The World of Physical Chemistry

The Hon. Robert Boyle (1627–1691)

The World of Physical Chemistry

KEITH J. LAIDLER

Professor Emeritus of Chemistry,
The University of Ottawa
Ontario, Canada

Oxford · New York · Toronto

OXFORD UNIVERSITY PRESS

Oxford University Press, Walton Street, Oxford OX2 6DP

Oxford   New York

Athens   Auckland   Bangkok   Bombay
Calcutta   Cape Town   Dar es Salaam   Delhi
Florence   Hong Kong   Istanbul   Karachi
Kuala Lumpur   Madras   Madrid   Melbourne
Mexico City   Nairobi   Paris   Singapore
Taipei   Tokyo   Toronto

and associated companies in
Berlin   Ibadan

Oxford is a trade mark of Oxford University Press

Published in the United States
by Oxford University Press Inc., New York

First published 1993
First published in paperback (with corrections) 1995
Reprinted 1995 (twice, with corrections)

A catalogue record for this book is available from the British Library

Library of Congress Cataloging in Publication Data
Laidler, Keith James, 1916–
The world of physical chemistry / Keith J. Laidler.—1st ed.
Includes bibliographical references (p.   ) and index.
1. Chemistry, Physical and theoretical—History.   I. Title.
QD452.L35   1993   541.3—dc20   92–41635

ISBN 0 19 855919 4 (Pbk)

Printed in Great Britain by Bookcraft (Bath) Ltd
Midsomer Norton, Avon

Preface

This book might have been entitled 'A History of Physical Chemistry', which to some extent it is. I am not, however, a historian, and my object has been somewhat different from that of a historian. I wanted primarily to give scientists some insight into how one important branch of physical science has developed. At the same time I have tried to write the book in such a way that historians of science who are not also scientists may find the book useful, even though they may well want to skip over some of the early material with which they will be familiar, and perhaps some of the more mathematical parts.

A science like physical chemistry is not necessarily best taught in the first instance by a purely historical approach. A historical approach to science does, however, have important uses. For one thing, the teaching of science can be more effective if the teacher has some knowledge of its history. An account of some personal incident that occurred during the course of scientific work can greatly arouse the interest of students. I have found, for instance, that the story of Arrhenius's difficulties with his Ph.D. examiners always seems to go down well with students, and seems to make them think that electrolytic dissociation is perhaps not such a dull subject after all.

There are benefits of a more basic kind. Ideas sometimes seem obvious to us simply because we have been brought up to believe them. Teachers are inclined to think it self-evident that the second law of thermodynamics is a statistical law, and may be impatient with students who can not see at once that this is so. It is helpful to know that Clausius, who did as much as anyone to establish the second law, believed all his life that it is a purely mechanical law, and that for a year or so Boltzmann thought the same. Lord Kelvin, who was also a pioneer in formulating the second law, never understood the idea of entropy at all. Appreciating these facts must help us to teach the second law more effectively to our students.

When scientists write the history of their subject, historians often complain that they write 'Whig' history, meaning that they write the history from the standpoint of the present day, failing to put themselves in the position of the scientists who were making the discoveries. Perhaps in a book like the present one a certain amount of Whiggery may be justified, but I have nevertheless been at some pains to discuss a number of ideas that eventually led to dead ends—but which were often fruitful in leading to further work which sometimes had the object of disproving them.

In order not to distract the reader with superscript references throughout the text I have simply included an alphabetically arranged bibliography at the end, sometimes with explanatory notes. It should always be easy to relate the references to the names and dates in the text.

Brief biographies of many of the scientists mentioned in the book are to be found at the end, and longer biographies of a selected few are distributed throughout the text. It will be obvious that I have not chosen, for the more detailed treatments, those who have made the most important contributions. Instead I have included people whose lives are of particular interest, and a few whose contributions to science seem to have been underrated.

Many readers of this book will feel that I have omitted topics that ought to have been included. To avoid excessive length I have kept to what in my view are the main trends in the overall development of the subject. Some topics, although in themselves of great importance, have been omitted since they seemed to me to be branches rather than the main growth itself. Different views of the subject are indisputably of equal validity.

Ottawa   K. J. L.
June, 1992

Acknowledgements

I am greatly indebted to Dr John Shorter, of the University of Hull, and to Mr A. V. Simcock of Oxford's Museum for the History of Science, for much help during the writing of this book. Both have been kind enough to read the entire manuscript, often in more than one version, and their comments have always been extremely positive and helpful. In addition to their careful reading they have provided me with much additional information. Without their painstaking help this book would contain many more errors that it now does. Mrs Heather Hoy has also been kind enough to read much of the manuscript, and to discuss with me some aspects of the presentation of the more historical material.

Many others have been generous with assistance. Professor C. J. Ballhausen provided me with information about some of the Danish chemists, and the late Dr Erik Bohr wrote me interesting reminiscences about his father and about Professors Brönsted, Bjerrum and Christiansen. To Professor Mansel Davies of the University of Aberystwyth I am grateful for discussions about the early work on electronic configurations, especially that of C. R. Bury. Professor E. T. Denisov has been of help in connection with some of the Russian contributions.

Correspondence with Mrs Janet Howarth, Fellow of St Hilda's College, Oxford, has greatly clarified my appreciation of the development of the physical sciences in the British universities. Professor W. Jaenicke, of the University of Erlangen, and Professor R. Haul, of the University of Hannover, have kindly provided me with not easily accessible information about some of the German scientists. Professor W. E. A. McBride of the University of Waterloo, and Professor J. T. Edward of McGill University, have made helpful comments relating to the development of physical chemistry in Canada. Professor Hitoshi Ohtaki, of the Institute for Molecular Science at Myodaiji, Okazaki, sent me his ideas on the introduction of chemistry into Japan in the nineteenth century. Professor Lord Porter read passages from the manuscript dealing with high-speed kinetics, and made suggestions regarding the classification of the different techniques.

I owe a special debt to the late Dr Christine King, with whom I had many valuable discussions on the history of physical chemistry, especially of kinetics. She had a deep insight into physical chemistry and its history, and her tragic and untimely death in a car accident was a great loss to many of her friends and colleagues.

Portraits

I am grateful to a number of people who at various times have given me portraits of themselves, which are reproduced in this book: the late Henry

Eyring, the late Sir Cyril Hinshelwood, and the late Professor R. G. W. Norrish; and Dr Gerhard Herzberg, Professor John Polanyi, and Professor Lord Porter.

For the portrait of William Allen Miller I am indebted to the Royal Society of Chemistry.

A number of the portraits reproduced in this book are taken from paintings and photographs protected by copyright, and arrangements have been made with the following:
The *Beckman Center for the History of Chemistry*, for use of the portraits of Bunsen, Draper, Laplace and Agnes Pockels.
The Earl of Porsmouth, for use of the painting of Isaac Newton by Sir Godfrey Kneller.
Godfrey Argent Studio, London, for use of the photograph of Walter Heitler.
The Hunterian Art Gallery of the University of Glasgow, for use of the painting of Joseph Black by Henry Raeburn. I am grateful to the Hon. John Warrender, whose family owned the portrait for many years, for informing me of its present location.
The Royal Society, for use of the painting of Robert Boyle by Johann Kerseboom. I am particularly grateful to Sandra Cumming, Information Officer of the Royal Society, for much help with locating some of the other portraits and for interesting information about them.
The Scottish National Portrait Gallery, for use of the portrait of Mary Somerville by Thomas Phillips. Miss Helen Smailes of the Gallery has been particularly helpful in providing information about the extensive portraiture of Mrs Somerville, involving paintings, busts and medallions.

As far as is known the remaining portraits used in the book are not protected by copyright. Every effort has been made to get in touch with persons and organizations who might hold the copyright on portraits used, but these efforts were not always successful. If notified, the publishers will be pleased to rectify any omissions in future editions.

Contents

CHAPTER 1

The origins of physical chemistry

It is sometimes said that the year of birth of physical chemistry was 1887. In that year there were indeed several important events relating to the subject, one being the launching of the *Zeitschrift für physikalische Chemie*, the first journal devoted exclusively to physical chemistry. That year also saw, in the first volume of the *Zeitschrift*, Svante Arrhenius's famous paper on electrolytic dissociation, and one of van't Hoff's more important papers on the thermodynamics of solutions, work that was later to win him the Nobel Prize for chemistry in the first year that the prizes were awarded.

Towards the end of the nineteenth century there was certainly something of a change in attitude towards physical chemistry, which then began to be recognized as a distinct branch of chemistry. However, a good deal of physical chemistry had been done previously. For example, two centuries earlier Robert Boyle had been carrying out physico-chemical investigations, and a good case can be made for regarding him as the first physical chemist. His approach to chemistry is well described by his own statement that he hoped 'to beget a good understanding 'twixt the chymists and the philosophers'. His famous book *The Sceptical Chymist; or Physico-Chymical Doubts and Paradoxes . . .* even contains the expression 'physico-chemical' in its full title. His approach to chemistry had a great influence on others, including Isaac Newton, some of whose work can certainly be called physical chemistry. In the eighteenth century Joseph Black in Scotland and Antoine Lavoisier in France also did much that can be classed as physical chemistry. In the nineteenth century Robert Bunsen in Germany, Michael Faraday in England, and many others whose work is considered later in this book, were also contributing to the development of physical chemistry. The first volume of an important book published in 1855 by William Allen Miller was entitled *Chemical Physics*; it might equally well have been called *Physical Chemistry*.

Perhaps, then, we should consider 1887 not as the year of birth of physical chemistry but rather as the year in which it reached a certain maturity.

1.1 The meaning of 'science', 'philosophy', 'physics' and 'chemistry'

It is useful to consider the meaning of some of the general terms that have been used to designate what is now called science and some of its branches. It is

1

important to be alert to the fact that the meaning of all of these words has shifted over the centuries, and that even today the terms are not entirely unambiguous. An additional complication is that similar words in other languages, such as the German *Physik* and the French *chimie*, do not necessarily mean quite the same as their English equivalents. Also, many scientists are individualists, and a few are highly eccentric; often they use a word in a sense that is different from the usually accepted one.

The word 'science' derives from the Latin scientia, and was for centuries regarded as including all kinds of knowledge. Gradually, however, the meaning narrowed, and the word, at least in certain contexts, began to mean knowledge of the universe that is derived from observation and experimental investigation. Even today, however, other meanings are encountered, in expressions such as 'economic science'.

The broader meaning of the English word 'science' was the prevalent one up until the seventeenth century and it still persisted into the nineteenth century. For that reason many of the earlier scientists—as we now call them—tended to avoid calling their work science. That word did, however acquire an official status in 1831 with the founding of the British Association for the Advancement of Science. Until late in the nineteenth century, however, many preferred to use the term 'natural philosophy', usually condensed to 'philosophy'. In the seventeenth century, when Boyle and Newton were active, this word meant more or less what we mean by science today; it certainly included physics, chemistry, geology, biology, and astronomy, as is evident from a perusal of the early volumes of the *Philosophical Transactions*.

Gradually, however, the meaning of the English word 'philosophy' shifted in such a way that, in the context of natural or experimental philosophy, the word tended to mean only physics as we now understand it, and this restricted meaning is still sometimes encountered; some ancient universities still have professors of natural or experimental philosophy, by which is usually meant physics.

The change in meaning of the word 'science' occurred in France rather earlier than in England. The name occurs in the title of the Académie des Sciences, founded in 1666, only six years after the founding of the Royal Society, which would have felt it inappropriate to use the word in connection with its activities. Conversely, the Académie members would not have used the word 'philosophie' to describe their work. The German word 'Wissenschaft' has changed its meaning in a similar way to the English word 'science', and is often qualified as 'exacte Wissenschaft' or 'Naturwissenschaft' to have much the same meaning as the English word science has today.

The meaning of the word 'physics' has also varied over the years. The word derives from the Greek (*physica*), meaning natural, and is now understood to refer to investigations of the general laws of nature, in contrast to investigations of specific substances, which would be called chemistry. It was not, however, until the latter part of the nineteenth century that the word acquired its present more restricted meaning, and even today the expression 'physical science' is interpreted in different ways by different people. In earlier times the word encompassed more than it does today. At one time, but not since the eighteenth century, the word 'physics' was sometimes used to refer to the science of physic, i.e., medicine, and there is still a vestige of that usage in the modern word 'physician'; it is interesting that the equivalent French word *physicien* means a physicist, not a medical practitioner.

Until well into the nineteeenth century the word 'physics' was commonly regarded as meaning much the same as natural philosophy, and therefore to include chemistry, biology and geology as well as what we now understand as physics. For example, in the Prospectus of the newly-formed University of London, issued in 1828, the discipline of physics is discussed as follows:

It is a matter of considerable difficulty to ascertain the distribution of PHYSICS, a vast science or rather class of sciences, which consists in the knowledge of the most general facts observed by the senses in the things without us. Some of these appearances are the subject of calculation, and must, in teaching, be blended with the Mathematics; others are chiefly discovered by experiment; one portion of physical observation relates to the movements of conspicuous masses, while another relates to the reciprocal action of the imperceptible particles or agents which we know only by their results; and a great part are founded on that uniformity of structure, and those important peculiarities of action, which distinguish vegetable and animal life.

Later in the Prospectus, physics is subdivided into

Mathematical Physics
Experimental Physics
Chemistry
Geology and Mineralogy
Botany and Vegetable Physiology
Zoology and Comparative Anatomy
Applications of Physical Sciences to the Arts

It seems curious today to include zoology as a branch of physics, but this was certainly the practice at the time. For example, the Royal Physical Society of Edinburgh was founded in 1854 for the 'promotion of zoology and other branches of natural history'!

A classification consistent with that in the University of London Prospectus is to be found in Robert Walker's *Textbook of Mechanical Philosophy*, published in 1850. In his opinion the physical sciences consist of physiology and physics, according to whether or not they involve living systems. Physics in turn is subdivided into chemical philosophy, which we would call chemistry, and mechanical philosophy, which we would call physics.

A change of meaning similar to that of the English word *physics* is found with the equivalent word in other languages. When the German journal *Journal der Physik* was founded in 1790 (its title being changed to *Annalen der Physik* in 1799) it was clearly understood that it would include chemistry as well as physics, and this is reflected by its contents; later, as the meaning of 'Physik' narrowed, changes of title were required (Section 2.1).

Although the words 'science' and 'physics' have been used for a long time, the corresponding words 'scientist' and 'physicist' were invented only in the nineteenth century. A word favoured by the seventeenth-century scientists to refer to themselves was 'virtuoso' (plural virtuosi). This word is related to the special meaning of 'virtue' as an inherent property, the virtuosi being concerned with the investigation of the specific properties of materials; the word 'virtuoso' did not acquire its present meaning of a skilled musical performer until the eighteenth century. Robert Boyle was particularly fond of this word, one of his books, published in 1690, having the title *The Christian Virtuoso*; in it he defined virtuosi as 'those that understand and cultivate natural philosophy'.

The modern words 'scientist' and 'physicist' were suggested by the Revd William Whewell (1794–1866), who in 1841 became Master of Trinity College, Cambridge. He was a man of wide interests, being highly proficient in Latin and Greek, mathematics, mechanical philosophy (physics), mineralogy, geology, engineering, German architecture, philosophy and theology. In a long essay *On the language of science*, which was part of the 1840 edition of his *The Philosophy of the Inductive Sciences*, Whewell first suggested the words 'scientist' and 'physicist'. In view of their origin it is amusing to find that in England these two words were for some time condemned as 'Americanisms'. Faraday, among many others, did not like either word and continued to call himself a philosopher. Even towards the end of the century many prominent scientists still disliked the word 'scientist', usually preferring to be called philosophers.

Even stronger objection was taken to the word 'physicist'. Lord Kelvin disapproved of it and sometimes called himself a naturalist, relying on Dr Samuel Johnson's definition of it in his famous *Dictionary* as 'a person well versed in natural philosophy'—a definition that was a rather eccentric one and never proved popular! Soon after Whewell proposed the word 'physicist', Faraday wrote to him saying that 'physicist is both to my mouth and ear so awkward that I think I shall never be able to use it. The equivalent of three separate sounds of "i" in one word is too much.' *Blackwood's Magazine* was even more scathing: 'The word physicist, where four sibilant consonants fizz like a squib. . . .' The same can, of course, be said of the word 'criticism'. Today the word 'physicist' does not seem so bad, perhaps because we are so used to it.

The words 'chemist' and 'chemistry' were used commonly in the seventeenth century, and probably derived from the Arabic word *al-kimiya*, which in turn came from the Greek (*chemia*), which meant the art of transmuting metals. Like the word 'physics', the word 'chemistry' has varied in meaning over the centuries, and even at a given time has meant different things to different people. In the seventeenth century much of chemistry was what we would now call alchemy, having as its objectives the conversion of base metals into gold and the discovery of an elixir that would prolong life; both Robert Boyle and Isaac Newton engaged in a certain amount of alchemy, but did some scientific chemistry as well.

Chemistry is more concerned with the properties of individual substances, in contrast to physics which is more concerned with general properties. Two types of chemist began to emerge in the eighteenth century. Some chemists were concerned with the preparation of new compounds, particularly those that would have useful properties, often for medicinal purposes. Well into the nineteenth century many chemists of this kind were iatrochemists or pharmaceutical chemists, interested primarily in the medicinal aspects of the subject.

1.2 The meaning of 'physical chemistry'

At the same time there were a number of others who called themselves chemists but whose approach to science was the same as that of the physicists: although interested in the properties of individual substances, they were primarily concerned with understanding the workings of nature and were only incidentally interested in preparing new compounds. Among eighteenth- and nineteenth-century investigators who called themselves chemists, and who approached their subject in this way, may be mentioned Joseph Black, J. F. Daniell, Michael Faraday, W. Allen Miller, John Draper, William Grove, Robert Bunsen and Victor Regnault; all of these held appointments as professors of chemistry rather than of physics.

Chemists of this type felt free to work on heat, light and electricity, which we now think of as physics rather than chemistry. Until the latter part of the nineteenth century these topics were in fact generally regarded as part of chemistry, which was in turn a branch of physics. Michael Faraday was professor of chemistry and not of physics at the Royal Institution, and when he carried out his important investigations on electricity and magnetism these topics were accepted as quite appropriate for a chemist. Faraday himself would not have thought about these distinctions; to him what he was doing was natural philosophy.

Mention should also be made of a number of people who were primarily physicists but who also did important work on chemical problems; Lord Kelvin, Clerk Maxwell and Edmund Becquerel are obvious examples.

Chemistry carried out with the primary object of investigating the workings of nature is what we now call physical chemistry. It can be defined as that part of chemistry that is done using the methods of physics, or that part of physics that is concerned with chemistry, i.e., with specific chemical substances. Since the distinction between physics and chemistry is not sharp and has varied over the years, physical chemistry cannot be precisely defined. Nobel Prizes in chemistry have sometimes gone to people (such as Ernest Rutherford and Gerhard Herzberg) who thought that they were physicists, and vice versa. The distinction between the sciences is, after all, no more than a matter of administrative convenience.

Some further discussion of the definition of physical chemistry seems appropriate. At various times the distinguished American chemist G. N. Lewis expressed himself on this matter, his ideas changing with the development of the subject and with his moods. On one occasion he defined physical chemistry as that which is done by a physical chemist, and it is hard to think of a better definition! At another time he said that there are two branches of science, physical chemistry and nuclear physics. In other words, the nuclear physicist deals with the nature and behaviour of the atomic nucleus, while the physical chemist deals with everything that is related to the extranuclear electrons. This remark was made before the technique of nuclear magnetic resonance was developed; today the physical chemist has made considerable inroads into nuclear physics!

Lewis's comment that science is either physical chemistry or nuclear physics is reminiscent of Rutherford's celebrated statement that all of science is physics or stamp collecting. Lewis's comment would make all of chemistry, biology, geology and astronomy, and much of physics, a branch of physical chemistry, and like Rutherford's is apt to produce cries of outrage. If not taken too seriously Lewis's definition has some merit in calling attention to the undoubted fact that physical chemistry is relevant to a wide variety of disciplines. Chemical kinetics and electrochemistry are related to many aspects of biology, particularly physiology. Mineral deposits, of special interest to geologists, were dealt with in the pioneering physico-chemical investigations of van't Hoff, and flow processes in geology have also been studied by physical chemists. Properties such as viscosity and diffusion, normally regarded as lying in the domain of physics, have often been investigated by chemical kineticists. The invasion of astronomy by physical chemistry is exemplified by spectroscopic studies which have provided evidence for the existence of elements and of species such as free radicals in the stars. Sometimes physical chemistry even strays outside the physical sciences; chemical kinetics, for example, has been applied to psychological problems, in that subjective time appears to obey the Arrhenius equation for the dependence of rate on temperature.

The fact that physical chemistry has so expanded and has merged with other scientific disciplines has led some to suggest that the subject can no longer be regarded as a separate discipline. This point of view was expressed by G. N. Lewis as early as 1922, when his mood was different from when he made his other statements:

The fact is that physical chemistry no longer exists. The men who have been called physical chemists have developed a large number of useful methods..., and as the applications of these methods grow more numerous, it becomes increasingly difficult to adhere to our older classification.

This argument admittedly has some merit. Today all undergraduates are required to include physical chemistry in their studies as a condition for obtaining a degree in chemistry, and physical chemistry is recognized as one of the tools of trade of every kind of chemist. Many universities require some knowledge of physical chemistry for degrees in physics, biology and geology.

Research in almost every branch of chemistry involves the use of much physical chemistry. Publications by organic chemists who have used the technique of nuclear magnetic resonance, for example, often read like papers in physical chemistry.

In practice, however, it still seems useful to regard physical chemistry as a separate discipline. Universities usually have departments or subdepartments of physical chemistry, and even if they do not they tend to specify the discipline of physical chemistry in making certain appointments. Some universities even contain their physical chemists in a separate building, but this may be undesirable since it seems better to let the physical chemists come into close contact with other chemists, especially now that the discipline has so expanded into other branches of science.

The question of what physical chemistry is has only been touched on in a roundabout way in the above discussion. I think it was Mr Gladstone who said in the British House of Commons, when contrary to parliamentary rules an alleged religious argument had been put forward, that he could not define a religious argument but that he knew one when he heard it. Although I cannot precisely define what physical chemistry is I can recognise it when I see it, and I am sure most of us can. Perhaps I can do no better than say that anything covered in this book, and much else besides, is physical chemistry! Alternatively we have the nice definition of G. N. Lewis, when he was not in the mood to say that physical chemistry does not exist, that physical chemistry is 'anything that is interesting'.

The nature of a physical chemist

The question of how a physical chemist spends his or her day is just as impossible to answer precisely. The difficulty is that there are vast differences between physical chemists and in the way they approach their subject. I can best illustrate this by considering the careers of two physical chemists, Henry Eyring and Ronald Norrish; I knew them both personally, and have written their biographies. Although both of them made outstanding contributions in the same branch of physical chemistry, chemical kinetics, it is hard to think of two people who were more different in their personalities and in their approaches to scientific research.

First as to their personalities. Eyring was a deeply religious man, while as far as I know Norrish had no religious beliefs; certainly they did not show. Eyring was brought up a member of the Church of the Latter-Day Saints, or Mormons, the beliefs of which are strongly fundamentalist. For a time he was an elder of the church and superintendent of its Sunday Schools. In accordance with the practice of his church Eyring was highly abstemious, eating only simple food and taking no stimulant, not even tea or coffee. Norrish can be summarized in a nutshell by saying that in this respect he was exactly the reverse!

In appearance Norrish was far from what is popularly regarded as a typical professor or scientist. He was powerfully built, with a close-cropped moustache, and was always neatly and formally dressed in a dark suit. Anyone trying to guess his occupation might well have taken him for an army officer or business executive. Eyring on the other hand was quite informal and easily identifiable as a scientist and professor.

Eyring had a friendly and open disposition. His way of doing research was often to look for someone, perhaps one of his graduate students, with whom he could have a friendly argument about some scientific topic that was troubling him. He was always brimming with ideas, many of them wrong but always stimulating, which he would pass on to anyone who would listen, and he had no objection to being contradicted. Norrish, on the other hand, was much more formal in his approach, and he was somewhat secretive about any ideas that he thought might be used by others. He made it clear to his graduate students that they were not to continue working on topics to which he had introduced them, since in that way they would become his competitors. He was aloof and uncommunicative with scientists in other institutions who were working in his field of flash photolysis, being unwilling to tell them much of what he was doing. Eyring never had such an attitude; he was pleased if his former students continued the work they had begun with him, always being conscious that there is a vast amount of research work to be done.

One thing that Eyring and Norrish had in common was that they were remarkable effective directors of research, although their approaches were completely different. The students of both of them often went on to careers of some distinction. In Eyring's case the influence arose from his great friendliness, so that his students felt that they were part of his intellectual family. Norrish's research students often found relations with him to be rather difficult, but they usually retained their admiration and respect for him.

Eyring was a skillful mathematician, and he approached scientific problems with great insight, having the ability to reduce them to mathematical terms and to solve the resulting mathematical relationships. At the same time he had a keen appreciation of experimental results, although he had little taste for carrying out experiments or for directing the experiments of others. Norrish was just the opposite. His main concern was with carrying out experiments, which he interpreted in an instinctive and intuitive way, without resort to formal theories. He did not have much skill in mathematics, and his theoretical background was no more that was necessary to work in his particular branch of chemical kinetics. He had no deep understanding of spectroscopy, of thermodynamics or of theories of reaction rates, but knew just enough for his own particular purposes. He once remarked to me with his usual cheerfulness that transition-state theory was 'high-falutin' stuff', and I doubt that he had ever bothered to find out much about it. It is at first sight surprising that one who was so successful (at the Nobel Prize level) in chemical kinetics and in certain aspects of spectroscopy should have been so uninterested in the theoretical treatments that lie behind those topics. The answer lies in his peculiarly intuitive approach to research. He had no objection to hypotheses based upon a particular set of experimental results, and framed many himself, but he regarded them as of temporary value, to be shot at and perhaps shot down by subsequent research. Eyring, too, was intuitive in his research—as indeed are all scientific pioneers—but his intuition was based on a firm understanding of the broad theoretical background to his particular area of research.

Neither Eyring nor Norrish was a good university lecturer in the conventional sense. Eyring would never prepare a lecture in a formal way, and as a result his lectures were considerably disorganized. He always knew what he was talking about, but he frequently went off on tangents as new ideas struck him. Undergraduate students would have found his courses impossible, but fortunately they were rarely exposed to them. Graduate students who were interested in what he was teaching, and had previous knowledge of it, found his lectures highly stimulating, although they often had to do much work to get the subject matter organized in their minds. Norrish's inadequacy as a lecturer sprang from a different cause; students who attended his university lectures have told me that they wondered whether he properly understood thermodynamics, kinetic theory, or the other subjects he was teaching.

This comparison of Eyring and Norrish makes it clear that the question of how a physical chemist carries out research is an unanswerable one. There is obviously a vast range of approaches. Some physical chemists are highly theoretical and mathematical, and little concerned with experimental results, but on the whole their contributions are not as significant as those of Eyring who maintained a good balance between theory and experiment. Linus Pauling's approach to science has been similar to that of Eyring; he too has a strong mathematical approach, but always with a keen regard for experimental results. At the other extreme there are some physical chemists whose competence in theory is even weaker than Norrish's was. The American chemist Wilder Bancroft comes at once to mind. He was forced to choose fields of research—the phase rule (Section 4.4) and colloid chemistry (Chapter 9)—in which he could work with the minimum of theory. Unlike Norrish, he was not very successful, and his work had little impact on the development of physical chemistry. He published many papers and achieved a certain prestige among chemists, but he became left behind as the subject developed, and his career finally went into a serious decline.

The majority of physical chemists who are successful in their work either follow a similar approach to that of Eyring, or are somewhat more on the experimental side and somewhat less on the theoretical and mathematical side. Commonly a physical chemist has a strong programme of experimental work which may be carried out largely or entirely by technicians, graduate students and postdoctoral fellows. The demands of teaching, writing and administration may be such that little time remains for work at the laboratory bench, but the research director must keep in close touch with the details of the work being done. It is interesting that this scenario applies well to Robert Boyle, who as I have suggested was perhaps the first physical chemist.

Mathematics and physical chemistry

The relationship between mathematics and physical chemistry has been touched on in the last section, but needs a little further discussion. A non-scientist looking over this book, or any textbook of physical chemistry, and seeing 'all those equations', would come away with the impression that the subject is a highly mathematical one. In one sense this is true, since the subject does have a highly mathematical superstructure. This superstructure has been built by an army of scientific pioneers of high mathematical skills, and has been brought to the attention of students of physical chemistry by the writers of textbooks and monographs. All of science has a mathematical superstructure. The mathematician David Hilbert said that 'Physics is far too difficult for physicists.' The physicist Einstein said that 'The trouble with chemistry is that it is too difficult for chemists.' A chemist might add that 'Biology is too difficult for biologists.' What this means is that there is a pyramidal hierarchy with mathematics at the top. The mathematician has to solve problems of the physicist, the physicist those of the chemist, and so along the line to the more descriptive sciences.

Those of us who teach physical chemistry must be well versed in its mathematical superstructure, at least in its broader aspects, even if we are not familiar with every detail. Those who lecture in physical chemistry have at some time had to work through all the mathematical details, but most of us would sometimes need to refer back to the texts if challenged to supply the mathematical proofs that lie behind the subject matter of our teaching.

At the same time it has to be recognized that competent and even outstanding research in physical chemistry can sometimes be done by people who are somewhat deficient in mathematical skills, and are not fully at home in the theoretical aspects of their subject. Michael Faraday is the obvious example; more recently Norrish was able to win a Nobel Prize in this way, and others with the same background have made significant contributions. Such people, however, are rather exceptional, and one cannot escape the conclusion that a sound education in physical chemistry, and indeed in all branches of chemistry, must have a strong mathematical basis.

Although mathematics now plays such an important role in physical chemistry, chemists on the whole were slow to recognize the usefulness of applying mathematics to their scientific work. There were several reasons for this. One factor that applied particularly to Britain but also to some extent to other countries, arose from too slavish adherence to Newton's form of the calculus. It later became clear that the calculus of Leibniz was much more powerful and versatile, and towards the beginning of the eighteenth century it began to be realized that British scientists were putting themselves at a disadvantage by their adherence to Newton's geometrical methods and his method of fluxions. Later the British work in both physics and chemistry became more successful.

Another factor was that there was in the minds of many chemists and even some physicists a conscious decision to exclude mathematics from science. This is well illustrated by a statement made by Charles Giles Bridle Daubeny (1795–1867), professor of chemistry at Oxford and a vigorous advocate of the extension of science education at the universities. In 1848 he issued a tract in support of a school of physical science at Oxford, and in it appears the following passage:

It would manifestly be quite foreign to the purpose, and fatal to the genius, of a School of Physical Science, to encourage the introduction of any subjects that are treated mathematically.

Daubeny could have pointed to many who had made outstanding contributions to physics and to the more physical aspects of chemistry without the use of any mathematics beyond simple arithmetic; examples that come to mind are Joseph Black, Antoine Lavoisier, John Dalton and Michael Faraday. Daubeny was not suggesting that mathematics should not be used in science, but rather that the mathematics should be left to the mathematicians while the scientists should go on with their experiments and concentrate on the descriptive aspects. There is one outstanding example where this was successful; Faraday made his remarkable contributions to electricity and magnetism without the use of any mathematics, while Clerk Maxwell developed the theory.

Until well after World War I even those educated in science at universities were often surprisingly deficient in their knowledge of mathematics. On the continent of Europe the situation was not too bad, but it was less satisfactory in Britain and worse in the United States. At Harvard, for example, it was possible until about 1930 to obtain a Ph.D. degree in physical chemistry without any knowledge of the calculus; Farrington Daniels, for example, did so in 1914, and had to learn the subject for himself before he could make his distinguished contributions to chemical kinetics. Irving Langmuir, who was later to receive a Nobel Prize for his work on the physical chemistry of surfaces (Section 9.3), had to withdraw from a course at Leipzig because his mathematical background was inadequate; later he had to abandon a piece of research under Nernst for the same reason.

Today we can see that an important factor in the development of physical chemistry was the introduction of mathematical methods into the subject. Some of the earlier books in physical chemistry tended to avoid mathematics, an extreme example being Ostwald's *Grundriss der allgemeinen Chemie* (1889), which contains not a single mathematical equation. In writing this book Ostwald was aware of the mathematical deficiencies of many of his readers but in his own research, and in his *Lehrbuch der allgemeinen Chemie* (1885–87) he made full use of mathematics, as did van't Hoff in his *Études de dynamique chimique*, which appeared in 1884. Although these more mathematical presentations of physical chemistry are easy reading for physical chemists today, they were quite beyond many chemists of the time, but both were important in leading chemists towards a greater appreciation of the value of mathematics.

1.3 Attitudes to science

The manner in which science developed, and the speed of its development, depended in an important way on the attitudes of society. Two factors are of particular importance, and they are somewhat interrelated: the official attitudes of the churches, and the general attitudes of society, as reflected by government policies and the establishment of official scientific societies.

The churches

In the countries where most of the new science was done—Italy, England, France, The Netherlands, and later Germany—the Christian churches were at first extremely powerful and exerted important control over all intellectual activity. Until the Middle Ages almost all learned men were clerics. Early Christian theology derived many of its ideas from the teaching of Aristotle, particularly as a result of the opinions put forward by St Thomas Aquinas (*c.*1226–1274), one of the founders of the scholastic movement. For many centuries scholastic philosophy dominated Roman Catholic theology, and to some extent the theologies of the other denominations. It is probably true to say that such opposition to science that came from the Christian churches was due more to the dogmatic acceptance of Aristotle's ideas than from any intrinsic feature of Christian theology.

The attitude of the churches to science was ambiguous and inconsistent, and varied widely from denomination to denomination. The rise of Protestantism brought about some liberalism, but some of the Protestant churches themselves created some difficulties for men of science. The Protestant mathematician and astronomer Johannes Kepler (1571–1630), who at one time had intended to study for the ministry of the Evangelical Lutheran Church, was so persecuted by his church that he had to seek refuge with the Jesuits!

There are many other examples of pressures to which scientists were subjected. In 1633 Galileo was summoned to abjure the 'heresies' that his ideas about the solar system were considered to be, and the Roman Catholic Church still officially condemned these ideas until 1922. In 1663 all of Descartes' writings were placed on the Roman Catholic *Index Librorum Prohibitorum* (List of Prohibited Books), despite the fact that Descartes had done all he could, and more than was acceptable from the scientific point of view, to make his opinions compatible with the teachings of the Church. Especially after the Revocation of the Edict of Nantes in 1685, the Roman Catholic clergy in France were particularly obstructive to science and to Protestantism, with the result that many Huguenots were forced to leave the country.

In England the situation was on the whole better than in Roman Catholic countries, since the established Church of England has no mechanism for enforcing uniformity and must tolerate both Catholic and Protestant beliefs. Many of the seventeenth-century 'virtuosi' were remarkably liberal in their religious ideas; Robert Boyle, for example, went to some trouble to study and appreciate other religions, even acquiring a knowledge of the Latin, Greek, Hebrew, Syriac and Chaldee languages for that purpose. Many of the Original Fellows of the Royal Society were devout Anglicans and a number were clergymen, but on the whole their scientific work was little affected by their religious beliefs; there were a few attacks by individual members of their church, but they had little effect. Some indirect consequences of the attitude of Anglican clergymen will be referred to later, and particular mention may be made of the bitter confrontation in 1860 between the biologist Thomas Henry Huxley (1825–1895) and the Bishop of Oxford, the Rt Revd Samuel ('Soapy Sam') Wilberforce, who took strong exception to Darwin's theory of evolution.

In contrast to the hostile actions by some church representatives is the undoubted fact that many clerics and devout laymen made substantial contributions to the development of science. Among laymen who took their religious beliefs very seriously may be mentioned Robert Boyle, John Dalton, Clerk Maxwell, Michael Faraday and Lord Kelvin. Among ordained clergymen of various denominations was the Rev. Roger Joseph Boscovich, a Jesuit priest whose atomic theory and other scientific contributions in the eighteenth century exerted a wide influence (Section 5.2), and who remained in good standing with his church. Another example is the Unitarian minister Joseph Priestley, who did pioneering work on the chemical properties of gases. It is significant that Anglican clergy played a prominent role in the founding in 1831 of the British Association for the Advancement of Science, and of its first fourteen Presidents, six were clergymen.

Adherents to Christianity can be classified according to their formal denominations. In England, where there is an established church, there are members of the Church of England (Anglicans), Dissenters (Protestants who oppose the views of the Church of England), Nonconformists (Protestants who do not oppose the Anglican Church but prefer other forms of worship), and Roman Catholics. Members of the Anglican Church can be classified as belonging to the High Church, the Broad Church, or the Low Church. Members of the High Church place emphasis on the more formal aspects of religion such as ritual, the significance of the priesthood, the importance of the sacraments, and revelation as a source of knowledge. The Roman Catholics have a somewhat similar approach.

At the other extreme are members of the Low Church, where the emphasis is more on a literal interpretation of the Bible, referred to as fundamentalism or Bibliolatry. Many Dissenters have essentially the same point of view.

In between the High Church and the Low Church is the Broad Church, founded by a group of seventeenth-century English clergymen known as Latitudinarians. Here there is a more liberal, tolerant and broadminded approach, with little fundamentalism or insistence on dogma and formal ritual.

While this classification of religious attitudes is convenient for present purposes it must be recognized that there is no sharp division between the three categories. A member of the High Church might, for example, have quite fundamentalist beliefs on some matters. The classification also applies to some extent to denominations other than the Anglican Church, and to countries other than England, with some differences of detail.

The members of the Broad Church saw no conflict between science and religion, and did not attempt to limit the activities of scientists. Those who were scientists themselves felt that their work strengthened rather than weakened their faith. Among church adherents, it is the Broad Church members who on the whole did the most important scientific work, although there are important exceptions, notably Michael Faraday who as a Sandemanian must be classified as very Low Church. The strongest opposition to the growth of science has come from members of the Low Church, who found some scientific conclusions to be inconsistent with their fundamentalist beliefs. The discovery in the sixteenth century of new stars and the observation of comets was sometimes condemned as contradicting the belief that God completed the work of creation in seven days. Those fundamentalists who were interested in geology—the 'Scriptural geologists'—interpreted their observations so as to make them consistent with the conclusion of Archbishop James Ussher (1581–1656) that the universe was created in 4004 BC and that catastrophic changes were later brought about by the Great Deluge, exactly as described by Moses in the Old Testament; they were greatly offended by geological and palaeontological evidence that indicated otherwise. Some other fundamentalist beliefs, not based on the Bible, are less easy to understand. Atomic or corpuscular views of matter were sometimes condemned, apparently because such a mechanical view of the universe seemed to leave no room for God, or sometimes because the existence of empty space between atoms seemed to mean that God was not omnipresent.

Members of the High Church usually did not object to scientific investigation as such, but they did tend to delay the advance of science in a less direct way. From the early days of Christianity most learned men were clerics who had been trained in the classics, ancient history and ancient philosophy, and this tradition survived until the nineteenth century. This had a powerful effect on the teaching at universities, as will now be discussed.

The universities

The example of England may be taken as broadly typical of the situation in Christian countries, although there are some significant differences, especially as regards the periods of time over which changes occurred. In 1800 in England there were two universities, Oxford and Cambridge, both steeped in the Anglican clerical tradition, particularly that corresponding to the Broad Church. Most of the teachers at the two universities had to be ordained Anglican clergymen, the main exceptions being a few specially appointed professors. At both universities a condition for obtaining a degree was subscription to the Thirty-Nine Articles of Religion, in order to demonstrate acceptance of the Church of England, a requirement that was not relaxed until the latter half of the nineteenth century.

All Roman Catholics and most Dissenters were thus effectively excluded from a university education in England. At Oxford until 1850 the only way to a degree was to take classics, ancient history and ancient philosophy (*Literae Humaniores*, later called 'Greats'). In 1807 a School of '*Mathematicae et Physicae*' was instituted for those who had already taken Greats, and it involved some knowledge of 'experimental philosophy'. Few students, however, took the examinations in this school.

At Cambridge the Mathematical Tripos was the requirement for a degree, but until the early 1800s the standards were extremely low, as they had to be if people like Wordsworth, Coleridge and Macaulay were to graduate. At both universities there were science professors who gave some lectures, but since a knowledge of science could in no way help a student to obtain a degree the lectures were often sparsely attended, and the professors were set apart from the rest of the university. In view of this unsatisfactory situation at the two English universities, it is not surprising that until later in the nineteenth century many English scientists had not attended a university at all.

When attempts were made in the middle of the nineteenth century to reform the curricula at Oxford and Cambridge, and to introduce the teaching of science and other subjects, there was considerable opposition from the clerics, who at Oxford were irreverently referred to as the 'black dragoons'. There was little fundamentalism in their position; they had no objection to science as such but believed that an education in the classics and ancient philosophy is a much better training for the mind than one in the sciences. This point is made well in an incident involving F. A. Lindemann shortly after being appointed professor of experimental philosophy at Oxford in 1919. At a dinner party he expressed concern about the serious shortage of physicists in England, but was told not to worry: 'Anyone with a First in Greats could pick up physics in a fortnight.'

One argument which seems curious today was concerned with the use of standard texts, such as the works of classical authors like Thucydides and Livy. Classical students could be examined on their knowledge of such texts, but in subjects like science, English literature and history, no such authoritative texts could be recognized.

In spite of this unsatisfactory situation in the two English universities, science did slowly develop in them and eventually flourished. In the seventeenth century, and for another century and a half, Oxford outshone Cambridge in science, largely owing to the activities of a few gifted people of whom Robert Boyle, who resided in the city from about 1655 to 1668, was the most distinguished. In 1683 the Ashmolean Museum was opened in Oxford, the first building in England and one of the first in the world devoted to scientific research. It was not so much a museum in the modern sense of a building in which materials are displayed; it was more a 'home of the Muses', a building devoted to education, study and research. For over a century and a half almost all of the lecturing and research in science was done at the Ashmolean, but hardly any practical instruction was given to students. Cambridge was less fortunate, for it had no university laboratories until nearly two centuries later; G. G. Stokes, for example, in the 1850s had to carry out his research on fluorescence in a narrow room behind the pantry of his Cambridge house.

It was partly as a result of the failure of Oxford and Cambridge to be universally accessible and to foster the teaching of science that other English universities came into being in the nineteenth century. The first of these was the University of London, begun in 1828 by a group of Dissenters, and soon a number of other London colleges became incorporated into it. Later a number of civic colleges were begun in Liverpool, Manchester and other cities, initially to meet local needs but eventually to develop into fully fledged universities with a wide range of faculties.

The situation in Scotland was very different from that in England. In 1800 the population of Scotland was only one-fifth that of England and its wealth one-fortieth, but the intellectual soil had become much more fertile to science. The Scottish Enlightenment was more luminous than the English, and the Universities of Edinburgh, Glasgow and Aberdeen were providing, in the sciences and other fields, a more liberal education than was being provided by the two English universities. The University of Edinburgh had one of the earliest chairs of chemistry to be established anywhere in the world, and Thomas Jefferson recommended it as the best university in Europe for the acquisition of pure science. When the American universities began to teach science in the late eighteenth century they often went to the Scottish universities for their professors. Students of Joseph Black at Edinburgh became the first professors of chemistry at two important American colleges: Samuel Mitchell in 1792 at Columbia College (which developed into Columbia University two years later), and John McLean in 1796 at the College of New Jersey (now Princeton University). In view of the superiority of science teaching in Scotland it is not surprising that a disproportionate number of British physical scientists came from there; Joseph Black, David Brewster, Lord Kelvin and Clerk Maxwell are obvious examples. The greater encouragement of science in the Scottish universities is due in part to less interference from the churches, which in Scotland tended towards the Broad Church and were therefore more tolerant of science.

On the continent of Europe the universities followed somewhat the same pattern as in England, although the teaching of science tended to come earlier. At first there was much clerical control, initially by the Roman Catholic Church, but in Germany and elsewhere a number of Lutheran and other Protestant universities were later established. In France the Université de Paris and other universities were first under the control of the Roman Catholic Church but nevertheless in the eighteenth century accomplished a good deal more science than did the English universities. As a consequence of the French Revolution (1789–1795) the French universities were secularized. In 1792 the Marquis de Condorcet (1743–1794) presented to the French National Assembly a report in which he made a strong case for the establishment of educational institutions which placed emphasis on science, and this was acted upon. In

1793 a decree of the Revolutionary Convention abolished all the universities and colleges of the *ancien régime*, and a number of new institutions were founded. The École Normale Supérieur and the École Centrale des Travaux Publics, which soon changed its name to the École Polytechnique, were established in 1794, with strong programmes in mathematics and the physical sciences. In 1808 Napoleon ordered a complete university reorganization in which the state took complete control, the professors and other university teachers becoming civil servants. Clerical control was therefore abolished, but later in the century religious authorities were allowed some freedom to establish universities under their own control.

In the German-speaking countries the universities also became emancipated from clerical control somewhat earlier than was the case in England. In the eighteenth century, Prussia developed an effective state system of higher education, and this played an important role in the German successes in science, particularly in the nineteenth century. During the early years of that century several technical schools were established and subsidized by the state, and by the end of the century the number of science graduates they were producing was much greater than from the British universities. Several German universities had particularly flourishing scientific institutes; in chemistry, for example, there were the institutes of Liebig at Giessen, of Bunsen at Heidelberg, and of Ostwald at Leipzig; in physics were the institutes of Helmholtz at Berlin and of Weber at Göttingen. Liebig's teaching laboratories at Giessen, established in 1832, were particularly effective. By the end of the nineteenth century the scientific output from Germany was of a similar quantity and quality to that of all the rest of the world.

Governments

As noted in the previous section, governments in different countries have supported scientific work to very different extents. The policies of the British and French governments offer a particularly sharp contrast. Until recently the British government gave little financial support to the universities or to scientific research. Although they had Royal Charters, Oxford and Cambridge were private universities, financed by student fees and endowments. When in the middle of the nineteenth century a Royal Commission was established with the object of reforming the two universities there was considerable opposition, a number of colleges refusing to cooperate. William Whewell, then Master of Trinity College, Cambridge, although himself an enlightened reformer, regarded the Commission as 'an unwarranted and undesirable intrusion into the affairs of the university'. The Royal Society was also a private institution, with a Royal Charter but supported entirely by the dues paid by the Fellows.

In France, on the other hand, the universities were, from the early years of the nineteenth century, entirely state institutions. The Académie Royale des Sciences, founded in 1666, was well supported by the Crown, and the Académiciens were provided with salaries and with laboratory facilities.

The invention of photography in 1839, almost simultaneously in France and England but involving different techniques, provides an interesting example of contrasting government policies (see also Section 8.4 where the scientific aspects of the two methods are outlined). In France in January, 1839, the technique invented by Louis Jacques Mandé Daguerre was presented to the Académie des Sciences by its perpetual Secretary, François Arago, who had already achieved great distinction in physics and astronomy and had personally made some daguerreotypes after witnessing Daguerre's demonstrations of the technique. Arago was also active in politics, having since 1830 been the elected *Député* from Paris in the Chambre des Députés, and in August of 1839 he presented Daguerre's invention to the Chambre, with a strong plea for the financial support of further work. This received approval, and Daguerre was awarded a state pension for the rest of his life. The intention at the time was that as a result of these arrangements the invention would become 'free to the world'—*à tout la monde*—but this was not to be so since France could not control the policies in other countries. In March, 1840, an application on behalf of Daguerre was made to the Board of Treasury in England, enquiring whether the British Government would be interested in purchasing the rights to daguerreotypy for 'the purpose of throwing it open in England for the benefit of the public and preventing this important discovery being fettered or limited by individual interest or exertion'. This proposal was at once declined.

In England Henry Fox Talbot presented his technique of 'Photogenic drawing' to the Royal Society early in 1839, and received much moral support from colleagues such as Michael Faraday and Sir John Herschel. There was no official support from the government, however, and he was entirely on his own as far as developments and applications were concerned. After some initial difficulties, and finding that his and Daguerre's techniques were being used widely, he took out an English patent on his improved technique, the 'calotype', in 1841.

The situation is greatly complicated by the different patent laws applying in different countries. English patents, or more correctly Royal letters patent, confer on a person the sole right to use and sell an invention for a specified period of time. In France at the time the law regarding patents (or *brévéts*) did not provide such protection and gave little more than a moral recognition of the priority for the invention. As a result few French patents were taken out for photographic techniques. Various patents were, however, secured in England with respect to both Daguerre's and Talbot's techniques and to subsequent refinements of the procedures. Since there were so many improvements in techniques during the 1840s and 1850s it was found that the patents gave little protection, and by 1852 they had been largely abandoned, leaving an open field to amateur photographers and not much restriction on professional ones.

English patents were effective in Wales and in the British colonies but not in Scotland; as a result there was a considerable flourishing of photography in Scotland during the 1840s. Although the English patents applied to British

North America they were not in fact enforced there, and as a result there was a good deal of photography done, particularly in Montréal and Québec. There was also an early flourishing of photography, mainly daguerreotypy, in the United States, where no patents were taken out.

1.4 Technology and scientific research

The relationship between technology and scientific research is an intimate and important one. It is well recognized that advances in technology owe much to basic research, and there are many examples that illustrate this. Michael Faraday's fundamental discoveries concerning electricity and magnetism soon led to electric motors and generators of electricity, and thence to the great development of the modern electrical industry. Conversely, technical advances have had a great impact on scientific research; the electrical devices that resulted from Faraday's work have found much application in research.

Photography provides another example of the two-way relationship between pure science and technology. The birth of photography in 1839 (Section 8.4) owed much to fundamental knowledge of chemistry and of the nature of light; Henry Fox Talbot, for example, had done a good deal of work in spectroscopy before arriving at his particular system of photography. The impact of photographic techniques on scientific research, as well as on all other aspects of life, has been overwhelming. Spectroscopy, in particular, was completely transformed by photography (Section 6.1), which has found applications in almost every area of science.

Prior to World War II (1939–1945) research had to be done with the simplest of apparatus, and there were no great changes in procedures during the previous hundred years. Chemical analysis, for example, was still done by classical chemical methods, little use being made of spectroscopic techniques since few commercial spectrophotometers were available.

During World War II there were a number of important innovations in technology which led to great improvements in techniques used in research in physics and chemistry. As a result of increased governmental support for practical science directed to particular objectives there were many practical advances, and some of these were later adapted for use in scientific work.

Table 1.1 summarizes some of the techniques that are commonly used in modern research laboratories in the physical sciences. Some of the techniques, such as spectroscopy and ultrasonics, were known before World War II, but the techniques became improved and refined so as to make them more convenient for laboratory use. The ultrasonic methods, for example, were used in the 1920s and 1930s, but they were not very effective in research. During World War II underwater research led to great advances in ultrasonics, and the technique then became of great value, particularly for investigating the kinetics of rapid processes occurring in solution.

Electron spin resonance and nuclear magnetic resonance were unknown

TABLE 1.1 Some experimental techniques in physical chemistry

Year	Author	Technique	Section
1752	Melvill	Flame spectroscopy	6.1
1839	Daguerre; Talbot	Photography	8.4
1903	Zsigmondy	Ultramicroscope	(p. 294)
1906	Tswett	Chromatography	(p. 292)
1912	von Laue	X-ray diffraction	
1919	Aston; Dempster (following work of J. J. Thomson, 1910)	Mass spectrometry	6.7
1923	Svedberg	Ultracentrifuge	9.1
1925	Heyrovsky	Polarography and electroanalysis	7.5
1925	Pierce	Ultrasonics	8.5
1927	Davisson and Germer; G. P. Thomson	Electron diffraction	10.2
1928	Raman	Raman spectroscopy	6.7
1932	Ruska	Electron microscopy	
1937	Tiselius	Electrophoresis	9.1
1940	Chance	Stopped flow	8.5
1944	Zavoiskii	Electron spin resonance	6.7
1946	Purcell; Bloch	Nuclear magnetic resonance	6.7
1949	Porter and Norrish	Flash photolysis	8.4
1954	Bull and Moon	Molecular beams	8.8
1954	Eigen	Relaxation techniques (e.g., temperature jump)	8.5
1954	Townes	Maser spectroscopy	6.7
1958	Mössbauer	Mössbauer spectroscopy	6.7
1960	Maiman	Laser spectroscopy	6.7

before World War II, and were a direct product of the radio and radar work that was done during that war. Within only a few years commercial instruments had become available, and the techniques are now used in many laboratories as a matter of routine, as well as in scientific research.

The technique of flash photolysis (Section 8.4) developed as a result of wartime research on the photography of missiles in flight. At first the duration of the flashes was about a millisecond, but over the next four decades there had been a reduction by eleven powers of ten, to about 10^{-14}s, or 10 femtoseconds. The technique has provided a powerful means of studying the fastest of chemical and physical processes. The improvements in flash photolysis would not have been possible without the invention of lasers. From 1960, when a laser was first devised, there has been remarkable development in this field, and lasers now have many applications in research laboratories.

Physical chemistry, like physics and other exact sciences, has also been transformed by the development of computers. Before World War II calculations had to be made step by step with the aid of slide rules and calculating machines that could do no more than add, subtract, multiply and divide; they could not even take square roots directly. Computers did exist, but they were rare, bulky and expensive, and generally not available to research workers. The quantum-mechanical calculations made in the 1930s, for example, had to be done on paper with the aid of slide rules and simple calculators, and took considerable time, so that the amount of exploration that could be done was extremely limited.

The introduction of the transistor in the early 1950s brought about a radical change in the situation, greatly reducing the size and expense of computers and their power requirements. In the 1960s integrated circuits, small structures containing a number of individual components, were developed, and by the early 1970s thousands of integrated circuits could be incorporated into a single silicon wafer or 'chip' that was smaller than a postage stamp. These chips eliminated the kilometres of wiring that had been necessary in the earlier computers.

In the early 1970s the microprocessor was introduced, essentially a computer incorporated into a silicon chip. The result of all these developments was the ready availability of inexpensive pocket-size computers as well as larger instruments capable of storing a great deal of information. It is a striking fact that tiny modern computers are vastly more effective than the massive computers that were in use during World War II.

It was originally thought that computers would be used only for mathematical calculations, but this is only one of their functions today. It was soon realized that computers are remarkably effective for storing, processing and retrieving information, and they are now used for cataloguing books in libraries and for storing scientific data.

All of these functions have been of great value in modern research in the physical sciences. Operations that formerly were extremely time-consuming can now be carried out much more rapidly. For example, the experimental study of complex molecules such as proteins by X-ray crystallography has been completely transformed. Although the general methods had been completely worked out before World War II, the computational procedures then available were so laborious that months of tedious calculations were required to obtain the structure even of a fairly simple molecule. With modern computers the results even for large molecules can be obtained in a short time.

Quantum-mechanical calculations, such as are involved in the construction of potential-energy surfaces for chemical reactions, involve a considerable amount of computation, and with modern computers can be carried out much more effectively than was previously possible.

One pitfall not always avoided in this computer age is thinking that a scientific problem can be solved simply by making appropriate computer calculations. Often the computations in themselves fail to provide the necessary insight, and it is always necessary that they be carried out only as an adjunct to a proper theoretical analysis.

1.5 Some schools and laboratories of physical chemistry

It is now of interest to consider various matters relating to the natural habitat of the physical chemist. What kinds of laboratories have been available to physical chemists, and how have they developed over the years? What schools of research have become established, and how strong has been the influence of some of the early physical chemists? What centres of excellence have been established? The topic is a vast one, and various aspects have been considered in a number of publications; a large book could be written from the point of view of physical chemistry alone. Only a brief account can be given here, with a few examples to illustrate the general trends.

England

If, as has been suggested earlier, Robert Boyle was the first physical chemist, the story can begin with his career. His first research of any importance was done in Oxford from about 1655 to 1668, and it was there that he made his famous investigations on the pressure–volume relationships of gases and on various other topics. He was never a regular student at any university, but he chose to set up his laboratory in a private house in which he lodged in Oxford, since its university then seemed to him to be one of the most active research centres in the world. He there became associated with a number of men, including Christopher Wren and Robert Hooke, who were later to be active in London in the early years of the Royal Society. Boyle transferred his research to his sister's house in London in 1668.

Oxford can thus be said to be the birthplace of physical chemistry, and it continued for many years to be pre-eminent in the physical sciences. An important event was the opening in 1683 of the Ashmolean Museum, which continued to be used for scientific research and teaching until about 1860. This was the first building in Britain, and one of the first in the world, to be devoted to scientific work: Cambridge had to wait for about two centuries to have much more than a few college laboratories. Another important event in 1683 was the establishment by Oxford of the first professorship of chemistry in Britain, with the appointment of Robert Plot (1640–1696) who did his teaching in the Ashmolean. For some years the professorship lapsed for lack of funds, but there was an almost continuous series of teachers of chemistry in the chemical laboratory. One of them was John Keill, whose physico-chemical work, carried out under Newton's influence, will be mentioned in Section 3.2. The Revd John Desaguliers was also carrying out physico-chemical work in Oxford, from

1710 to 1712, but he does not appear to have been associated with the Ashmolean.

Cambridge was not so fortunate in its early physical scientists or in its laboratories. Its first professor of chemistry was appointed almost 20 years after Oxford's, and almost by accident. In about 1680 an Italian, John Francis Vigani (c. 1650–1713), arrived in Cambridge and began to teach chemistry, especially in its medical aspects. Queens' College allowed him the use of a room for his lectures and experiments, and he appears to have been on good terms with Newton who besides writing his *Principia* and other publications was carrying out some chemical and alchemical work in his laboratory at Trinity College. The importance of Vigani's contributions became recognized by the university authorities, and in 1703 the title of professor of chemistry was conferred on him, but no stipend was awarded to him or to some of his successors. Appointments have been made to this chair continuously until the present day, so that this Cambridge professorship is the oldest existing one in Great Britain.

On the whole, however, the Cambridge professors of chemistry, and other Cambridge chemists, were not of any particular distinction until the end of the nineteenth century. An exception was the Revd Richard Watson, whose work on the depression of the freezing point will be mentioned in Section 4.4. His tenure of the chemistry chair was brief, from 1764 to 1771, and much of his chemical work was done while he was neglecting his episcopal duties as Bishop of Llandaff. The quality of the Cambridge chemists began to improve in 1861 with the appointment of G. D. Liveing to the chair. Even he was not of great distinction, but after 1875, when James Dewar secured the Jacksonian chair of chemistry at Cambridge, the two began to collaborate on a useful series of spectroscopic investigations (Section 6.5).

An important reason for the greater early success of Oxford in the physical sciences was undoubtedly the existence of suitable laboratories. Besides the Ashmolean, there was a group of college laboratories that played an important role in scientific research, particularly in physical chemistry. The first of these was at Christ Church, and it was used for teaching and research in chemistry from 1776 to 1941; Vernon Harcourt's pioneering research in kinetics (Section 8.1) was done there. In 1848 Charles Giles Bridle Daubeny, who was an extremely enterprising professor of chemistry from 1822 to 1854, persuaded Magdalen College to let him build at his own expense a laboratory in the Physick (or Botanic) Garden, across the High Street opposite to the College and belonging to it. A little later Balliol College built a laboratory in some cellars and soon these were extended in collaboration with Trinity College. Important physico-chemical work was done in these college laboratories from the second half of the nineteenth century onwards. A little of Frederick Soddy's work that led to the 1921 Nobel Prize for his work on isotopes was done in one of the Balliol cellar laboratories, while Hinshelwood's 1956 Nobel Prize was awarded partly for work that he did in a delapidated converted lavatory that was part of the Trinity College laboratories.

Cambridge also had some College laboratories, particularly at St John's, Downing, Gonville and Caius, Sidney Sussex, Newnham and Girton Colleges, the latter two being necessary since women were not at first allowed to work in the other laboratories. These Cambridge college laboratories did not, however, play as important a role in research as did those at Oxford.

Cambridge also lagged somewhat behind Oxford in the organization of the teaching of science, and there are significant differences in the way this came about in the two universities. In Oxford the initiative came from the scientists themselves, an important role being played by Daubeny. He and a few others pressed the university authorities to establish a school of natural science, and they were finally successful in 1850. Previously the only way to a degree was to take the final examinations in *Literae Humaniores*, or 'Greats', covering ancient history and philosophy. After the Honour School of Natural Science was established in 1850, Greats was no longer required, although the student had to pass a preliminary examination, called Moderations, to demonstrate a knowledge of Latin and Greek.

The situation at Cambridge was different, in that the requirement for a degree had been the Mathematics Tripos, which at first sight would have been thought to lead more easily into a career in science. This tripos did not, however, involve any experience in the laboratory, laboratory facilities not being available to undergraduates. Until the middle of the nineteenth century Cambridge produced a number of distinguished mathematical physicists, and a few men like Stokes, Clerk Maxwell and Kelvin who gained a proficiency in experimental physics through their own efforts, but it did not produce many experimental chemists, even physical chemists. The establishment by Cambridge of a Natural Sciences Tripos owed little to pressure from scientists within the university; it resulted in large measure from concern on the part of the university authorities that they were going to be outstripped by Oxford, which by 1850 had its Honour School and was planning to build a substantial laboratory to house a number of the sciences.

This Oxford laboratory, known as the University Museum and still in use today, was also a result of pressure from the Oxford scientists rather than a consequence of administrative action, which indeed was somewhat hostile to the idea. The requirement for laboratory space was no longer being satisfied by the Ashmolean and the college laboratories, and Daubeny and several other influential scientists pressed for a new building. John Ruskin exerted an important influence on its design, and his rather romantic ideas about the nature of chemistry led to the curious result that the chemistry laboratory was a replica of the Abbot's Kitchen at Glastonbury Abbey, a thirteenth-century octagonal structure with a roof towering some 50 feet above the ground. Those who have worked in the building have been unanimous as to its inconvenience. So strong was Ruskin's spell that other universities, including Glasgow, Durham and Toronto, built chemistry laboratories to the same design very soon afterwards. The Oxford Museum's foundation stone was laid in 1855, and the laboratories began to be used in 1858, their first occupant being Benjamin Collins Brodie (1817–1880) who had succeeded Daubeny as professor of chemistry in 1855.

These developments at Oxford produced some concern at Cambridge where it was realized that its lack of laboratory facilities was putting it at a disadvantage. In 1870 the Seventh Duke of Devonshire, who was the university's Chancellor, offered to finance the construction of a building for experimental physics and to provide the necessary equipment and apparatus. This building became known as the Cavendish Laboratory, Cavendish being the family name of the Dukes of Devonshire (one famous kinsman being the great chemist Henry Cavendish). With the appointment in 1871 of Clerk Maxwell as the first Cavendish professor, and with such distinguished successors as Lord Rayleigh, J. J. Thomson, and Lord Rutherford, Cambridge soon established a pre-eminent place in the world of physics.

The Cambridge chemists had to wait a few more years for a building, and this was erected mainly though the exertions of Professor Liveing. Some laboratory facilities had been provided in a building in the Botanic Garden, and in 1872 Liveing persuaded the university authorities to build an upper storey to it for the teaching of theoretical and practical chemistry. This was still inadequate, and Liveing continued to press for better accommodation. Complications arose from the fact that non-scientists seem convinced—not entirely without justification—that chemistry laboratories blow up with unfailing regularity, and even when not doing so emit noxious fumes. One popular solution is to provide chemists with the upper storey of a building, apparently with the idea that if the laboratory blows *up*, those in the lower floors will not notice. A second solution, favoured in some of the college laboratories at Oxford, is to hide the chemists in strongly constructed cellars. One offer made to Liveing was that the chemists should work in 'a series of vaults' and that their quarters should be 'capable of standing violent explosions and [be] as uninflammable as possible'. Liveing, however, would have none of this, and insisted on a well-designed and equipped separate building. This was finally provided for him on Pembroke Street, not far from the Cavendish Laboratory and other scientific laboratories. After World War II a new large chemistry laboratory was built a little further from the centre of Cambridge, on Lensfield Road.

In the meantime other universities had been founded in England and had established their laboratories. The institution that later had a distinguished career as University College, London, opened its doors in a building on Gower Street in 1828, calling itself the University of London. When in 1836 it received its Royal Charter it was required to use the name University College, and at the same time a new foundation, the University of London, was created as the organization that was licensed to grant degrees. University College was founded by a group of Dissenters, and unlike Oxford and Cambridge imposed no religious restrictions on its teachers or students. The year 1836 saw the appointment of Thomas Graham, a man already of great distinction, as professor of chemistry. Ten years later Graham was able to teach chemistry in a newly opened chemistry laboratory based to some extent on Liebig's laboratory at Giessen. For the first time in any English university students were able to learn chemistry not simply through lectures but by carrying out personal experiments in a laboratory.

King's College, London, was opened in 1831, three years after University College. Within the next few decades colleges giving courses approaching a university standard were opened in other cities, such as Manchester, Birmingham and Leeds. At first the laboratory facilities in these were largely improvised. At King's College, for example, John Frederic Daniell and later William Allen Miller had to carry out their researches in electrochemistry (Section 7.3) and spectroscopy (Section 6.1) in a storage room under the lecture theatre.

A quite creditable amount of physical chemistry was done in England during the nineteenth century. Following the discovery of electrolysis by Nicholson and Carlisle there was a good deal of work in electrochemistry (Chapter 7), notably by Faraday at the Royal Institution, by Daniell at King's College, London, and by Grove at the London Institution. W. Allen Miller's work in spectroscopy (Section 6.1) at King's College, was also a significant contribution, as were Thomas Graham's investigations of diffusion and of colloidal systems (Chapter 9). In his private laboratories in Manchester James Prescott Joule, sometimes in collaboration with Kelvin, made his important investigations relating to the new science of thermodynamics (Chapter 4). Pioneering contributions to kinetics were begun by Vernon Harcourt in the college laboratories at Oxford (Section 8.2), and were further pursued by Dixon in the Balliol laboratories and later at the University of Manchester (Section 8.5). Later in the century William Ramsay at University College, London, did his Nobel-Prize-winning work leading to the discovery of argon. In addition to this work done by men who were primarily chemists, there were many important physico-chemical investigations by physicists such as Kelvin and Clerk Maxwell.

These contributions by the English scientists were, however, individualistic in style, with little administrative organization and little recognition of physical chemistry as a distinct discipline. At no English university was there a chair of physical chemistry until after World War I. Also, since the English universities were late in awarding advanced degrees, the influence of English physical chemistry was slow to spread to other countries. Oxford, for example, did not award a Doctor of Philosophy (D.Phil.) degree until 1917 and even then did not give research students any special status, treating them the same as undergraduates, a policy not always appreciated by graduates of other universities.

Scotland

The situation in Scotland offers an interesting contrast to that in England. In view of its much smaller population and wealth Scotland would not have been expected to be a leader in scientific research and teaching, but a number of sociological factors enabled it to become so. One of these was the great emphasis placed there on education, in order to offset economic factors, and as a result the Scottish people tended to be more literate than elsewhere. Facilities for education were provided more generously in Scotland than in England or on the continent of Europe. Scotland had also retained its cultural links with other countries, particularly France, and there was some exchange of teachers between the two countries. William Davidson, a graduate of Marischal College, Aberdeen, was in 1648 appointed the first professor of chemistry at the Jardin du Roi, and he then became known as d'Avisonne. After the Glorious Revolution of 1688, when James II was replaced by William and Mary on the British throne, French cultural influence in Scotland was to some extent replaced by Dutch. The Calvinist religion and culture became more prominent, and with it a greater tolerance of science and an appreciation of the scientific methods of Boyle and Newton.

Four Scottish universities were ancient foundations: St Andrew's, Glasgow and Aberdeen were founded in the fifteenth century, and Edinburgh in the sixteenth. For about three centuries there were thus twice as many universities in Scotland as in England with its much larger population. By the early eighteenth century all of the Scottish universities, and particularly those at Glasgow and Edinburgh, had reasonably satisfactory facilities for research and teaching in chemistry, and in this regard were over a century ahead of the English universities. A particularly important contribution to chemical teaching and research was made by William Cullen (1710–1790), who took an MD degree at the University of Glasgow and then held a professorship there until 1755 when he moved to the University of Edinburgh. His main contributions were in medicine, particularly physiology, but he had a great interest in the physical aspects of chemistry, such as chemical reactivity and the nature of heat. One whom he inspired to continue work along these lines was Joseph Black, who transformed the whole field of thermochemistry (Section 4.1), first at the University of Glasgow and then at the University of Edinburgh. By the end of the eighteenth century he was lecturing in chemistry to classes of 200 students, as compared with the handful of students in the chemistry classes at Oxford and Cambridge even a century later.

Important roles in Scottish physical chemistry were played by a number of others, including John Robison, a student of Black who published Black's lectures on heat. Another chemist of particularly influence was Thomas Thomson (1773–1852), who taught at the University of Glasgow for many years and conducted the first school of practical chemistry in any British university. Numbered among his students were Thomas Andrews and William Thomson (later Lord Kelvin), both of whom held him in high esteem. Thomas Thomson worked on heat, then regarded as a part of chemistry rather than of physics, and on research relating to Dalton's atomic theory. He was remarkably effective in establishing a research school, and many owed their success in chemistry to his teaching and inspiration. In 1831 he persuaded the reluctant university authorities to build new laboratories, the 'Shuttle Street Laboratories', for the teaching of chemistry. In the same year Thomas Graham began to teach a highly successful course in practical chemistry at the Andersonian College in Glasgow, a private college without charter which had been founded as a result of a bequest by John Anderson (1729–1796).

One consequence of these successes of the Scottish universities in the teaching of chemistry was that graduates of the Scottish universities began to play a disproportionately prominent role in English institutions and industries. Thomas Graham, for example, later became a professor at London University, and was the first president of the Chemical Society of London.

North America

Until late in the nineteenth century relatively little physical chemistry, or indeed science of any kind, was done in the United States or Canada. It is true that there were a few contributions of major importance, such as those of Benjamin Franklin, Joseph Henry and John Draper, but there was even less organization of research than in England and much less than in the Scottish universities. Towards the end of the nineteenth century the situation began to change, and research in North America developed rapidly, the United States eventually becoming the leader in most branches of science.

It might have been expected that the change would have come as a result of the influence of the British scientists, in view of the strong cultural links, but this was not the case and the reason is not hard to find. Science in Britain, even to some extent in Scotland, had been rather individualistic in style, the English scientists in particular tending to have few contacts with scientists in other countries. In addition, as previously noted, the English universities were not organized in a way to attract research students, having no graduate schools and offering no advanced degrees.

As far as physical chemistry is concerned, the main influence can be clearly traced to the men referred to as the Ionists, van't Hoff, Wilhelm Ostwald and Arrhenius. Their work attracted considerable attention, and Ostwald was by far the most dynamic and charismatic of them. Van't Hoff was probably the greater scientist, but although of a kindly nature he tended to be reserved and uncommunicative. Arrhenius was dynamic, but his personality was a somewhat difficult one. Ostwald was from all accounts the ideal research director, being freely available to his students at all times and giving them every encouragement.

Ostwald was at Leipzig from 1887 until his retirement in 1906, and in 1897 he acquired a fine new building for his research. He attracted a considerable number of students, the majority of them from outside Germany since he had somewhat antagonized German chemists, most of them organic chemists, with his rather blunt comments; he told them, quite correctly but not too tactfully, that they should make use of physical methods, but most were not ready for such advice; Emil Fischer replied curtly 'I have no need of your methods.'

On the occasion of the celebration of Ostwald's fiftieth birthday he had already directed the research of 147 students, of whom 34 were professors and most had achieved scientific distinction. Over 40 American or Canadian chemists worked with Ostwald, and by the beginning of World War I many of these had become full professors at major universities, including Harvard, the Massachusetts Institute of Technology (MIT), the California Institute of Technology (at Pasadena), the University of California (at Berkeley), and the University of Toronto. For many years the bulk of the teaching of physical chemistry at the universities in North America was done by professors who had either worked with Ostwald or had come less directly under his influence. Perhaps the most influential of these in the early years was Arthur Amos Noyes, whose career is also referred to in Section 2.3 in connection with the physical chemistry journals.

It is of interest that Noyes went to work with Ostwald by something of an accident. He had gone to Europe in 1888 with the intention of working in organic chemistry with Adolf von Baeyer at the University of Munich. However, on arriving in Rotterdam he learnt that Baeyer had no room for him, and he decided to go to Ostwald instead. After working at Leipzig from 1888 to 1890, on a problem relating to deviations from van't Hoff's law of solution, Noyes went to MIT where he established what he called the Research Laboratory of Physical Chemistry. The agreement that he made with MIT. was an unusual one. The administration supplied a building to house the research laboratory, and in return Noyes supplied half of the operating costs, including the salaries, out of his own pocket. The research done in the laboratory was concerned with the properties of solutions, particularly at high temperatures, and the determination of free-energy changes in chemical reactions. Much of it was not of great originality; its importance lay in the fact that Noyes was planting the seeds of physico-chemical research that were to bear abundant fruit in the years to come. During this MIT period Noyes surrounded himself with a number of people who carried out work of great distinction. One of them was G. N. Lewis, much of whose work on activities (Section 4.4) and on atomic theory (Section 10.3) was done while he was at MIT. Another was R. C. Tolman, later well-known for his contributions to statistical mechanics (Section 5.4) and to other fields. The organic chemist C. S. Hudson, of 'Hudson's rule' fame, was also in Noyes's department.

Although Noyes served as acting President of MIT from 1907 to 1909 and had been offered but had refused its presidency, his relationship with the administration became somewhat strained, and in 1919 he resigned his position and moved to the Throop College of Technology in Pasadena, California, where he remained to the end of his life; the College later became known as the California Institute of Technology. At Pasadena he built up, for a second time, a strong research laboratory for physical chemistry. In 1922 the future Nobel Prize winner Linus Pauling began graduate research there, and Noyes did much to support his subsequent career in the department and to make possible his later great achievements in physical chemistry.

Another important school of physical chemistry was founded at the University of California at Berkeley in 1912. In that year G. N. Lewis, who had been Noyes's deputy director of research in physical chemistry at MIT, was appointed Dean of the College of Chemistry at Berkeley. In making the move he took with him several graduate students and three colleagues. One of them was the Canadian-born William C. Bray, who after taking his degree at the University of Toronto also worked with Ostwald at Leipzig, and did important work in thermodynamics and kinetics; another was Merle Randall, with whom Lewis wrote his authoritative book on thermodynamics; the third was R. C. Tolman. Bray and Randall remained for many years at Berkeley, but Tolman left in 1916. The College of Chemistry was greatly strengthened in 1913 by the appointments of the Englishman George Ernest Gibson, who had taken his Ph.D. degree at the University of Breslau with Lummer, and Joel C. Hildebrand who had taken his degree and taught at the University of Pennsylvania. For the next 20 years all further appointments were of people who had taken their degrees at Berkeley, two of them being William F. Giauque and Willard F. Libby, both future Nobel Prize winners. In 1935 the future Nobel Prize winner Melvin Calvin (b. 1911) came to the College from the University of Minnesota.

Both the Berkeley and Pasadena laboratories soon developed into two of the most distinguished physical chemistry laboratories in the world. It is interesting that although the research in both of them had come under the influence of Ostwald, who had done much work in kinetics, little work in that field was done in either Pasadena or Berkeley in the early years. Bray at Berkeley was the only one to do much work in kinetics, the main emphasis in both departments being on thermodynamics and the properties of electrolyte solutions. Lewis for the most part showed little interest in kinetics, but he did work on one interesting kinetics problem, relating to the radiation hypothesis (Section 8.3), in the 1920s.

One of the first of the American universities to begin a strong programme of research in kinetics was Princeton after the arrival of H. S. Taylor in 1914. Two Leipzig graduates, George A. Hulett and the Scottish-born Alan W. C. Menzies, were at Princeton for many years and worked on more classical aspects of physical chemistry. After taking a degree at the University of Liverpool, Taylor spent a year with Arrhenius and a year with Bodenstein before going to Princeton where he remained to the end of his life. He did important work on a variety of kinetic problems, dealing with homogeneous reactions, reactions on surfaces, and reactions induced by radiation. As Chairman of the chemistry department and Dean of the graduate school he exerted a wide influence, one of his more significant actions being the appointment in 1931 of Henry Eyring, who did much of his important work on theoretical kinetics while he was at Princeton.

The relationship between Canada and the United States is in some ways similar to that between Scotland and England. Canada's population is about one-tenth that of the United States and its *per capita* wealth is much smaller.

Canada's educational system is more uniformly satisfactory than that of the United States. Canada's role in scientific research has not always been evident, since many scientists educated in Canada have done much of their work in the United States.

The first Canadian universities to do much work in physical chemistry were the University of Toronto and McGill University in Montreal. The research in both of these universities was much influenced by the European laboratories. Mention has been made of Toronto's first chemistry building, in the style of the Abbot's Kitchen at Glastonbury, which was designed in 1856 and opened in 1859, one year after the similar building in Oxford. Physical chemistry at the University of Toronto was initiated by William Lash Miller (1866–1940) and Frank Boteler Kenrick (1874–1951). After graduating from the University of Toronto they both went to Europe, where Miller obtained two Ph.D. degrees, one from Munich and one from Leipzig, where he worked with Ostwald. Kenrick obtained a doctorate from Leipzig. Miller exerted a particularly strong influence on the Toronto chemistry department, organizing a physical chemistry sub-department in 1900, and a sub-department of electrochemistry in 1904. His own research was mainly in thermodynamics, and he played a role in interpreting the work of Willard Gibbs. Important work in electrochemistry was later done by Andrew Robertson Gordon (1896–1967) and Frank Ellsworth Waring Wetmore (1910–1963), and in kinetics by Donald James Le Roy (1913-1985). The work in molecular dynamics by John Polanyi, also done at the University of Toronto, is outlined in Section 8.8.

McGill University completed its first chemistry building in 1862; a later and more satisfactory laboratory, the Macdonald Building, was opened in 1898 and still stands, but the chemists moved to the new Otto Maass building in 1965. One of the early teachers of chemistry was the Scotsman J. W. Walker who had obtained his Ph.D. degree at Leipzig. A particularly important contribution to physical chemistry was made by Otto Maass (1890–1961), a McGill graduate who went to work with Nernst, narrowly escaping from Germany at the outbreak of World War I; later he took a Ph.D. degree at Harvard under T. W. Richards A hundred and thirty-one McGill students obtained their doctorates under Maass's direction. His work covered a wide range, including calorimetry and investigations of the critical conditions. Important work in kinetics at McGill was later done by Edgar William Richard Steacie (1900–1962), and by Carl Arthur Winkler (1909–1978), who after obtaining his Ph.D. degree with Maass at McGill obtained a D.Phil. degree at Oxford under Hinshelwood.

Steacie obtained his Ph.D. in 1926 from McGill University, and he later worked at the University of Frankfurt with K. F. Bonhoeffer. He taught at McGill until 1939 when he was appointed Director of the division of chemistry at the National Research Council in Ottawa, becoming its President in 1952. In spite of heavy administrative duties he succeeded in continuing his work in chemical kinetics, his main interest being in the kinetics of elementary reactions involving atoms and free radicals, which he investigated by photochemical techniques. Steacie's greatest achievement, however, was in establishing strong programmes of basic scientific research in Canada. He was instrumental in initiating a postdoctoral fellowship scheme which attracted many capable young scientists to Canada, and which had the result of greatly strengthening the scientific work not only in the National Research Council but in the universities.

Unfortunately, subsequent to his untimely death in 1962, Steacie's good influence has been severely eroded by the unwise policies of politicians and government administrators. This indeed has been a serious problem in many countries during the past few decades. One is struck by the contrast between modern governmental policies and the farsighted action taken by the French politicians in 1839 on the occasion of the invention of photography (Section 1.3). An important factor then was that a distinguished scientist, Arago, was playing an active role in politics, and could ensure that advances in photographic techniques were properly encouraged. Now that politics has become so professional and time-consuming it is no longer possible for a scientist to become a politician and remain active in science. We scientists must do everything we can to ensure that appropriate policies are applied to our work. Whether or not we accept Einstein's remark that chemistry is too difficult for chemists, we can hardly doubt that politics is too difficult for politicians.

Suggestions for further reading

For accounts of the development of the physical sciences, particularly in their social context, see

H. Butterfield, *The Origins of Modern Science*, 1300–1800, G. Bell and Sons, London, 1965.

D. S. L. Cardwell, *The Organisation of Science in England, Heinemann*, London, 1957, 1972.

Sir William Dampier, *A History of Science, and its Relations with Philosophy and Religion*, Cambridge Univ. Press, 1938.

Mary Hesse, *Science and the Human Understanding*, SCM Press, London, 1954.

A. Rupert Hall, *From Galileo to Newton*, Harper & Row, New York, 1963; Dover Publications, New York, 1983.

B. W. G. Holt, Social aspects in the emergence of chemistry as an exact science: the British chemical profession, *British Journal of Sociology*, **21**, 181–199 (1970).

Janet Howarth, Science education in late-Victorian Oxford: a curious case of failure, *English Historical Review*, **102**, 334–371 (1987).

Trevor H. Levere, Elements in the structure of Victorian science, in *The Light of Nature* (Eds J. D. North and J. J. Roche), Martinus Nijhoff, Dordrecht, The Netherlands, 1985.

J. T. Merz, *A History of European Thought in the Nineteenth Century*, Blackwood, London, 1904–1912; Dover Publications, New York, 1965. Especially Volume 1, Part 1, *Scientific Thought*.

J. R. Partington, *A History of Chemistry*, 4 vols., Macmillan, London 1961.

H. T. Pledge, *Science Since 1500: A Short History of Mathematics, Physics, Chemistry, Biology*, H. M. Stationery Office, London, 1939, 2nd edition 1966; Harper & Brothers, New York, 1939; reprint with Prefatory Note, 1959.

For discussions of the application of mathematics to science see

W. W. Rouse Ball, *A Short Account of the History of Mathematics*, London, 1908; Dover Publications, New York, 1960.

Harvey W. Becher, William Whewell and Cambridge mathematics, *Hist. Stud. Phys. Sci.*, **11,** 1–48 (1980).

John W. Servos, Mathematics and the physical sciences in America, 1880–1930, *Isis*, **77,** 611–629 (1986).

The relationship between science and the churches is discussed in the books by Butterfield and Hesse and also in

James R. Jacob and Margaret C. Jacob, The Anglican origins of modern science: The metaphysical foundations of the Whig constitution, *Isis*, **71,** 251–267 (1980).

Jack Morrell and Arnold Thackray, *Gentlemen of Science: Early Years of the British Association for the Advancement of Science*, Clarendon Press, Oxford, 1981.

Richard S. Westfall, *Science and Religion in 17th Century England*, Archon Books, Hamden, Conn., 1970.

Some other books that are helpful in providing background material in connection with the subject matter of this chapter are

T. K. Derry and Trevor I. Williams, *A Short History of Technology*, Clarendon Press, Oxford, 1960.

A. Hellermans and B. Bunch, *The Timetables of Science. A Chronology of the Most Important People and Events in the History of Science*, Simon and Schuster, New York, 1988; updated edition, 1991.

Mary Jo Nye, *Science in the Provinces: Scientific Communities and Provincial Leadership in France, 1890–1930*, Berkeley, 1986.

Robert Fox and George Weisz (Eds.), *The Organization of Science and Technology in France, 1808–1914*, Cambridge, 1980.

Trevor I. Williams, *The History of Invention*, Facts on File Publications, New York & Oxford, 1987.

Trevor I. Williams, *Science. A History of Discovery in the Twentieth Century*, Oxford University Press, 1990.

A useful reference book is

R. M. Gascoigne, *A Historical Catalogue of Scientists and Scientific Books from the Earliest Times to the Close of the Nineteenth Century*, Garland, New York, 1984.

Informative and detailed accounts of the development of chemistry departments in Great Britain and Ireland are given in a series of articles, with the general title 'Schools of chemistry in Great Britain and Ireland', in the *Journal of the Royal Institute of Chemistry*, Volumes 77 to 82 (1953–1958); for example
Cambridge: W. H. Mills, *JRIC*, **77**, 423–431, 467–473 (1953).
Oxford: Sir Harold Hartley, *JRIC*, **79**, 116–127, 176–184 (1955).
King's College, London: D. H. Hey, *JRIC*, **79**, 305–315 (1955).

For accounts of chemistry in some of the English universities see in addition
E. J. Bowen, Chemistry in Oxford: The development of the university laboratories, *Chemistry in Britain*, **1**, 517–520 (1965).
F. M. Brewer, The place of chemistry. I. At Oxford, *Proc. Chem. Soc.*, 185–189 (1957).
F. M. Brewer, Oxford as a home of chemistry and industry, *Chem. and Ind.*, 845–853 (1961). On the article the author's name is given as J. M. Brewer, but the author was Frederick Mason Brewer (1902–1963).
J. P. Earwaker, Natural science at Oxford, *Nature*, **3**, 170–171 (1870). This gives a good account of the situation at the time.
Negley Harte, *The University of London, 1836-1986*, The Athlone Press, London, 1986.
Negley Harte and J. North, *The World of University College, London*, University College, London, 1978; revised edition, 1991.
Janet Howarth, Science education in late-Victorian Oxford: a curious case of failure?, *English Historical Review*, **102**, 334–371 (1987).
K. J. Laidler, Chemical kinetics and the Oxford college laboratories, *Arch. Hist. Exact Sci.*, **38**, 197–283 (1988).
F. G. Mann, The place of chemistry. II. At Cambridge, *Proc. Chem. Soc.*, 190–193 (1957).
D. R. Oldroyd and D. W. Hutchings. The chemical lectures at Oxford (1822–1854) of Charles Daubeny, M.D., F.R.S., *Notes and Records of the Royal Society*, **33**, 217–259 (1979).
A. V. Simcock, *The Ashmolean Museum and Oxford Science*, Museum of the History of Science, Oxford, 1985.
Tom Smith, The Balliol-Trinity laboratories, in *Balliol Studies* (Ed. John Prest), Leopard's Head Press, London, 1982.
F. Sherwood Taylor, The teaching of science in Oxford in the nineteenth century, *Annals of Science*, **8**, 82–112 (1952).

For accounts of chemical research in Scotland see some of the above-mentioned articles in the *Journal of the Royal Institute of Chemistry*, for example
Edinburgh: E. L Hirst and M. Ritchie, *JRIC*, **77**, 509–517 (1953).
Glasgow: J. W. Cook, *JRIC*, **77**, 561–572 (1953). See also
David V. Fenby, Chemistry during the Scottish enlightenment, *Chemistry in Britain*, 1013–1016 (November, 1986).

Andrew Kent, The place of chemistry. X. In the Scottish universities, *Proc. Chem. Soc.*, 109–113 (1959).

Ostwald's laboratories in Leipzig and their influence are discussed in
K. J. Laidler, Chemical kinetics and the origins of physical chemistry, *Arch. Hist. Exact Sci.*, **32**, 43–75 (1985).
J. W. Servos, *Physical Chemistry from Ostwald to Pauling: The Making of a Science in America*, Princeton University Press, 1990.

For an account of physical chemistry in Germany see
W. Jost, The first 45 years of physical chemistry in Germany, *Annual Review of Physical Chemistry*, **17**, 1–14 (1966).

The establishment of physical chemistry laboratories in the United States is discussed in
Melvin Calvin and Glenn T. Seaborg, The College of Chemistry in the G. N. Lewis era: 1912–1946, *J. Chem. Education*, **61**, 11–13 (1984).
John W. Servos, G. N. Lewis: The disciplinary setting, *J. Chem. Education*, **61**, 5–10 (1984).
J. W. Servos, *Physical Chemistry from Ostwald to Pauling*, op. cit.

For information about early physical chemistry in Canada see
J. T. Edward, McGill chemistry, *Canadian Chemical News*, 12–17 April 1984.
M. Christine King, *E. W. R. Steacie and Science in Canada*, University of Toronto Press, 1989.
C. J. S. Warrington and R. V. V. Nicholls, *A History of Chemistry in Canada*, Pitman, New York, 1949.

CHAPTER 2

Communication in the physical sciences

Progress in the sciences is closely linked to the manner in which scientists communicate with each other and with the general public. Over the years there have been important changes in the extent to which different kinds of communication have been used. In earlier times scientists tended to be secretive about their work, but were faced with the dilemma of not being able to claim priority for a discovery if anyone else announced it later. Newton was involved in a number of difficulties of this kind, and often became engaged in acrimonious controversy. One way around the dilemma was to summarize a discovery in the form of an anagram so complicated that it could only be deciphered by the author himself. In its early days the Royal Society was a repository for anagrams.

Much of the communication of early science was in the form of personal letters. Robert Boyle, for example, described many of his experiments in the form of letters to friends and relatives, often publishing them later as pamphlets. He also communicated many of his findings orally to the numerous visitors to his laboratories in Oxford and London. Even later the same kind of informal communication was still used to some extent. In the nineteenth century Maxwell, Kelvin and P. G. Tait were frequent correspondents on matters relating to thermodynamics and the kinetic theory. Very little of Maxwell's important work on thermodynamics, including his 'Maxwell demon', was published in a formal way, and our knowledge of much of it has come to us indirectly from the many private letters (and often postcards!) he wrote, and from his university lectures.

Since it is now recognized that priority in scientific work can only be established by a proper publication in the form of a book or an article in a scientific journal, the informal methods of communication are only used to supplement formal publication.

Robert Boyle published some of his work in the *Philosophical Transactions*, but most of it appeared in privately-published books and pamphlets. Since he was a wealthy man this presented no problem, but it is obviously not a satisfactory procedure for the most part. Little original work in the physical sciences is now published in that way, perhaps the last important example being van't Hoff's famous pamphlet of 1874 on the 'tetrahedral carbon atom'.

Today practically all research appears in scientific periodicals, of which there

36

are two kinds: the institutional, which are issued officially by scientific societies, and the non-institutional, which are not. The history of scientific journals, whether institutional or not, is inextricably bound up with the history of scientific societies, and the situation has varied a good deal from country to country. The present account does no more than outline the main trends; some details about the journals referred to in this book are included in the Appendix.

2.1 General scientific periodicals

The first scientific society to be formed was the Italian Accademia del Cimento, or Academy of Experiments, which was set up in Florence in 1657. It was both a scientific institute and a scientific society, and it survived for only about ten years. The second scientific society to be formed, the Royal Society of London, was founded in 1660 and survives to this day. It was founded by a group of men who for some time had met informally from time to time at Robert Boyle's lodgings in Oxford, at Gresham College in London and in a few other places. In 1660 they formally banded together as the Royal Society of London for Improving Natural Knowledge—to give it its little-used full title. In 1662 the Society was granted its first Charter by King Charles II. It has met regularly from then until the present day, and except for a period in the eighteenth century has maintained high standards. Its aim was to 'improve the knowledge of naturall things, and all useful Arts . . . by experiments—(not meddling with Divinity, Metaphysics, Moralls, Politicks, Rhetorick, or Logick)'. The work was to 'advance the glory of God, the honour of the King . . . and the general good of mankind'. At first the Royal Society was essentially a gentlemen's club with a membership of enthusiastic and highly talented amateurs. Membership was open to any who wished to join, and the minutes of the early years do not reveal that any applicant was rejected; it may well be, however, that men who might have been rejected, such as the philosopher Thomas Hobbes (1588–1679), who was an avowed atheist, knew better than to apply.

The initial success of the Royal Society sprang from the fact that its Original Fellows, of whom there were 48, were enthusiastic men of outstanding ability. Important to its success was the appointment of Henry Oldenburg (c.1617–1677) as one of the first Secretaries of the Society. Oldenburg was a rather unusual individual who on the face of it would not seem to have been a suitable candidate for the position. He was a German, born in Bremen, and held the degree of Master of Theology. Having a good knowledge of English he was sent to England by the Free City of Bremen as agent and envoy. He became a friend of the poet John Milton and studied for some time at Oxford where he became acquainted with Robert Boyle. It was Boyle who recommended Oldenburg for election to the Fellowship of the Royal Society and for the post of Secretary. Oldenburg entered very conscientiously into the duties of his position, engaging in much correspondence with Fellows and with persons abroad, and organizing the discussions that took place at the regular meetings of the Society.

The Philosophical Transactions

Probably the most important action he took was to launch in 1665 the *Philosophical Transactions* (see Figure 2.1), which except for brief periods has been published regularly until the present day. It is the oldest scientific publication of any importance, and nearly the oldest of any; the oldest, the *Journal des Sçavants*, which survived until 1790, was started only two months before the *Philosophical Transactions*. It is often said that the *Transactions* was from the start a publication of the Royal Society, but this statement requires qualification. The Society passed a resolution 'that the *Philosophical Transactions*, to be composed by Mr Oldenburg, be printed the first Monday of every month if he have sufficient matter for it; and that the tract be licensed by the Council of the Society, being first reviewed by some of the members of the same; . . .'. However, the Society took no financial responsibility for the journal, which was a private speculation on the part of Oldenburg, who hoped to make a profit from it.

The *Philosophical Transactions* was an immediate success not only in Britain but on the continent of Europe, although it does not appear that Oldenburg's annual profit was ever more than £50, and it was often much less. That it was such a success is a tribute to its quality, since at the time the English language was not well known outside Britain and the American continent. Various translations of the journal appeared, including Latin translations of the early volumes, published in Amsterdam even while The Netherlands was at war with England (1664–1676).

Besides editing, and sometimes translating, articles submitted to the *Transactions*, Oldenburg reprinted articles that had appeared elsewhere. He also wrote numerous reviews of books and excellent accounts of the proceedings at meetings of the Society and of demonstrations that had been carried out at them. He succeeded in getting the monthly issues out on time, except for delays resulting from the Great Plague of London in 1665, and arising from his two-month imprisonment in the Tower of London in 1667, due apparently to an indiscreet letter he had written to someone on the Continent criticizing the conduct of the war with the Dutch.

A perusal of the early volumes of the *Transactions* reveals a surprising number of names of famous men whose work is described or who had contributed papers— Robert Boyle, Robert Hooke, Edmund Halley, the Revd John Flamsteed, Christiaan Huygens, and Christopher Wren among many others. Two interesting papers were from John Winthrop (1606–1676), Governor of Connecticut, who was one of the Original Fellows. Newton published a few papers in it between 1672 and 1676, but when his paper on light and colour received some criticism he characteristically sulked, and never published in it again or had much to do with the Society until he served as its President from 1703 until his death in 1727.

Another reason for the early success of the Royal Society was that Robert Hooke was appointed its Curator of Experiments, with the onerous duty of

PHILOSOPHICAL
TRANSACTIONS:
GIVING SOME
ACCOMPT
OF THE PRESENT
Undertakings , Studies , and Labours
OF THE
INGENIOUS
IN MANY
CONSIDERABLE PARTS
OF THE
WORLD.

Vol I.

For *Anno* 1665, and 1666.

In the *SAVOY*,
Printed by *T. N.* for *John Martyn* at the Bell, a little with-
out *Temple-Bar* , and *James Allestry* in *Duck-Lane*,
Printers to the *Royal Society*.

FIGURE 2.1   The title page of the first volume of the *Philosophical Transactions*.

performing an original experiment or demonstration at each one of the Society's weekly meetings. Hooke, who had worked with Robert Boyle in Oxford and had designed a highly efficient air pump that was effective for the pressure–volume studies, was an ingenious experimentalist who also had a sound grasp of theory. Some of the demonstrations performed were rather bizarre, such as the autopsy on a man who had swallowed a musket ball with the object of clearing his gut, but many played a valuable role in advancing basic knowledge. Like Newton, Hooke had a difficult and quarrelsome personality, and the two men were often at loggerheads, particularly on matters of priority. Oldenburg also disliked Hooke and did what he could to foster the disagreements between him and Newton. He also caused trouble between Hooke and Huygens regarding the invention of a spring-controlled watch. Also, in his reports of Society meetings Oldenburg tended to neglect Hooke's contributions, and as a result Hooke refused to publish in the *Transactions*.

When Oldenburg died in 1677 Hooke was unanimously elected to succeed him as Secretary, and he held the position until 1682. Not surprisingly, he declined to edit the *Philosophical Transactions*, and its publication lapsed for five years. In 1679, after it had ceased to appear for two years, the Council of the Society requested Hooke to produce some kind of a news-sheet, and he edited what was called the *Philosophical Collections* from 1679 to 1682. His successor as Secretary was Robert Plot (1640–1696) who had been a close friend of Boyle's at Oxford, and who in 1683 became the first Keeper of the Ashmolean Museum and the first professor of chemistry at Oxford, retaining these positions until 1690. Plot at once revived the *Philosophical Transactions*, and the Secretaries who succeeded him continued its publication.

Since for many years the *Philosophical Transactions* was edited by successive Secretaries of the Society, contained detailed reports of the activities of the Society, and included articles written for the most part by Fellows and approved by a Society committee, it was widely assumed that it was the official publication of the Society. The Society was, however, at pains to deny this, and on several occasions issued statements to the effect that the journal was the sole responsibility of its editor; the successive editors were naturally glad to take the credit for what from the beginning had been an outstandingly successful publication. In 1753, however, the Society finally bowed to the inevitable and accepted responsibility for the publication, establishing a committee to oversee it. In 1776 the title of the journal was officially changed to the *Philosophical Transactions of the Royal Society*.

The decline in the quality of British science during the eighteenth century is strikingly confirmed by an examination of the *Philosophical Transactions* during that period; few great names appear, and few articles are of much importance.

Other British journals

Following the revival of British science in the early nineteenth century the Royal Society began another publication, the *Transactions of the Royal Society*, first issued in 1832. Its first six volumes consisted of reports of the business transacted at the meetings of the Society, and brief accounts, usually two to three pages in length, of scientific work presented at those meetings. Many of these papers later appeared in the *Philosophical Transactions*. In 1854, with the issue of Volume 7, the format was changed; full papers then began to be published, and the title of the publication became the *Proceedings of the Royal Society*. After that, the *Philosophical Transactions* modified its character somewhat and began to concentrate on longer articles covering wider fields of research and including more review material.

The year 1797 saw the appearance in England of a non-institutional journal, the *Journal of Natural Philosophy, Chemistry and the Arts*. This journal was founded by William Nicholson (1753–1815) and came to be known as *Nicholson's Journal*. Nicholson was primarily a journalist and science writer, with a sound knowledge of science. In 1800, together with Anthony Carlisle (1768–1840), a fashionable London surgeon who was later knighted and became President of the College of Surgeons, Nicholson electrolyzed water for the first time, making use of the 'pile' that the Italian Alessandro Volta (1745–1827) had just invented. Nicholson reported this work in the first volume of his journal and, as further discussed in Section 7.1, it attracted considerable attention, leading to the construction of Voltaic piles in many places and contributing greatly to the success of the new journal.

In the following year the *Philosophical Magazine*, another non-institutional journal, was founded by Alexander Tilloch (1759–1825), who was primarily a journalist. This was not at first as successful as *Nicholson's Journal* which attracted better scientific publications. Nicholson did not, however, gain much financial return from his journal, and when in 1813 Thomas Thomson (1773–1852) brought out yet another journal, the *Annals of Philosophy*, Nicholson sold his journal to Tilloch and it became absorbed in the *Philosophical Magazine*. In 1819 the Scottish physicist David Brewster (1781–1868) founded the *Edinburgh Philosophical Journal*, which was often known as *Brewster's Philosophical Journal*, and he followed it in 1824 with the *Edinburgh Journal of Science*. In 1832 the latter also became absorbed in the *Philosophical Magazine*, which became greatly strengthened by these two incorporations, its title officially becoming the *Philosophical Magazine and Journal of Science*. Two of its editors, Richard Taylor (1781–1858) and the chemist William Francis (1817–1904) set up a publishing firm, Taylor and Francis, which to this day publishes the *Philosophical Magazine* and the *Proceedings of the Royal Society*, as well as many other periodicals and books.

In the early years of the nineteenth century there was some concern among scientists in Britain that the Royal Society was not being as effective as it could be; as already noted, the Society went through something of a decline in the eighteenth century, as did scientific and other intellectual work in Britain. In 1831 a group of prominent scientists began a new scientific society, the British Association for the Advancement of Science. This was much less exclusive than the Royal Society, to which by this time only a limited few could be elected. Whereas the Royal Society held all of its meetings in London, the Association held its annual meetings in different British cities, and later held some abroad. One of its aims was to communicate science to the general public. It issued an annual Report on what occurred at each annual meeting. The Association was successful in that it was well supported by prominent scientists, and its meetings were lively and contributed much to the progress and popularization of science. One important function that the Association performed effectively was to face squarely the opposition to science from both High Churchmen and Low Churchmen. Of the 20 founders of the Association, 18 were Anglicans and 9 of them were clergymen; all were members of the liberal Broad Church and were in a strong position to defend science effectively against attacks from church-men holding opposing views.

The year 1869 saw the founding of an important non-institutional journal, *Nature*, which still flourishes today. It was begun by Joseph Norman Lockyer (1836–1920), who was its editor for 50 years, almost until his death. At the time of the journal's foundation Lockyer was a civil servant with an amateur interest in astronomy; he later made important contributions to astronomy (Section 6.3) and became Director of the Solar Physics Laboratory in South Kensington.

French journals

In France the Académie Royale des Sciences was established in 1666, just four years after the Royal Society received its Charter. There are some important differences between the Académie and the Royal Society, which although it had a Royal Charter was essentially a private club, receiving no financial support from the state and relying on the Fellows' subscriptions of a shilling a week, which were often in arrears. The French Académiciens, of whom there were originally 16, were well supported by the Crown, being given salaries, and soon after the founding of the Académie they were provided with excellent laboratory facilities in the newly-built Paris Observatory.

At first the Académie did not have a serial publication, adopting instead the policy of publishing in book form collections of its members' scientific papers. This procedure, however, proved unsatisfactory, and in 1699 the Académie began its *Histoire et Mémoires*. Each issue consisted of a small section, the *Histoire*, containing news of the activities of the Académie and summarizing the papers that appeared in the later part, the *Mémoires*, which contained full accounts of the communications that had been made at the weekly meetings of the Académie.

That the *Histoire et Mémoires* was a great success was largely due to a remarkable man, Bernard Le Bovier de Fontenelle (1657–1757), who was Secretary of the Académie from 1697 to 1741. Fontenelle died one month before his hundredth birthday, and was an active writer on a number of topics for nearly 80 years, which perhaps constitutes a record for literary longevity. He began as a poet and playwright—not a very good one—and was well versed in philosophy. Although not really a scientist at all he became a great popularizer of science, about which he wrote in an elegant and lively style. In 1686 appeared his *Entretien sur la pluralité des mondes* (A discourse on the plurality of worlds), which presented science in a form that could easily be understood and which was read avidly by fashionable ladies of the Court of Versailles. In the remaining 71 years of Fontenelle's life the book ran to 31 editions. His contribution to the *Histoire* included writing annual reviews of the activities of the Académie, and preparing his famous *Éloges* (funeral or memorial orations) for deceased Académiciens—including a memorable one of Newton who had been elected a foreign associate of the Académie in 1699.

In contrast to the situation in Britain, French science developed strongly during the eighteenth century, particularly in mechanics, applied mathematics and chemistry. During that time the organization and achievements of French science were much admired and widely imitated. In 1771 the Abbé François Rozier (1734–1793) started a journal which in 1794 changed its name to the *Journal de physique, de chimie, et d'histoire naturelle*. It was published in Paris, the first of the French non-institutional journals, and was usually known as *Rozier's Journal* or as the *Journal de physique*. For a few years it was very successful, becoming larger than the *Philosophical Transactions* or the Académie's *Mémoires*. From 1785 until his death its editor was Jean Claude de Lamétherie (1743–1817), who was a strong supporter of the phlogiston theory of chemistry and who refused to publish papers by Antoine Lavoisier and others who opposed that theory. The phlogiston theory became discredited by about 1800 and the *Journal de physique* then suffered a great loss of prestige because of its policies, ceasing publication six years after Lamétherie's death.

During the French Revolution (1792–1795) the Académie des Sciences was abolished, but in 1795 it was recreated in a new form as part of the Institut National de France. The *Mémoires* of the old Académie were carried on as the *Mémoires de l'Institut National de France*. The École Polytechnique was founded at the same time as the Institut de France and in 1795 it began the *Journal de l'École polytechnique*, which was largely mathematical in nature and accepted papers only from members of the École. Clapeyron's famous mathematical formulation of Carnot's ideas (Section 4.2) was published in that journal. In 1807 there was founded at Arcueil, near Paris, the Société d'Arcueil, which issued its *Mémoirs de physique et de chimie*. This journal only lived for ten years, but it played an important role in its time.

A weakness of the *Mémoires* was that it was often seriously in arrears, publication delays of as much as six years being encountered. To overcome this problem the Académie began in 1835 to publish the *Comptes rendus hebdomadaire des séances*, which as its name implies appeared weekly like the meetings of the Académie. This journal included short papers by members of the Académie as well as papers from non-members subject to approval by a committee. The publication became so successful that it became necessary to restrict members to ten papers annually of not more than four pages each;

non-members were allowed five papers. This rule no doubt contributes to the tendency of French scientists to write with greater brevity than often seems desirable.

German journals

An important German publication, founded in 1682 in Leipzig, was the *Acta Eruditorium*, the first volume of which contained contributions from Boyle and Leibniz. It included articles in law and theology as well as in natural philosophy. In 1700 the Societas Regia Scientiarum (Royal Society of Science) became established in Berlin, largely at the instigation of Leibniz. Ten years later it began to publish the *Miscellanea Berolensis ad incrementum scientiarum*, its language being Latin, which was then more widely understood by educated people than any other language except possibly French. Neither the Society nor its journal was at all successful, and in 1745 there was a complete reorganization. The official language was changed from Latin to French, the society being renamed the Académie Royale des Sciences de Berlin, and the publication became the *Histoire, avec les Mémoires*. In 1804 it was renamed the *Abhandlungen des könglige (*later *könglige preussische) Academie der Wissenschaft*.

In 1790 there was founded in Germany a non-institutional journal called the *Journal der Physik*. As discussed in Section 1.1, at that time the word 'physics' still meant much the same as natural philosophy, and therefore included chemistry, biology and geology as well as what we now understand as physics.

Over the years the *Journal der Physik* underwent a number of confusing changes of title which are understandable in the light of the shift in the meaning of the word 'physics' and the tendency of the journal to become more specialized. In 1799 the title was changed to the *Annalen der Physik*, the intention still being that chemistry should be included. To make this explicit the title in 1819 became *Annalen der Physik und der physikalischen Chemie*, the word 'physikalischen' indicating that papers on purely descriptive aspects of chemistry were not encouraged. By 1824 it was apparently thought unnecessary to specify this restriction and the title was shortened to *Annalen der Physik und Chemie*. The editor from 1824 and for the next 53 years was Johann Christian Poggendorff (1796–1877) and the journal was commonly referred to as *Poggendorff's Annalen*. From 1877 it was edited by Eilhard Ernst Gustav Wiedemann (1853–1928) and was officially renamed *Wiedemann's Annalen der Physik und Chemie*. In 1900 the words 'und Chemie' were dropped; by then the emphasis was on physics as we understand it today, and the chemists had their own *Annalen der Chemie* and other journals such as the *Zeitschrift für physikalische Chemie*, to be considered later.

In addition, a number of regional scientific societies were formed in Germany, particularly during the nineteenth century. Most of these published journals, and a number of them played an important role in the communication of science. Some of these regional journals are included in the Appendix.

Other countries

Until well into the nineteenth century the societies and journals of England, France and Germany exerted by far the greatest influence on the progress of science. Academies had been established in a number of other countries but they did not become of much significance until later.

In Russia the Academia Scientiarum Imperialis Petropolitana (Imperial Academy of Science of St Petersburg) was founded by the Emperor Peter the Great in 1724; St Petersburg was then the capital of Russia. At first a number of people from other countries were invited to be members. The Academy was organized in three divisions, concerned with mathematics, the physical sciences, and the humanities. Its journal was first called *Commentarii* and later the *Bulletin*. In 1934 the headquarters of the Academy was transferred from Leningrad (as St Petersburg was then called) to Moscow. The most distinguished and influential of the early members of the Academy was the chemist and astronomer (and poet and playwright) Mikhail Vasilevich Lomonosov (1711–1765), who is regarded as the founder of modern Russian science. He was professor of chemistry at the University of St Petersburg from 1745, and although he published little he is known to have had advanced ideas on a number of topics. He opposed the phlogiston theory at an early date, he recognized the true nature of heat and light, and he believed in the conservation of energy almost a century before this had been established by Meyer, Joule and others.

Science in the United States remained in an embryonic state until the middle of the nineteenth century, most of its science and many of its science professors being imported from Europe. The American Academy of Arts and Sciences, essentially a regional society based in Boston, was founded in 1780 and published its *Memoirs* and, after 1846, its *Proceedings*. The most influential of the early American journals was the *American Journal of Science*, founded in 1818 by Benjamin Silliman (1775–1864) who was trained as a lawyer but after studying science in Europe became professor of chemistry, mineralogy and geology at Yale. He was succeeded as editor by his son Benjamin Silliman Jr (1816–1885) and the journal was generally known as *Silliman's Journal*. It covered all the sciences with somewhat of an emphasis on geology. The journal *Science*, still highly influential today, was begun in 1883 and from 1895 was edited by J. M. Cattell who made it the official publication of the American Association for the Advancement of Science which had been founded in 1848. The Association had published its *Proceedings* from the beginning, but from 1901 the proceedings appeared in the journal *Science*.

Like American science, Russian and Japanese science became very strong during the twentieth century, but little of any importance was done until the nineteenth century when science was introduced from western Europe. No important Russian or Japanese societies or journals devoted to science in general were founded during the nineteenth century, but some chemical societies are mentioned later.

2.2 Chemical journals

The journals considered so far were begun for the most part as general journals covering various aspects of science. Since many of these journals tended more and more to concentrate on what we now call physics, there was an increasing need for journals covering other branches of science. The first chemical journals were founded in the late eighteenth century by Lorenz von Crell (1744–1816) who had been a student of Joseph Black in Edinburgh and who was a professor of chemistry in Helmstadt. The first of his journals, founded in 1778, was the *Chemische Journal für die Freunde der Naturlehre, Arzneygelahrheit, Haushaltung-skunst und Manufacturen*. This changed its name several times and finally ceased publication in 1803. Its lack of lasting success was due to the fact that most of its papers pertained to the phlogiston theory, von Crell being a staunch opponent of the new ideas of Lavoisier.

The first chemical journal of lasting importance was the French *Annales de Chimie*, which was founded in 1789 by a group of distinguished chemists, including Lavoisier, who were unable to publish their papers in the *Journal de physique, de chimie, et d'histoire naturelle* (*Rozier's Journal*) because of its editorial policies under its editor Lamétherie. In the first issue of the *Annales de chimie* it was announced that its editorial policy would be a liberal one, but in fact little appeared that supported the phlogiston theory, chemists who favoured that theory publishing in *Rozier's Journal*. In 1815 the editors of the *Annales de chimie* decided to accept papers on physics and changed the title to *Annales de chimie et physique*. This title remained until 1914 when the *Annales* split into two journals, the *Annales de chimie* and the *Annales de physique*.

Several German chemical journals started in the early nineteenth century, the most important being the *Annalen der Pharmacie*, a non-institutional journal which was first published in 1832; at the time the pharmaceutical aspects of chemistry were the most important. In 1838 this journal began to be edited jointly by the two distinguished chemists Justus von Liebig (1803–1873) and Friedrich Wöhler (1800–1882). Two years later, at Wöhler's suggestion, the title was changed to the *Annalen der Chemie und Pharmacie* to recognize the growing importance of other aspects of chemistry.

At about the time of this change of title two amusing satirical pastiches were published in the *Annalen*. The first was written jointly by Liebig and Wöhler and published anonymously as a note following a synopsis (*Ann. der Pharmacie*, 29, 93–100 (1839)) of an article by Pierre Turpin that had appeared in the *Comptes Rendus*. In this article Turpin had argued that fermentation can only be brought about by living organisms. Liebig and Wöhler, on the other hand believed that fermentation was a purely chemical process, and they satirized Turpin's paper by claiming to have carried out fermentation experiments in which they had detected tiny animals, some remarkable features of which they described in detail.

The second satire appeared in the following year and was written by Wöhler, who had worked with the great Swedish chemist Jons Jakob Berzelius (1779–

1848) and like him disliked the substitution theory that had been put forward by the French chemist Jean-Baptiste Dumas (1800–1884). Wöhler wrote Berzelius a letter in which he claimed to have made successive substitutions of chlorine atoms into an organic compound, finally obtaining a substance whose molecule contained 22 atoms, all of which were chlorine! The letter was in French, and its author named as S. C. H. Windler (*Schwindler* being the German for swindler). Much amused, Berzelius showed the letter to Liebig who published it in the *Annalen* in 1840 without telling Wöhler; the article appears in French but with a German title. The article is listed in the Name Index of the *Annalen* and in the *Royal Society Catalogue of Scientific Publications*.

Two years after Liebig's death in 1873 the name of the *Annalen* was officially changed to *Justus Liebig's Annalen der Chemie und Pharmacie*, the last two words being deleted a year later. Other important German chemical journals were the *Journal für praktische Chemie*, begun in Leipzig in 1834, and the *Berichte der deutschen chemischer Gesellschaft*, begun in Berlin in 1868.

The need for a British chemical journal was not as great as in other countries, as the British chemists had the *Philosophical Transactions* and the *Philosophical Magazine* in which to publish their research. The Chemical Society of London was founded in 1841, the qualification 'of London' being dropped when the Society was granted its Royal Charter in 1848. This society, the first President of which was Thomas Graham, is the oldest chemical society in the world that has had a continuous existence until the present day. As noted below, in 1972 the Chemical Society became merged with the Faraday Society, which was mainly concerned with physical chemistry. In 1980 the Society became unified with the Royal Institute of Chemistry, with the new title of the Royal Society of Chemistry. The Society now has a number of divisions, the one concerned with physical chemistry being called the Faraday Division. The Society's journal, first called the *Memoirs of the Chemical Society*, was begun in 1858, the title being changed in 1862 to the *Journal of the Chemical Society* one series of which, as noted later, is now devoted to physical chemistry.

A national chemical society of France was established in 1858 with the title Societé chimique de Paris, changed to Societé chimique de France in 1907. Its *Bulletin de la Societé chimique de Paris* (later *de France*) appeared in 1858 and it gradually overtook the *Annales de chimie* in size. In Germany the Deutsche Chemische Gesellschaft was founded in 1867 and its *Berichte* was first issued in the following year, rapidly becoming a publication of great importance. At about the same time chemical societies were started in many other countries. In 1869 in St Petersburg the *Russkoe Khimicheskoe Obshchestvo* (Russian Chemical Society) was formed, and in 1873 it changed its name to the *Russkoe Fiziko-Khimicheskoe Obshchestvo*, which is best translated as the Russian Society for Physics and Chemistry since it was concerned with both physics and chemistry rather than with physical chemistry. Its *Zhurnal* was begun in 1869. The Tokyo Chemical Society was founded in 1878 and its journal, the *Tokyo kwangaku kwai shi*, appeared two years later.

The American Chemical Society was founded in 1876, and the *Journal of the American Chemical Society* began publication in 1879. In the same year Ira Remsen (1848–1927) launched the *American Chemical Journal*, which for some years was more successful than the Society's *Journal*. In 1914 the *American Chemical Journal* became incorporated into the *Journal of the American Chemical Society*, which became one of the world's leading chemical periodicals. One of its distinguished editors was William Albert Noyes (1857–1941), who was editor from 1902 to 1917 and who also edited *Chemical Abstracts* and *Chemical Reviews*; another was Arthur B. Lamb of Harvard University.

In 1895 the *Review of American Chemical Research* was founded by Arthur Amos Noyes (1866–1936), who is sometimes said to have been a distant cousin of W. A. Noyes; however I was told by the latter's son, W. Albert Noyes Jr, the distinguished photochemist, that he knew of no such relationship. This journal greatly expanded its scope in 1907 and became *Chemical Abstracts*, which has played a role of great importance in the dissemination of research.

2.3 Journals for physical chemistry

The first journal devoted to physical chemistry was the *Zeitschrift für physikalische Chemie, Stöchiometrie und Verwandtschaftslehre*, begun in 1887 by Ostwald and van't Hoff and published in Leipzig, where Ostwald was professor. Although the articles were required to be in German the journal had a strong international flavour, as is indicated by its board of associate editors which included most of the well-known physical chemists of the day: for example, M. Berthelot, H. Le Chatelier and F. M. Raoult from France, C. M. Guldberg and P. Waage from Norway, D. Mendeleev and N. Menschutkin from Russia, W. Ramsay and T. E. Thorpe from Britain, Julius Thomsen from Denmark, as well as Lothar Meyer and Victor Meyer from Germany. A good fraction of the papers in the early issues of the *Zeitschrift* came from laboratories outside Germany.

Another important journal for physical chemistry was the *Zeitschrift für Elektrochemie*, which was begun in Leipzig in 1894 by the Deutsche Elektrochemische Gesellschaft. This society and its journal interpreted electrochemistry in a very broad way, as including the whole of physical chemistry, and in 1902 the society changed its name to the Deutsche Bunsen-Gesellschaft für angewandte physikalische Chemie (Bunsen Society of Germany for applied physical chemistry). In 1904 the title of the journal became the *Zeitschrift für Elektrochemie und angewandte physikalische Chemie*, and in 1963 it became the *Berichte der Bunsen-Gesellschaft für physikalische Chemie*, which accepts papers in either German or English. In 1890 the *Elektrochemische Zeitschrift* was started in Berlin; it ceased publication in 1922.

In 1896 the American chemist Wilder Dwight Bancroft (1867–1953) founded the *Journal of Physical Chemistry*. The rather chequered career of this journal deserves some detailed attention, since the circumstances are of considerable significance in connection with the early development of physical chemistry in the United States. Problems arose from the rather eccentric and complex personality of Bancroft. A member of a distinguished and affluent New England family, he was educated at Harvard and then worked in the laboratories of Ostwald and van't Hoff, obtaining his Ph.D. in 1892 at the University of Leipzig. In 1895 he went to teach at Cornell University, where he remained to the end of his life. During his first few years at Cornell he declined to accept any salary, his private means being sufficient for his needs. Bancroft was highly eccentric in manner, at a time when unconventional behaviour was regarded as evidence if not of insanity at least of an artistic temperament. Van't Hoff, after visiting Bancroft at Cornell and finding him wearing knickerbockers and shoes instead of boots (!), said that he reminded him of an impressario. Bancroft had a keen sense of humour but it was often misplaced. Once at a Ph.D. oral examination he asked the candidate 'What in water puts out fires?'; after rejecting some of the student's answers Bancroft revealed that the answer he wanted was 'A fireboat'.

From one point of view Bancroft's career was a highly successful one. At Cornell he attracted attention as an enthusiastic and eloquent lecturer, and he devoted much energy to his research. His reputation spread nationally and internationally, and he received a number of honours. He was twice President of the American Electrochemical Society, he served as President of the American Chemical Society and he was elected a member of the National Academy of Sciences. He was made an honorary member of the Polish Chemical Society and of the Société Chimique de France, and an honorary fellow of the Chemical Society (of London). He received an honorary degree from Cambridge, but it is significant that no major American university accorded him a similar honour.

His personality also had a darker side. Along with his enthusiasm and eccentricity came much intolerance and acerbity of manner. He was autocratic and outspoken, and sometimes seemed to go out of his way to antagonize those who could have been his allies. As a result he was held in higher regard by those who did not come into personal contact with him, such as physical chemists in other countries and American organic chemists, than by those who did. By and large the leading American physical chemists kept their distance from him. The most sympathetic obituary notice for him was written by the British chemist Alexander Findlay, and even he made reference to a number of Bancroft's weaknesses.

Bancroft's mathematical ability was very limited; he had taken only one course at Harvard in the calculus and had obtained a grade of C. As a result his teaching and research concentrated on what might be called qualitative physical chemistry. At first his research was on applications of Le Chatelier's principle and the phase rule (Section 4.4); later he did much work in colloid chemistry (Chapter 9). In these fields he did some competent work. Unfortunately, in editing his journal he discouraged contributions on other aspects of physical chemistry, so that it included few papers on those topics that were becoming particularly important: applications of quantum mechanics to atomic and molecular structure, solution chemistry, X-ray diffraction, spectroscopy, and chemical kinetics. Besides being devoted mainly to the phase rule and colloids—many of the papers being by Bancroft himself and his students—his journal was patronized by cranks such as Louis Kahlenberg (1870–1941) who although teaching electrochemistry at the University of Wisconsin never accepted the existence of ions in solution (Section 7.2).

During the early years of the life of the *Journal of Physical Chemistry* one of the most distinguished American physical chemists was Arthur Amos Noyes, whose New England background and early career were somewhat similar to those of Bancroft. Noyes graduated from the Massachusetts Institute of Technology (MIT) and then worked with Ostwald, receiving his doctorate from Leipzig in 1890. In 1903 he established and became Director of the Research Laboratory in Physical Chemistry at MIT, and there he founded a distinguished school of research. He served for a period as acting President of MIT but declined to accept the Presidency, moving instead in 1919 to the Throop Institute of Technology, which later became famous as the California Institute of Technology ('Cal. Tech.').

Noyes was a highly competent chemist, who did some significant research, but his main strength was as a teacher and as a director of research. Unlike Bancroft he had a clear appreciation of which scientific problems were of greatest importance. He was an excellent judge of the research potential of students, whom he inspired by his own devotion to science. At both MIT and California he built up excellent departments and played a great part in fostering the distinguished careers of such men as G. N. Lewis, R. C. Tolman, and Linus Pauling.

It is hardly surprising that such men as these held Bancroft's *Journal of Physical Chemistry* in some contempt, preferring to publish in the *Journal of the American Chemical Society*. A. A. Noyes did not even allow the *Journal of Physical Chemistry* in the library of his department at the California Institute of Technology. Inevitably the journal suffered greatly in scientific quality, and at no time was it financially self-supporting. W. Albert Noyes, during his editorship of the *Journal of the American Chemical Society*, proposed to Bancroft a merger of the two journals, but only received a rebuff. For a time Bancroft contributed to the publication of his journal from his private resources, but by the 1920s this had become impractical. For a few years he was able to arrange for his journal to be published under the joint auspices of the American Chemical Society, the Chemical Society (of London) and the Faraday Society. He also managed to obtain some financial support from the Chemical Foundation, but when this was discontinued in 1932 Bancroft had no alternative to giving up the editorship and allowing the journal to be taken over entirely by the American Chemical Society. The new editor was Samuel C. Lind, a Leipzig graduate who had worked with Bodenstein on the kinetics of the hydrogen–bromine reaction, and who had later done distinguished work in radiation chemistry at the University of Minnesota. His editorial policies were more reasonable than those of Bancroft, and the journal soon became a more important medium for the publication of research of high quality.

The various negotiations and discussions that led to the wresting of the editorship from Bancroft also led to the founding of an important new journal, the *Journal of Chemical Physics*, which appeared in 1933. Its first editor was Harold Clayton Urey (1893–1981), who was soon to receive the 1934 Nobel Prize for his isolation of deuterium. This journal received strong support from the American Chemical Society, and the editor of the *Journal of the American Chemical Society*, Harvard's Arthur B. Lamb, agreed to divert to the new journal those papers submitted to him that seemed suitable for it. The journal thus began under excellent auspices, and its first volume included papers from almost all of the physical chemists and chemical physicists of the time that one can think of: J. D. Bernal, Farrington Daniels, Henry Eyring, R. H. Fowler, William D. Harkins, Gerhard Herzberg, George B. Kistiakowsky, Irving Langmuir, Gilbert N. Lewis, Robert S. Mulliken, Linus Pauling, John C. Slater, Hugh S. Taylor, John H. van Vleck, and many others. Several of these had also agreed to serve as associate editors of the journal. The third volume, of 1935, contained the classic paper of Henry Eyring on transition-state theory, and the journal has maintained its high standards over the years.

In 1903 the Faraday Society was founded in London, its stated object being to 'promote the study of electrochemistry, electrometallurgy, chemical physics, metallography and kindred subjects'; later the wording was modified to 'promote the study of sciences lying between chemistry, physics and biology'. From the outset the Society was actively supported by many eminent physicists and chemists, including Lord Kelvin, Lord Rayleigh and Sir J. J. Thomson, and has been extremely effective. One of its important functions has been to organize discussions on lively topics in physical chemistry, always at opportune times, so that contributions made at the meetings have often led to important advances. The procedure at Faraday Society Discussions is somewhat unusual. The formal papers are printed and distributed ahead of time to those intending to be present at the meeting; these preprinted papers serve a double purpose, since they are also page proofs for the subsequent publication. At the meeting itself these papers are taken as 'read' and the author is allowed only a few minutes to present a summary. Because of this procedure plenty of time remains for discussion; those who participate can submit written accounts which are published along with the formal papers.

The *Transactions of the Faraday Society* was first published in 1905, and it included reports of the various Society Discussions as well as regular papers. In 1947 the Society began to publish the Discussion papers in a separate publication, the *Discussions of the Faraday Society*. In 1972, after the Faraday Society had merged with the Chemical Society (later called the Royal Society of Chemistry) the *Transactions* was split into two publications entitled the *Journal of the Chemical Society. Faraday Transactions I*, and the *Journal of the Chemical Society.*

Faraday Transactions II, with a different emphasis of subject matter in the two series; in 1990 these two series were merged. The *Discussions of the Faraday Society* became the *Journal of the Chemical Society. Faraday Discussions.*

2.4 Some difficulties of communication

It is evident that from 1662 onwards there were a number of journals available for the publication of scientific papers. There were nevertheless some difficulties of communication. Sometimes, for example, editorial policies made it impossible for authors of good papers to have them published, and examples of this have been noted in connection with the phlogiston theory. Another notorious case relates to a paper by John Waterston (1811–1883) on the kinetic theory of gases (Section 5.3). This was submitted to the Royal Society in 1846 but not published until 1892, nine years after the author's death, the paper having been disinterred from the Society's files by Lord Rayleigh. It is impossible to know how many papers of value have been rejected for publication and never saw the light of day. The famous paper by Henry Eyring (1901–1981) on transition-state theory (Section 6.3) nearly met that fate; it was originally rejected flatly by the editor of the *Journal of Chemical Physics*, and was only published in 1935 through the intervention of others.

There were other problems. Clausius, for example, complained in 1857 about the tendency of British scientists to publish in journals not easily available to him; he particularly mentioned a paper that Joule had published in the *Proceedings of the Manchester Literary and Philosophical Society*. This publication, commonly known as the *Manchester Memoirs*, had at the time established something of a reputation as a national scientific journal, and Joule in his reply said that copies of it were sent to the leading scientific societies in Europe. It is understandable, however, that Clausius should have missed it.

Another type of communication difficulty came from the inability or unwillingness of some scientists to express themselves clearly. Clausius and Boltzmann in particular were very difficult to understand, even by Clerk Maxwell who fortunately was able to work out their theories for himself and explain them with great clarity. Parts of Clausius's very verbose papers still defy interpretation, and Boltzmann's style of writing was particularly difficult; it is almost impossible to translate even the titles of some of his papers into acceptable English.

Perhaps the most obscure writer in the physical sciences was that great American genius Josiah Willard Gibbs (1839–1903), who was obscure in several ways: he chose an obscure medium of publication—the *Transactions of the Connecticut Academy*—he chose titles for his publications that in no way suggested their contents, and he expressed himself so badly that few could or still can understand him. The medium of publication was the least of the problems, since Gibbs widely distributed reprints of his papers, sending them to Clerk Maxwell and Lord Rayleigh among others. In 1892 Rayleigh wrote to Gibbs politely suggesting that perhaps Gibbs might consider publishing something that would be more understandable; he commented that one of Gibbs's papers was 'too condensed for most, I might say all, readers'. Gibbs, however, rejected this suggestion, and he also declined an invitation from Ostwald to write an article, suitable for chemists, in the first volume of the *Zeitschrift für physikalische Chemie*.

It is unfortunate that Gibbs presented his thermodynamics in this way, because although by the 1870s he had become a master of the subject his work at first exerted little influence, and everything he did was worked out by others a decade or so later, particularly by Helmholtz and van't Hoff. The early development of thermodynamics would have been just as rapid if Gibbs had not done anything, so that his contributions are only important from a historical point of view. This was particularly unfortunate for physical chemistry, since Gibbs was the first to apply thermodynamics to chemical problems, although he did so in a purely abstract way, with only rare and passing references to experimental results.

This episode emphasizes more than any other the great importance of communication in the progress of science.

Suggestions for further reading

For further details about scientific societies and publications see

E. N. da C. Andrade, The birth and early days of the *Philosophical Transactions*, *Notes & Records Roy. Soc.*, **20**, 9–27 (1965).

Gwen Averley, The 'social chemists': English chemical societies in the eighteenth and early nineteenth century, *Ambix*, **33**, 99–128 (1986).

Charles A. Browne, *A History of the American Chemical Society*, Washington, D.C., 1952.

Mansel Davies, A century of physical chemistry: The foundation document: Zeitschrift für physikalische Chemie, Stöchiometrie und Verwandtschaftslehre: Volume 1, pp. 1–678, 1887', *Z. physikal. Chem.*, **170**, 7–14 (1991).

R. M. Gascoigne, *A Historical Catalogue of Scientific Periodicals, 1665-1900*, Garland, New York, 1985.

R. M. Gascoigne, *A Historical Catalogue of Scientists and Scientific Books from the Earliest Times to the Close of the Nineteenth Century*, Garland, New York, 1984. This useful compilation contains the names of over 15 000 scientists, with dates, nationalities, and biographical references, and lists any books written, with date and place of publication.

Marie Boas Hall, The Royal Society's role in the diffusion of information in the seventeenth century, *Notes & Records Roy. Soc.*, **29**, 173–192 (1975).

D. A. Kronick, *Scientific and Technical Publications of the Seventeenth and Eighteenth Centuries: A Guide*, Scarecrow Press, Metuchen, N. J., 1991.

Wyndham Miles, Early American chemical societies, *Chymia*, **3**, 95–113 (1950).

T. S. Moore and J. C. Philip, *The Chemical Society, 1841-1941*, The Chemical Society, London, 1947.

John W. Servos, *Physical Chemistry from Ostwald to Pauling: The Making of a Science in America*, Princeton University Press, 1990. This book discusses the American journals in considerable detail.

J. W. Stout, *The Journal of Chemical Physics*: the first 50 years, *Annual Review of Physical Chemistry*, **37**, 1–23 (1986).

Arnold Thackray, Jeffrey Sturchio, P. Thomas Carroll, and Robert Bud, *Chemistry in America, 1876-1976: Historical Indicators*, Dordrecht, 1985.

D. H. Whiffen, *The Royal Society of Chemistry; the First 150 Years*, The Royal Society of Chemistry, London, 1991.

Virginia E. Yagello, Early history of the chemical periodical, *J. Chem. Education*, **45**, 426–429 (1968).

For translations of the satirical pastiches in the *Annalen der Chemie* see

P. de Mayo, A. Stoessl and M. C. Usselmann, The Liebig/Wöhler satire on fermentation, *J. Chem. Education.*, **67**, 552–553 (1990). This includes a translation of the anonymous satire, actually written by Liebig and Wöhler and entitled 'Das enträthselte Geheimniss der geistigen Gährung' (The solved mystery of spirituous ferment), in *Annalen der Pharmacie*, **29**, 100–104 (1839).

Ralph E. Oesper, *The Human Side of Scientists*, Univ. of Cincinnati Press, 1975. This includes a translation of S. C. H. Windler, (Liebig's) *Ann. der Chemie und Pharmacie*, **33**, 308–310 (1840). The paper was actually written by Wöhler.

CHAPTER 3

The growth of the physical sciences

There is a strong link between modern physical chemistry and the ancient science of mechanics. From the time of the ancient Greeks, who devoted much attention to mechanics, until the middle of the nineteenth century, progress in the physical sciences was to a large extent progress in the understanding of mechanical principles. The scope of mechanics is very wide. It deals with the motions of bodies ranging in size from massive stars to submicroscopic particles. It has application to the biological sciences and to engineering. It forms a basis for the kinetic theory of gases, for thermodynamics, and for molecular statistics. Even when quantum mechanics was introduced it by no means superseded classical mechanics, which still applies satisfactorily to all but submicroscopic particles.

3.1 The mechanical universe

Until the later years of the nineteenth century it was generally thought that physics was nothing more than an application of the principles of mechanics; as already noted (Section 1.1), physics was often referred to as 'mechanical philosophy'. The realization that there is more to physics than mechanics came with the statistical theories of matter, with Maxwell's theory of electromagnetic radiation, and with the quantum theory. The present section outlines the development of 'mechanical philosophy', with special reference to its significance for physical chemistry.

Early science

The Greeks, a few centuries BC, took the first steps in the creation of science as it is known today. Before them the Babylonians and Egyptians had made many systematic observations and measurements, but it was left to the Greeks to introduce theoretical ideas in the form of abstraction and generalization. The dominant figures were Plato (427–347 BC) and his pupil Aristotle (384–322 BC), whose work dominated scientific thought until after the Renaissance. Aristotle worked principally in biology, in which he did excellent work, Charles Darwin commenting that other great biologists were 'mere schoolboys to old Aristotle'. He also did some work in the physical sciences, where he was less successful but highly influential. His physical science was a logical system in which experimental observations were interpreted as deductions from a number of fundamental principles. Its great weakness was that these principles were arrived at on the basis of religious or metaphysical arguments, by intuition rather than from experimental observations. For example, one of Aristotle's principles was that circular motion is the 'natural' type of motion, and he interpreted the motion of the planets in that way. Another principle was that every body has a tendency to reach its 'natural' place, the natural movement being towards the centre of the universe, which to Aristotle was the centre of the earth. Motion that is not towards the earth required a moving influence, which he believed must act during the whole of the motion. To explain why a stone continues its motion after it has left the hand of the thrower, Aristotle concluded that propulsion is exerted by the air, which is compressed in front of the stone and rushes round to the back to prevent a vacuum, another of his principles being that nature abhors a vacuum.

Today we can see clearly that where Aristotle went wrong was not in his idea of deducing observations from general principles, but in his method of arriving at the principles on the basis of metaphysical ideas. There are two methods of inference: *deduction*, which is the drawing of conclusions in particular instances from general premises, and *induction*, which is the reverse process of arriving at a general rule from particular instances. Aristotle made good use of deductive methods, but he neglected induction.

The beginnings of modern science

What is conveniently called 'modern science' began in the late sixteenth and early seventeenth centuries, mainly in Italy, England, France and the Netherlands. Earlier Germany had been the leader in mining and technology, but it was devastated by political and religious strife, particularly the Thirty Years War (1618–1648), and did not make much contribution to pure science until later. The two great contributions of the twentieth century, Planck's and Einstein's quantum theory and Einstein's theory of relativity, were made by Germans.

Three men who did more than any other to establish modern science were the Englishman Francis Bacon (1561–1626), the Italian Galileo Galilei (1564–1642) and the Frenchman René Descartes (1596–1650). The overthrow of the Aristotelian system was the particular contribution of Francis Bacon, whose work was done in the first quarter of the seventeenth century. Throughout his active career as a statesman and lawyer he gave much thought to philosophy and science. As a student at Trinity College, Cambridge, he had become aware of the Aristotelian method of investigation, and decided that an entirely new approach must be made. Much later, in 1620, he published his great work, the *Novum Organum*, which expounded his scientific philosophy in which fundamental principles were arrived at not by intuition based on metaphysics, but by inductive reasoning based on experimental evidence.

A good example of Bacon's inductive method of arriving at general principles is his procedure for discovering the nature of heat. He drew up a list of examples of heat production, and then a list of situations in which heat is not produced. He then eliminated factors that were common to his two lists, and finally decided that the only remaining conclusion, that heat is a form of motion of the constituent particles, must be the correct one. He was ahead of his time in drawing this conclusion; Boyle and Newton accepted it, but over a century later Joseph Black, Antoine Lavoisier and many others thought that heat was a substance.

The modern scientist makes much use of both inductive and deductive reasoning, and moves freely and perhaps unconsciously from one type of inference to the other. In some branches of science, deduction is the more important; in others induction plays a greater role. Once general principles have become established, as in many branches of physics and in quantum chemistry, it is deduction that is used for the most part. On arriving at a reaction mechanism from kinetic evidence, on the other hand, one proceeds inductively. The problems of chemistry have proved very complex, and progress still has to be made to a great extent by reviewing a range of experimental evidence and proceeding inductively.

One of Bacon's weaknesses was that although he did appreciate the importance of applying mathematics to physical problems, he did not have much facility with mathematics. Galileo was proficient in mathematics, and Descartes was a distinguished mathematician who was the founder of coordinate geometry, which is important for handling problems of dynamics. He too made a decisive break with the philosophy of Aristotle, emphasizing the importance of getting rid of principles that are not based on experimental evidence.

Descartes often Latinized his name as Renatus Cartesius, which is why the adjective Cartesian is often applied to his work. He first explained his philosophy of science in a small book entitled *Discours de la méthode pour bien conduire la raison et chercher la vérité dans les sciences* (A discourse on the method of best reasoning and of seeking the truth in the sciences), which appeared in 1637. He expanded his ideas in books that appeared in 1641 and 1644. The weakness of his method is that he underestimated the importance of experimental results in arriving at general principles, and in particular he did not properly appreciate the importance of Galileo's experimental work in mechanics. Also, whereas he criticized Aristotle's general principles, he himself sometimes arrived at principles on the basis of metaphysical or theological arguments. On the assumption that God is omnipresent, for example, he concluded that atomic theories must be incorrect as the existence of empty space implies regions where God does not exist.

Until well into the eighteenth century Descartes' methods exerted a wide influence in France. In England, largely owing to the influence of Bacon and Newton, the emphasis was more on inductive methods. In the middle of the eighteenth century, however, the French began to pay much more attention to Bacon's methods, and indeed held Bacon in even higher esteem than he was held in England. A little later the British scientists began to pay more attention to Descartes, particularly to his mathematics and to the system of calculus that had been developed by Leibniz and which proved more powerful and versatile than that of Newton.

Galileo's mechanics

The first important steps towards overcoming the confusions of Aristotelian mechanics were taken by Galileo, who can be said to be the founder of modern scientific mechanics. Important advances were also made by Christiaan Huygens, and in particular Isaac Newton brought the subject to a certain degree of completion, giving us a system that came to be called Newtonian mechanics. Table 3.1 summarizes some of the more important contributions to Newtonian and post-Newtonian mechanics.

TABLE 3.1 Highlights in classical mechanics

Author	Date	Contribution	Section
Galileo	1589–1604	Laws of falling and sliding bodies	3.1
Leibniz	1684–1686	Differential and integral calculus	3.1
Newton	1687	Laws of motion	3.1
Newton	1710–1717	Theory of matter	3.2
Bernoulli	1738	Hydrodynamics; kinetic theory of gases	3.2
d'Alembert	1743	Principle of equilibrium	3.2
Euler	1740–1750	Analytical dynamics	3.2
Maupertuis	1751	Principle of least action	3.2
Lagrange	1788	Generalized coordinates, q and $\dot{q}$	3.2
Laplace	1799–1825	Interparticle forces; 'astronomical' view of matter	3.2
Coriolis	1829	Significance of $\frac{1}{2}mv^2$	3.3
Hamilton	1834	Principle of varying action	3.2
Rankine	1853	Principles of energetics; kinetic and potential energy	3.3

At Pisa, and later at Padua, Galileo carried out many experiments on moving bodies and developed mathematical equations to interpret their motion. His conclusions are described in his two great books, *De motu* (on motion), which was written in about 1590, and *De mechanicce* (on mechanics), written in about 1600. He allowed a ball to roll on a table with uniform speed until it came to the edge and fell in a curved path to the floor. He showed that the path is a semi-parabola, and concluded that the path of a projectile is a full parabola. He realized that the horizontal and vertical motions are independent of one another, and that after the ball left the table its horizontal motion remained unchanged.

As far as the vertical motion of a falling body is concerned, Galileo concluded that its speed v is proportional to the time t of fall:

$$v = at, \tag{3.1}$$

where the proportionality factor a is the acceleration. Since the speed after time t is at, the average speed during the time t is $\frac{1}{2}at$. The distance fallen is thus

$$s = \frac{1}{2}at{\cdot}t = \frac{1}{2}at^2. \tag{3.2}$$

Galileo confirmed this conclusion that the distance is proportional to t^2 by allowing a ball to roll down an inclined plane, so that there was a reduction in the acceleration a. These investigations were of great importance, being about the first in which mathematics and experiment had been combined.

Galileo considered also the case of a ball rolling down an inclined plane and then along a horizontal plane. He concluded that the ball tends to continue along the horizontal plane without coming to rest, and this he attributed to its 'inertia'. This was the first clear break with Aristotle's idea that motion requires a moving influence.

Newton's mechanics

Newton's experiments and mathematical analysis, carried out at Trinity College, Cambridge, led him to formulate three laws of motion. The first of these had already been arrived at by Galileo and by Descartes in his *Principia Philosophiae* (1644). It was expressed by Newton in his *Principia Mathematica* (1687) as follows:

I. Every body perseveres in its state of resting or moving uniformly straight on, except inasmuch as it is compelled by impressed forces to change that state.

The version of Newton's second law that is commonly used today is quite different from that given by Newton, who stated that

II. Change in motion is proportional to the motive force impressed and takes place following the straight line along which that force is impressed.

Today Newton's second law is commonly expressed as

$$F = m\,\frac{\mathrm{d}v}{\mathrm{d}t}. \tag{3.3}$$

where F is the force and m the mass; that is, force is mass times acceleration $\mathrm{d}v/\mathrm{d}t$. This equation was first formally stated in 1700 by Pierre Varignon (1654–1722), and in 1750 Euler stated that the equation represented a 'general and fundamental principle of all mechanics', and that it provides a definition of force as mass times acceleration.

Newton's third law is comparatively straightforward, and is the only one that did not give rise to extensive discussion:

III. To any action there is always a contrary, equal reaction; in other words, the actions of two bodies each upon the other are always equal and opposite in direction.

Sir Isaac Newton

(1642–1727)

Newton was born in Lincolnshire, the son of a yeoman farmer who died shortly before Newton was born. His mother remarried three years later, to a clergyman whose Christian principles did not stretch to bringing up a stepson; instead Newton was raised by his grandparents. This rejection as a child probably contributed to his paranoid behaviour throughout his life; he was vain, manipulative, humourless and excessively religious. In many respects his approach to problems was quite unscientific, with much recourse to metaphysical and theological arguments, but his achievements were nevertheless outstanding, and he has been regarded as the greatest scientist of all time.

It was intended by his family that he should become a farmer, but showing no inclination for such a career he was sent to Trinity College, Cambridge, where he excelled in mathematics. Because of the Great Plague he spent 1665–1666 at home where he began to develop his theory of gravitation and his method of fluxions which was his form of calculus (Section 3.1). In 1667 he was elected a Fellow of Trinity, and took up permanent residence in the college. Two years later Isaac Barrow, the Lucasian professor at Cambridge, resigned his chair in favour of Newton, whose superior abilities he freely acknowledged. Newton then carried out investigations on a wide range of problems, including the theory of equations, optics, the lunar and planetary orbits, gravitation, alchemy and chemistry. His chemical experiments were carried out in a laboratory which he established in the garden behind his college rooms.

Newton was often reluctant to publish, being exceedingly sensitive to criticism, but at the same time he was always eager to establish his own priority, which he sometimes did by depositing anagrams decipherable only by himself. His famous

It was on the basis of these three laws that Newton made his great contributions to understanding the movement of the moon and the planets. In order to do so he invented a new branch of mathematics which he called 'fluxions' and which is a version of differential calculus. Suppose that distance x is a function of time; then the ratio of a small change in x to a corresponding small change in the time t was called by Newton the fluxion; he gave it the symbol $\dot{x}$. Newton arrived at his calculus by geometrical rather than analytical reasoning; thus if x is plotted against t the fluxion is given by the slope at any point. In the alternative form of the calculus given independently by Leibniz, use was made

book *Philosophiae Naturalis Principia Mathematica* (Mathematical Principles of Natural Philosophy), written in Latin and published by the Royal Society in 1687, was prepared only at the insistence of the astronomer Edmund Halley (1656–1742), who even covered much of the cost of the printing and put his own research aside to help with the publication. Newton's *Opticks*, written in English, was first published in 1704.

Some of Newton's early work was alchemical in nature, as he believed in the transmutation of the elements. Some of his later work is physico-chemical, and is expounded in a short tract entitled *De Natura Acidorum* (Concerning the Nature of Acids), which appeared in 1710, and in the second edition of his *Opticks*, published in 1717. He believed light to be corpuscular in nature, and since it can pass through some material he concluded that matter consists of exceedingly small particles, with much empty space between them (Section 3.2). He assumed the existence of attractive and repulsive forces between the particles, and he and others attempted to develop a comprehensive theory of matter in terms of these forces.

In 1696, after becoming afflicted with a severe nervous condition, Newton became Warden of the Royal Mint, being promoted to Master in 1699, with a substantial salary that enabled him to live in some luxury in London till the end of his life. He resigned his Cambridge chair in 1701, became President of the Royal Society in 1703, and was knighted in 1705. He is buried in Westminster Abbey.

References: I. B. Cohen, DSB, 10. 20–101 (1974); Andrade (1950a); Ball (1908, 1960); King-Hele (1988); Koyré and Cohen (1972); More (1934); Thackray (1970); Westfall (1971); Westfall (1980); Whitrow (1989).

of differential quantities, which today we write as dx and dt; Newton's fluxion $\dot{x}$ was the same as dx/dt. Although Newton's calculus was perhaps more rigorous than that of Leibniz it was less versatile and powerful, and the great advances made in the eighteenth century, mainly by the French mathematicians, were based on Leibniz's form of calculus rather than that of Newton.

Newton's mechanics was developed in detail in his great book *Philosophiae Naturalis Principia Mathematica* which appeared in 1687. Although much more work in mechanics remained to be done, the mechanics described in this book brought the subject to the end of an important phase. The expression 'Newtonian mechanics' is often used to refer not just to the mechanics developed by Newton but to any used before the introduction of quantum theory; this is more appropriately called classical mechanics.

Analytical dynamics

Also of great significance was the work of a group of mathematicians on the continent of Europe, who extended Newton's mechanics in many ways and whose work had a strong influence on the development of science, including physical chemistry, in the nineteenth and twentieth centuries.

Whereas Newton developed his mechanics on the basis of geometrical arguments, the Continental mathematicians used analytical methods, representing the relationships of mechanics by means of equations. Also, instead of using Newton's fluxions, they preferred the system of calculus that had been developed independently by Leibniz. Since in this system Newton's $\dot{x}$ appears as dx/dt, the differential quantities dx and dt could be separated. Leibniz's calculus proved to be more versatile and powerful than that of Newton, and the fact that the Continental mathematicians used it put them at an advantage over the British, who clung to Newton's fluxions until the early nineteenth century.

The branch of mechanics that is concerned with motion is conveniently referred to as *dynamics*, as was suggested by Leibniz in his book *Specimen Dynamicum* (1695). The mechanics of stationary systems is often called *statics*. This distinction is convenient, but it should be noted that other definitions have been given. Thus William Thomson (Lord Kelvin) and P. G. Tait in their *System of Natural Philosophy* (1867) defined dynamics as the science of force, whether it produces motion or not, and subdivided dynamics into statics (systems at equilbrium) and kinetics (systems in motion).

Others who made notable advances in dynamics were the Swiss mathematician John or Johann Bernoulli (1667–1748) and his son Daniel Bernoulli (1700–1782). The latter was the first to suggest the usefulness of resolving the compound motion of a solid body into translational and rotational components.

Mention has already been made of the Swiss mathematician Leonhard Euler, who first defined force as mass times acceleration. He also developed methods of dealing with systems of several mass points, and formulated a treatment of hydrodynamics. Particularly important contributions were made by Joseph Louis Lagrange (1736–1813) who was born in Italy but did most of his work in Berlin and Paris. His treatise *Mechanique analytique*, which appeared in 1788, proposed a completely new system of dynamics. It was based on generalized coordinates for each component of motion, one for position q and the other for velocity $\dot{q}$. Lagrange introduced the *concepts* of kinetic energy T and potential energy V, although these terms were suggested later. If T and V are expressed in terms of the generalized coordinates q and $\dot{q}$, their difference L is now known as the Lagrange function or Lagrangian:

$$L(q,\dot{q}) = T(q,\dot{q}) - V(q,\dot{q}). \tag{3.4}$$

Lagrange's differential equations of motion are of a form particularly suited to complicated mechanical systems. Lagrange also introduced the idea of *potential*, which was later developed by Laplace and by Poisson.

Laplace's Theory of Matter

Undoubtedly the most influential of the Continental mathematicians, although perhaps not the greatest, was Pierre-Simon Laplace (1749–1827). He made important contributions to pure mathematics, and in addition developed a new style of physics that has been referred to as the astronomical view of matter. He developed the transform that is known by his name, and the expansion used in the theory of determinants that is also known by his name. He developed the concept of *potential*, which Lagrange had previously introduced. For a period he entered into collaboration with Antoine Lavoisier on the theoretical and experimental study of heat. Laplace not only contributed to the mathematical theory but was also mainly responsible for the design of an ice-calorimeter that was used extensively in the experiments. Laplace's success in science was no doubt partly due to his interest and experience in experimental science as well as in mathematics.

Laplace also made important contributions to the theory of matter. Just as the movements of the planets can be interpreted in terms of gravitational forces, Laplace considered that the behaviour of matter could be understood in terms of interparticle forces, which may be attractive or repulsive. In his time it was generally believed that heat, light, electricity and magnetism are imponderable fluids of a particulate nature; heat was not regarded by Laplace as a form of energy, and the wave properties of light were not generally recognized. Each of the imponderable fluids was thought to consist of particles that are mutually repulsive but which are attracted by the particles of ordinary (ponderable) matter.

It was Laplace's view that a treatment of the short-range forces between particles would bring the physics of materials to the same level of understanding that Newton's laws had achieved for celestial physics. He built on the Newtonian tradition, developing the principles in a mathematical form, and by treating the short-range forces developed a comprehensive physics of matter. In the first edition of his *Exposition du système du monde*, published in 1796, he gave a preliminary treatment of the cohesion of solids and even of chemical reactions. He assumed the interparticle forces to be gravitational in nature, although he had to invoke deviations from the inverse-square law, attributing them to the fact that the particles are not necessarily spherical.

Laplace's theories were expounded in his five-volumed *Traité de méchanique céleste* (Treatise on celestial mechanics), which appeared from 1799 to 1825 and deals with much more than the title suggests since it covered his theories of matter as well as astrophysics. When the book appeared it was generally regarded as the second most influential scientific book ever published, the first being Newton's *Principia*. Mary Somerville (1780–1872) gave a popular

Pierre-Simon Laplace, Marquis de Laplace

(1749–1827)

Laplace was born in Normandy and attended the University of Caen where he was originally intended to be a cleric. His mathematical gifts were soon recognized, and he was appointed professor of mathematics at the École Militaire. He was elected a member of the Académie des Sciences at the early age of 24.

He was one of the most versatile and influential scientists of all time. He made valuable contributions to a wide variety of problems in mathematics and physics, he developed a highly productive philosophy of science, and he played an important part in establishing the modern scientific disciplines. He was interested in experimental science as well as in pure and applied mathematics.

From 1768 Laplace worked on the integral calculus, astronomy, cosmology, the theory of games, probability, and causality. During the next ten years he developed the transform that is known by his name, and also the expansion used in the theory of determinants that is also known by his name. He made contributions to the theory of potential (Section 3.3). During this period he entered into a collaboration with Lavoisier on theoretical and experimental investigations of heat, being largely responsible for the design of the ice-calorimeter used in the work.

During the revolutionary period in France, from 1789 to 1805, his reputation was at its height. Always an opportunist, he shifted his political opinions as occasion demanded, and always managed to retain a position of influence. He was an active member of Lagrange's commission that introduced the metric system in 1799, and

exposition of the ideas in the book in her *Mechanism of the Heavens* (1831). She had been invited to prepare a translation of the book but decided instead to write her own version of the ideas in it, and also introduced her own diagrams to make the arguments easier to understand.

Laplace's theory of matter appeared in the fourth volume of his treatise, which was published in 1805. By this time he had received support for his ideas from the distinguished chemist Claude Louis Berthollet (1748–1822), with whom he had been friendly since the early 1780s. Berthollet's chemical ideas were much the same as Laplace's; in his *Essai de statique chimique* (Essay on chemical statics), which appeared in 1803, he had said that 'the forces that bring about chemical phenomena all derive from the mutual attraction between the molecules of bodies'. In Laplace's 1805 volume he treated refraction of light and capillary attraction in terms of short-range forces between densely

he played a dominant role in the creation of the Institut de France and the École Polytechnique. The first four volumes of his comprehensive *Traité de méchanique céleste* (Treatise on celestial mechanics) appeared from 1799 to 1805, with later instalments appearing from 1823 to 1825; it was these later volumes that contained much of Laplace's physics with its 'astronomical' theory of matter. Laplace presented a copy of the book to the Emperor Napoleon I who commented that, although dealing with the universe, the book made no mention of its creator. To this Laplace stiffly and bluntly replied: 'Je n'avais pas besoin de cette hypothèse-là' (I had no need for that hypothesis). Napoleon, much amused, told the story to many, including Lagrange whose comment was: 'Ah! c'est une belle hypothèse; ça explique beaucoup de choses!' (Ah! that is a nice hypothesis; it explains many things!).

For a short time Laplace served as minister of the interior under Napoleon; after six weeks Napoleon was dissatisfied and dismissed him, but appointed him to the Senate. When in 1814 the monarchy was restored in France, Laplace hastened to offer his services to the Bourbons, and was rewarded with the title of marquis. He was vain and self-seeking in some matters, but could be generous and considerate under other circumstances.

References: Gillispie *et al.*, DSB, 15, 273–403 (1978); Ball (1908, 1960); Fox (1974).

distributed particles, and gave a lengthy mathematical treatment of these properties. Laplace claimed that if he were given the initial positions and motions of every particle in the universe he could in principle predict the future course of history.

Although Laplace's comprehensive theory of matter was undoubtedly of great importance and exerted a wide influence, it had only a short life and had become more or less obsolete by about 1825. The main problem was with the treatment of imponderable fluids. The idea that heat is such a fluid was becoming increasingly untenable (Section 4.1) and the experiments of Fresnel on the diffraction of light contributed greatly to the acceptance of the idea that light has wave properties. Some of Laplace's work therefore lost its validity. Furthermore, the particular form of the atomic theory that was put forward by John Dalton required important modifications to the theories of Laplace and

Mary Somerville

(1780–1872)

Mary Fairfax, the daughter of Lieutenant William Fairfax who later became an Admiral and was knighted, was born in Jedburgh, just north of the border between England and Scotland. As was common with females at the time, Mary received little education, her only schooling being for a period of one year when she was aged ten. Aside from that she had occasional lessons on such important matters as dancing and needlework. She was, however, inspired to educate herself, and became perhaps the most remarkable of all autodidacts. At the age of about 15 she accidentally heard of algebra and geometry, and surreptitiously obtained a copy of Euclid and a textbook on algebra. Her studies had to be concealed from her parents, who were convinced that intellectual effort would seriously unbalance a female mind: 'We shall have Mary in a straightjacket one of these days' said her kindly but misguided father, on learning something of her activities.

In 1804 Mary married a cousin, Samuel Grieg, who died three years later; this can hardly be regretted, since he had a low opinion of female ability and no interest in anything intellectual. Mary was left with a comfortable income, and she used the independence of widowhood to educate herself further, reading Newton's *Principia* and studying higher mathematics and physical astronomy. In 1812 she remarried, and in her second husband, her first cousin William Somerville, she was much more fortunate since he was generous and intellectual, supporting his wife's activities in every way. He was an army doctor, and in 1816, when he was appointed to the Army Medical Board, the family moved to London. There they soon became well acquainted with a number of prominent scientists, and Dr Somerville was elected a Fellow of the Royal Society. Their close friends in London included Sir Humphry and Lady Davy, Thomas Young, William Hyde Wollaston, Michael Faraday and Sir John Herschel; Jane Marcet, whose *Conversations in Chemistry* had so inspired Faraday when he was a bookbinder's assistant, became a particular friend. On a trip to the Continent in 1817 the Somervilles met Laplace, Arago, Gay-Lussac, Cuvier and others, and later maintained a friendly correspondence with them.

All of these prominent scientists held Mary Somerville in high esteem, her considerable beauty and charming personality doing her no harm. In her early days she had been known as the 'Rose of Jedburgh' and her beauty was preserved into her old age. She is described as petite, shy, modest and vivacious. In addition to her studies and scientific investigations she devoted much time to her many children and had an active social life.

Mary Somerville published three experimental papers in what can be called physical chemistry. The first, communicated to the Royal Society by her husband and published in 1826 in the *Philosophical Transactions*, was on the magnetizing effects of sunlight; it stimulated much later work, but her conclusions were later disproved. Ten years later Arago presented to the Académie des Sciences an abstract of one of her letters to him, and this was published in the *Comptes Rendus*; it dealt with the passage of solar rays through various media. Her third paper appeared in 1845 in the *Philosophical Transactions*, in the form of a letter to Sir John Herschel about the effects of light on plant materials. This was in the early days of photography, and Herschel himself had done much work on plant pigments, with a view to achieving colour photography (Section 8.4).

More important were her several books. In 1827 Henry (later Lord) Brougham, who had been a student of Joseph Black, suggested that she should prepare a translation of Laplace's *Mécanique céleste*. In the end she prepared not a translation but a version of her own, entitled *The Mechanism of the Heavens*, published in 1831. In it the mathematics were expressed in a more explicit form, for the benefit of British readers who at the time were well behind in their understanding of mathematics. This book was an instant success, probably because her own struggles to learn mathematics made her more appreciative of the difficulties that others would face. The book was used for courses at Cambridge and elsewhere. In 1834 her second book, *On the Connexion of the Physical Sciences*, was published, and it was also a great success, running to ten editions many of which were revised by her.

A later book, *Physical Geography*, published in 1848 when she was 68, was an even greater success. When she was 89, her final work, *On Molecular and Microscopic Science*, appeared in two volumes. Sadly, this was something of an embarrassment, since she had not kept up with developments in a field that was advancing rapidly. Critics treated it kindly and with deference to its venerable author, and used tact in pointing out its shortcomings.

Mary Somerville received many honours and awards, from several countries including the United States. As a woman she could not be elected a Fellow of the Royal Society, but the Society voted a more signal honour for her by commissioning a bust of her to be displayed in its Great Hall. Somerville College, one of the first two colleges for women in Oxford, was named in her honour. She gave powerful support to the cause of the education and emancipation of women, not only by her always temperate advocacy but by her personal example as a conscientious wife and mother who also had an active intellectual career.

References: Elizabeth C. Patterson, DSB, 12, 521–525 (1975); Patterson (1983).

Berthollet, which had not required that each element has its particular type of atom.

In spite of this collapse of Laplace's comprehensive treatment of matter, many of the ideas he originated were of great importance and are still useful today. For example, he developed the idea of potential which had first been introduced by Lagrange. He gave an equation for the electric potential V, but this was corrected in 1811 by Simeon Denis Poisson (1781–1840) who showed that if within a closed surface the density of electric charge is ρ, the equation for the electric potential is

$$\nabla^2 V = \frac{\delta^2 V}{\delta x^2} + \frac{\delta^2 V}{\delta y^2} + \frac{\delta^2 V}{\delta z^2} = -4\pi\rho. \tag{3.5}$$

Willard Gibbs later gave the name del to the operator ∇, and ∇^2, which is defined in eqn 3.5 is called del squared. A number of other useful relationships originated with Laplace.

Laplace and Lagrange were both actively concerned with the introduction by the French of the metric system. In 1793, during the French Revolution, a decree was enacted ordering all foreigners to leave France, but a specific exception was made for Lagrange, an Italian, who was appointed President of the commission for the reform of weights and measures. Laplace was an active member of the commission, and he and Lagrange had much to do with the final recommendations. The kilogram was chosen as the standard unit of mass and the metre as the standard of length. These units, together with the second, which is a much older unit, formed the basis of the centimetre–gram–second (cgs) system, which was formally adopted by the French Legislature on 2 June 1799. The cgs system was later adopted throughout the world as the most logical and convenient system for scientific purposes; modifications of it are now used by most countries for commercial purposes also. It formed the basis for the more extensive Système Internationale (SI) which is used in science today.

Hamilton's mechanics

One important contribution made by the eighteenth-century mathematicians was the calculus of variations, which has been adapted for use in modern quantum-mechanical calculations. In 1751 Pierre Louis Maupertuis (1698–1759) proposed his *principle of least action*. His theory was later defended and elaborated by Euler, and Lagrange finally expressed the principle in its exact form in his *Méchanique analytique* (Analytical mechanics), which was published in 1788. Later an important extension of the principle was made by the Irish mathematician Sir William Rowan Hamilton (1805–1865). He had great admiration for Lagrange's formulation of the principle, which he described as a scientific poem, being himself something of a poet—although not a very good one! Hamilton's treatment appeared in a paper published in 1834, but it was expressed very obscurely and our understanding of it has to rely on later formulations of it, such as that given by Clerk Maxwell in his *Treatise on Electricity and Magnetism* (1873). Hamilton referred to his treatment as the *principle of varying action*. In the formulation of Maxwell and others the kinetic energy T is expressed in terms of the coordinates of position q and momentum p of each of the particles constituting a system. (By contrast Lagrange used q and $\dot{q}$ as variables.) When q and p are used the velocity is given by $\dot{q} = dT/dp$ and the momentum by $p = dT/dq$. Hamilton's treatment emphasized the importance of the sum $T + V$ of the kinetic and potential energies, expressed in terms of the generalized coordinates q and p, and this sum is now known as the Hamiltonian. It can be converted into the Hamiltonian operator which is convenient for generating the Schrödinger wave equation (Section 10.2). Maxwell made use of both Lagrange's and Hamilton's coordinates, and often found the latter to be more convenient for his work in optics, magnetism and electricity.

3.2 Early physical chemistry

Early physical chemistry was essentially an attempt to apply the principles of mechanics to chemical problems. This is true of the work of Robert Boyle who in Oxford from about 1655 to 1668, with a number of assistants notably Robert Hooke, investigated the properties of gases; this work is described in more detail in Section 5.1. In 1668 Boyle moved to London and continued this work in a laboratory he established in his sister's house in Pall Mall. He had made a careful study of the work of Francis Bacon and was convinced by Bacon's arguments in favour of the atomic nature of matter. Like Bacon he recognized heat as motion of the fundamental particles of matter, and he attributed the behaviour of gases to the movement of these particles. He recognized that the difference between fluids and solids is due to the different amounts of motion in the particles, the particles of solids being relatively quiescent. Boyle was a prolific writer, and his chemical ideas appeared in numerous books and pamphlets, and articles in the *Philosophical Transactions*.

Important work in physical chemistry was also done by Robert Hooke, who began his scientific work as an undergraduate at Christ Church, Oxford, acting as an assistant to Boyle from about 1655 to 1662. His construction of a highly efficient air pump made possible the pressure–volume experiments carried out in Boyle's laboratory in Oxford (Section 5.1), and he also participated in Boyle's investigations on combustion and respiration. Hooke continued these studies after 1662 when he became the Royal Society's Curator of Experiments, and he succeeded in elucidating the essential features of combustion and respiration. He also made important contributions to physics but much of his work in mechanics was eclipsed by the greater contributions of Newton. Hooke was a distinguished microscopist, and his book *Micrographia* (1665) contained remarkable drawings of microscopic observations.

Newton was 15 years younger than Boyle, holding him in high esteem, and

The Honourable Robert Boyle

(1627–1691)

Robert Boyle, whose ancestors were English rather than Irish, was born in Lismore Castle, County Waterford, in the south of Ireland. His father, Richard Boyle, lived for a period in Ireland, and in 1620 was created the Earl of Cork, of the Irish peerage, being commonly referred to as the 'great' earl. Robert Boyle inherited property which yielded a substantial income and never had to work for a living. He was educated mainly by private tutors, not attending a university but travelling on the continent of Europe where he gained a considerable interest in philosophy and science.

In about 1655 Boyle took lodgings in Oxford where after a period of study he embarked on his investigations of the properties of gases (Section 5.1). He also began studies on combustion and respiration, and he continued them after 1668 when he moved to London to lodge with his sister Katherine, Lady Ranelagh, who had a substantial house on Pall Mall. Boyle established a laboratory in the house and there he received many visitors, spending much time demonstrating his experiments and exhibiting the famous air pump designed in Oxford by his talented assistant Robert Hooke. In the London house Boyle continued the work he had begun in Oxford, and worked on some other problems, including the properties of acids and alkalies and the purity of salts. After the discovery of phosphorus in about 1669 he established many of its properties, particularly its reaction with air accompanied by light emission.

Boyle made no outstanding discoveries (he merely confirmed and publicized 'Boyle's law'), and his fame rests on his entirely new approach to experimental science. He was probably the first to apply to scientific investigation the inductive

his chemical work was greatly influenced by Boyle. Some of Newton's experiments, carried out in a garden shed behind his rooms at Trinity College, Cambridge, were partly alchemical, concerned with trying to transmute elements, but he also made important contributions to the theory of matter. Some of his ideas on this are expressed in his short tract *De Natura Acidorum* (On the nature of acids) which was published in 1710 but which was passed round among his colleagues for 20 years or so previously. His theory of matter is also explained in the second edition of his *Opticks*, which appeared in 1717. The *Opticks* consisted of a number of queries, each one of which is discussed at length. Because of this format it is often difficult to know precisely what Newton's opinions

methods that had been advocated by Francis Bacon. Until Boyle made his investigations, chemistry had been pursued haphazardly and empirically, with much emphasis on attempts to convert base metals into gold. Boyle's great contribution was to transform chemistry from a subject in which there was much mysticism and charlatanism into an effective branch of natural philosophy, based firmly on the experimental evidence. If Boyle was not the father of chemistry, as he has often been called, a good case can be made for regarding him as the first physical chemist.

Boyle had a very fine character, being invariably kind, gentle and courteous to everyone. He was what today would be called charismatic, having a wide circle of friends who were devoted to him, and he exerted a great influence on Newton among many others. His investigations were carried out in collaboration with a team of technicians and research assistants some of whom, notably Robert Hooke, became distinguished scientists in their own right.

Boyle died on December 31 1691, a week after his sister, and both were buried in the chancel of the famous Church of St Martin's-in-the-Fields in London. Characteristically he specified in his will that his funeral should be a simple one, and so it was. In 1720 the church was demolished, and the remains were presumably buried elsewhere by members of the Boyle family. In spite of much investigation, however, no trace of the remains has been discovered. There was no memorial to Boyle in the old church, nor is there one in the present church which was completed in 1726.

References: Marie Boas Hall, DSB 1. 377–382 (1970); Boas (1958); Conant (1957); Hall (1965); Maddison (1969); More (1944); Webster (1966).

were, especially as he expressed different opinions at different times. Also, Newton's theories of matter are, as with his other theories, mingled with theological and metaphysical arguments. Newton's situation was complicated by the fact that although he was nominally an Anglican—which he was required to be to hold his professorship at Cambridge—he held some unconventional theological opinions; he secretly did not believe in the Trinity, and today would be called a Unitarian. In this respect he was in contrast to Boyle, who was firm although liberal in his Anglican beliefs and found no difficulty in reconciling his religion with his science.

Newton's belief in the particulate nature of matter was supported by his optical experiments. His view was that light is a stream of corpuscles, and in order to explain the fact that some material is transparent he assumed that some of the corpuscles pass through matter without encountering the particles of which it is composed. Matter must therefore consist mainly of empty space, and he made estimates of the size of the particles. His conclusion was that the particles must be so extremely small that if all of the particles in the solar system came together the total volume would be that of a nut. Later Joseph Priestley (1733–1804), in his *Disquisitions Relating to Matter and Spirit* (1777), first used the expression 'matter in a nut-shell' to describe Newton's ideas.

To explain the coherence of matter Newton assumed that there are attractive forces between particles. He thought that these are gravitational in character, and that they would obey the inverse square law which he had found to be successful in explaining the movement of the planets. To explain the behaviour of gases Newton assumed that there are also repulsive forces between particles. He was never able to clarify this question of attractive and repulsive forces, or to explain when one or the other would be predominant.

Even before Newton published his ideas about the theory of matter, some of them had already been put forward by associates to whom he had communicated them. Because of his great sensitivity to criticism Newton was reluctant to publish, although he was always eager afterwards to claim credit for what had been in his mind but was unpublished. One device he used was to have a selected colleague publish his ideas; if they proved wrong the blame would not be directed to Newton, but if they were right he would claim the credit!

One who was used in this way by Newton was John Keill (1671–1721), an Edinburgh-educated Scot who in 1699 became deputy to the Sedleian professor of natural philosophy at Oxford; he later became Savilian professor of astronomy at Oxford. There he gave a series of lectures on Newtonian mechanics, in which he developed the theory that matter consists of particles held together by short-range attractive forces. In 1701 with Newton's support he published his major work, *Introductio ad veram physicam* (Introduction to true physics) based on these lectures, and a paper in 1708 dealt with the laws of attraction between particles. Keill was particularly useful to Newton because of his theological position. Most of Newton's support was coming from Low Churchmen, but Keill was a High Churchman who was well versed in theology, and was able to balance the Low Church influences. John Keill's younger brother James (1673–1719) was a physician who also gave support to Newton's ideas about matter in various anatomical and physiological works.

Newton's theories of matter were also publicised by John Freind (1675–1728) who as an undergraduate at Christ Church, Oxford, had learnt Newtonian mechanics from John Keill. In 1704 Freind gave, by invitation, nine lectures on chemistry in the Ashmolean Museum at Oxford, and in 1709 these were published as *Praelectiones chymicae*, with an English translation as *Chymical Lectures* in 1712. In these lectures Freind interpreted chemistry on the basis of Newtonian mechanics.

Aside from these theoretical treatments of matter and its behaviour, based on Newtonian principles, a number of related experimental investigations were carried out during the early eighteenth century. Some of the more interesting of these were performed by the Revd John Theophilus Desaguliers (1683–1744), the son of a Huguenot refugee to England. After being an undergraduate at Christ Church, Oxford, he held a Church appointment near London and being a man of great experimental skill he was invited by Newton to carry out a number of experiments on heat, optics and electricity. He was elected a Fellow of the Royal Society in 1714 and became the Curator of Experiments for the Royal Society, a post that Hooke had held for so many years. The ultimate objective of his experiments was to quantify the short-range forces between the particles of matter, but in this he was unsuccessful. He wrote an excellent book, *A Course in Experimental Philosophy*, the two volumes of which appeared in 1734 and 1744.

A little later the Revd Stephen Hales (1677–1761) made even more significant contributions on the basis of Newton's theory of matter with its attractive and repulsive interparticle forces. Hales was educated at Cambridge where, although Newton had left to become Master of the Royal Mint, he was exposed to the Newtonian ideas. In 1709 he became perpetual curate—or, as we would now say, vicar—of Teddington, near London. Although he published nothing until he was in his fiftieth year, he made outstanding contributions, especially to plant and animal physiology. He performed important experiments on 'fixed air' (carbon dioxide), using ingeniously designed apparatus such as the pneumatic trough, which he invented, and he laid the foundations of the chemistry of gases. His work had an important influence on the later great contributions of Joseph Black, Henry Cavendish, Antoine Lavoisier and Joseph Priestley. In 1727 he published his important book *Vegetable Staticks*, and in the preface he acknowledged his great indebtedness to Newton; indeed this indebtedness is apparent throughout the book. In explaining the rise of sap in trees, for example, he applied the idea of attractive and repulsive forces between particles.

On the whole, notwithstanding these important contributions, the British investigations in the eighteenth century that owed their origins to Newton's theories were rather disappointing. The work in Scotland of Joseph Black during the latter part of the eighteenth century was of particular significance for physics and physical chemistry, and is considered in Section 4.1.

The work in France of Antoine Lavoisier (1743–1794), who is sometimes referred to as the father of modern chemistry, was also of outstanding importance. His great contribution was to introduce quantitative methods into chemistry, and he placed particular stress on the use of the chemical balance. His important calorimetric studies in collaboration with Laplace were referred to on p. 63. By his quantitative studies on oxidation reactions he was able to provide convincing evidence against the phlogiston theory, and to explain oxidation and reduction in what is now known to be essentially the correct manner. His book *Traité élémentaire de chimie* (Elementary treatise on chemistry), published in 1789 in two volumes, presented a unified picture of his new ideas, and exerted a wide influence. It contained the first list of the chemical elements.

3.3 Concepts in the Physical Sciences

Over the centuries a number of fundamental concepts and theories have gradually developed and have played a vital role in the growth and success of the physical sciences. Many of them lie in the field of physics but at the same time are essential to the development of chemistry. A few of them are largely chemical, such as the atomic theory in the form proposed by John Dalton (1766–1844) in the early years of the nineteenth century. The most important of the concepts and theories that have been so essential to the growth of physical chemistry are

1. The fundamental laws of mechanics, as formulated particularly by Newton; these have already been discussed in Section 3.1.
2. The atomic theory of Dalton (Section 5.2).
3. The elucidation of the nature of energy and heat (Section 4.1).
4. The elucidation of the nature of light and other forms of radiation.
5. The realization that matter is electrical in nature (Sections 6.7 and 7.1).
6. The appreciation of the fact that purely mechanical models of the universe are unsatisfactory, and that operational definitions of physical quantities are required.
7. The idea that energy and radiation are quantized (Section 10.1).

Most of these concepts are dealt with in more detail, in the sections indicated; here some general comments will be made.

The atomic nature of matter

The idea that matter is not continuous but is particulate, corpuscular or atomic in nature can be traced back to the Greeks, but for a long time it was no more than speculation. One of the first to put the idea on a firm foundation was Francis Bacon, who used the principle of induction to conclude that an atomic hypothesis is more consistent with the experimental results than the assumption that matter is continuous. Newton's theories were also based on his conclusion that matter is corpuscular in nature. The atomic theory did not, however, have much impact on chemistry until 1803 when John Dalton showed that many of the results relating to chemical composition could be interpreted in terms of the idea that different chemical elements contain different kinds of atoms, which form the building blocks for the various chemical compounds.

The concept of energy

A concept that became clarified rather late is that of energy. This property is so important in modern science and technology that it is hard to realize that it did not emerge until the middle of the nineteenth century. This was due to an historical accident. The idea of energy was implicit in the treatments of Galileo, Descartes and Newton, but they chose to formulate their mechanics in terms of time, distance, velocity, acceleration and force; momentum and energy did not emerge until later. Until about the middle of the nineteenth century a number of words were used indiscriminately and interchangeably to mean energy, force, power and momentum. All these concepts, as we now understand them, were variously referred to by words such as vis viva, impetus, power and force. The word 'force', in particular, could sometimes be understood to mean any property that could produce some effect; this was stated explicitly, for example, by W. R. Grove in his book *Correlation of Physical Forces* (1846), and he even included heat as 'a force capable of producing motion'. In the famous memoir *Über die Erhaltung der Kraft* (1847) by Hermann von Helmholtz, the word Kraft, usually translated as force, must be understood to mean what we now call energy.

The failure to recognize energy and work was a conceptual one and not a mathematical one. For a constant acceleration a Galileo arrived at eqns 3.1 and 3.2, and since the speed v is equal to at it follows that

$$as = \tfrac{1}{2}(at)^2 = \tfrac{1}{2}v^2. \tag{3.6}$$

Multiplication by the mass m gives

$$mas = \tfrac{1}{2}mv^2. \tag{3.7}$$

The force F is equal to ma and thus

$$Fs = \tfrac{1}{2}mv^2. \tag{3.8}$$

The product Fs is what we now call the work, w, and $\tfrac{1}{2}mv^2$ we call the kinetic energy. Thus the work done on a body is the kinetic energy produced in it. Galileo, Newton and others could easily have arrived at these relationships, and perhaps did so, but would not have seen any significance in the concepts involved.

Towards the end of the seventeenth century a vigorous controversy raged about the significance of a quantity that was supposed to represent the capacity a body has to set another body in motion when a collision occurs. This quantity was variously called the impetus, *potentia* (power), *vis* (force), *vis motus* (moving force) or *vis viva* (vital force). The controversy was about how the quantity should be defined. One view was that it is the mass m multiplied by the speed v, and it was considered by some that for all the bodies involved in a collision the sum Σmv remains constant. It is true that if mv, which we now call the momentum, is treated as a vector there is conservation in a collision, even in an inelastic one. At the time, however, the concern was not with vectors, which were introduced only in the early nineteenth century; it was thought that the sum Σmv is conserved, which is not the case.

The view that the total mv is conserved in a collision was maintained by Descartes, but characteristically his arguments were of a religious or metaphysical nature. The great Dutch physicist and astronomer Christiaan Huygens

(1629–1695), on the other hand, carried out many important and ingenious experiments on colliding bodies, some of which he demonstrated at meetings of the Royal Society, and these led him to the correct conclusion that it is Σmv^2 and not Σmv that is preserved in an elastic collision. He also established that if motion *in one direction* is considered, the quantity Σmv remains constant, which is also correct. Huygens had essentially reached the idea of a *vector* quantity, a quantity that was introduced formally only in the nineteenth century. Later, in 1686, Leibniz drew the same conclusion, although his argument was metaphysical and quite unsound. By 1743 the constancy of Σmv^2 seems to have been generally accepted, being taken for granted in that year in d'Alembert's authoritative book *Traité de dynamique*.

Leibniz was aware that when two bodies collide there may be a decrease in the apparent Σmv^2, and concluded that the *vis viva* that appeared to be missing had gone into the particles of which the colliding bodies are composed. If he had gone only a little further and had realized that the missing *vis viva* was actually the heat developed in the colliding bodies, he would have anticipated the ideas developed a century and a half later. Similar ideas about the missing *vis viva* were suggested in 1742 by Émilie, Marquise du Châtelet (1706–1749), a woman of remarkable ability who is noteworthy for having been Voltaire's mistress, and for her excellent translation of Newton's *Principia*.

The *vis viva* was first defined as Σmv^2, but in 1829 Gustave Gaspard de Coriolis (1792–1843), pointing out that the work done in giving a velocity v to a mass m is $\frac{1}{2}mv^2$ (eqn 3.8), introduced the practice of defining the *vis viva* as $\frac{1}{2}mv^2$ rather than mv^2. A little earlier the British physicist Thomas Young (1773–1829) had suggested replacing *vis viva* by 'energy', but this word, so common today, was little used until half a century or so later.

In 1853 a comprehensive treatment of energy was developed by William John Macquorn Rankine (1820–1872), a Scotsman who until then had worked mainly as a civil engineer, with a particular interest in naval architecture. He had also done some interesting work in thermodynamics and atomic theory, to be considered in Sections 4.3 and 5.2. In his 1853 paper Rankine for the first time distinguished clearly between *potential energy* and what he called 'actual or sensible energy', renamed *kinetic energy* by Kelvin. Rankine has been called the founder of the science of energetics. In 1855 he became regius professor of civil engineering and mechanics at the University of Glasgow. In spite of his original contributions Rankine was very conservative, and strongly resisted attempts to replace the British system of measurements—which now, paradoxically, is used mainly in the United States, almost all other countries having adopted the metric system. Rankine wrote a number of songs, one of which included the stanza:

> Some talk of millimetres and some of kilograms,
> And some of decilitres to measure beer and drams;
> But I'm a British workman, too old to go to school,
> So by pounds I'll eat and by quarts I'll drink and
>   I'll work by the three-foot rule.

Rankine was of course opposed to the replacement of the Fahrenheit scale of temperature by the Celsius or centigrade scale, and he devised an absolute scale based on the Fahrenheit scale; on this Rankine scale the freezing point of water is 492 R and the boiling point 672 R.

The use of the word work also requires some discussion. The concept of work had been used by James Watt (1736–1819) in his pioneering development of steam engines, but it had not been clearly defined. Work is now defined in a simple way: if a constant force F acts through a distance s the work done is Fs, and if the force is variable the work is $\int F ds$. It is not certain how this definition arose, but the first important contribution to the concept of work was made by Charles Augustin Coulomb (1736–1806), who in his earlier years worked as a military engineer. In that capacity he investigated with great psychological insight the efficiency with which work is performed by humans and animals. He went carefully into the question of fatigue, and found that frequent rest periods result in higher output. It was not for more than a century that much more was done in this important field, which is now known as ergonomics. The results of Coulomb's investigations on this problem were described in an essay entitled 'Théorie des machines simples', which appeared in 1781 and was published as a book in 1809; it attracted much attention.

Jean Victor Poncelet (1788–1867) also made important contributions to the understanding of mechanical work. He was the founder of the field of projective geometry, but in 1824 he agreed after some hesitation to become professor of 'mechanics applied to machines' at the École d'Application de l'Artillérie et de Génie at Metz. He then abandoned his work on geometry and entered enthusiastically into research and teaching in applied mechanics. Notes on his highly successful lectures were widely distributed, sometimes without his knowledge, and they spread the understanding of work done by machines. In 1829 he specifically suggested the use of the word 'travail' (work) to mean what it means today.

The nature of heat

Since it took so long for the idea of energy to be clarified, it is not surprising that the same is true of heat, which is now known to be a form of energy. By 1620 Francis Bacon had correctly concluded that heat is a form of motion of the constituent particles of matter. This idea was expressed very clearly by Newton's contemporary the physician, chemist and philosopher John Locke (1632–1704):

Heat is a very brisk agitation of the insensible parts of the object, which produces in us that sensation from which we denominate the object hot; so that what in our sensation is heat, in the object is nothing but motion.

A similar view was taken by Newton.

The alternative view of heat is that it is a substance, and this idea persisted long after the time of Bacon. Lavoisier was one who always believed heat to be a substance, and he even listed 'calorique' as one of the chemical elements. It is of interest to note that the exemplary experiments of Joseph Black in the late eighteenth century seemed to lead to this incorrect concept. It was Black who first clearly distinguished between temperature and heat, and who showed how temperature measurements can be used to determine the quantity of heat. Black was struck by the fact that mercury, which is very dense, has a lower heat capacity than an equal volume of water. He thought, not unreasonably at the time, that more motion should be possible in a material of greater density; therefore, he concluded, heat cannot be motion. At the time there was no knowledge of the atomic structures of mercury and water; we know now that the higher heat capacity of water arises from its more complicated structure than that of mercury so that more modes of motion are possible, and that for a more detailed understanding quantum effects must be taken into account. This is an instructive example of how the most careful experiments and intelligent reasoning can lead one astray if the time is not ripe.

The nature of light

Establishing the nature of light was another problem that proved very difficult, and for years there was much controversy. Newton's experiments on light, described in his *Principia* (1687) and in his *Opticks* (1704) convinced him that light was a stream of 'corpuscles'. Christiaan Huygens in 1690 rejected Newton's corpuscular theory and proposed instead a wave theory. This theory seemed, however, to have the difficulty of failing to explain why light travels in straight lines. Although Huygens was able to show how a wave theory could explain this, and could be consistent with the laws of reflection and refraction, his theory was largely ignored for over a century, particularly as a result of Newton's great prestige. In 1803 Thomas Young demonstrated the existence of diffraction bands, a phenomenon that could not be explained in terms of the corpuscular theory. In about 1814 the French physicist Augustin Jean Fresnel (1788–1827) carried out similar and more extensive investigations on diffraction, and these provided further evidence in favour of the wave theory. In 1850 Jean Bernard Léon Foucault (1819–1868) and Armand Hippolyte Louis Fizeau (1819–1896), who had begun working together but reached their conclusions independently at about the same time, demonstrated that light travels faster in air than in water, a result that can only be interpreted on the basis of the wave theory.

With the advent of the quantum theory, however, it became clear that a dual theory is necessary, light behaving in some experiments as a stream of particles, in others in the manner of waves (Section 10.1).

The electrical nature of matter

The idea that matter is electrical in character developed from Faraday's work on electrolysis, which suggested that electricity as well as matter is discontinuous, and that there is an intimate relationship between atoms and the fundamental particles of electricity. In the 1870s both Clerk Maxwell and Johnstone Stoney called attention to this relationship, and a particularly clear statement about it was made in 1881 by Helmholtz. Ten years later Stoney suggested that there is a fundamental particle of electricity, to which he gave the name electron. He also made a number of interesting suggestions about atomic structure, even suggesting the idea of orbital electrons, and he attempted to explain spectra in terms of electronic motions. These ideas were, however, rather ahead of their time, and did not attract much attention. Only after the direct establishment of the existence of electrons, particularly by J. J. Thomson (Section 6.7), and the formulation of the quantum theory (Section 10.1), was it possible for the electronic structure of the atom to be properly formulated.

Decline of mechanical models: operational definitions

The mechanical view of the universe that was outlined in Section 3.1 dominated the physical sciences until about the middle of the nineteenth century. Over the next few decades three entirely new concepts in physics were developed which provided a view of the universe entirely different from that of the mechanical theories. These are the electromagnetic theory of radiation, the theory of relativity, and the quantum theory. The quantum theory has had particularly important implications for chemistry.

The electromagnetic theory of radiation was developed by the Scottish physicist James Clerk Maxwell to explain the propagation of light. He developed the theory in a series of papers and in a book, *Treatise on Magnetism and Electricity* which appeared in 1873. By that time the wave theory of light had become accepted, and it had been assumed that the waves occur in a medium, called the aether, which was supposed to pervade all space. There were unsatisfactory features about this aether, and an important feature of Maxwell's theory was that he had decided that it was not necessary for a theory to be based on a mechanical model. Instead he developed a mathematical theory of radiation which represented the situation not in terms of a model that could be visualized but rather as a mathematical analogue. This was one of the first physical theories that had this characteristic, and Maxwell's approach has become a prototype for modern theories of physics.

Some physicists found themselves unable to appreciate Maxwell's theory and later theories that could not be visualized in terms of a model. Lord Kelvin, for example, always insisted on a mechanical model; in his own words,

I never satisfy myself unless I can make a mechanical model of a thing. If I can make a mechanical model I can understand it.

Because of this attitude Kelvin could never understand the concept of entropy, which Clausius had developed in connection with the second law of thermodynamics (Section 5.5). Although Kelvin had himself made such an important contribution to the second law, he always explained it in terms of the dissipation of heat, a concept by no means as powerful or satisfactory as that of entropy.

One can visualize the dissipation of heat, but entropy is a mathematical concept which can not be visualized in terms of a mechanical model. The emergence of quantum theory and quantum mechanics in the twentieth century confirmed the view that a purely mechanical approach to scientific problems is not satisfactory.

The second of the entirely new theories of physics was the theory of relativity, first put forward in 1905 by Albert Einstein and developed in later years. The theory of relativity has had a less direct impact on chemistry than the electromagnetic theory and the quantum theory, but it has important indirect implications. One consequence of relativity theory that is important in chemistry is that a modification to the principle of conservation of energy is required. According to relativity theory, energy and mass can be interconverted, the relationship between the two being that the energy E is equal to mc^2, where m is the mass and c the speed of light. When nuclear transformations occur this relationship assumes great practical importance, in that large amounts of energy can be released at the expense of a small loss of mass. For ordinary chemical processes, however, the effects are undetectable.

The theory of relativity is also important to chemistry in less direct ways. For one thing, it gave further support to the conclusion that the purely mechanical theories are not satisfactory. With the relativity theory a new principle entered physics and had a profound influence on our attitude to scientific theories. This principle, called the *operational definition* of physical quantities, states that

physical quantities must be defined in terms of the operations that are used to measure them.

For example, the length of an object is not an absolute property, but only relates to the manner in which the length is measured. This type of operational definition is in complete contrast to the ideas of Newton, who stated for example that 'time flows equably without relation to anything external'. This is equivalent to saying that time moves with a constant speed, which from the modern, operational point of view is meaningless since time is involved in the measurement of speed.

Quantization

The third of the entirely new concepts came from the quantum theory, which was born in 1900 with the analysis by Max Planck of blackbody radiation (Section 10.1). It followed from this analysis that the energy of an oscillator is not continuous but comes in packets, called quanta. One of the most important implications of quantum theory came in 1905 with Einstein's proposal that radiation itself is quantized. This idea took a few years to be accepted, since so much evidence had accumulated in favour of the wave theory of radiation. Einstein resolved the difficulties by pointing out that those properties that suggest wave motion—diffraction and interference—are concerned with time averages. When there is a one-to-one relationship between radiation and atoms or molecules, as in the absorption and emission of radiation, the particle nature of radiation becomes evident. There thus must be a dual theory of radiation, a wave theory to explain properties like diffraction and interference, and a particle theory to explain properties where there is a one-to-one relationship between radiation and matter.

These ideas had far-reaching implications. Just as light has particle as well as wave properties, particles such as electrons also have wave properties. This realization led to the development of a comprehensive wave mechanics or quantum mechanics which has dominated the thinking of physical scientists since the 1920s. One important aspect of quantum mechanics is the uncertainty principle proposed in 1927 by the German physicist Werner Heisenberg. This principle is firmly based on the operational principle, inherent in the theory of relativity. If a measurement is made on a particle such as an electron, the act of making the measurement disturbs the particle, so that there is inevitably an uncertainty in the measurement of the quantity.

The implications of this principle are profound. In particular, it discredits the purely deterministic theory of the universe. Laplace had said that from a complete and detailed knowledge of the universe the future could be predicted, but the uncertainly principle shows that it is impossible to have such a detailed knowledge. The whole deterministic hypothesis therefore collapses; human free will can no longer be denied.

Quantum theory, and its development as quantum mechanics, has completely transformed our thinking about chemical problems. Experimental results that were previously incomprehensible can now be explained. For example, in the nineteenth century it seemed impossible to reconcile the spectra of atoms and molecules with the specific heats of gases (Section 10.1). The spectra indicated that many types of motion are possible in atoms and molecules, while the specific heats could not be explained if so much motion were involved. For a diatomic molecule the specific heats indicated only two degrees of rotational freedom in addition to three of translational motion, and apparently no vibrational motion. The idea that energy is quantized resolves this dilemma; at ordinary temperatures the vibrations are not excited, and are not revealed in the specific heats, because of wider spacing of the energy levels.

Another matter that could not be understood at all was the nature of photochemical reactions, in which light induces chemical reaction. The earlier workers had to invoke mysterious properties of light, such as 'actinism' and 'tithonicity', which merely described the effects and explained nothing. The photochemical reaction between hydrogen and chlorine was particularly baffling; a flash of light of extremely short duration and not particularly high intensity can induce rapid reaction, the process being complete in an instant. Einstein's suggestion in 1905 that light is a beam of quanta, later called photons, and that a photon can interact with a single molecule, resolved part of the difficulty, and led to the ultimate solution in terms of a chain reaction proposed in 1919 by Walther Nernst (Section 8.4).

Suggestions for further reading

For general accounts of the history of mechanics see

Sir William Dampier, *A History of Science*, Cambridge University Press, 1929 and later editions.

R. J. Forbes and E. J. Dijksterhuis, *A History of Science and Technology*, Penguin Books, 1963, Chapter 17.

P. J. Harman, Newton to Maxwell: The *Principia* and British physics, *Notes & Records Roy. Soc.*, **42**, 75–96 (1988).

M. von Laue, *History of Physics*, translated by P. Oesper, Academic Press, New York, 1950, Chapter 2.

A. E. E. MacKenzie, *The Major Achievements of Science*, Cambridge University Press, 1960), Vol. 1, Chapter 3.

Some articles dealing with special aspects of mechanics are

M. Crossland and C. Smith, The transmission of physics from France to Britain: 1800–1840, *Hist. Stud. Phys. Sci.*, **9**, 1–61 (1978).

R. Fox, The rise and fall of Laplacian physics, *Hist. Stud. Phys. Sci.*, **4**, 89–136 (1974).

T. Hankins, Eighteenth-century attempts to resolve the *vis viva* controversy, *Isis*, **56**, 281–297 (1965).

L. L. Landau, The *vis viva* controversy: a post mortem, *Isis*, **59**, 131–143 (1968).

J. R. Partington, *A History of Chemistry*, Macmillan, London, 1962; Vols. 2 and 3 contain numerous references to mechanics, including Newton's theory of matter (Vol. 2, pp. 468–485).

For accounts of early theories of matter see Partington's book and also

A. Thackray, Matter in a nut-shell: Newton's *Opticks* and eighteenth-century chemistry, *Ambix*, **15**, 29–53 (1968–69).

A. Thackray, *Atoms and Powers: An Essay on Newtonian Matter-Theory and the Development of Chemistry*, Harvard University Press, 1970.

The development of the concepts of energy, force and work is treated in

A. E. Bell, The concent of energy, *Nature*, **151**, 519– (1943).

D. S. L. Cardwell, Some factors in the early development of the concepts of power, work and energy, *British J. Hist. Sci.*, **3**, 209–224 (1967).

P. M. Harman, *Energy, Force and Matter. The Conceptual Development of Nineteenth-Century Physics*, Cambridge University Press, 1982.

Y. Elkana, Helmholtz's 'Kraft': an illustration of concepts in flux, *Hist. Stud. Phys. Sci.*, **2**, 263–298 (1970).

CHAPTER 4

Thermodynamics

The word 'thermo-dynamic' was first used in 1849 in a paper presented by William Thomson (1824–1907) to the Royal Society of Edinburgh. Thomson was then a young professor of natural philosophy at the University of Glasgow, where he was to remain for over half a century. Later he was created Baron Kelvin of Largs, and although all his thermodynamic work was done as Professor William Thomson it seems best to refer to him as Kelvin, if only to avoid confusion with various other Thomsons, Thomsens and Thompsons who worked in similar areas of research.

When he wrote his 1849 paper Kelvin had been grappling for a few years with the ideas presented in a remarkable book which had been published in 1824 by Sadi Carnot, and with a paper by Clapeyron that had appeared in 1834. Carnot's book, which can be said to have inspired the whole subject of thermodynamics, was concerned with the performance of heat engines, and Kelvin was one of the first to appreciate its great importance. Carnot had based his treatment on the concept that heat is an 'imponderable fluid', but by 1849 Kelvin was just beginning to realize, mainly because of the work of James Prescott Joule, that heat is instead a form of energy. The idea was developing in Kelvin's mind that in heat engines, such as the steam engines that were being used more and more, there is actually a conversion of heat into another form of energy, mechanical work. His word 'thermo-dynamic' was invented by him to refer to processes of this kind, and today we define thermodynamics as the science that deals with the interconversion of heat and work and all its applications.

There are several reasons why the science of thermodynamics could not have arisen until about the middle of the nineteenth century. In the first place, there was much confusion between temperature, the intensive quantity, and heat, which is extensive; this matter is discussed briefly in Section 4.1. Even after the clear distinction between temperature and heat had been made, there still remained much confusion about the nature of heat. During the first half of the nineteenth century there were three theories of heat, all with their strong proponents, and these are touched upon briefly at the end of Section 4.1. The realization in the middle of the nineteenth century that heat is neither a substance nor a manifestation of wave motion, but is a form of energy, soon led to a consideration of the interconversion of heat and mechanical work, and the subject of thermodynamics was born.

TABLE 4.1 Highlights in thermodynamics

Author	Date	Contribution	Section
Black	1760–1766	Calorimetry	4.1
Mayer	1842	Interconversion of heat and work	4.2
Joule	1843–1852	Interconversion of heat and work	4.2
Thomson (Kelvin)	1848–1849	Absolute temperature scale	4.3
Clausius	1850	Second law	4.3
Thomson (Kelvin)	1851	Second law; dissipation of energy	4.3
Clausius	1854–1865	Concept of entropy	4.3
Gibbs	1873	Chemical thermodynamics	4.4
Gibbs	1876–1878	Chemical potential; phase rule	4.4
Helmholtz	1882	Theory of equilibrium; free and bound energy	4.4
van't Hoff	1884–1887	Chemical thermodynamics; theory of equilibrium constant; solutions	4.4
Nernst	1906	Heat theorem (third law)	4.5
Simon	1927	Improved version of third law	4.5

Table 4.1 shows some of the more important contributions to the science of heat and thermodynamics.

4.1 Temperature and heat

That the science of heat was much slower to develop than the science of mechanics is well illustrated by the fact that specific gravities were being measured, by the ancient Greeks, two thousand years before specific heats were first measured, by Joseph Black, in the eighteenth century.

Some advance in the understanding of temperature, the intensity of heat, came with the invention of thermometers. for which Galileo is usually given the credit; he was certainly one of the first to use such an instrument, even if the idea was not his. The first written record of a thermometer, however, was made in 1612 by the Italian physiologist Santorio Santorii (1561–1636), who is usually known as Sanctorius. Another who devised a thermometer at about the same time was the English physician Robert Fludd (1574–1637), one of whose instruments, reproduced from his book *Meteorologica Cosmica* (1626) is illustrated in Figure 4.1. The type of thermometer used by Galileo, Sanctorius and Fludd consisted of a bulb containing air at less than atmospheric pressure; the tube attached to the bulb is immersed in a liquid, usually water. The height of the liquid in the tube gives a rough estimate of the temperature, but the instrument was not very reliable, since the level depends also on the atmospheric

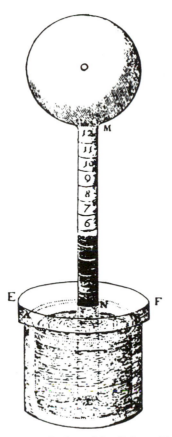

FIGURE 4.1   An air thermometer designed by Robert Fludd, as illustrated in his *Meteorologica Cosmica* (1626).

pressure. Modifications of this device were used throughout the seventeenth century. Because of constant minor variations in temperature and pressure there was usually an incessant movement of the liquid in the tube, and the instrument was sometimes regarded as exhibiting perpetual motion.

In about 1700 a much more reliable air thermometer was devised by the French physicist Guillaume Amontons (1663–1705), who used it to make studies of the expansion of gases. A little later a great advance was made by Daniel Gabriel Fahrenheit (1686–1734), a German who resided in Holland, with his invention of a reliable liquid thermometer. Fahrenheit acquired fame as a manufacturer of such thermometers, and he devised the scale known by his name. The scale used widely today, and universally in scientific work, is essentially that proposed in 1742 by Anders Celsius (1701–1744), a Swede.

It was Joseph Black (1728–1799) who first clearly recognised the distinction between temperature and heat, and who showed how temperature measurements

Joseph Black

(1728–1799)

Black, of Scottish descent, was born in Bordeaux, France. He was educated first in Belfast and later at the Universities of Glasgow and Edinburgh, receiving an MD degree from Edinburgh in 1754.

Two years later he became professor of chemistry at the University of Glasgow where he carried out investigations on a variety of problems in physics, chemistry and medicine. In 1766 he became professor of chemistry at Edinburgh where he remained until his death. Besides giving many highly successful lectures and carrying out much research, Black, who never married, had an extensive and demanding medical practice and carried out many university duties. Many of his students attained distinction, some becoming professors of chemistry. One of his students was Henry Peter Brougham (1778–1868), later the first Baron Broughham and Vaux, who was active in law and politics. He is best remembered as a law reformer—he was concerned with the abolition of slavery—and as one of the founders of the University of London. In 1838 he designed the one-horse closed carriage that is named after him; it provided the pattern for the coachwork of the first Rolls Royces to be made.

As noted in Section 4.1, Black's work was the first to lead to a fundamental understanding of heat in its experimental aspects, most of his work on heat being done between 1760 and 1766 when he was at Glasgow. Although he did much research on a variety of problems he published a total of only three papers, and nothing on heat. After his death his former student John Robison (1739–1805), who had become professor of natural philosophy, prepared a book based on notes taken by some of Black's students, including himself, and on Black's own notes. This book, entitled *Lectures on the Elements of Chemistry*, appeared in 1803, four years after Black's death. It was dedicated to James Watt (1736–1819), who had been employed as an instrument maker at the University of Glasgow and whose friendly association with Black had played a part in some of his many technical successes, particularly in his famous steam engine.

Black was active in the Royal Society of Edinburgh, but was never elected a Fellow of the Royal Society.

References: H. Guerlac, DSB, 2, 173–183 (1970); Boyer (1943); Cardwell (1971); Roller (1957).

can be used to determine the quantity of heat; he can be said to have been the founder of the field of calorimetry. At the University of Glasgow, where he became professor in 1756, Black carried out important investigations on a variety of problems in physics, chemistry and medicine. Since Black published little, our knowledge of his research comes from the book *Lectures on the Elements of Chemistry*, edited by John Robison, which appeared in 1803.

Black's careful experiments on heat were of the greatest importance, and were the first to lead to a proper understanding of the experimental aspects of the subject. He applied Newton's law of cooling and used a dynamical method of measuring heat, in much the same way as is done in modern calorimetrical studies. He measured the amounts of heat required to melt solids and to vapourize liquids, and he invented the term 'latent heat' to describe these amounts. He also measured, for a number of substances, what he called the 'capacity for heat' and is now called the heat capacity or (for unit mass of substance) the *specific heat*.

It was noted in Section 3.3 that Black's meticulous experiments led him to reject the idea that heat is a form of motion and to believe that it is a substance.

Carnot and heat engines

The book published in 1824 by Sadi Carnot (1796–1832) probably had more influence than anything else on the founding of the science of thermodynamics. Although Carnot's ideas required much further development, they represented an important step forward, and stimulated others to make further advances.

Sadi Carnot was 28 years of age, and residing in Paris after a brief military career, when he published his 118-page book, *Réflections sur la puissance motrice du feu et sur les machines propres à dévelloper cette puissance*. This book, Carnot's only publication, was of great importance; Lord Kelvin described it as an 'epoch-making gift to science', and Sir Joseph Larmor as 'perhaps the most original in physical science'. It was clearly written but by no means a popular account; it presupposes some knowledge of steam engines, physics and basic mathematics.

In his book Carnot developed his highly original treatment of heat engines on the basis of his belief that heat is a substance, called caloric, and he thought that when in an engine heat flows from a higher to a lower temperature and work is done, the heat is actually conserved. Carnot discussed the analogy of a waterfall causing a wheel to turn (Figure 4.2); there is no loss of water, the wheel being turned by the force of the water. In the same way he thought that the force of falling heat would cause the piston in an engine to move and perform work. We now know that when work is done as a result of a flow of heat, some of the heat is converted into an equivalent amount of work.

Carnot realized that a steam engine cannot function if only a single temperature is involved. His great contribution to science was to consider ideal types of engines operating between a higher temperature which we will call T_h and a lower one T_c. It is convenient to consider one of these 'Carnot engines' with

Nicolas Leonard Sadi Carnot

(1796-1832)

Sadi Carnot was born in Paris and was a member of a distinguished family. His father, Lazare Nicolas Marguerite Carnot (1753–1823) is important in the history of mechanics as the author of *Essai sur les machines en général* (Essay on machines in general), published in 1783, in which he discussed the principle of conservation of energy. He was also prominent in political and military spheres, being called the 'organisateur de la victoire' because of his activities under the First Republic of France. Sadi Carnot's nephew, Marie François Sadi Carnot (1837–1894) became President of the Third Republic of France in 1887.

Sadi Carnot was educated at the École Polytechnique as a military engineer, and saw active service in 1814. For a few years he held various routine military positions, but being frustrated by the work obtained permanent leave of absence and took up residence in Paris where he undertook study and research in science and engineering. For some years he worked on his famous book *Réflexions sur la puissance motrice du feu . . .* , which was published in 1824. Although this book was of the greatest importance (Section 4.1) it soon became difficult to obtain a copy.

In 1827 Carnot was required to return to active duty, with the rank of captain, but after less than a year's service he was able to return to Paris. He continued his studies on the theory of heat and the design of engines, but made no further publications. His surviving papers show that he was beginning to abandon the view, assumed in *Réflexions*, that heat is a substance, and was beginning to favour the idea that heat is a form of motion.

Carnot's health was always fragile. An attack of scarlet fever in June, 1832, further undermined his constitution and in August he fell victim to a cholera epidemic, dying within a day at the age of 36.

References: J. F. Challey, DSB, 3, 79–84 (1971); Barnett (1958); Fox (1969–1970), Kerker (1957); Wilson (1981).

reference to the pressure–volume diagram shown in Figure 4.3(a), although such a representation was not known to Carnot. In one of the Carnot cycles he considered a gas represented initially by point A on the diagram, the temperature being T_h. He first allowed the gas to expand at constant temperature (isothermally) so that it follows the curve AB represented by the equation of Boyle's law (pV = constant). The state of the gas at the end of this proces is represented by point B, and the gas is then placed in an insulating container and

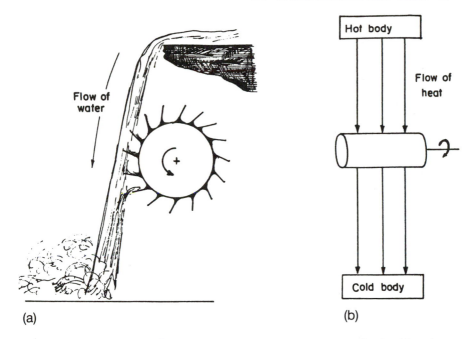

FIGURE 4.2 Falling water can perform work by turning a water wheel, without loss of water. Carnot thought that heat 'falling' from a higher to a lower temperature could produce work, without loss of heat.

allowed to expand to point C. Such an expansion, with no heat allowed to enter or leave is called adiabatic, and when it occurs there is necessarily a lowering of temperature, the final temperature being T_c.

In the third stage the gas is compressed isothermally, the temperature remaining at T_c, until point D is reached. Finally, in the fourth stage, the gas is compressed adiabatically until the temperature is T_h again, and the initial point A has been reached.

One important contribution made by Carnot in his book was to use the idea of thermodynamic reversibility for the first time. (He did not use the *word* 'reversibility', which appears to have been first used in its thermodynamic sense by the Scottish physicist Peter Guthrie Tait in 1876.) For a process to be thermodynamically reversible it must occur infinitely slowly and be reversible at every stage of the process; for example, if a substance is being cooled reversibly the external temperature must at all times be infinitesimally lower than the temperature of the substance. A thermodynamically reversible process can only occur in our imagination; any process that really occurs must be irreversible in the thermodynamic sense, although if it is slow it may be close to reversible. To visualize the first step A–B in the Carnot cycle we may imagine first holding the gas at the initial pressure P_1 by placing a weight at the top of a piston (Figure 4.3(b)).

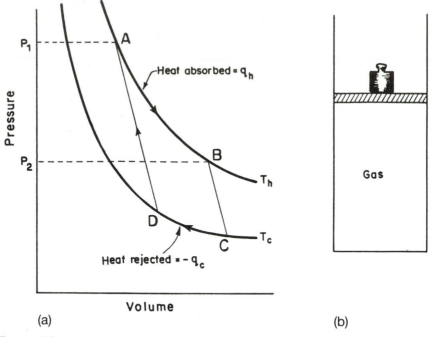

FIGURE 4.3 (a) An indicator diagram (pressure against volume) for a Carnot cycle. Such a diagram was first used by James Watt; it was unknown to Carnot, but used by Clapeyron. (b) A gas at equilibrium supported by a weight placed on a piston.

We can then slice off an infinitesimal piece of the weight so that the piston rises by an infinitesimal amount. The process is continued by taking off infinitesimal slices, until the final weight on the piston corresponds to the pressure P_2. The piston has then raised the maximum weight possible, and therefore has done the maximum possible amount of work.

Carnot considered going round the cycle completely reversibly, and also going round it by some irreversible processes. He realized that the work done in the completely reversible process is the maximum possible, and that a consideration of the completely reversible engine is useful in giving the maximum amount of work that could be done when an engine consumes a given amount of fuel such as coal. Modern treatments of the Carnot cycle usually deal with the efficiency of the system, but this is something that Carnot was not in a position to consider, for two reasons. In the first place, the efficiency is most easily expressed in terms of the absolute temperature, and it was only later that this concept was applied to the Carnot cycle. Secondly, the efficiency is the fraction of the heat absorbed at the higher temperature that is converted into work; Carnot, believing in the caloric theory, did not consider that the work was done at the expense of heat that disappeared.

Instead, Carnot considered the maximum 'duty' of an engine, which is the amount of work it does for a given amount of fuel. He expressed the work done as the mass of water that could be lifted multiplied by how high it is lifted. One important conclusion he reached is that if heat falls from a higher temperature T_h to a lower temperature T_c, the duty is larger the larger the difference between the two temperatures, just as in a waterfall (Figure 4.2(a)) the greater the fall of the water the more work is done by the water wheel. Carnot also found that for a given drop in temperature, $T_h - T_c$, the work is greater the smaller is T_h; thus a drop from 1 °C to 0 °C will produce more work than one from 100 °C to 99 °C. These relationships were later clarified in terms of the absolute temperature. If T_h and T_c are the absolute temperatures, the efficiency of the reversible engine, defined as the ratio of the net work done in the cycle to the heat absorbed at T_h, is given by

$$\text{efficiency} = \frac{T_h - T_c}{T_h}.$$

Carnot discussed his conclusions with reference to some of the steam engines of his time. He was able to explain why a high-pressure steam engine is more efficient than a low-pressure one; in his own words (in translation):

It is easy to see why the so-called high-pressure steam engines are better than the low-pressure ones; their advantage lies essentially in their ability to utilize a greater fall of caloric. Steam generated at a higher pressure is also at a higher temperature and as the temperature of the condenser is nearly always the same, the fall of caloric is evidently higher.

In terms of efficiency and absolute temperature this can be understood as follows. If the lower temperature is 27 °C (300 K) and the higher is 100 °C (373 K), the efficiency of a reversible engine is (373–300)/373 = 0.2 = 20%; in practice the efficiency will be substantially less. If the higher temperature is 400 K, because of the use of high pressure, the efficiency is raised to (400–300)/400 = 25%. Towards the end of the nineteenth century this important conclusion of Carnot was put into practice by Rudolf Diesel (1858–1913), who designed an engine, named after him, in which T_h is much higher than in a steam engine. In designing his engine Diesel gave full attention to thermodynamic principles, which he expounded in his *Theorie und Konstruktion eines rationallen Warme-Motors* (1893).

Another important conclusion reached by Carnot in his book is now referred to as Carnot's theorem. He considered two engines, both working between two particular temperatures. One of the engines worked reversibly, and the question he asked was whether the other could be designed in such a way as to produce more work from the same amount of fuel. He answered the question by first postulating that there could be such an engine, and he caused it to drive the first one backwards. He then showed that if this were to occur there would be a net flow of heat from the lower to the higher temperature. This, he pointed out, is contrary to experience; if it could happen, perpetual motion machines could be constructed. Today we express his conclusion by the statement that the efficiencies of all reversible engines working between two given temperatures must be the same, namely $(T_h - T_c)/T_h$.

This theorem had great practical implications, which Carnot pointed out. Previously it had been thought that engines could be improved without changing the working temperature, by changing to different materials; perhaps a steam engine could be improved by changing from water to alcohol or some other material. Carnot had shown, however, that attention must be directed to the working temperature and not to the materials used.

Carnot's great contributions to science and engineering may be summarized as follows:

1. His discussion of engines in terms of simple cycles.
2. His conception of reversible processes, as giving an upper limit to the usefulness of an engine.
3. His conclusion as to the dominance of the working temperatures in determining the effectiveness of engines.

Some of Carnot's other conclusions were incorrect, but this was inevitable in view of his acceptance of the caloric theory of heat. His main conclusions were correct, and this is surprising since the caloric theory might well have led him astray. His intuition and sound common sense led him for the most part to the truth. In August 1832 Carnot fell victim to a cholera epidemic and died within a day at the age of 36. In accordance with the custom of the time with cholera victims, all of his personal effects that could be found were burned. Most of his papers were therefore destroyed, but those that survived showed that he was abandoning the view that heat is a substance, and was beginning to favour the 'mechanical theory of heat'. Had he lived more than his 36 years he would probably have made further important advances; perhaps he would even have been led to the idea of entropy.

Carnot's book did not at first exert much influence on engineers or scientists. It soon went out of print, and copies were almost impossible to obtain. In 1845 the 21-year-old Kelvin went to Paris to work with Henri Victor Regnault (1810–1878), and tried to find a copy of it. To his surprise there was no copy in the library of the École Polytechnique, and no Paris bookseller had heard of it or its author. Kelvin had first learned of Carnot's work from a paper published in 1834 by Clapeyron. Later, when Clausius developed the concept of entropy he also was unable to see Carnot's book and had to base his knowledge of the work on the papers of Clapeyron and Kelvin. Today Carnot's book is easily available, through reprints and translations.

Benoit Pierre Emile Clapeyron (1799–1864) was born in Paris and had known Carnot when they were both at the École Polytechnique. He became an engineer of some distinction, specializing in the construction of metal bridges and of locomotives. He published little in pure science, but the 1834 paper in which he developed Carnot's ideas caused his name to be remembered in two equations, the Clapeyron equation and the Clapeyron–Clausius equation, concerned with the effect of temperature on vapour pressure. But an even more important contribution made in that paper was that Carnot's treatment was restated in the much more precise language of the calculus. Also, Clapeyron clarified the treatment by representing the Carnot cycle by means of a pressure–volume diagram (Figure 4.3(a)). This type of diagram, known as an indicator diagram, had first been used by James Watt in his work on steam engines, but he had kept the idea a close secret. It is not known whether the idea of an indicator diagram had been leaked to Clapeyron or whether he had thought of it himself.

The nature of heat

After Clapeyron's paper appeared, not much further progress could be made until the question of the nature of heat had been settled. Both Carnot and Clapeyron accepted the caloric theory, and this proved unproductive. The full significance of Carnot's ideas could only be realized when heat had been recognized as a mode of motion, that is, as a form of energy capable of being converted into other forms of energy such as mechanical work.

Ideas as to the nature of heat shifted considerably over the years. It was mentioned in Chapter 3 that in the seventeenth century Francis Bacon concluded that heat is a form of motion, and that the same point of view was taken by many others including Boyle, Newton and many of their followers. During the eighteenth century, however, this concept of heat was displaced in the minds of many investigators by the idea that heat is instead a substance, often referred to as an 'imponderable fluid'. It was considered that when a body is heated some 'heat matter' is added to it; when it is cooled 'heat matter' is withdrawn. We have seen that Joseph Black was convinced by some of his calorimetric experiments that heat is a substance, and the same conclusion was drawn by Lavoisier who had carried out important investigations of heat in collaboration with Laplace. Although we no longer accept this view, we still find it convenient to use language that implies that heat is a substance. We speak of heat 'flowing' from one body to another, and we refer to the 'quantity of heat' in a body and to the 'latent heat' of fusion or vapourization (these latter terms, however, are not recommended by IUPAC; one should say 'enthalpy of fusion' and 'enthalpy of vapourization'). The old-fashioned terms are, however, harmless as long as the true nature of heat is recognized, and it would be awkward to express the flow of heat or the quantity of heat in any other way.

A third theory of heat, which had only a short life from about 1830 to 1850, was that heat is due to vibrations in the aether that was supposed to pervade all space. This theory arose largely as a result of work on the transmission of heat by radiation, which had been studied in particular by Sir William Herschel (1738–1822). It was later realized that the way in which heat is transmitted by radiation is not direcly relevant to the nature of heat itself. Clerk Maxwell expressed this succinctly: 'I do not think that radiant heat is heat at all as long as it is radiant'.

A considerable number of investigations led to the conclusion that heat is a form of motion. Some of these were carried out by the American-born Benjamin Thompson (1753–1814), who acquired the title of Count Rumford while serving in the Bavarian army. While in Munich he was concerned with the boring of cannon, and in experiments in which a horse operated a dull drill he was impressed by the fact that large amounts of heat, capable of boiling ice-cold water, were continuously produced for periods of up to 2½ hours. He established that no weight change occurred during the process, and that the metal shavings had the same properties (such as heat capacity) as the unbored metal. From these investigations he concluded that heat could not be a substance but that there was a conversion of mechanical work into heat, and he obtained a value—not a very accurate one—for the 'mechanical equivalent of heat'.

A more persuasive contribution was made by the German physician Julius Robert Mayer (1814–1878). In 1840 he sailed to Java on a Dutch vessel, and noticed that venous blood has a much redder colour in the tropics than in a colder climate. He concluded that this is due to a lower metabolic rate in warmer weather, so that there is smaller consumption of oxygen with the result that there is less contrast in colour between venous and arterial blood. He speculated ingeniously on the relationship between food consumption, heat production and work done, and arrived at the conclusion that heat and work are interconvertible. He considered both to be different forms of what he called force (by which he meant energy) and which is indestructible. On his return home Mayer prepared a paper on the subject and submitted it the *Annalen der Physik*. Unfortunately for Mayer, his paper was couched in metaphysical terms, and worse still he was ignorant of mechanics and made many elementary errors; as a result his paper was rejected by Poggendorff, the editor of *the Annalen*. Mayer was very angry and frustrated at this treatment, but he later became aware of the paper's limitations and revised it extensively, submitting it for publication in the *Annalen der Chemie und Physik*; its editor, Liebig, accepted it, and it appeared in May, 1842. Mayer later claimed that this paper established his priority for the principle of conservation of energy.

Mayer published further books and articles, paying particular attention to the physiological aspects of the problem, and in a book that appeared in 1845 he was critical of Liebig's ideas about animal metabolism, which often involved a vital force which Mayer considered to be unsatisfactory. For a time Mayer's ideas were either ignored or ridiculed, and this had a serious effect on his mental stability; he attempted suicide and was confined several times in mental institutions, sometimes in a straightjacket. In his later years he received due recognition, receiving in 1871 the Copley medal of the Royal Society and being elected a corresponding member of the French Académie des Sciences.

James Prescott Joule (1818–1899), a prosperous English amateur scientist, took a rather different approach to the problem of the interconversion of heat and work. He began his investigations in the early 1840s by following up Faraday's experiments on electrolysis (Section 7.1) and on the mechanical generation of electricity. He found that the heat produced by an electric motor was the same as the heat developed if the work done in driving the motor were used instead to produce frictional heat. He later carried out many further experiments on heat, and published a number of papers on the subject. In some of his experiments he produced heat in water by stirring it with large paddles (Figure 4.4). In 1847 he lectured in a Church hall in Manchester and argued that 'the hypothesis of heat being a substance must fall to the ground'. Joule's work was in marked contrast to that of Mayer, who had reached his conclusions on the basis of imaginative intuition and had based his arguments mainly on the experimental evidence obtained by others. Joule's conclusions were careful inferences from skilfully designed and meticulously conducted experiments. For good measure, however, he added some metaphysical and religious arguments: 'Believing that the power to destroy belongs to the Creator alone ... I affirm ... that any theory that demands the annihilation of force [i.e. energy] is necessarily erronious.' The trouble with this type of argument is that another person could apply it to heat rather than total energy.

Since he neither was a university graduate nor held a recognized scientific appointment, Joule at first met with difficulty in having his views accepted. He was fortunate in gaining the interest and support of Kelvin, who shortly after returning from his year in Regnault's laboratory in Paris met Joule at the 1847 meeting in Oxford of the British Association for the Advancement of Science. Kelvin was impressed by Joule's paper at that meeting, and had some private discussions with him. By a curious chance, two weeks later Kelvin was on a

FIGURE 4.4 A diagram given by Joule of apparatus he used to measure the heat generated when a liquid is stirred. The rotation of handle *f* raises two weights *e*. When the handle is released the weights descend, and cause large paddles to stir the liquid in the container. The heat generated can then be related to the work done by the falling weights.

James Prescott Joule

(1818–1889)

Joule, a member of a wealthy brewing family, was born in Salford, near Manchester. He received much of his education at home; from 1834 to 1837 he was taught mathematics and science by John Dalton. He lived on his private income, never playing much part in the management of the family brewery, and carried out his investigations in laboratories established at his own expense. Later, after experiencing some financial losses, he received some subsidies from scientific societies, and during his final years he received a government pension.

From 1837, when he was 19, until about 1847, Joule carried out investigations that led him to conclude that work can be done at the expense of heat, a form of energy, and that the total energy is conserved (Section 4.2). His initial experiments, concerned with the development of heat from an electric current, led him to the discovery that the heat is proportional to the square of the current and to the first power of the resistance. In 1845 he measured the heat absorbed or produced in the expansion or compression of air, and in 1847 he determined the heat produced by large paddle wheels rotated in water.

At first Joule's work met with a chilly reception; after rejection of papers by the Royal Society he presented some of his results to the Manchester Literary and Philosophical Society, where the response was more sympathetic. His work became well known after he gave a public lecture in the reading room of St Ann's Church in Manchester; the text was reported in detail in a newspaper, the *Manchester Courier*. In 1847 he met William Thomson (later Lord Kelvin) who after some initial scepticism became enthusiastic about Joule's conclusions and gave them much support. Three years later Joule was elected a Fellow of the Royal Society, and began to enjoy a position of great scientific authority.

In 1848 Joule made an application of the kinetic theory of gases that had been proposed in the previous year in a book by John Herapath, making a calculation of the speed of a gaseous hydrogen molecule and of the specific heat of hydrogen gas. Between 1852 and 1862 Joule collaborated with Kelvin on experiments relating to what has come to be called the Joule–Thomson effect. They measured the temperature changes that occur when a gas expands through a porous plug, experiments that demanded Joule's great practical skill.

Joule had a shy and modest disposition, and his health was always delicate. He married in 1847 but his wife died in 1854 leaving him with two children. He always resided in the neighbourhood of Manchester, and died in Sale, near Manchester, after a long illness.

References: L. Rosenfeld, DSB, 6, 180–182 (1972); Cardwell (1971, 1983, 1989); Crowther (1935); Reynolds (1892).

walking tour in Switzerland and unexpectedly ran into Joule who was carrying a large thermometer; although on his honeymoon, with his bride waiting patiently in a carriage not far away, the enthusiastic Joule was making temperature measurements at the top and bottom of a large waterfall. These meetings, and Joule's papers, finally convinced Kelvin that Joule was correct; heat is a mode of motion, not a substance, and in a heat engine such as had been considered by Carnot there is an actual conversion of heat into mechanical work. Kelvin's interpretation of the Carnot cycle in terms of this idea is considered in Section 4.3.

By the middle of the nineteenth century this view of heat had become fairly universally accepted. The theory was often referred to as the 'dynamical theory of heat' or the 'mechanical theory of heat'. The latter expression seems inappropriate, since mechanical, with its usual connotation, is just what heat is not. Perhaps what was implied was that 'mechanical' should be understood to mean 'relating to the science of mechanics'.

4.2 The first law

Deciding that heat is a form of energy and that heat and work can be interconverted is not quite the same as deciding that energy is conserved. It is possible to conceive of a universe in which heat and not energy is conserved; for a time this was Kelvin's opinion, and such a universe would not be greatly different from our own. We have seen in Section 3.3 that a number of earlier investigators had decided that energy is conserved in purely mechanical systems involving collisions between spheres. Huygens' experiments had suggested that this is the case, and similar ideas had been expressed by Leibniz and by Émilie, Marquise du Châtelet. The situation, however, is a good deal more complicated for engines in which heat and work are also involved. An early suggestion that energy is conserved in such systems was made in 1783 by Sadi Carnot's father, Lazare Nicolas Marguerite Carnot (1753–1823), in his *Essai sur les machines en géneral*. Convincing evidence was not, however, possible until it had been established that heat is not a substance but is a form of energy that can be converted into mechanical work.

Mayer and Joule's work on the interconversion of heat and work also provided good circumstantial evidence that there is conservation of energy. In 1848 the two men became involved in a bitter priority dispute, carried out mainly through the French Académie des Sciences, with other scientists taking sides; Mayer was supported by Helmholtz, Clausius, John Tyndall (1820–1893) and others, Joule by other British scientists. The question of priority seems of little importance today and is impossible to resolve since there were no crucial experiments that led inevitably to the conclusion that energy is conserved; the conclusion was seen by both Mayer and Joule to be a probable one in view of the results of experiments relating to the interconversion of heat and work. Mayer's paper of 1842 does give him chronological priority. However, Joule's evidence, based on his numerous careful experiments carried out shortly afterwards, was more thorough and reliable than the less direct evidence adduced by Mayer from the experiments of others.

Perhaps the most convincing expositions of the principle of conservation of energy were those of William Grove and of Helmholtz. Grove for a time practised as a lawyer and later became a judge, but also had an active scientific career, being for a period professor of experimental philosophy at the London Institution. In 1846 he published his book *On the Correlation of Physical Forces* in which he made a clear statement of the principle of conservation of energy. Helmholtz, who later became a close friend of Kelvin, did outstanding work in both physiology and physics. In 1847 he published his book *Über die Enhaltung der Kraft* in which he dealt with the conservation of energy in a very comprehensive way. He discussed in detail the dynamical theory of heat, and showed that in inelastic collisions the energy that is apparently lost is converted into heat. This publication probably did more than any other to lead investigators to accept the modern position with regard to energy, heat and work.

The first law of thermodynamics is simply a statement of the conservation of energy. Today it is conveniently expressed as

$$\Delta U = q + w, \tag{4.1}$$

where ΔU is the change in the internal energy of a system, q is the heat added to it, and w is the work done on the system. In terms of infinitesimal quantities the law can be expressed as

$$dU = đq + đw. \tag{4.2}$$

The symbol đ, which originated with Carl Gottfried Neumann (1832–1925) in his *Vorlesungen über die mechanische Theorie der Wärme* (1875), indicates that q and w are not state functions, their values depending on how the processes are carried out. The modern practice is to use the joule, named in honour of Joule, as the unit of energy, heat and work. One joule is one newton-metre, or one kilogram metre squared per second per second:

$$J = N\ m = kg\ m^2 s^{-2}.$$

In scientific work today heat as well as other forms of energy are measured in joules, the use of the calorie being no longer recommended.

As a postscript to this section, it should be mentioned that a qualification to the principle of conservation of energy is required in view of Einstein's general theory of relativity. According to this, mass and energy are interconvertible, the relationship between the two being $E = mc^2$ where c is the speed of light. When nuclear transformations occur this relationship assumes great importance, but in ordinary chemical reactions the mass changes are too small to detect. Suppose, for example, that a mole of hydrogen reacts with half a mole of oxygen: $H_2 + \frac{1}{2}O_2 = H_2O$. The heat evolved is 241.75 kJ, which corresponds to a decrease in mass of 2.7×10^{-9} g, which could not be detected even by the most modern techniques.

4.3 The second law

It took a few years for Kelvin to accept Joule's suggestion that heat is a form of energy which can be converted into other forms of energy such as mechanical work. Paradoxically, the reason for his initial reluctance was that he had made a deeper study than Joule had of the works of Carnot and Clapeyron. Kelvin knew that if there was an interconversion of heat and work there was something not straightforward about it. Work could be converted into heat without any apparent complications, as in Rumford and Joule's experiments, but there were some restrictions on the conversion of heat into work. Kelvin grappled with this problem for some time, as did Clausius from a somewhat different standpoint, and both were led to a deeper understanding of the restrictions and to what is now known as the second law of thermodynamics.

Kelvin's absolute temperature scale

In 1848 Kelvin presented to the Cambridge Philosophical Society a paper in which he proposed constructing an absolute temperature scale based on the Carnot cycle. Previously such a scale could only be based on the expansion of gases. The French scientists Guillaume Amontons, Jacques Alexandre Charles (1746–1823) and, more satisfactorily, Joseph-Louis Gay-Lussac (1778–1850) had shown that at constant pressure the volume V of a gas is related to the Celsius temperature Θ by the equation

$$V = \text{constant} \, (a + \Theta), \tag{4.3}$$

where a is a constant, equal to about $273\,°C$. The Celsius temperature can therefore be adjusted by the addition of the constant a to give an absolute temperature T; the volume is then proportional to T. One weakness of this procedure is that the absolute scale is being based on the behaviour of particular substances. Another weakness is that the procedure depends on the assumption that temperature is reliably measured on the basis of the variation of the length of a column of mercury in a thermometer; even if the assumption of a linear variation of volume of mercury with temperature is valid, there is still a question as to whether the linear dependence holds down to the absolute zero. John Dalton, in particular, in his *A New System of Chemical Philosophy* (1808), had proposed a temperature scale very different from that obtained in the conventional way by a thermometer reading.

Accounts of Kelvin's work on the absolute scale of temperature usually state that he based his scale on the efficiency of a Carnot engine, but this is not really the case. The situation is a little complicated, especially as Kelvin changed his mind as to how best to proceed. In his 1848 paper he pointed out that since Carnot's theorem shows that the 'duty' (work output per unit quantity of fuel) depends only on the two temperatures used, one could base a more reliable temperature scale on the Carnot cycle. When he wrote this paper he had not been able to see Carnot's book and perhaps had not fully understood the argument (or made a slip of the pen) since he wrote:

A unit of heat descending from a body A at the temperature T° of this scale to a body B at the temperature (T − 1)° would give out the same effect, whatever be the number T. This may be termed an absolute scale.

This statement is not, however, correct; Carnot had shown that the work *diminishes* as *T* increases. In any case, the scale that Kelvin finally advocated, and the one which we use in principle today, is not the one advocated in this 1848 paper, based on the efficiency of a Carnot engine.

In a paper that appeared in 1849 Kelvin gave tables of values of work done over a range of temperatures, the values having been calculated from data obtained by Regnault on the densities of steam at various temperatures and from thermal data. These values show clearly that the work done when a given amount of heat falls from T° to (T − 1)° decreases as T increases. Kelvin did not say what value of the absolute zero he deduced from the data; it appears to be much below −300°C. In 1854, in a paper with Joule in which they investigated what has come to be called the Joule–Thomson effect, they concluded that, if 273.7° is added to the Celsius temperature, the work is proportional to the temperature so obtained. Kelvin had not constructed a scale based on a Carnot engine, as is often stated; subsequent to writing his 1848 paper he had no doubt realized that it would be quite impractical to do so. What he had done was to show that calculated values of work done were proportional to temperatures obtained by adding a constant (now known to be 273.15°) to the Celsius temperatures. This conclusion was important since it showed that Dalton had been wrong in proposing his modified temperature scale, which was very different from the conventional one based on the length of the column of mercury in a thermometer.

Kelvin's statement of the second law

Until about 1850 Kelvin was ambivalent as to the nature of heat, but afterwards had become convinced by the evidence presented by Mayer and Joule that heat is a form of energy. In a paper presented to the Royal Society of Edinburgh in March, 1851, he made a specific reference to Mayer's paper of 1842 and to Joule's statement of the dynamical theory of heat to the British Association in 1847. He then proceeded to analyse Carnot's treatment in the light of that theory. In this paper he quoted a statement that Clausius had made in the previous year, and which can be said to have been the first statement of what came later to be called the second law of thermodynamics. Kelvin himself expressed the second law in an alternative way as follows:

It is impossible, by means of inanimate material agency, to derive mechanical effect from any portion of matter by cooling it below the temperature of the coldest of the surrounding objects.

William Thomson, Baron Kelvin of Largs

(1824–1907)

Of Scottish descent, Thomson was born in Belfast, Ireland. He was educated at the University of Glasgow and at St Peter's College (now called Peterhouse), Cambridge, graduating as Second Wrangler in the Mathematical Tripos in 1845, and being elected a Fellow of St Peter's. He then spent a year in Paris working with Victor Regnault. In 1846, when he was 22, he was appointed professor of natural philosophy at the University of Glasgow, retaining the post for over half a century. In 1871 he declined an invitation to become the first Cavendish professor at Cambridge, a position that went to his friend Clerk Maxwell.

Thomson was a prodigious worker on a wide range of mathematics, physics and chemistry, much of his work being on problems of fundamental significance. His interests included electrostatics, magnetism, thermodynamics, hydrodynamics, and geophysics. He exerted a wide influence in scientific and educational circles, and under his influence the teaching of both experimental and theoretical physics at the University of Glasgow helped to establish it as one of the leading educational institutions in the world.

His contributions to thermodynamics (Sections 4.2 and 4.3) were made for the most part in the early years of his career. Later he developed theories of atomic structure (Section 5.2). Besides his work in basic science he was also greatly interested in practical applications. In 1857–1858 he played an active personal role in the laying of the first submarine cable across the Atlantic Ocean, and he devised a mirror galvanometer for improving telegraphic transmission. He improved ships' compasses and invented numerous instruments, including a tide predictor and many electrical devices.

He was knighted in 1866, and in 1892 he was created Baron Kelvin of Largs, the title coming from the River Kelvin at Glasgow and from the small town of Largs on the Firth of Clyde, near to which his house was built. He was greatly honoured in his time, and was buried next to Newton in Westminster Abbey.

References: J. D. Buchwald, DSB, 13, 374–388 (1976); Crowther (1935); Larmor (1908); Macdonald (1964); Smith and Wise (1989); Thompson (1901).

The inclusion of the phrase 'inanimate material agency' is interesting; Kelvin was leaving open the possibility that living systems might be able to defy the second law, a possibility that is no longer accepted.

In later papers Kelvin discussed what he called the 'dissipation of energy'. If two bodies at temperatures T_h and T_c are placed in contact, heat flows from the warmer to the colder body with no work being done. There is no loss of energy, but the energy has become dissipated, and there has been a loss of capability to do work. In the early 1850s Kelvin began to transfer his interests to a variety of other scientific problems, and his further contributions to thermodynamics were only spasmodic. For a number of years it was Clausius who made the important contributions to the subject.

Clausius's approach to the second law

Rudolf Julian Emmanuel Clausius (1822–1888), of German nationality, carried out the most important of his thermodynamic work in Switzerland, at the University of Zürich, where he was professor of mathematical physics from 1855 to 1867. Like Kelvin, Clausius only knew about Carnot's work at second-hand, gaining his information from the papers of Clapeyron and Kelvin. His 1850 paper, already mentioned, stated the second law in a form that was expressed by Kelvin in his 1851 paper as follows:

If an engine be such that when it is worked backwards the physical and mechanical agencies in every part of its motion are all reversed [i.e., if it operates reversibly], it produces as much mechanical effect as can be produced by any thermodynamic engine, with the temperature of source and refrigerator, from a given quantity of heat.

Willard Gibbs later said that with Clausius's 1850 paper the 'science of thermodynamics came into existence'.

The concept of entropy

In 1854 Clausius published another paper in which he presented a detailed analysis of the Carnot cycle and showed that if the engine operated reversibly between two absolute temperatures T_h and T_c (Figure 2.3(a)) the ratio of the heat absorbed at T_h to the heat rejected at T_c is the ratio T_h/T_c. If q_h^{rev} is the heat absorbed at T_h and $-q_c^{rev}$ the heat rejected at T_c (the convention is that q is always the heat absorbed by the system and may be positive or negative), then

$$\frac{q_h^{rev}}{-q_c^{rev}} = \frac{T_h}{T_c} \quad \text{or} \quad \frac{q_h^{rev}}{T_h} + \frac{q_c^{rev}}{T_c} = 0. \tag{4.4}$$

He then extended the argument to a cycle in which there is a range of temperatures, and showed that then, for the entire cycle,

$$\int \frac{dq^{rev}}{T} = 0. \tag{4.5}$$

He realized that this relationship gives to the quantity dq^{rev}/T a special significance. There are only certain properties for which this type of relationship is true. Volume is one of them; if a body has a certain volume and we take it through a series of operations finally returning it to its original condition, the volume returns to its original value, and this can be expressed by saying that $\int dV = 0$ for a complete cycle. The same is true for temperature: $\int dT = 0$.

Rudolf Julius Emmanuel Clausius

(1822–1888)

Clausius was born in Köslin, Prussia (now Koszalin, Poland) and was educated at the Stettin Gymnasium and the University of Berlin, obtaining his doctorate at the University of Halle in 1847. In 1855 he became professor of mathematical physics at the Polytechnicum in Zürich, moving to the University of Würzburg in 1867 and the University of Bonn in 1869. He remained at Bonn to the end of his life, serving as Rector of the University in his later years.

Clausius was not an experimentalist, although much of his work had significant practical implications. He carried out research on a variety of topics in theoretical physics, making particularly important contributions to thermodynamics and to the kinetic theory of gases. His work on thermodynamics, carried out in the 1850s and 1860s, is described in some detail in Section 4.3, his main contributions being to make a detailed analysis of the Carnot cycle and to introduce the concept of entropy. In 1865–1867 he published an important book, *Die mechanische Wärmetheorie* which was later translated into English as *The Mechanical Theory of Heat*.

His work on the kinetic theory of gases was also begun in the 1850s and is described in Section 5.3. Perhaps his most important contribution was to introduce the concept of the mean free path. A weakness of his work on kinetic theory was that although he recognized that there is a distribution of velocities he made no use of Maxwell's distribution equations, dealing only with average velocities. Also, he never recognized that the second law of thermodynamics is a statistical law, believing it to be a consequence of the principles of mechanics.

From time to time he became particularly interested in electricity and magnetism and in electrokinetic effects. In 1857 he developed the theory that electrolyte molecules in solution are constantly interchanging, and that the effect of an imposed electromotive force is to influence the interchange and not to cause it (Section 7.2); this has been referred to as the Williamson–Clausius hypothesis, but this does not seem justified as Williamson did not, in fact, consider electrolytes. After 1875 Clausius's main work was on electrodynamic theory, but in this he was not very successful.

Clausius was a man of fine character and personality who exerted a wide influence. In the unpleasant controversy with P. G. Tait, who was unnecessarily aggressive (Section 4.3), Clausius behaved with exemplary restraint and dignity. He received many honours both at home and abroad.

References: E. E. Daub, DSB, 3, 303–311 (1971) ; Brush (1958, 1976); Fitzgerald (1891); Gibbs (1899); Klein (1969).

For heat, on the other hand, this type of relationship is not true; $\int dq$ is not equal to zero for a complete cycle, since some heat has been converted into work. The integral $\int dq^{\mathrm{rev}}/T$ therefore has a special significance, being equal to zero for a complete cycle in which a system is taken through some transformations but is finally returned to its initial state. It also follows that if a system is changed from a state A to a state B the integral

$$\int_{B}^{A} \frac{dq^{\mathrm{rev}}}{T}$$

represents the change in a particular property of the system. In his 1854 paper Clausius called the property the 'equivalence value of transformation' (in his later 1865 paper he called it the *entropy*). At the end of the 1854 paper Clausius briefly discussed irreversible processes, and showed that for them the corresponding integral round a complete cycle must be negative:

$$\int \frac{dq^{\mathrm{irr}}}{T} < 0. \tag{4.6}$$

This relationship has come to be called the 'inequality of Clausius'. In the 1854 paper he also expressed another version of the second law of thermodynamics:

Heat can never pass from a colder to a warmer body without some change, connected with it, occurring at the same time.

During the next few years Clausius published a number of papers on the same general subject. In 1865 appeared a paper of particular importance in which he introduced the symbol S, defined by the relationship

$$dS = \frac{dq^{\mathrm{rev}}}{T}. \tag{4.7}$$

For a complete cycle, therefore, $\int dS = 0$. He gave the name entropy to S, and his reason for the choice of the word is of some interest:

... schlage ich vor, die Grösse S nach dem griechischen Worte η τροπη [trope], die Verwandlung, die *Entropie* des Körpers zu nennen.
(... I propose to name the quantity S the *entropy* of the system, after the Greek word [trope], the transformation.)

The word entropy also involves the Greek prefix εν (en), meaning 'in'. The German word Verwandlung, here translated as 'transformation', could also be translated as a turn-about, or as a change of direction, and the Greek word can be translated in the same way. Clausius evidently had in mind that entropy is the property concerned with giving a direction to a process. His next sentence, however, shows that he was thinking not only of etymology:

Das Wort *Entropie* habe ich absichtlich dem Worte *Energie* möglichst ähnlich gebildet, den die beiden Grössen, welche durch diese Worte benannt werden sollen, sind ihren physikalischen Bedeutungen nach einander so nahe verwandt, dass eine gewisse Gleichartigkeit in der Benennung mir zweckmässig zu seyn scheint.

(I have deliberately chosen the word *entropy* to be as similar as possible to the word *energy*: the two quantities to be named by these words are so closely related in physical significance that a certain similarity in their names appears to be appropriate.)

In the same paper Clausius also made his famous concise statement of the first two laws of thermodynamics:

Die Energie der Welt ist constant: die Entropie strebt einem Maximum zu.
(The energy of the universe is constant; the entropy tends towards a maximum.)

This 1865 paper of Clausius makes difficult reading. The type of mathematics in it is rather strange, and even accomplished mathematicians, such as Kelvin and Clerk Maxwell, have had difficulty with it. Clausius used a considerable amount of intuition in arriving at his conclusions, and the arguments are not presented at all clearly. For our understanding of Clausius's great contribution we must be grateful to later writers such as Clerk Maxwell, Lewis and Randall, Guggenheim, and many others. In his later papers Clausius introduced a new concept, 'disgregation', which was supposed to be a heat contribution arising from molecular arrangements. It was shown, however, particularly by Maxwell, that this splitting of heat into two types cannot be justified.

Rankine's thermodynamics

Mention should also be made of the work of Rankine in the field of thermodynamics. Early in 1850, the year in which Clausius published his first paper on the second law, Rankine presented to the Royal Society of Edinburgh a paper which gave a similar analysis to that in Clausius's 1850 and 1854 papers. A year previously Rankine had started to develop a comprehensive theory of matter in terms of what he called 'molecular vortices' (Section 5.2). According to his hypothesis an atom consists of a tiny nucleus surrounded by an atmosphere consisting of vortices, or circulating streams, of elastic matter. Unfortunately Rankine's thermodynamics was so entangled with his hypothesis of molecular vortices, and was expressed so obscurely, that it was largely incomprehensible and exerted little influence. Later Clerk Maxwell, in his 1878 review of Tait's book on thermodynamics, described Rankine's writing as 'inscrutable', and quoted one sentence from his *Manual of the Steam Engine* (1859):

If the absolute temperature of any uniformly hot substance be divided into any number of parts, the effects of these parts in causing work to be performed are equal.

Maxwell's comment on this sentence was as follows:

The student who thinks that he can form any idea of the meaning of this sentence is quite capable of explaining on thermodynamic principles what Mr. Tennyson says of the great Duke [of Wellington]—
    'Whose eighty winters freeze with one rebuke
    All great self-seekers trampling on the right'

When deciphered, Rankine's mathematics turned out to be similar to that of Clausius, so that Rankine deserves some credit, but certainly not for clarifying the problem!

Although the idea of entropy is of great value to scientists today, it took some time for the idea to be properly appreciated. Van't Hoff, who made important advances in chemical thermodynamics in the latter part of the nineteenth century, made no use of the concept in his own research and made only passing references to it in his writings; thermodynamic relationships like the Gibbs–Helmholtz equation (eqn 4.26), which today we derive using entropy, were obtained by him in other ways. Kelvin, who had made such important contributions to the second law of thermodynamics, never appreciated the idea of entropy. For one thing, although he was a highly proficient mathematician, he could not follow Clausius's mathematics, and in this he was not alone. Also, Kelvin's philosophy of science was that everything must be explained in terms of a mechanical model, and entropy cannot be explained in this way. Whereas properties like volume, pressure and temperature can be measured with simple intruments and can be appreciated even by people who do not know much about science, entropy is elusive; no instrument can directly measure an entropy change, which has to be calculated from data relating to reversible processes. Clausius's conception of entropy was, in fact, something like Maxwell's electromagnetic theory, involving a mathematical analogue rather than something that can be visualized.

In Kelvin's view the idea of entropy was unnecessary, since everything could be explained satisfactorily in terms of the dissipation of heat which is more easily visualized. However, the dissipation of heat is not as satisfactory as entropy in leading to an understanding of why processes occur. Clerk Maxwell pointed this out in later editions of his *Theory of Heat*, and showed that there is not an exact correlation between the dissipation of heat and the change in entropy, which is the property that does give a precise understanding of the tendency of processs to occur. Only in some special cases is there an exact correlation between the two factors. If a Carnot engine operates at two temperatures T_h and T_c, and heat q is absorbed at T_h, the available energy is $q(T_h - T_c)/T_h$. If, on the other hand, heat q flows irreversibly from T_h to T_c this energy is lost, and this is the heat dissipation. The entropy change in the irreversible process is $q(T_h - T_c)/T_hT_c$, and there is a correlation. For a general cycle, with a range of temperatures, this correlation is no longer found. A second reason for rejecting Kelvin's argument that heat dissipation gives a satisfactory interpretation of processes is that it does not at once provide an understanding of mixing processes at a constant temperature. Mixing processes, rarely considered in the early days of thermodynamics, are readily interpreted in terms of the idea of entropy, as is further considered in Section 5.3.

Today we think of thermodynamics as a rather stolid and established discipline, with little scope for controversy, and it comes as a surprise that in the latter part of the nineteenth century there were fundamental and acrimonious disagreements about it, with harsh words written on both sides. The unpleasantness was precipitated by P. G. Tait, a close friend of both Kelvin and Maxwell. For a time Tait exerted a considerable influence through his textbooks of physics, particularly the famous 'Thomson and Tait'. Unfortunately, Tait was the worst kind of chauvinist. His view appeared to be that the best work had to be done by a Scot (e.g. Kelvin and Maxwell), that the work of a man from northern England might be tolerable, but that little good could come from a Frenchman (e.g. Carnot and Clapeyron) and certainly not from a German (e.g. Clausius). He gave Kelvin the entire credit for the second law of thermodynamics, and discounted Clausius's idea of entropy change, which he said was the same as Kelvin's dissipation of heat. Unfortunately for him his exposition of this in his *Sketch of the History of Thermodynamics* (1868) was quickly demolished by Clausius.

Worse still, Tait reversed the meaning of entropy. Believing (not quite correctly) that Clausius's entropy was a measure of the unavailability of energy, he perversely defined entropy as the availability of energy, and his version of the second law was that entropy should decrease in a spontaneous process. In so doing he even confused Maxwell, who in earlier editions of his *Theory of Heat* defined entropy in Tait's way but discussed it in Clausius's way. For spreading such confusion Maxwell apologized in the preface to the fourth edition (1875) and in a mild and amusing note to Tait chided him for getting him into such a mess.

Modern textbooks of thermodynamics often make much of the fact that the subject can be developed without any regard to the existence of atoms and molecules. The earlier workers, however, by no means adopted this position; Joule, Kelvin, Clausius, Rankine and Maxwell all thought about thermodynamics in terms of molecular motions. Section 5.5 will deal with how the second law is interpreted in terms of molecular distributions.

4.4 Chemical thermodynamics

The investigations described so far were concerned with fundamental principles of thermodynamics—with the circumstances under which heat and work can be interconverted. By the 1870s, in spite of some confusions and controversies, these principles were beginning to be fairly well understood and accepted. During that decade attempts began to be made to apply the principles to chemical problems, since it was realized that thermodynamics could be of great value in leading to an understanding of the factors determining the direction of chemical change. Previously chemical processes had been considered with reference to the rather ambiguous concept of affinity, which sometimes related to the question of whether a reaction would occur at all, and sometimes to the speed of a chemical process.

The story of how chemical thermodynamics developed is a somewhat tangled one, since several investigators worked along different lines and quite independently of one another. From the modern point of view the most successful of these investigators was the American theoretical physicist Josiah Willard Gibbs (1839–1903), who between 1873 and 1878 developed the subject in a masterly and rigorous fashion. However, Gibbs was not able to make his ideas understandable, and he used a complicated notation with much use of Greek letters. As a result, only a few, such as Lord Rayleigh and Clerk Maxwell, had much idea of what Gibbs was saying, and only in the twentieth century did his work become deciphered and receive recognition. As a result, it was the less rigorous thermodynamics of van't Hoff that exerted a greater influence until the 1930s.

Because of this lack of appreciation of Gibbs's work, much of the same ground was covered independently by others, particularly by van't Hoff, Helmholtz and Max Planck. Of these three, Helmholtz and Planck worked along fairly conventional lines, but van't Hoff's approach was a more chemical one and was more practical but less rigorous than that of the others. He posed various chemical problems, such as the way a chemical equilibrium is shifted by various factors, and he derived simple equations that explained the results. Although he was well aware of the work of Clausius, and made good use of some of it, he derived these equations without making explicit use of the concept of entropy.

The person who deserves the credit for having done the first work in chemical thermodynamics is August Friedrich Horstmann (1842–1929), of the University of Heidelberg. Investigations on the determination of molecular weights by the vapour density method led him in 1868 to begin to investigate the effect of temperature on the equilibrium constant for a dissociation process. He began with the so-called Clapeyron–Clausius equation, which is an equation originally given in 1834 by Clapeyron and modified by Clausius; it related the vapour pressure p of a liquid to the temperature T and in modern notation can be written as

$$\frac{\mathrm{d}p}{\mathrm{d}T} = \frac{\Delta H_{\mathrm{m}}}{T \Delta V_{\mathrm{m}}}, \tag{4.8}$$

where ΔH_{m} is the heat that must be given to one mole of a liquid at constant pressure to bring about vapourization, and ΔV_{m} is the volume change when one mole of liquid is converted into vapour. Horstmann extended this expression to apply to a substance that is dissociating, and obtained the equation

$$\frac{1}{p}\frac{\mathrm{d}p}{\mathrm{d}T} = \frac{q}{RT^2} \quad \text{or} \quad \frac{\mathrm{d}\ln p}{\mathrm{d}T} = \frac{q}{RT^2}, \tag{4.9}$$

where p is now the change in pressure resulting from the dissociation and q is the heat bringing about the dissociation. Van't Hoff later acknowledged the importance of Horstmann's work, which he said had inspired some of his own.

The thermodynamics of Willard Gibbs

Willard Gibbs's work in thermodynamics, carried out while he was a professor of theoretical physics at Yale, was described in three papers all published in the

Josiah Willard Gibbs

(1839–1903)

Gibbs was born in New Haven, Connecticut, his father being a noted philologist who was professor of sacred literature at Yale. The younger Gibbs graduated from Yale in 1858 and continued there as a student of engineering, in 1863 obtaining one of the first Ph.D. degrees awarded in the United States. He spent a year at each of the universities of Paris, Berlin and Heidelberg, studying mathematics and physics. After his return to New Haven he rarely left the city and never again travelled abroad. He lived all his life in the house in which he had grown up, and never married.

In 1871 he was appointed professor of mathematical physics at Yale, holding the position for the first nine years without salary and living on his inherited income. After he had been offered a position at Johns Hopkins University a salary was provided by Yale, where he remained until his death.

After some work in engineering Gibbs turned his attention to thermodynamics in the early 1870s. His first published paper was in that field, and appeared in 1873 when he was 34. It was at once followed by another in the same year and by a lengthy memoir, in two parts, a few years later. These papers, discussed in some detail in Section 4.4, demonstrate that he had a complete mastery of the subject, but because of their obscurity he exerted little influence for a number of years, until they had been interpreted by more lucid writers.

During the 1880s Gibbs concentrated on optics and particularly on Maxwell's electromagnetic theory, on which he lectured. He became interested in Hamilton's theory of quaternions, and as an alternative to them he worked out a simpler and more useful system of vector analysis. In about 1890 he began to develop Boltzmann's work into a system of statistical mechanics, and in 1902 his important book *Elementary Principles in Statistical Mechanics Developed with Special Reference to the Rational Foundations of Thermodynamics* appeared (see Section 5.4).

Because of his retiring nature and his inability to communicate clearly, Gibbs gained little recognition during his lifetime. It is said that for a time Clerk Maxwell was the only one who understood Gibbs's thermodynamics, which he did by working it out himself. Some prominent scientists found little use for the work; Kelvin, for example, in a letter to Lord Rayleigh wrote 'I find no light for either chemistry or thermodynamics in Willard Gibbs.' The story is told that the President of Yale told Maxwell that Yale was looking for a professor of theoretical physics since it did not have one; Maxwell mentioned Gibbs, but the President had not heard of him.

References: M. J. Klein, DSB, 5, 386–392 (1973); Garber (1969); Klein (1969, 1983); Rukeyser (1964); Wheeler (1951); Wilson (1901).

1870s. His contributions were all based on those of Clausius, and he made no mention of Kelvin in any of his papers. He realized at the outset that entropy is a property equal in importance to the more familiar properties, such as energy, temperature, pressure and volume. In his first paper which appeared in 1873, uninvitingly entitled 'Graphical methods in the thermodynamics of fluids', he combined eqn 4.2 with the equations $dq = dS/T$ and $dw = -PdV$ to give

$$dU = TdS - PdV. \tag{4.10}$$

The significance of this equation is that it involves state variables, unlike eqn 4.2 which involves heat and work which are not state variables. In this first paper Gibbs limited himself to what can be done with two-dimensional representations of thermodynamic relationships, such as entropy diagrams.

In his second paper, which appeared later in the same year, Gibbs extended his geometrical discussion to three dimensions. He particularly emphasized diagrams in which entropy, energy and volume are used as three orthogonal coordinates. He pointed out that it follows from eqn 4.10 that

$$T = \left(\frac{\partial U}{\partial S}\right)_V \tag{4.11}$$

and

$$P = -\left(\frac{\partial U}{\partial V}\right)_S. \tag{4.12}$$

He showed that one can use surfaces in the entropy–energy–volume diagrams to discuss the coexistence of various phases (liquid, solid and gas) of a pure substance, and to discuss the stability of such phases at various temperatures and pressures. He discussed in particular the critical point, which had been discovered experimentally for carbon dioxide only a few years earlier by the Irish scientist Thomas Andrews (1813–1885).

An important contribution in Gibbs's second paper was his suggestion of a function that is concerned with the approach to equilibrium when a system is maintained at constant temperature and pressure. This function is now known as the *Gibbs energy* (formerly as the Gibbs free energy) and is defined by

$$G = U + PV - TS \tag{4.13}$$
$$= H - TS. \tag{4.14}$$

The quantity H which appears in this second equation, defined by

$$H = U + PV, \tag{4.15}$$

was recognized by Gibbs as playing an important role for processes occurring at constant pressure, and was called by him the 'heat content'. Some years later, in 1909, the Dutch physicist Heike Kamerlingh Onnes (1853–1926) gave H the name enthalpy, from the Greek εν (en), in, and θαλπος (thalpos), heat, or from the single Greek word ενθαλπος (enthalpos), to warm within.

Gibbs recognized in this paper that the function G has a special significance, in that the condition for equilibrium at constant pressure and temperature is

$$dG = 0 \qquad (4.16)$$

In other words, the condition for a process to occur spontaneously is that the value of G must decrease, until finally at equilibrium it reaches a minimum value. Modern textbooks give clear accounts of G and its significance, but even those who are already thoroughly familiar with the subject find Gibbs's argument quite incomprehensible.

Gibbs sent reprints of his first two papers to a number of scientists, including Clerk Maxwell who found them impossible to follow but was able to work out the theory for himself. He was sufficiently impressed that he started to extend Gibbs's methods to mixtures of substances, but he soon abandoned this work when he heard of Gibbs's own extensions to mixtures.

Gibbs's third paper is entitled 'On the equilibrium of heterogeneous substances'; it is in two parts and runs to some 300 pages. It was directed to the problem of the conditions for physical and chemical equilibrium. For a homogeneous system containing n independent chemical species $B_1, \ldots B_n$, of variable amounts $n_1, \ldots n_n$, Gibbs defined the chemical potential of a component i as

$$\mu_i = \left(\frac{\partial U}{\partial n_i} \right)_{V, n_j}, \qquad (4.17)$$

where the subscript n_j means that the amounts of all components except B are held constant, as well as V.

This procedure led Gibbs to his famous phase rule, which expresses the number of independent variations f in a system of r independent phases having n independent components:

$$f = n + 2 - r. \qquad (4.18)$$

Gibbs also obtained the conditions for chemical equilibrium. A chemical reaction such as $H_2 + Br_2 = 2HBr$ can be written as

$$2HBr - H_2 - Br_2 = 0, \qquad (4.19)$$

and for a general reaction the stoichiometric equation can be written as

$$\Sigma \, v_j \, B_j = 0, \qquad (4.20)$$

where the vs are the stoichiometric coefficients (positive for products, negative for reactants). Gibbs showed that the equilibrium condition for such a reaction is

$$\Sigma \, v_j \, \mu_j = 0, \qquad (4.21)$$

where the chemical symbol B_j has been replaced by the corresponding chemical potential μ_j.

It is clear to anyone reading Gibbs's papers today that he had a thorough mastery of the basic principles of thermodynamics. His failure to exert more influence in his time was due in part to the obscurity of his presentation of the subject but to a greater extent because he derived his treatment as an austere and logical extension of the ideas of Clausius. He made only passing references to experimental results, and unlike van't Hoff never presented tables of data to show how his theories applied to them. Although the concept of the equilibrium constant had become useful to chemists by then, Gibbs had perhaps never encountered it; he certainly never mentioned it.

Our present understanding of Gibbs's thermodynamics is due largely to the interpretations given to it by others. Maxwell presented clearly the main features of it in his *Theory of Heat*. Ostwald later became enthusiastic about Gibbs's work, and expounded it in his *Lehrbuch*. Accounts of it were also given by Roozeboom in his *Die heterogenen Gleichgewichte vom Standpunkt der Phasenlehre* (1901), and later in the well-known books of Lewis and Randall (1923) and of Guggenheim (1933).

The thermodynamics of Helmholtz

In the 1880s Helmholtz, in ignorance of what Gibbs had done, carried out quite similar work. After writing his book on the conservation of energy in 1847 (Section 4.4), Helmholtz turned his attention to other areas of research, such as various aspects of physiology including sensory perception. He later returned to thermodynamics, and in 1882 published an important paper on the thermodynamics of chemical change, paying particular attention to the conditions under which a reaction occurs spontaneously. Previously Julius Thomsen (1826–1909) and Pierre Eugène Marcellin Berthelot (1827–1907) had suggested that the tendency of a reaction to occur was determined by the heat evolved in the reaction, but the occurrence of a number of endothermic reactions showed that this could not be the case. To overcome this difficulty Helmholtz distinguished between two forms of energy, 'free' energy and 'bound' energy. Only the free energy can be converted into other forms of energy; the bound energy is obtainable only as heat. On the basis of Clausius's treatment Helmholtz deduced that for an energy U and an entropy S the free energy is defined as

$$A = U - TS. \tag{4.22}$$

The bound energy is thus TS. This free energy A is now known as the Helmholtz energy (formerly as the Helmholtz free energy) and is to be distinguished from the Gibbs energy G which is defined by eqn 4.13. Helmholtz showed that if a system is maintained at constant temperature and constant volume, reaction can only occur spontaneously if there is a decrease in the Helmholtz energy A. The Helmholtz and Gibbs energies thus differ in that the former provides the condition for spontaneity at constant volume, the latter at constant pressure.

Hermann Ludwig Ferdinand von Helmholtz

(1821–1894)

Helmholtz was born in Berlin, and in 1837 he obtained a government stipend for medical studies at the Friedrich-Wilhelm-Institüt in Berlin; in return he committed himself to eight years' service as an army surgeon. He obtained his MD degree in 1842, and in 1848 secured his release from military duty to become professor of physiology at Königsberg. In 1858 he became professor of physiology at Heidelberg, and in 1871 he moved to the Univerity of Berlin as professor of physics. He had a remarkably original mind and covered a wide range of science, including physiological acoustics, physiological optics, thermodynamics, hydrodynamics and electrodynamics.

Helmholtz's early studies in physiology led him to consider how chemical energy, heat and work are interconverted in living systems. He rejected the idea of a 'vital force' that was popular at the time, and in a pamphlet *Über die Erhaltung der Kraft*, published in 1847 shortly after the work of Mayer and Joule, he discussed in detail various aspects of energy, presenting a convincing argument in favour of the principle of the conservation of energy.

For some years after 1847 he devoted his attention mainly to physiological problems and to some aspects of electrochemistry. In 1851 he invented the ophthalmoscope and worked out a mathematical theory of it. During the next few years he developed important theories of vision. In 1858 he published a treatment of vortex motion which excited interest in connection with theories of the atom (Section 5.2). In the late 1870s he considered electrochemical cells, and developed a theory of the electric double layer at an electrode in contact with an electrolytic solution (Section 7.5).

In about 1880 he returned to the subject of thermodynamics and in 1882 he distinguished between 'bound' and 'free' energy, introducing a function that has come to be called the Helmholtz energy. He also developed an equation usually known as the 'Gibbs–Helmholtz equation', although Gibbs never gave such an equation.

By the 1880s Helmholtz had achieved a position of high esteem and prestige, similar to that of his close friend William Thomson (later Lord Kelvin). In 1887 Helmholtz was appointed President of the newly formed Physikalische-technische Reichanstalt, which carried out both pure and applied research. During his lifetime he received many honours.

References: R. S. Turner, DSB, 5, 241–253 (1972); Maxwell (1877); Reinold (1896).

From Clausius's equations Helmholtz also deduced that the entropy S is given by

$$S = -\left(\frac{\partial A}{\partial T}\right)_V, \qquad (4.23)$$

and from eqn 4.23

$$A = U + T\left(\frac{\partial A}{\partial T}\right)_V. \qquad (4.24)$$

This equation is mathematically equivalent to

$$\left(\frac{\partial (A/T)}{\partial T}\right)_V = -\frac{U}{T^2}. \qquad (4.25)$$

The corresponding equation for the Gibbs energy is

$$\left(\frac{\partial (G/T)}{\partial T}\right)_P = -\frac{H}{T^2}. \qquad (4.26)$$

These equations have become known as Gibbs–Helmholtz equations, although neither of them appears in any of Gibbs's papers. As Wilder Bancroft said in 1927 in a book review: 'This [eqn 4.26] is an equation which Helmholtz did deduce and which Gibbs could have, and perhaps should have, deduced but did not. The error goes back to Ostwald who was advertising Gibbs at that time.'

The thermodynamics of van't Hoff

The thermodynamic work of van't Hoff, carried out for the most part at the University of Amsterdam where he had a formidable teaching load, was in marked contrast to that of Gibbs and Helmholtz. His approach to thermodynamics was that of a practical chemist interested in understanding chemical reactions, and he was less concerned with scientific rigour than in arriving at simple relationships applicable to laboratory work. He was a modest and unassuming man who never made priority claims; indeed in his later books he sometimes gave credit to others for things—like the 'Arrhenius equation' and the 'Le Chatelier principle'—that he had first discovered himself. He tended to proceed rather independently; although later recognizing the importance of the work of Clausius, Helmholtz and Gibbs, he preferred to use the equations of thermodynamics that he had derived himself. He never, for example, made use of the concept of entropy, and instead of free energy he preferred to think of the work done by a chemical reaction.

Much of van't Hoff's thermodynamics is contained in his book *Études de dynamique chimique*, which appeared in 1884, with a second edition in 1896. With regard to the title, it should be noted that at the time the expression 'chemical dynamics', now applied to the study of the details of chemical change, included also the study of chemical equilibrium. van't Hoff's book was therefore a textbook of chemical thermodynamics as well as of chemical kinetics. It was, in fact, the first book to cover either of these topics.

The part of the book concerned with chemical kinetics is considered in Chapter 8. Perhaps the most important contribution to thermodynamics in the book was the treatment of equilibrium constants. Van't Hoff was one of the first to recognize the dynamical nature of chemical reactions, and it was he who introduced, in this book, the symbol $\rightleftarrows$ to replace the previously used equals sign in a chemical equation. (The modified symbol $\rightleftharpoons$, now more commonly used, was proposed by H. Marshall in 1902.) The idea that when a chemical system is at equilibrium, chemical change is still occurring, taking place at equal rates in the two opposite directions, did not originate from van't Hoff. It had been first suggested by the British chemist Alexander William Williamson (1824–1904) and had been accepted previously by Clausius and by the Austrian physicist Leopold Pfaundler (1839–1920). Prior to their work it had been thought that chemical equilibrium is reached when the *forces* in the two opposite directions have become equal—as is the case in a purely mechanical system—and that all reaction has ceased at equilibrium.

When van't Hoff wrote his book the idea of an equilibrium constant was fairly new. In 1864 the Norwegian chemist Peter Waage (1833–1900) and his brother-in-law the mathematician Cato Maximilian Guldberg (1836–1902) had arrived at the idea of an equilibrium constant, but in a somewhat unsatisfactory way. They argued not in terms of the rates of reactions in forward and reverse directions but in terms of forces. Thus for a reaction

$$aA + bB \rightleftharpoons yY + zZ,$$

they argued that the force in the left to right direction would be proportional to $[A]^p[B]^q$, where p and q are non-integers to be determined empirically. Similarly, the forces from right to left are proportional to $[Y]^s[Z]^t$. By equating the forces they obtained an equilibrium equation, but the indices bore no relationship to the stoichiometry. Only in a later paper which appeared in 1879 did Guldberg and Waage obtain the relationship

$$K = \frac{[Y]^y[Z]^z}{[A]^a[B]^b}, \tag{4.27}$$

where the indices a, b, y and z are the stoichiometric coefficients in the equation. This indeed is the correct relationship, but Guldberg and Waage had not arrived at it in anything like a satisfactory way. Neither did they make any contribution to kinetics, since they worked in terms of forces and not of rates, although they did tentatively suggest that the rates might be proportional to the forces. At the time of the appearance of van't Hoff's book equations such as 4.27 had a good empirical foundation, as was shown by some of his own experiments, but their theoretical basis had yet to be provided.

Jacobus Henricus van't Hoff

(1852–1911)

Van't Hoff was born in Rotterdam and studied in The Netherlands, Germany and France. He later obtained his Ph.D. degree in 1874, his research being on a rather mundane problem in organic chemistry. In the meantime he had published his now famous pamphlet on the 'tetrahedral carbon atom', but he shrewdly made no mention of this in his thesis, realizing that in view of its then controversial nature it would do him no good. He first suffered a period of unemployment for a year and a half, and then held a rather unsatisfactory position at the State Veterinary School in Utrecht. In 1878 he was appointed professor of chemistry, mineralogy and geology at the University of Amsterdam, where he embarked on his research in physical chemistry. In 1896 he moved to the University of Berlin, and in 1901 he was awarded the first Nobel Prize in chemistry, 'in recognition of the extraordinary value of his discovery of the laws of chemical dynamics and of the osmotic pressure in solutions'.

During the period he was in Amsterdam van't Hoff carried out a comprehensive programme of research in thermodynamics and kinetics, described in some detail in Sections 4.4, 8.1 and 8.2. He published many papers in various journals including the *Zeitschrift für physikalische Chemie* which he and Ostwald founded in 1887; the work is conveniently summarized in his *Études de dynamique chimique*, which appeared in 1884, and in the second edition, in German, which was published in 1896. His thermodynamic work was less rigorous than that of Gibbs and Helmholtz, his concern being more with arriving at relationships that would be useful in the interpretation of experimental results.

His appointment at the University of Berlin did not require him to give more than the occasional lecture, and he was able to devote himself entirely to research. He embarked on a completely new line of research, a study of the equilibria involved in the marine salt deposits at Stassfurt, Germany. In this work, further described in Section 4.4, the problem was for the first time treated in a systematic manner in terms of Gibbs's phase rule. This work established van't Hoff's position as the founder of the science of petrology.

Van't Hoff always emphasized the importance of imagination in scientific research. He was a man of remarkable kindness and modesty, and he exerted a strong influence on the many students who worked with him. He was a man of few words, and it is said that if a student asked him a question the result was sometimes silence for several days, until finally a well thought out response would be revealed.

References: H. P. M. Snelders, DSB, 13, 565–581 (1976); Donnan (1912); Holleman (1952); Eugster (1971); MacCallum and Taylor (1938); McBride (1987); Root-Bernstein (1980); Servos (1990); Van Kloster (1952); Veibel (1970); Walker (1913).

In his book van't Hoff accepted the idea of the equilibrium constant, and he only later obtained it theoretically. He stated without proof the equation

$$\frac{d \ln K}{dT} = \frac{q}{RT^2}.$$ (4.28)

This expresses the variation with temperature T of the equilibrium constant K; R is the gas constant and q the heat absorbed by the system if the reaction occurs at constant pressure; it is the change of enthalpy $\Delta H°$. It is not clear how van't Hoff originally arrived at this equation, but in view of his later statements of indebtedness to Horstmann it seems likely that he obtained it by extending Horstmann's eqn 4.9 for the dissociation pressure p. For two temperatures T_1 and T_2 this equation integrates to

$$\ln \frac{p_1}{p_2} = \frac{q}{R}\left(\frac{1}{T_2} - \frac{1}{T_1}\right)$$ (4.29)

At a given temperature T this equation becomes

$$\ln p = -\frac{q}{RT} + \text{constant}.$$ (4.30)

For a chemical reaction at equilibrium the pressure p can be replaced by the equilibrium constant K, with q taken to be the heat absorbed in the chemical process:

$$\ln p = -\frac{q}{RT} + \text{constant}.$$ (4.31)

If q is constant this may be differentiated with respect to T to give

$$\frac{d \ln K}{dT} = \frac{q}{RT^2} = \frac{\Delta H°}{RT^2}.$$ (4.32)

It is typical of van't Hoff's approach to problems that this derivation is by no means rigorous, and that a good deal of intuition is involved. He had nevertheless given for the first time a correct relationship of great practical importance. He recognised that q is not necessarily independent of temperature, and considered various possible relationships for its temperature dependence.

In papers published in 1886, van't Hoff derived eqn 4.32 in another way in terms of an infinitesimal Carnot cycle, but the treatment is rather laborious. The same proof was repeated in the second edition of the *Études*.

Since van't Hoff had a constant volume system in mind when he obtained eqn 4.32 it has come to be known as the van't Hoff isochore. This description, however, is misleading, since the way in which an equilibrium constant varies with temperature in no way depends on the conditions under which the reaction has occurred. The replacement of q by $\Delta H°$ makes eqn 4.32 exactly correct provided that K is K_p, the equilibrium constant expressed in terms of pressures

If q is replaced by $U°$ the equation is valid if K is K_c, the equilibrium constant expressed as a ratio of concentrations.

In the *Études* van't Hoff included a criticism of the conclusion reached by Julius Thomsen and Berthelot that the heat evolved is the driving force of a chemical reaction. He realized that the sign of q influences only how the equilibrium constant K changes with the temperature, in accordance with eqn 4.32; it does not control its actual magnitude. At the absolute zero, van't Hoff noted, the Thomsen–Berthelot principle is correct, but not at other temperatures. In his discussion of eqn 4.32 he presented an important qualitative discussion of the way in which K is affected by temperature. If heat is evolved when reaction occurs from left to right (q is negative) the equilibrium constant will decrease as the temperature is raised. Conversely, if q is positive (heat is absorbed) a rise in temperature will increase K. Later in 1884, a few months after the *Études* appeared, the French chemist Henri Louis Le Chatelier (1850–1936) quoted these conclusions of van't Hoff, which have come to be referred to as the *Le Chatelier principle*. Characteristically, van't Hoff never raised any objection to this incorrect attribution.

In the *Études* and in his later papers van't Hoff made much use of the concept of the work, w_A, of chemical affinity, which is the work done by the system when a chemical reaction occurs under a specified set of conditions. In introducing this concept he took an important step forward, since previously affinity had been treated in an unsatisfactory manner; sometimes the term applied to rates, sometimes to the extent to which a reaction occurred. In concentrating attention on the work of chemical affinity van't Hoff was at last providing a precise definition of the concept. For a process occurring at constant pressure and temperature the work of chemical affinity can be shown to be the negative of the change in Gibbs energy; in other words, if when a reaction occurs there is a decrease in Gibbs energy by a certain amount, this is the work done by the system. In his own investigations van't Hoff made no use of the free energy concept, preferring to deal with the work of chemical affinity.

In papers published in 1886 van't Hoff derived an expression for the work of affinity for a reaction

$$aA + bB \rightarrow yY + zZ,$$

where A and B are initially at pressures p_A and p_B, and are converted into Y and Z at pressures p_Y and p_Z. He did this by making use of a device that came later to be called *van't Hoff's equilibrium box*. He imagined a box in which the system was present at equilibrium, the components having the pressures p_A^{eq}, p_B^{eq}, p_Y^{eq} and p_Z^{eq}. He then considered the work done when, for example, the pressure p_A of a moles of gas A is changed reversibly to pressure p_A^{eq}; this work is

$$w_A = -aRT \ln \frac{p_A^{eq}}{p_A}, \tag{4.33}$$

and there are similar expressions for the other three components. Proceeding in this way he showed that the work of affinity, which is the work of converting a moles of A and b moles of B into y moles of Y and z moles of Z, at the appropriate pressures, is

$$w_A = RT \ln \left(\frac{p_Y^y p_Z^z}{p_A^a p_B^b}\right)_{eq} - RT \ln \left(\frac{p_Y^y p_Z^z}{p_A^a p_B^b}\right). \tag{4.34}$$

Since w_A cannot depend on the particular pressures that correspond to the equilibrium condition, it follows that the first of the ratios in this equation must be a constant, which is the equilibrium constant for the reaction:

$$K_p = \left(\frac{p_Y^y p_Z^z}{p_A^a p_B^b}\right)_{eq}. \tag{4.35}$$

In this derivation van't Hoff had given the first satisfactory proof, based on thermodynamics, of the existence of an equilibrium constant. Equation 4.35 may be written as

$$w_A = RT \ln K_p - RT \ln \left(\frac{p_Y^y p_Z^z}{p_A^a p_B^b}\right), \tag{4.36}$$

which is equivalent to the equation we write today as

$$\Delta G = -RT \ln K_p + RT \ln \left(\frac{p_Y^y p_Z^z}{p_A^a p_B^b}\right). \tag{4.37}$$

In his *Études* and in many later publications van't Hoff expressed the work of affinity at a given temperature T as

$$w_A = q \frac{T_t - T}{T_t}, \tag{4.38}$$

where q is now the heat evolved and T_t is what he called the transition temperature, and can be interpreted as the temperature at which the equilibrium constant is unity. His later papers suggest that he obtained this equation from arguments presented by Clausius, who had shown that the efficiency of an engine working between two temperatures T_h and T_c is $(T_h - T_c)/T_h$. This efficiency is the work of affinity w_A divided by the heat q absorbed at the higher temperature:

$$\frac{w_A}{q} = \frac{T_h - T_c}{T_h}. \tag{4.39}$$

If T_h is identified with the transition temperature and T_c is the working temperature, eqn 4.38 at once follows.

Equation 4.39 can be related to an analysis of the situation in terms of Gibbs energy changes and entropy changes, as would be done today. The standard Gibbs energy change is

$$\Delta G° = \Delta H° - T\Delta S°, \tag{4.40}$$

and is zero if the equilibrium constant is unity. If ΔH° and ΔS° are independent of temperature, which is often true to a good approximation, the temperature at which the equilibrium constant is unity, which may be identified with van't Hoff's transition temperature, is given by

$$T_t = \frac{\Delta H^\circ}{\Delta S^\circ}. \tag{4.41}$$

Elimination of ΔS° between eqns 4.41 and 4.42 gives

$$\Delta G^\circ = \Delta H^\circ \frac{T_t - T}{T_t}. \tag{4.42}$$

Since $\Delta H^\circ = -q$ and $\Delta G^\circ = -w_A$, this is equivalent to eqn 4.39. Thus, without using the concepts of Gibbs energy and entropy, van't Hoff had obtained a useful if not rigorously derived expression for the work of chemical affinity.

Van't Hoff also differentiated eqn 4.38 to obtain

$$\frac{dw_A}{dT} = -\frac{q}{T_t}, \tag{4.43}$$

and elimination of T_t between this equation and eqn 4.38 gives

$$\frac{dw_A}{dT} = -\frac{w_A - q}{T} \tag{4.44}$$

or

$$w_A = q + T\frac{dw_A}{dT}. \tag{4.45}$$

Since $w_A = -\Delta H^\circ$ and $q = -\Delta H^\circ$, this is equivalent to

$$\Delta G^\circ = \Delta H^\circ + \frac{d\Delta H^\circ}{dT}, \tag{4.46}$$

which is a form of the Gibbs–Helmholtz equation (eqns 4.24–4.26). Today this equation is usually obtained by making use of the entropy concept, and it is interesting to see that van't Hoff obtained it without any reference to entropy.

Some mention should also be made of the thermodynamic work of Max Planck. During Planck's earlier years, at the Universities of Munich, Kiel and Berlin, most of his efforts went into thermodynamics, including the theory of solutions, and his work that led to the quantum theory in 1900 was thermodynamic in character. His achievements in this field, although always of high quality, were a considerable disappointment to him since time and time again he found that what he did had already been done by others, particularly by Gibbs, Helmholtz and van't Hoff. He devoted some attention to chemical problems, but having been trained as a physicist he lacked the chemical insights that van't Hoff in particular applied to such great advantage. As a result Planck's research in thermodynamics is today rarely mentioned, and is in any case eclipsed by his great work leading to the quantum theory (Section 10.1). His book *Vorlesungen über Thermodynamik* (Lectures on thermodynamics), first published in 1897, did exert a considerable influence, giving an exceptionally clear and systematic treatment of the subject. It appeared in many editions (the 11th edition in 1966) and was translated into many languages, including English, Russian and Japanese.

Solutions and phase transitions

Solutions have always been of great interest to chemists, and the study of them has played an important part in the development of thermodynamics. A number of important investigations were made quite early and inspired much later work. The investigations in the eighteenth century by the Abbé Nollet on osmotic pressure, and by Richard Watson on freezing-point depression, are referred to later. In 1803 William Henry (1774–1836) described investigations in which he had shown that the amount of a gas dissolved in a liquid is proportional to its pressure, and this is known as Henry's law. Henry, who held an MD degree from Edinburgh, superintended the chemical manufacturing business that his father had established in Manchester, where he was a close friend of John Dalton.

Van't Hoff's development of thermodyamics embraced solutions as well as gases, and it was his work on solutions that played the major part in securing for him the Nobel Prize in chemistry in 1901, the year the prizes were first awarded. Much of his work on solutions is contained in the *Études*, in papers he published in 1886, and in an article that appeared in the first volume of the *Zeitschrift für physikalische Chemie* that Ostwald and van't Hoff launched in 1887 (Section 2.3). This volume also contained an article by Max Planck on solutions of electrolytes and Arrhenius's famous paper on electrolytic dissociation (Section 7.2).

A number of different aspects of solution theory were treated by van't Hoff. One of them was concerned with osmotic pressure, which had first been studied in a systematic way in 1748 by the Abbé Jean Antoine Nollet (1700–1770) who was professor of experimental philosophy in the Collège de Navarre in Paris and is better known for his work on electricity. He tied a piece of bladder over the mouth of a glass cylinder containing wine and immersed the cylinder in water, finding that the bladder became distended and eventually burst, owing to the passage of water into the alcohol. The behaviour later came to be applied to the strengthening of wine and brandy.

Work of this kind continued for many years. Particular mention should be made of the investigations of René Joaquim Dutrochet (1776–1847) who studied what he called *endosmose*, from the Greek ενδον (endon), within, and ὠσμος (osmos), impulse, for the inward flow of liquid through a membrane;

he used the term exosmose for passage outwards, and the modern word 'osmosis' derives from these words. The most important quantitative investigations of osmosis are those of Friedrich Philipp Pfeffer (1845–1920) who began his career as an apothecary, later becoming a chemist and finally a botanist; his work on osmotic pressure was done at Bonn and Tübingen where he held professorships. He made the important practical innovation of depositing a copper ferrocyanide membrane in the walls of a porous clay pot, in this way producing a satisfactory 'semipermeable membrane' which was permeable to the solvent—usually water—but not the solute.

Van't Hoff's work on osmotic pressure was mainly directed towards analysing the results obtained by Pfeffer. He found that for many solutes the osmotic pressure π obeys the equation

$$\pi V = nRT \quad \text{or} \quad \pi = cRT, \tag{4.47}$$

where n is the molar amount of solute, V the volume, R the gas constant, T the absolute temperature and c ($= n/V$) is the concentration. He was struck by the fact that this equation is analogous to the equation for an ideal gas,

$$PV = nRT. \tag{4.48}$$

In other words, the osmotic pressure is the pressure which the dissolved substance would exert if it existed as an ideal gas in the volume occupied by the solution, with all the solvent removed. Van't Hoff's explanation was that the pressure is due to the 'impacts of the dissolved molecules on the semipermeable membrane. The molecules of the solvent present on both sides of the membrane, since they pass freely through it, need not be taken into consideration.' Today we prefer to look at the situation a little differently: in a pure liquid the concentration of the solvent is greater that in a solution, so that there is a greater flow of solvent from the pure liquid than from the solution.

For electrolytes there were anomalies, the osmotic pressures being greater than predicted by eqn 4.47. In his 1887 paper van't Hoff mentioned having heard from Arrhenius about his theory of electrolytic dissociation (Section 6.3) and he realized that the abnormally high values could be understood in terms of the increase in the number of particles present in the solution. He modified eqn 4.47 by the addition of a factor i,

$$\pi V = inRT, \tag{4.49}$$

which may be written as

$$\pi = icRT, \tag{4.50}$$

where c, equal to n/V, is the molar concentration. Values of i obtained from Arrhenius's data and in other ways were found to provide a satisfactory interpretation of the osmotic pressure results.

The accurate measurement of osmotic pressures is difficult. Interesting and important investigations, with the attainment of high osmotic pressures, were made in the first decade of the twentieth century by Randal Thomas Mowbray Rawdon Berkeley, the Eighth Earl of Berkeley (1865–1942), in collaboration with Ernald George Justinian Hartley (1875–1947). Berkeley was one of the last of the amateur physical scientists, having had no formal education in science. After a short career in the Royal Navy he attended some lectures in chemistry in London and Oxford, and established a research laboratory beside his house on Boar's Hill, just outside Oxford. Some of his osmotic pressure work with Hartley, who taught chemistry at Oxford, was done in this laboratory and some in a cellar laboratory at Balliol College.

In his 1887 paper van't Hoff also considered other properties that are referred to as *colligative*. This word, from the Latin *colligare*, to bind together, refers to several properties of solutions that are closely related in the sense that if one of them has been measured the others can be found by calculation, the reason being that these properties are all determined by the proportions of the different molecules present and not by their nature. Besides osmotic pressure, other colligative properties are the depression of the freezing point, the elevation of the boiling point and the lowering of the vapour pressure. An important relationship was first deduced thermodynamically by Guldberg in 1870, who showed that the molar lowering of the freezing point ΔT_f is related to the latent heat of fusion l_f by the equation

$$\Delta T_f = \frac{RT^2}{l_f}.$$

(4.51)

The corresponding relationship for the molar elevation of the boiling point,

$$\Delta T_b = \frac{RT^2}{l_b},$$

(4.52)

where l_b is the latent heat of evaporation, was given in 1889 by Arrhenius.

The first experiments on the freezing points of salt solutions were carried out by the Revd Richard Watson (1737–1816) whose career was a remarkable one. He was professor of chemistry at Cambridge from 1764 to 1771, in spite of initially having no knowledge of the subject; in his own words, in his auto-biography,

At the time ... I knew nothing at all of Chemistry, had never read a syllable on the subject; not seen a single experiment in it.

He did quickly learn some chemistry and lectured and carried out some research until 1771 when he was appointed regius professor of divinity, a subject of which he was also ignorant. He retained that position for 45 years until his death, and from 1782 was also Bishop of Llandaff. For many years, however, he resided at Windermere and rarely visited either Cambridge or his diocese. His original work on freezing points was done while he was at Cambridge, his paper appearing in 1770. He found that 'in salts of the same kind, the resistance to congelation is in the direct simple proportion to the quantity of salt dissolved'. The same result was later found by Charles Blagden (1748–1820), who held an MD degree from Edinburgh and who had worked with Henry Cavendish (1731–1810), later serving as Secretary to the Royal Society. Blagden's 1788 paper on the freezing points does not mention Watson. The conclusion that the depression of the freezing point is proportional to the concentration is often called Blagden's law.

Particularly important work on the depression of the freezing point and other colligative properties was done by the French chemist François Marie Raoult (1830–1901), who taught at the Université de Grenoble from 1867 until his death. In his first paper, which appeared in 1878, Raoult confirmed the thermodynamic relationship between the colligative properties that had been deduced by Guldberg, and in an important paper that appeared in 1882 he provided extensive data for a variety of solutes in water and showed that the proportionality coefficient was approximately the same for different solutes. Later publications showed that the same is true for other solvents, each solvent having a different molecular constant. Raoult's work at once attracted particular attention, and within a few years the freezing-point method was being used widely for the determination of molecular weights.

Raoult also did much work on the lowering of the vapour pressure, and he published a series of papers on the subject between 1886 and 1888. He showed that to a good approximation in many systems the difference between the vapour pressure p_0 of a pure solvent and that of a solution containing an involatile solute, p, is proportional to the mole fraction x_2 of the solute:

$$p_0 - p = p_0 x_2. \qquad (4.53)$$

Then, since the mole fraction x_1 of the solvent is equal to $1 - x_2$, it follows that

$$p = p_0 x_1. \qquad (4.54)$$

Important work on the elevation of the boiling point as well as other colligative properties was done by Ernst Otto Beckmann (1853–1923) who published a number of papers on the subject beginning in about 1888. He designed ingenious apparatus for measuring the properties, the best known being the Beckmann thermometer in which the amount of mercury in the stem can be varied so as to change the temperature range. It could measure temperatures with an accuracy of about $0.001\,°C$, which is important for measuring the small differences in temperature that are found in the boiling-point and freezing-point work. He also designed a device for measuring small vapour pressure differences, useful for the vapour pressure studies.

All of this work on the colligative properties was considered very carefully in van't Hoff's papers and books. He did not make original contributions other than to the understanding of osmotic pressure, but he went to great pains to show that the thermodynamic equations were consistent with the experimental results. His approach to thermodynamics from the experimental side is in contrast to Gibbs's completely theoretical attack, with little appeal to experiment.

For some years the colligative properties provided the most useful methods for the determination of molecular weights. Since the boiling-point and freezing-point methods involve small differences between two temperatures they can only be used for substances of fairly low molecular weight. Differences between the vapour pressure of a solvent and a solution can be measured directly, as first done by Beckmann, and it is possible to measure molecular weights up to about 10 000 in this way. The osmotic pressure method can be used for molecular weights of up to 3 000 000. A difficulty with this technique is that it often takes a long time for equilibrium to be established; Berkeley and Hartley and others overcame this by measuring rates of flow with different applied pressures.

Willard Gibbs proposed his phase rule in 1876 but it was some years before any attention was paid to it. The first experimental work of any importance related to the phase rule was carried out by the Dutch chemist Hendrik Willem Bakhuis Roozeboom (1854–1907), whose work was begun at the University of Leyden and continued in Amsterdam where he succeeded van't Hoff as professor of chemistry in 1896. The phase rule was brought to Roozeboom's attention by van der Waals, who was professor of physics at Amsterdam, and from about 1886 to 1902 Roozeboom carried out a great many investigations on a large number of systems, at the same time making important theoretical studies. He also inspired a number of other Dutch chemists to work on problems of a similar kind.

Important work related to the phase rule was also carried out by van't Hoff. In 1896 he left Amsterdam to become professor at the University of Berlin, and there he devoted almost his entire research efforts to a study of the mineral equilibria involved in the marine potash and magnesite deposits at Stassfurt, which is near Magdeburg. His decision to work in this field, which was rather different from anything he had done before, was partly due to the influence of Wilhelm Meyerhoffer (1864–1906), who had been his student in Amsterdam after working with Ostwald, and had previously done phase-rule work. Together with some 30 collaborators and students van't Hoff for the first time treated the problem of marine deposits in a systematic way, and they published over 50 papers on the subject. As a result, van't Hoff is today regarded by petrologists as one of the pioneers in their subject.

Extensive work on the phase rule was also carried out by Wilder Bancroft. His book *The Phase Rule* appeared in 1897, two years after he had joined Cornell University. Bancroft had little mathematical ability, and his book did not even contain a derivation of the phase rule, being merely a systematic compilation of data from many sources. In the course of his work he was the first to use the word 'solute' but insisted that there was a fundamental difference between a solvent and a solute, a view that few others could appreciate.

Non-ideal systems

Soon after van't Hoff formulated his theory of osmotic pressure and of equilibrium constants in solution it became clear that the simple equations that had been developed were not always obeyed exactly. Difficulties were also encountered with rate equations for reactions in solution, and in an 1889 paper Arrhenius suggested that it was better to use osmotic pressures rather than concentrations in such equations.

The first systematic attempt to deal with deviations from ideal behaviour was made by the American chemist Gilbert Newton Lewis (1875–1946). After taking his doctorate at Harvard in 1899 Lewis worked in Germany with both Ostwald and Nernst. His treatment of the thermodynamics of solutions is described in papers he published from 1900 to 1907, three pairs of papers appearing simultaneously: in English in the *Proceedings of the American Academy of Arts and Sciences* and in German in the *Zeitschrift für physikalische Chemie*. In his 1900 papers he introduced the concept of the escaping tendency of a gas, which in his 1901 papers he called the fugacity. He developed in some detail the thermodynamic equations that involve the *fugacity*, which replaces the pressure in such equations as eqn 4.37.

For a substance in solution Lewis introduced the concept of the *activity*, which for solutions that deviate from ideal behaviour replaces the concentration. His new procedures were expounded in some detail in English and German papers which appeared in 1907, and in his influential book with Merle Randall, *Thermodynamics and the Free Energy of Chemical Substances*, which was first published in 1923.

It is of interest that in these early treatments of non-ideal solutions Lewis made only a passing reference to the thermodynamics of Willard Gibbs, his procedures instead following the course that had been established by van't Hoff and expounded by him and by Ostwald.

4.5 The third law, or Nernst's heat theorem

In the early years of the twentieth century there was great interest in equilibrium constants, especially from a practical point of view. It is a great convenience to be able to calculate equilibrium constants from thermal data without the necessity of measuring them directly, which is sometimes a matter of experimental difficulty. According to van't Hoff's equation 4.32, the value of $\Delta H°$, which can often be estimated from thermal data, determines the way in which the equilibrium constant K changes with temperature, but it does not determine the actual magnitude of K. The equilibrium constant can be expressed as

$$K = e^{\Delta S°/R}\, e^{-\Delta H°/R}. \tag{4.55}$$

In order to calculate K one needs to know not only $\Delta H°$ but $\Delta S°$. From thermal data one can calculate the entropy of a substance relative to its value at the absolute zero, but there still remains the problem of knowing the entropy change for the reaction at the absolute zero. Expressed differently, the integrated form of eqn 4.32 is

$$\ln K = \int \frac{\Delta H°}{RT^2} \, dT + I. \tag{4.56}$$

The value of the constant of integration I is needed for the equilibrium constant K to be calculated.

This difficulty had been clearly recognized in 1884 by Le Chatelier, and it was again emphasized in 1905 by Fritz Haber (1868–1934) in his book on the thermodynamics of technical gas reactions. Attempts by them and by van't Hoff to calculate the constant of integration, and hence the equilibrium constant, were unsuccessful.

This was the problem attacked by Walther Nernst (1864–1941), who after acting as an assistant to Ostwald at the University of Leipzig was professor of physical chemistry at the University of Göttingen from 1891 to 1905. While at Leipzig and at Göttingen he worked mainly on electrochemistry and thermodynamics, and became particularly interested in the calculation of chemical equilibrium constants from thermal measurements. His first paper on this topic appeared in 1906, by which time he had moved to the University of Berlin. It was concerned particularly with the chemical aspects of the problem; the paper contains no applications to physics, and no mention of the quantum theory, of which Nernst was still unconvinced. These points are significant, since today the main importance of Nernst's work is very different from his original conception of it.

Nernst approached the problem from Helmholtz's equation 4.24, which in terms of changes occurring in a chemical process may be written as

$$\left(\frac{\partial \Delta A}{\partial T}\right)_V = \frac{\Delta A - \Delta U}{T}. \tag{4.57}$$

It follows that the value of $(\partial \Delta A / \partial T)_V$ at any temperature T is given by

$$\left(\frac{\partial \Delta A}{\partial T}\right)_{V,T} = -\int \frac{1}{T} \frac{\partial \Delta U}{\partial T} \, dT + \left(\frac{\partial \Delta A}{\partial T}\right)_{V,0}. \tag{4.58}$$

The last term is the value of $(\partial \Delta A / \partial T)_V$ at the absolute zero, and is related to the unknown constant of integration in eqn 4.56. Nernst realized from this equation that if the temperature coefficient of ΔU does not become zero at the absolute zero, the temperature coefficient of ΔA must be infinite. For a gas reaction $(\partial \Delta U / \partial T)_V$ is not zero at the absolute zero, and this had been the source of the difficulties in previous attempts to calculate the constant of integration. Nernst therefore decided to transfer his attention to condensed systems.

On the basis of experiments on condensed systems Nernst postulated that the value of $(\partial \Delta U / \partial T)_V$ at the absolute zero is zero, in this way ensuring that the temperature coefficient of ΔA does not become infinite at the absolute zero. This was the original form of what has come to be known as the Nernst heat theorem or as the third law of thermodynamics. At the absolute zero $A = U$ and therefore $d\Delta U/dT = d\Delta A/dT$. Since $-d\Delta A/dT$ is equal to the change in entropy ΔS, an alternative statement of the theorem is that entropy changes become zero at the absolute zero. Nernst himself, however, always disliked thinking in terms of entropy changes. In 1912 he formulated his theorem in another way, in terms of the unattainability of the absolute zero.

During the next few years much work was done, in Nernst's laboratory and elsewhere, to establish the validity of the theorem. Many investigations were made on the properties of substances at very low temperatures, and were useful quite apart from their relationship with Nernst's theorem. Some of the results turned out to be inconsistent with the theorem in its original form, and a number of scientists of high reputation, including the British theoretical physicist Ralph Howard Fowler (1889–1944) and the German physical chemist Arnold Eucken (1884–1950), concluded that the theorem was incorrect. It was evident that if the theorem in its original form had any validity at all it had to be modified.

The most successful of the new formulations were due to the German-born physicist Franz Eugen (later Sir Francis) Simon (1893–1956). Simon had been a student of Nernst and later worked in the Clarendon Laboratory at Oxford where he did pioneering work in low-temperature physics. Simon realized that the reason for the deviations from the Nernst heat theorem in its original form was that substances are often not in a state of true equilibrium at low temperatures; glasses, for example, are in a state of disorder and are not at equilibrium. In 1927 Simon therefore proposed the following modification of the original theorem:

At absolute zero the entropy differences disappear between all those states of a system between which reversible transitions are at least possible in principle.

In 1930 he expressed the same idea in a modified form:

The entropy differences disappear between all those states of a system that are in internal thermodynamic equilibrium.

Nernst himself was never willing to accept these modified versions of his theorem. His original theorem and its subsequent modifications are often referred to as the third law of thermodynamics, although it may be questioned whether such a description is justified, since the theorem seems to lack the status of the first and second laws.

The main value of the heat theorem today is different from its original purpose, which was the calculation of equilibrium constants from thermal data. It is now possible to calculate entropy changes by the methods of statistical mechanics, based on quantum theory, so that the evaluation of the constant of integration by Nernst's theorem is no longer of practical value. The main applications of the heat theorem are now in connection with low-temperature solid-state physics.

Suggestions for further reading

For accounts of the measurement of temperature and heat, and general accounts of heat theories and of thermodynamics see

C. B. Boyer, History of the measurement of heat, *Scientific Monthly*, **57**, 442–452, 546–554 (1943).

S. G. Brush, *The Kind of Motion that we Call Heat*, North-Holland Publishing Co., 1976.

D. S. L. Cardwell, *From Watt to Clausius*, Manchester University Press, and Cornell University Press, Ithaca, New York, 1971.

D. V Fenby, Heat: its measurement from Galileo to Lavoisier, *Pure & Applied Chemistry*, **59**, 91–100 (1987).

R. J. Forbes and E. J. Dijksterhuis, *A History of Science and Technology*, Penguin Books, 1963, especially Vol. 2, Chapters 17 and 18.

A. E. E. McKenzie, *The Major Achievements of Science*, Cambridge University Press, 1960, especially Chapters 3 and 13.

D. Roller, The early developments of the concepts of temperature and heat, in *Harvard Case Studies in Experimental Science* (Ed. J. B. Conant), Harvard University Press, Cambridge, Mass., 1957.

F. Sherwood Taylor, The origin of the thermometer, *Annals of Science*, **5**, 129–156 (1942).

The following publications deal more particularly with the second law:

M. Barón, With Clausius from energy to entropy, *J. Chem. Education*, **66**, 1001–1004 (1989).

E. E. Daub, Atomism and thermodynamics, *Isis*, **58**, 293–303 (1967).

E. E. Daub, Entropy and dissipation, *Hist. Stud. Phys. Sci.*, **2**, 321–354 (1970).

R. Fox, Watt's expansive principle in the work of Sadi Carnot and Nicholas Clément, *Notes and Records Roy. Soc.*, **24**, 233–253 (1969–1970).

Elizabeth Garber, James Clerk Maxwell and thermodynamics, *Amer. J. Phys.*, **37**, 146–155 (1969).

M. J. Klein, Gibbs on Clausius, *Hist. Stud. Phys. Sci.*, **1**, 127–149 (1969).

T. S. Kuhn, Carnot's version of 'Carnot's cycle', *Amer. J. Phys.*, **23**, 91–95 (1955).

C. Truesdell, *The Tragicomedy of Classical Thermodynamics*, Springer-Verlag, Vienna and New York, 1971.

For accounts of pre-thermodynamic ideas about affinity and of chemical aspects of thermodynamics see

E. N. Hiebert, Developments in physical chemistry at the turn of the century, in *Science, Technology and Society in the Time of Alfred Nobel* (Eds. C. G. Bernhard, E. Crawford and P. Sörbom), Pergamon Press, Oxford, 1982, pp. 95–105.

F. L. Holmes, From elective affinities to chemical equilibria: Berthollet's law of mass action, *Chymia*, **8**, 105–145 (1962).

T. H. Levere, *Affinity and Matter: Elements of Chemical Philosophy*, 1800–1865, Clarendon Press, Oxford, 1971.

M. W. Lindauer, The evolution of the concept of chemical equilibrium from 1775 to 1923, *J. Chem. Educ.*, **39**, 384–390 (1962).

V. V. Raman, The permeation of thermodynamics into nineteenth-century chemistry, *Indian J. Hist. Sci.*, **10**, 16–37 (1975).

C. A. Russell, *The History of Valency*, Leicester University press, 1970.

W. G. Palmer, *A History of the Concept of Valency to 1930*, Cambridge University Press, 1965.

R. B. Root-Berstein, The ionists: founding physical chemistry, 1872–1890, Ph.D. Thesis, Princeton University Press, 1980.

For accounts of the history of Nernst's heat theorem (the third law) see

E. N. Hiebert, Hermann Walther Nernst, *Dict. Sci. Biog.*, **10**, 432–453 (1974).

Sir Francis Simon, The third law of thermodynamics: an historical survey, *Yearbook of the Physical Society*, 1956, pp. 1–22.

CHAPTER 5

Kinetic theory and statistical mechanics

The kinetic theory of gases also played an important role in the development of physical chemistry. This theory, which evolved over a period of many decades, is concerned with understanding the behaviour of gases in terms of molecular motions. In its more advanced formulations the theory developed into statistical mechanics, which treats the properties of matter in a comprehensive way.

Not much progress could be made with the kinetic theory of gases until the pressure–volume–temperature relationships had become established and until more was known about the nature of atoms. The gas laws were only properly understood by the middle of the nineteenth century, the final stumbling block being the problem of determining relative atomic masses. Kinetic theory and thermodynamics developed at the same time, through the efforts of many of the same investigators, and there was strong mutual support between the two subjects.

Before the kinetic theory is discussed it will be useful to give brief accounts of the gas laws and of the development of ideas about the atomic nature of matter.

5.1 The gas laws

Every beginner in physics or chemistry becomes acquainted with two gas laws. One of them states that at constant temperature the product of the pressure p and the volume V of a fixed amount of gas is a constant:

$$pV = \text{constant}. \tag{5.1}$$

The other states that at constant pressure the volume of a given amount of gas varies linearly with the temperature Θ:

$$V = \text{constant} + \text{constant} \times \Theta. \tag{5.2}$$

If absolute (Kelvin) temperatures are used the volume is proportional to the temperature T:

$$V = \text{constant} \times T. \tag{5.3}$$

In English-speaking countries the first of these relationships, that pV is constant, is usually referred to as Boyle's law, while in France it is known as

Mariotte's law. The second relationship, that volume is proportional to the absolute temperature, is known in France as Gay-Lussac's law, in English-speaking countries usually as Charles's law.

Historians of science today, of all nationalities, are almost all agreed that the credit for the publication, if not the discovery, of the first relationship should go to Robert Boyle, and that Mariotte did little more than republicize it a few years after Boyle announced it. As to the second relationship, it is now generally agreed that the French attribution to Gay-Lussac is the only reasonable one, the contribution of Charles being of less importance.

Boyle's Law

There is little doubt that the first statement (as opposed to publication) of the pressure–volume relationship was made in 1660 or 1661 by two amateur investigators, Richard Towneley (1629–1668) and Henry Power (1623–1668), the experiments being carried out in Towneley Hall, a fortified fourteenth-century house, still in existence, near Burnley, Lancashire. Power, who was the Towneley family physician, began to collaborate with Towneley in about 1660, and soon afterwards they reached the conclusion that the product of the pressure and the volume of a gas is constant. In 1661 Power wrote to Boyle mentioning the relationship, and Boyle at once discussed the matter with his assistant Robert Hooke who said that he had already made observations that were consistent with the constancy of pressure times volume.

In the second edition (1662) of his book *New Experiments physico-mechanical, touching the Spring of the Air*, Boyle reported the relationship for the first time. His own experiments had been made possible by an air pump that had been devised by Robert Hooke, who served as Boyle's laboratory assistant in the house in High Street, Oxford, in which Boyle lodged from about 1655 to 1668. In his book Boyle gave credit to Hooke, who had previously confirmed the relationship through his own experiments, and to Richard Towneley and Viscount Brounker. He did not mention Power, but this was evidently due to a misunder-standing; in communicating to Boyle the results of the Towneley–Power experiments Power had modestly failed to make it clear that he had been concerned with them himself.

It is not known what contributions to the problem were made by William, Viscount Brounker (1620?–1684). In his book Boyle referred to 'some Tryals made about the same time by that Noble *virtuoso* and eminent Mathematician the Lord *Brounker*, from whose further enquiries into this matter, if his occasions will allow him to make them, the Curious may well hope for something very accurate'. Probably Boyle was doing no more than paying an effusive tribute to the man who had just been made the first President of the Royal Society.

There are several reasons for concluding that Hooke played a more import-ant role than Boyle in confirming the pressure–volume relationship. In his *Micrographia* (1665) Hooke himself said that he had first performed the experi-ments, claiming no priority for the relationship but merely saying that the results were consistent with 'Mr. Townly's (*sic*) hypothesis'; this statement was entirely consistent with what Boyle had said in his second edition. Hooke always had the greatest admiration for Boyle, with whom he remained friendly for many years, and would hardly have made a claim that would have caused Boyle pain or annoyance. Another argument is that Boyle's interest was largely in qualitative science, whereas Hooke was more concerned with quantitative relationships such as Hooke's law which relates stress to strain. Another compelling argument is that Newton, in his *Principia*, gave the credit to 'Hooke and others'. Newton detested Hooke, with whom he had had many acrimonious arguments over priority, and was generally reluctant to give him credit for anything. On the other hand, Newton held Boyle in high esteem, and it seems hardly likely that he would have attributed the relationship to Hooke if he had not been convinced that Hooke deserved the credit.

The claims of Edmé Mariotte (*c.* 1620–1684) are hardly to be taken seriously. In his book *De la nature de l'air* (On the nature of air), published in 1679, 17 years after Boyle's book appeared, Mariotte stated the pressure–volume relationship, with no reference to Boyle. He made no claim of originality, mentioning the relationship as if it were one of many well-known properties of air. Mariotte was an influential man in France, playing a central role in the work of the Académie des Sciences shortly after its formation, and no doubt it was partly his prestige that led the French to attribute the law to him. Also, Boyle was a very long-winded writer, with a tendency to write around a subject instead of coming to the point. The passage in which he introduced the Towneley suggestion that pressure times volume is constant is a good example of this; it runs to about 350 words and covers much irrelevant material (such as a comment on the distance between Towneley's and Boyle's places of residence) but never states the relationship explicitly. Mariotte, on the other hand, was much more readable and precise, and thus a better publicist for the law.

The situation regarding the pressure–volume relationship has been nicely summed up by I. Bernard Cohen in an article that appeared in 1964:

This is a law discovered by Power and Towneley, accurately verified by Hooke, verified again by Boyle (aided in some degree by Hooke), first published by Boyle, but chiefly publicized by Mariotte.

Gay-Lussac's law

The situation regarding the volume–temperature relationship is much simpler. The French attribution of the law to Joseph Louis Gay-Lussac (1778–1850) is the most reasonable, but much credit should also go to Guillaume Amontons (1663–1705) who had suggested the relationship, on the basis of more limited experiments, over a century earlier, and had even hinted at the existence of an absolute zero. The claim on behalf of Jacques Alexandre César Charles (1746–1823) has little basis. Charles, who became professor of experimental physics at the Conservatoire des Arts et Métiers in Paris, was particularly interested in ballooning, and invented the first hydrogen balloon in 1783, making several ascents in it. It was in connection with ballooning that he made some rudimentary studies on the expansion of gases. He never published this work, and concluded that the linear relationship did not apply to certain gases, which Gay-Lussac found to be incorrect.

Gay-Lussac's work was much more thorough. His work on the expansion of gases was published in 1802, and in it he scrupulously made reference to Charles's unpublished work. This reference was seized upon by P. G. Tait, who in his influential textbooks of physics introduced the designation 'Charles's law' which has persisted to this day in English-speaking countries. One is tempted to suspect that Tait, a highly nationalistic Scot, preferred the name Charles because it had belonged to two Scotsmen who had become English kings. Generations of British students have assumed Charles to be British, which they would hardly have done with Amontons or Gay-Lussac.

Avogadro's hypothesis

The two relationships expressed by equations 5.1 and 5.3 can be combined in the one equation

$$pV = \text{constant} \times T. \tag{5.4}$$

The proportionality constant in this equation, which applies to a given amount of gas, is different for each gaseous substance.

A further important relationship, due to Avogadro, developed from the discovery in 1808 by Gay-Lussac that gases combine in simple proportions by volume; two volumes of hydrogen, for example, and one of oxygen, produce two volumes of steam. Three years later the Italian physicist Amadeo Avogadro (1776–1856) proposed the hypothesis that equal volumes of all gases, at the same temperature and pressure, contain equal numbers of molecules. Unfortunately Avogadro expressed his ideas clumsily, in poor French with many irrelevant details, and his work was ignored until 1858 when his fellow-countryman Stanislao Cannizzaro (1826–1910), at a chemical congress in Karlsrühe, distributed a pamphlet explaining how Avogadro's hypothesis would clear up all the confusion that was still reigning about atomic weights. After reading the pamphlet Julius Lothar Meyer (1830–1895) wrote in 1864: 'It was as though scales fell from my eyes; doubt vanished, and was replaced by a feeling of peaceful certainty.' Avogadro, who had succeeded to his father's title of Count of Quaregna, was dead by the time his hypothesis was accepted.

The consequence of Avogadro's hypothesis is that the constant in eqn 5.4 is the same for all gases if one takes the same number of molecules for each gas. This is taken care of by defining a base unit called the *mole*, which is the amount of substance of a system which contains as many elementary entities as there are atoms in 0.012 kilogram of carbon-12. The unit mole is always used to express amount of substance, and if n is the amount of substance eqn 5.4 can be replaced by a single equation that applies to all gases, often to a good approximation:

$$pV = nRT. \tag{5.5}$$

Here R, known as the gas constant, is a universal constant. It is the IUPAC recommendation that n should be called the 'amount of substance', but many people feel that this expression is not sufficiently explicit, and prefer to call it the 'molar amount' or the 'chemical amount'; it is often called the 'number of moles', but this is frowned upon as it is not a number but has the unit of mole. In honour of Avogadro, the ratio of the number of molecules present in a given volume to the amount of substance is called the Avogadro constant, and an IUPAC-approved symbol for it is L (for Joseph Loschmidt (1821–1895), who did important work relating to this quantity). Another acceptable symbol for it is N_A.

Non-ideal gases

It was realized early that gases sometimes show marked deviations from the laws of Boyle and Gay-Lussac. The Irish physical chemist Thomas Andrews (1813–1885) carried out detailed investigations of the pressure–volume relationships, with particular reference to the conditions under which a gas can be liquified by the application of pressure. For carbon dioxide he found that above 31.1°C, which he called its *critical temperature*, the gas cannot be liquified. Below this temperature, when pressure is gradually applied to the gas, there is a sudden decrease in volume, the gas having been converted into liquid. His results were first announced in 1863 in the third edition of W. A. Miller's *Chemical Physics*, and his further work is described in a number of later publications, particularly in Bakerian lectures given to the Royal Society in 1869 and 1876.

A molecular explanation for this behaviour of gases was put forward in 1873 by the Dutch physicist Johannes Diderik van der Waals (1837–1923); the treatment was in his Ph.D. dissertation, and as Maxwell remarked 'this at once put his name among the foremost in science'. For one mole of gas eqn 5.5 becomes

$$pV_m = RT, \tag{5.6}$$

where V_m is the molar volume. Van der Waals's suggestion was that this equation should be modified to

$$\left(p + \frac{a}{V_m^2}\right)(V_m - b) = RT, \tag{5.7}$$

where a and b are constants. The constant b is related to the volume occupied by the molecules themselves, the idea being that the molecules do not have access to the entire volume V_m because of the presence of the other molecules. The constant a is related to the existence of attractive forces between the molecules. Van der Waals carried out a number of further investigations on gases, and a few years later proposed his *law of corresponding states*; one defines a reduced pressure, temperature and volume, in terms of which there is a universal equation of state. Van der Waals was professor of physics at the University of Amsterdam from 1877 to 1907, and he received the Nobel Prize for physics in 1910.

The intermolecular attractive forces that van der Waals envisioned have come to be called *van der Waals forces*, although he had little knowledge of their nature. It was later shown that several types of force are involved in the van der Waals forces. If the molecules are electric dipoles, there are electrical attractions between them. Even if they are not, they attract one another by what are called *dispersion forces*, which can only be understood on the basis of quantum mechanics. A quantum-mechanical theory of them was given in 1931 by Fritz London (1900–1954), and developed in later publications.

The van der Waals equation gives a reasonably good interpretation of the behaviour of gases, but a number of refinements to it have to be made for more exact agreement with experiment, and some quite different types of equations have also been proposed.

5.2 Atomic theories

The idea that matter is not continuous but consists of tiny particles known as atoms has been discussed for many centuries. The founder of the atomic theory was the Greek philosopher Democritus ($c.$470–$c.$380 BC), whose views were in opposition to those of Anaxagoras ($c.$500–$c.$428 BC) who had propounded the doctrine of continuity. The Greek word ατομος (atomos) derives from the Greek prefix α- (a-), not or without, and τομος (tomos), to cut, so that the word means 'indivisible'.

The atomic theory was expounded and developed by the Roman philosopher Titus Lucretius Carus ($c.$95–$c.$55 BC) in his *De Rerum Natura* (Concerning the nature of things), which appeared in 56 BC. His idea was that atoms are nothing more than extremely small particles of matter, having essentially the same properties as the bulk material. This Lucretian view was widely held until the nineteenth century, when it began to be realized that atoms must have other important features, including some electrical character.

Until the early years of the present century there has been some opposition to the atomic theory, on various grounds. Some of these were metaphysical or religious. Aristotle strongly opposed the idea of atoms, which did not fit into his scheme of things, and because his influence was so strong atomic theories did not play an important role for many centuries. Descartes rejected the atomic theory on the grounds that the existence of atoms implied a limitation to the power of God who, in his view, must retain the ability to subdivide matter to any extent. A more recent objection to the real existence of atoms, due principally to Mach and Ostwald, is considered later.

It has been seen in Chapter 3 that Francis Bacon, Boyle and Newton had an approach to science that was firmly based on an atomic theory. Newton in particular developed a detailed theory of matter ('matter in a nut-shell') which was applied extensively by many of his followers to the interpretation of physical and chemical processes. To these investigators atoms were assumed to be like tiny billiard balls except that they were infinitely hard.

Boscovich's atomic theory

An alternative theory of atoms was put forward in 1758 by the Croatian scientist Roger Boscovich (1711–1787), a Jesuit priest who wrote prolifically on a variety of scientific and philosophical problems. In 1763 he published in Venice a remarkable book entitled *Theoria naturalis philosophiae* (Theory of natural philosophy), which contained an atomic theory which was to have a profound influence on the course of science until the middle of the nineteenth century. Previously Boscovich had accepted the Newtonian attitude to atoms, but began to realize that it had some serious difficulties. It had previously been assumed that a collision between two atoms would be much the same as a collision between two billiard balls. To Boscovich the situation must be more complicated. When two billiard balls collide there is elastic deformation, made possible by the fact that the atoms within the ball can move relative to each other. With the infinitely hard atoms that had been postulated, however, such deformation is impossible.

Boscovich's solution to the dilemma was to abandon the idea that atoms are particles, and to regard them as centres of force, acting from a single point; his theory has been referred to as a theory of point atoms. As shown in Figure 5.1, he postulated a rather complicated law of force, oscillating several times between repulsion and attraction as the distance from the point atom increases. At short distances there is increasing repulsion as the distance decreases, the repulsion being infinite at zero separation, so that the atoms can not interpenetrate. At distances greater than those corresponding to point T in the diagram there is attraction which he assumed to correspond to Newton's law of gravitation, following the inverse-square law. At shorter distances the curve passes through several maxima and minima, so that there is repulsion at certain distances and attraction at others. At certain distances, such as those represented by points G, I and L in the diagram, the attractive and repulsive forces balance; bonds between atoms may therefore be formed corresponding to these equilibrium distances. By postulating a rather complicated law of force Boscovich was able to explain the solid, liquid and gaseous states of the same substance in terms of different equilibrium distances.

Boscovich's theory gave a general explanation of the nature of chemical combination. A difficulty with Newton's theory of atoms, which attracted one another according to a universal law of gravitation, was that no explanation could be given for the fact that certain atoms will combine chemically and

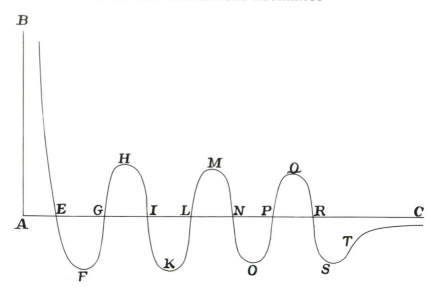

FIGURE 5.1 A copy of part of a figure in Boscovich's *De lege virium in natura existentium* (On the laws of force existing in nature), Rome, 1755; the complete diagram is complicated and confusing. The ordinate AB represents repulsive force, the abscissa AC the distance from the centre of force.

others will not. Boscovich's explanation was that the various atoms had complicated laws of force, and that bonding could only occur if the force patterns interlocked in an appropriate way.

Reaction to Boscovich's theory was mixed. Some rejected it outright, and for a period anyone who adhered to the theory was regarded with some suspicion, especially by chemists. It is now known that both Humphry Davy and Michael Faraday were favourable towards the theory, but thought it prudent not to say so. Faraday, for example, first revealed his support of the theory in a paper in the *Philosophical Transactions* of 1844. Davy and Faraday have often been regarded as purely empirical scientists, who were little interested in theory, but this is not the case; they were cautious in expressing their theories.

Dalton's atomic theory

Dalton's atomic theory was of particular importance for chemistry, which before it was put forward was no more than a mass of observations, incapable of any coherent explanation.

John Dalton (1766–1844), the son of a Cumberland hand-weaver, became a teacher of mathematics in the New College of Manchester. At first his primary interests were in mathematics and meteorology, but in his spare time he carried out chemical experiments, such as collecting marsh gas in the Lake District where he spent many holidays. He suggested various hypotheses regarding the nature of atoms, basing them on Newton's idea of repulsive particles but accepting the caloric theory of heat. Some of his ideas did not prove fruitful, but when in about 1803 he began to consider chemical problems he made advances of great importance. It was known that chemical substances appear to be of constant composition, and he realized that this can be explained if the different elements consist of different kinds of atoms, all atoms of the same element being identical. The atoms therefore form the building blocks for the various chemical compounds.

Dalton then attacked the problem of determining the relative masses of the different kinds of atoms, from the masses of substances that reacted chemically to form other substances. His first paper on this subject was presented orally to the Literary and Philosophical Society of Manchester in 1803, and it appeared in print in 1805; his ideas were further elaborated in his book *A New System of Chemical Philosophy*, the first volume of which appeared in 1808.

For example, Dalton thought that seven parts by weight of oxygen combine with one part of hydrogen to form water (eight parts is actually more correct). He assumed, as the simplest hypothesis but as we now know incorrectly, that one atom of hydrogen combines with one of oxygen (i.e., water is HO), and therefore concluded that an atom of oxygen is seven times as heavy as an atom of hydrogen. Proceeding in this way he compiled his first table of atomic weights, based on hydrogen with an atomic weight of unity, and this was included in his 1805 paper.

This example illustrates the difficulty that Dalton encountered, and which he could never overcome: he had no way of knowing how many atoms of each element come together to form a molecule, and as a result there were errors in his atomic weight tables. He never accepted as significant the discovery by Gay-Lussac that gases combine in simple proportions by volume, believing that the results were accidental and of no significance compared to the mass relationships that he emphasized. With appropriate modifications, such as could be made later with the help of Gay-Lussac's law and Avogadro's hypothesis, Dalton's theory was of far-reaching significance for chemistry. Dalton deduced from his theory the law of multiple proportions: if two elements combine to form more than one compound, the masses of one element that combine with a fixed mass of the other are in a simple ratio of whole numbers. It is sometimes stated that Dalton used this law to help him to arrive at his atomic theory; he did use the law of constant composition as a basis for his theory, but he deduced the law of multiple proportions as a necessary consequence of his theory.

Dalton was a crude experimenter, and much of the credit for developing the consequences of his atomic theory must go to the Swedish chemist Jöns Jakob Berzelius (1779–1848), who exerted a profound influence on chemistry. He was a great systematizer of knowledge, and was able to obtain results of lasting significance using simple equipment that he kept in his house. He established the law of multiple proportions without any doubt, and in 1814 introduced our modern chemical symbols in which the initial letter, or the first two letters, of the Latin name is used. Dalton himself reacted strongly against these new symbols, preferring his own rather clumsy system.

An interesting atomic theory was also put forward by the eccentric and irascible Irish chemist William Higgins (1763–1825). In 1789 he published a book which was mainly concerned with supporting Lavoisier's views as opposed to the phlogiston theory, but which also briefly suggested an atomic theory similar to Dalton's; he did not, however, deduce any atomic weights. After Dalton's theory had appeared he wrote another book in 1814 in which he strongly implied that Dalton had stolen his idea: 'I cannot with propriety or delicacy say that Mr. Dalton is a plagiarist, although appearances are against him.' Dalton later said that he had not read Higgins's earlier book, and in any case his theory was much better developed. Higgins was fond of accusing others of plagiarism; in the same 1814 book he accused Mrs Fulhame of stealing his ideas about what was later called catalysis, and this claim also had little foundation (Section 8.5).

On the whole the chemists of the early nineteenth century, while successfully applying Dalton's atomic theory to problems of chemical composition, did not give much consideration to the precise nature of atoms and molecules. and tended to retain Newton's idea that atoms were hard spheres. Some physicists, however, regarded the chemists' ideas as naive, in that they were incapable of explaining some of the physical evidence that was emerging, such as the details of spectra which were then beginning to attract attention. In 1867, for example, in a paper given to the Royal Society of Edinburgh, Kelvin spoke rather harshly of the 'monstrous assumption of infinitely strong and infinitely rigid pieces of matter' which he said had been put forward by 'some of the greatest modern chemists in their rashly-worded introductory statements'. At the time he was enthusiastic about the vortex theory of the atom which, as will now be seen, did not meet with much greater success.

The vortex atom

The vortex theory of the atom had a considerable vogue among physicists for several decades during the nineteenth century, but was soon forgotten. It seems to have originated with the Scottish engineer W. J. M. Rankine whose contributions to thermodynamics were referred to in Section 4.3. Rankine's papers were never easy to understand, particularly as his atomic theory was deeply entangled with his thermodynamics, and it is not quite clear what evidence he had for his vortex theory. In one of his papers, published in 1850, he does give a lucid statement of his idea, stating that:

Each atom of matter consists of a nucleus, or central point, enveloped by an elastic atmosphere which is retained in its position by attractive forces, and that the elasticity due to heat arises from the centrifugal force of these atmospheres, revolving or oscillating about their nuclei or central points.

This passage is of some interest as one of the first in which an atomic nucleus was mentioned. Rankine regarded the internal energy of molecules as residing in these atomic vortices.

At first Rankine's ideas attracted little attention. even among his friends Kelvin and Clerk Maxwell, neither of whom quite understood what he was proposing. Interest was, however, revived in 1858 when Helmholtz developed a theory of homogeneous incompressible fluids of zero viscosity, such as the aether was supposed to be. Helmholtz showed that for such a fluid closed vortex rings are formed having some remarkable characteristics. Once formed they are indestructible, and if they are in a particular rotational state they cannot change from that state. In 1867 P. G. Tait prepared a translation of Helmholtz's paper and also devised a spectacular and convincing simulation of the vortex theory by the use of smoke rings. Although smoke rings in air are by no means the same as vortex rings in an incompressible fluid of zero viscosity, they do appear to behave in a somewhat similar way and to provide an interesting analogy. Tait demonstrated, for example, that two smoke rings projected towards one another at an oblique angle would never come into contact but would glance off each other and begin to vibrate. The smoke rings could not be cut with a knife, but instead moved away from it.

A month after seeing this demonstration Kelvin read to the Royal Society of Edinburgh a paper entitled 'On vortex rings' in which he argued enthusiastically that atoms are vortices in the aether that was then believed to pervade all space. He later published several more papers on the properties of vortices, and for a period developed interpretations of atomic and molecular structure in terms of vortices. However, he later began to realize that this approach was unfruitful and in his Baltimore Lectures, delivered at Johns Hopkins University in 1884 and first published in 1904, he jokingly remarked that the most important aspect of Rankine's hypothesis of molecular vortices was its name.

Clerk Maxwell also had a considerable interest in the vortex atom. He wrote a detailed item entitled 'Atom' which appeared in 1875 in the *Encyclopaedia Britannica* in which he discussed the vortex theory in some detail. As he often did, he greatly overestimated the intellectual capacity of most of his intended readers, as the article contains Latin quotations (without translation) and a number of differential equations. It must have been one of the most highbrow articles that ever appeared in an encyclopedia intended for the general reader, but it is useful today in giving a clear idea of the scientific thinking of the time. The vortex theory is treated explicitly and for the most part with approval. There is also much about the kinetic theory of gases, and about the difficulties of understanding spectra in terms of atomic and molecular theories.

The reality of atoms

By the end of the nineteenth century the existence of atoms had become generally accepted among scientists, but there were some strong pockets of resistance. During the last decade of the century several influential voices were raised against the real existence of atoms, notably those of the Austrian physicist and philosopher Ernst Mach (1836–1916) and the chemist Wilhelm Ostwald. Their reasons, however, were different. Mach was a leader of the positivist school of thought, according to which concepts such as the atomic theory should not be accepted since the evidence for them must always be indirect, based on inference. The atomic theory might be of practical use, but one should not conclude that atoms really exist.

Ostwald based his objections to atoms on energetic grounds, with which Mach did not agree. His criticisms of the atomic theory, which he referred to as a 'mere hypothesis', were clearly stated in London in 1904 in his Faraday Lecture to the Chemical Society; he admitted to some temerity in giving this lecture as the Chemical Society had the previous year been holding celebrations in honour of the birth of Dalton's atomic theory. In this lecture Ostwald asserted that

It is possible to deduce from the principles of chemical dynamics all the stoichiometric laws: the law of constant proportions, the law of multiple proportions and the law of combining weights [his italics].

He then argued in some detail that the phase rule, together with dynamical arguments, leads to the stoichiometric laws without any need to invoke the atomic hypothesis. Boltzmann had previously pointed out to him on a number of occasions, and especially at a meeting in Lübeck, the fallacy of this argument: the results of chemical dynamics and the phase rule themselves require the atomic hypothesis. Ostwald would have argued, for example, that hydrogen iodide always has the same composition because when hydrogen and iodine react together they are required by dynamics to do so in a particular way; he failed to see that the reason this is so is that the reaction involves a rearrangement of atoms.

Ostwald remained sceptical of the real existence of atoms until 1909, when he was finally convinced by the experiments of John Joseph Thomson (1870–1942) on cathode rays and related matters (Section 6.7), and of Jean Baptiste Perrin (1870–1943) on the Brownian movement (Section 9.1). Until that time Ostwald's very influential textbooks of physical chemistry had given no explanations in terms of atoms. However in the 1909 edition of his 'grosse Ostwald' he finally agreed that the evidence for atoms was now overwhelming. He added, however, that one could still explain the laws of chemical composition without the assumption of atoms. Mach never conceded the existence of atoms.

5.3 The kinetic theory of gases

Some of the more important contributions to the kinetic theory of gases, and to statistical mechanics, are listed in Table 5.1.

Even before Dalton had proposed his atomic theory a number of theories had been suggested to explain the gas laws, especially the pressure–volume relationship, the most important being those of Newton and Daniel Bernoulli. In his *Principia* (1687) Newton briefly discussed various hypotheses about inter-

TABLE 5.1 Highlights in kinetic theory and statistical mechanics

Author	Date	Contribution	Section
D. Bernoulli	1738	Kinetic theory of gases	5.3
Waterston	1843	Detailed kinetic theory	5.3
Clausius	1857	Improved kinetic theory	5.3
Clausius	1858	Theory of mean free path	5.3
Maxwell	1860, 1867	Velocity distribution; transport properties	5.4
Maxwell	1867	Statistical basis of the second law	5.4
Boltzmann	1868	Distribution of energy	5.4
Boltzmann	1871	Thermodynamic properties from distribution functions	5.5
Boltzmann	1872	Theory of approach to equilibrium	5.3
Boltzmann	1877	Entropy and probability	5.4
Gibbs	1902	General treatment of statistical mechanics	5.4

atomic forces. He considered a static model for a gas in which the atoms repel one another with forces that are inversely proportional to the square of the distance, and he showed that such a model would explain the fact that pressure is inversely proportional to volume. He did not suggest that this theory is necessarily the correct one for a gas, simply that it led to the right kind of behaviour.

Half a century later Daniel Bernoulli (1700–1782) proposed a model for a gas that is very similar to the one accepted today. One section of his important treatise *Hydrodynamica*, which appeared in 1738, is entitled, in translation, 'On the properties and motions of elastic bodies, especially air', and it included the diagram reproduced as Figure 5.2. His view was that heat is nothing but atomic motion and that the pressure of a gas is due to the bombardment of the particles on the walls of the vessel. In presenting this theory he was a century ahead of his time, but his treatment had little effect since the caloric theory of heat was then universally accepted, and few believed that the motion of particles was of any significance.

For over a century after Bernoulli developed his theory hardly any progress was made on the theory of gases. During the interval, however, there were two contributions, made by Herapath and Waterston, which would have exerted much influence if their efforts had been treated more sympathetically by the scientific establishment of the time.

John Herapath (1790–1868) received little formal education, but gained some competence in science while working in the business of his father, who was a maltster, and while still in his teens he was recognized locally as

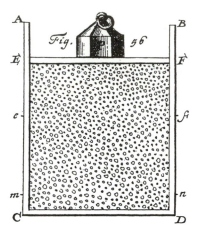

FIGURE 5.2   Daniel Bernoulli's diagram relating to his kinetic theory of gases (from his *Hydrodynamica*, Strasbourg, 1738).

something of a mathematical prodigy. He made attempts to extend Newton's methods, and in 1820 he submitted to the Royal Society a manuscript entitled 'A mathematical enquiry into the causes, laws and principal phenomena of heat, gases, gravitation, etc.' It included among other things a comprehensive treatment of the kinetic theory of gases but at the same time there were a number of errors. The Royal Society rejected the paper, largely as a result of Humphry Davy's opinion that the material was too speculative, and Herapath withdrew it and sent it to the *Annals of Philosophy*, where it appeared in 1821. He made a number of subsequent attempts to publicize his ideas; he wrote letters to *The Times*, and rather inappropriately he published articles in the *Railway Magazine* which he later edited himself. In 1847 he published a book, *Mathematical Physics*, which exerted some influence on Joule. On the whole, however, his work was neglected, for various reasons. One was that the caloric theory of heat was still the most popular one, so that Herapath's idea that the movement of particles is related to heat and pressure was suspect. Also, his treatment of elastic collisions was unsatisfactory, and he gave an unacceptable definition of absolute temperature, reaching the conclusion that pV is proportional to T^2 rather than to T. His treatment therefore required a good deal of modification to become entirely satisfactory, but the ideas were essentially correct and important, and with a little help from others he might have been able to produce a correct formulation of the kinetic theory.

John James Waterston (1811–1883) received a good education and while apprenticed as a civil engineer attended lectures at the University of Edinburgh. He subsequently practised as an engineer in London and in India, where he was a naval instructor. It has been suggested that his success with kinetic theory was partly due to the fact that his leisurely life in India allowed him to become a proficient billiard player. In 1846 he submitted to the *Philosophical Transactions* a paper which gave an essentially correct and comprehensive kinetic theory, better than that of Herapath. It included for the first time a statement of equipartition of energy among molecules of different masses, it treated the velocity of the molecules as proportional to the square root of the temperature and it gave a satisfactory treatment of specific heats.

The paper was rejected for publication by the Royal Society, but an abstract appeared in the *Proceedings of the Royal Society*. The Secretary of the Royal Society at the time was the physiologist and lexicographer Peter Mark Roget (1779–1869), now best remembered for his *Thesaurus of English Words and Phrases* (1852), and he sent Waterston's paper to two referees. One was the Revd Baden Powell (1796–1869), the Savilian professor of geometry at Oxford, who had done some experimental work in the field of optics; he was the father of the founder of the scouting movement. The other was Sir John Lubbock, Bt. (1803–1865), a banker and amateur scientist who had done some good work in astronomy and the theory of the tides. Neither referee had been at all concerned with the subject-matter of Waterston's paper, and both rejected it outright. Baden Powell's objection was that the basic assumption that pressure is due to molecular bombardments is 'very difficult to admit, and by no means a satisfactory basis for a mathematical theory'. Lubbock said that 'the paper is nothing but nonsense, unfit even for reading before the Society'. The paper was, however 'read'—in the formal sense that its title was read out at a Society meeting—and Roget prepared the abstract that was later printed in the Society's *Proceedings*.

The policy of the Royal Society was that a paper that had been read before the Society became its property and could not be returned to the author. In Herapath's case he had elected not to have his paper read, and therefore it was returned to him and he published it elsewhere. Waterston, however, perhaps because he was in India at the time and out of touch with the situation, allowed his paper to be read and therefore could not have it returned. He had no copy, and the paper did not see the light of day for over 40 years.

It was in 1892, nine years after Waterston's death, that the paper was finally resurrected. In that year John William Strutt, 3rd Baron Rayleigh (1842–1919), discovered it in the Royal Society archives and had it printed in the *Philosophical Transactions*, with an introductory note by himself in which he expressed regret that this valuable paper had not been printed earlier. By the time the paper appeared the work had all been done again, particularly by Clerk Maxwell and Clausius, so that the paper was then only of historical interest. The comment has sometimes been made that it seems surprising that when others developed the same theory those who knew of Waterston's work did not call attention to it. However, those who did know of his work, Roget, Baden Powell and Lubbock, probably did not follow the developments in kinetic theory, their own scientific interests being elsewhere.

Waterston did present his theory to the British Association for the Advancement of Science, at its 1851 meeting at Ipswich. The only person who seems to have paid any attention to it was Rankine, who briefly referred to it in three of his papers (1853, 1864 and 1869), but was too much concerned with his own theory of molecular vortices to pay much attention to it.

Another contribution to kinetic theory was made in 1848, when Joule presented to the Literary and Philosophical Society of Manchester a paper in which he used Herapath's theory to calculate the velocity of a hydrogen molecule, and to calculate the specific heat at constant volume. Joule's mathematical abilities were, however, limited, and this paper was not of much significance. It did not appear in print until 1851 and it did not attract much attention, although it was referred to in Clausius's important paper of 1857. In that paper Clausius remarked that it was to be regretted that Joule did not publish his memoir in a more widely circulated journal, and Joule then had his paper reprinted in the *Philosophical Magazine*.

Clausius's kinetic theory

In 1856 the German chemist August Karl Krönig (1822–1879) published a short memoir on the kinetic theory which was not of great significance but which did have the important effect of inspiring the later work of Clausius and Clerk Maxwell. Clausius's first venture into kinetic theory resulted in a paper that appeared in 1857 and went far beyond Krönig's treatment. Clausius considered not only the translational motions of molecules but also their rotations. He pointed out that one could not assume that translational energy would be conserved in a collision, since there might be transfer of one type of energy into another. He suggested that there must be a distribution of velocities and gave an explanation for the evaporation of a liquid, resulting in cooling: those molecules with higher than average velocities would be the ones that would escape the attractive forces of the other molecules, and the loss of energy would result in a fall in temperature. However, Clausius never took this important idea any further, and it was Maxwell who developed the idea of the distribution of velocity.

In this paper Clausius also derived the fundamental equation which gives an expression for the product of pressure and volume, pV, in terms of the mass m of each molecule, the number N of molecules, and the square of the mean velocity u^2. His proof is essentially the same as that now given to students of introductory chemistry and physics, and need not be repeated here; it led to the important equation

$$pV = \frac{1}{3} Nmu^2. \tag{5.8}$$

(As was later shown by Maxwell this equation is not quite correct; it ought to involve not the square of the mean velocity but the mean square velocity, as in eqn 5.15.) Since the total translational kinetic energy E_k is $\frac{1}{2}Nmu^2$ it is therefore given by

$$E_k = \frac{3}{2}pV. \tag{5.9}$$

Clausius deduced that for a given amount of gas the difference $C_p - C_v$ between the specific heat at constant pressure and that at constant volume is pV/T. The total heat q in the gas is therefore given by

$$q = C_v T = \frac{C_v \, pV}{C_p - C_v}. \tag{5.10}$$

The ratio of the translational energy E_k to the total heat is therefore

$$\frac{E_k}{q} = \frac{3}{2}pV\frac{C_p - C_v}{C_v \, pV} = \frac{3}{2}(\gamma - 1), \tag{5.11}$$

where γ is the ratio C_p/C_v. For simple gases γ was known to be about 1.4 and the ratio E_k/q is thus about 0.6. Motions other than translational must therefore occur. For a monatomic gas Clausius concluded that the molar specific heat at constant volume, $C_{v,m}$, should be $(3/2)R$. That at constant pressure, $C_{p,m}$, should be $(3/2)R + R = (5/2)R$, so that the ratio $C_{p,m}/C_{v,m}$ (which is equal to C_p/C_v for any given amount of gas) should be 5/3 or 1.67. There were then no known monatomic gases, and for other gases Clausius pointed out that there should be contributions for other types of motion, so that the ratio C_p/C_v would be reduced. He could not, however, estimate them since he was not convinced that there must be equipartition of energy. These were important conclusions, and for the first time a connection had been made between kinetic theory and thermodynamics.

In the same paper Clausius also produced an argument from kinetic theory in support of Avogadro's hypothesis that equal volumes of gases at the same temperature and pressure contain equal numbers of molecules. Clausius argued that if it were assumed that all molecules have the same average kinetic energy at a given temperature, then at a given temperature and volume equal numbers of them would exert the same pressure.

In the following year, 1858, Clausius published another paper of great importance in which he first introduced the idea of the mean free path, which is the average distance a molecule moves between successive collisions. This paper owed its origin to an objection raised by the Dutch meteorologist Christoph Hendrick Diederik Buys-Ballot (1817–1890), who had argued that if it were true, as Joule and Clausius claimed, that the molecules of a gas move at speeds of several hundred metres a second, gases should mix much more rapidly than they do. The odour of hydrogen sulphide released in one corner of a room should be detected elsewhere in the room in a fraction of a second, which is not the case. Also, it was known that carbon dioxide, being heavier than air, would remain at the bottom of a vessel for some period of time; it appeared to Buys-Ballot that it could not remain more than an instant if the molecular speeds were so high.

Clausius realized that this argument would only be true if molecules had no size. In reality there will be collisions between them, and a molecule cannot travel very far in a straight line. He derived an expression for the mean free path on the basis of a model for a gas in which all the molecules except one were in fixed lattice positions. If d is the average distance between the stationary molecules in the lattice, and r is the radius of the repulsive collision sphere round each molecule, Clausius found that the mean free path λ was given by

$$\lambda = \frac{d^3}{\pi r^2}. \tag{5.12}$$

Clausius then considered the true situation in which all of the molecules are in motion, and obtained a relationship that he thought highly significant. He stated, without proof, that the relative velocity is 4/3 times the actual velocity, and concluded that the mean free path is reduced by the factor 3/4:

$$\lambda = \frac{3d^3}{4\pi r^2} \tag{5.13}$$

or

$$\frac{\lambda}{r} = \frac{d^3}{(4/3)\pi r^3}. \tag{5.14}$$

In other words, the ratio of the mean free path to the radius of the collision sphere is equal to the ratio of the average space between molecules to the volume of the collision sphere for each molecule. However, Maxwell showed that the ratio 4/3 is not correct, a conclusion that Clausius was never able to accept.

These contributions of Clausius inspired later work of Clerk Maxwell. A weakness of Clausius's approach was that he was content to work with the average energy, without considering the distribution of velocities, which has a small but significant effect on the final conclusions. One of the only references Clausius made to the distribution of velocities was in his 1857 paper in which he discussed evaporation, recognizing that the more energetic molecules are more likely to leave the liquid, which therefore cools. Clausius's equations for the mean free path should be modified to take account of the distribution of velocities, and in his eqn 5.8 the square of the mean velocity u^2 should be replaced by $\overline{u^2}$, the mean square velocity:

$$pV = \frac{1}{3} Nm\overline{u^2}. \tag{5.15}$$

According to Maxwell's distribution (eqn 5.22) the mean square velocity is greater than the square of the mean velocity by the factor $3\pi/8 = 1.178$.

Equation 5.15, and Clausius's theory of the mean free path as modified by Maxwell, provided a sound basis for the treatment by kinetic theory of a variety of important problems. For example, they lead in a relatively simple way to expressions for the frequency of molecular collisions, a matter of importance in chemical kinetics (Section 8.3). The fundamental equations also readily led to treatments of viscosity, thermal conductivity and diffusion.

Clausius made few further contributions to kinetic theory. He showed little appreciation of Maxwell's theory of the distribution of velocities, and was strangely uninterested in the later important developments of Boltzmann and Gibbs. One important contribution that he did make in 1870 was to derive the *virial theorem*, which states that the mean value of the kinetic energy of a number of particles is equal to the mean value of a quantity he called the *virial*, a word he coined from the plural *vires* of the Latin word *vis*, which means force or energy. The virial is defined by the equation

$$\Sigma \; \tfrac{1}{2}m \; <u^2>_{av} = \tfrac{1}{2}\Sigma(Xx + Yy + Zz).$$ (5.16)

Here x, y and z are the Cartesian coordinates of each molecule, and X, Y and Z are the components of the forces acting on each molecule. The average value $<u^2>_{av}$ is taken over time. The virial finds its main application in the virial coefficients introduced in 1901 by the Dutch physicist Heike Kammerlingh Onnes (1853–1926), who suggested that the deviations from the ideal gas law shown by a real gas can be expressed by the equation

$$\frac{pV}{nRT} = 1 + \frac{Bn}{V} + \frac{Cn^2}{V^2} + \frac{Dn^3}{V^3} + \dots$$ (5.17)

where B, C, D, etc., are called the virial coefficients.

The distribution of velocity and energy

It may seem obvious today that molecules in a gas should be moving with a range of speeds, but that is because we have been brought up to believe so. The fact is by no means self-evident, and during the nineteenth century many who had devoted much thought to the matter had convinced themselves that if at any time there was a range of velocities these would soon become equalized as a result of the collisions between the molecules. Clausius did not believe this, but even after Maxwell had developed his theory of the distribution of velocities he remained convinced that it was satisfactory to deal with average velocities and to ignore their distribution for most purposes.

Maxwell's first work on the kinetic theory of gases was done while he was a professor at Marischal College, Aberdeen, and it led to an important paper which appeared in 1860. This paper, in which Maxwell first gave an expression for the distribution of velocities, has an interesting background. A letter that Maxwell wrote to George Gabriel Stokes (1819–1903) on 30 May 1859, reveals that he had undertaken the work as an 'exercise in mechanics', and that he had begun it with serious doubts as to the validity of a kinetic-theory

James Clerk Maxwell

(1831–1879)

Maxwell was born in Edinburgh to a family of comfortable means, and was educated at the Edinburgh Academy and the University of Edinburgh. In 1850 he went to St Peter's College (now called Peterhouse), Cambridge, but after a term transferred to Trinity College, graduating in 1854 as Second Wrangler. He was at once elected a Fellow of Trinity.

In 1856 Maxwell became professor at Marischal College, Aberdeen, but did not enjoy lecturing and was not much of a success at it. When in 1860 there was some reorganization at the college there was no longer room for him, and he accepted an appointment at King's College, London. He remained there until 1865 when he retired to his family estate in south-west Scotland. There he enlarged his family house and worked on his celebrated *Treatise on Magnetism and Electricity*, which appeared in 1873. In 1871 he somewhat reluctantly agreed to become the first Cavendish professor at Cambridge, the post having previously been refused by William Thomson (later Lord Kelvin).

Maxwell is particularly noted for his treatment of the distribution of molecular speeds in a gas, and for his theory of electromagnetic radiation. He also made important contributions in other fields, including thermodynamics, colour vision, colour photography, geometrical optics, photoelasticity, viscoelasticity, relaxation processes, and the theory of Saturn's rings; it was work on the latter that led to his kinetic theory. He was also a skilful experimentalist, and designed some of the instruments used in the Clarendon Laboratory. He and his wife made the first reliable measurements of viscosities of gases, using apparatus of his design.

interpretation of the properties of gases. He thought it likely that his work would show the theory to be inadequate, and perhaps quite wrong. In that letter he mentioned that he had obtained 'one curious result', namely that the viscosity η of a gas is independent of the pressure or density, being given by the equation

$$\eta = \frac{mu}{\sqrt{2}\pi d_m^2},\tag{5.18}$$

where d_m is the molecular diameter. At the time there were no reliable data on the viscosities of gases, and Maxwell thought that his conclusion was 'certainly very unexpected'. He asked Stokes, the leading expert in hydrodynamics, if he had 'the means of refuting this result of the hypothesis', namely that the viscosity should be independent of the pressure. Stokes replied that 'the only

His work on the theory of electromagnetic radiation began in about 1855 and culminated in a comprehensive paper published in 1856, followed by his *Treatise*. His work on kinetic theory, described in Section 5.3, was begun when he was at Aberdeen, and his first paper on the subject appeared in 1860; it was continued at King's College and at Cambridge.

His work on thermodynamics was done over a period of years but he wrote no major paper on the subject; he communicated some of it in correspondence with Kelvin and P. G. Tait, and he played a role in reconciling the views of Clausius, Kelvin and Tait. His book *Theory of Heat*, which first appeared in 1870 and ran to 11 editions, gave a lucid account of the subject. Since Maxwell was able to think about problems in a simple way he was able to clarify many matters that had been put forward obscurely by Clausius, Boltzmann and Gibbs.

In 1858 Maxwell married Katharine Mary Dewar, the daughter of the Principal of Marischal College and seven years his senior; they had no children. By all accounts Mrs Maxwell was a 'difficult' woman who resented his scientific interests and who would have preferred him to live the life of a country gentleman. This being so, it is a little surprising that he was able to persuade her to help him with the viscosity experiments at King's College. Maxwell died in Cambridge of abdominal cancer and is buried in Corsock Churchyard near his estate in Scotland.

References: C. W. F. Everitt, DSB, 9, 198–230 (1974); Anon. (1872); Brush (1965, 1976); Everitt (1983); Goldman (1982); Klein (1970); Macdonald (1964).

experiment I have met with on the subject does not seem to confirm' the prediction that the viscosity is independent of the pressure. Maxwell himself later undertook some viscosity experiments, with results that did confirm the prediction, as is mentioned below.

His 1860 paper gave his theory of the viscosity of gases, which was based on the idea that the friction between two parallel planes in a gas is due to the transfer of molecules from one plane to the other. The effect is that the more rapidly moving plane is slowed down by molecules jumping from the more slowly moving plane. Maxwell interpreted the independence of the viscosity from pressure by pointing out that at higher pressures more molecules jump from one plane to the next but do not have as far to jump; the effects approximately balance each other.

Perhaps the most important feature of his 1860 paper was his treatment of the distribution of velocities. In his derivation he paid no attention to the mechanics of collisions between molecules, instead basing his arguments on probability theory such as the theory of errors that had been given by Laplace. Maxwell's derivation was in brief as follows. Let the components of molecular velocities along three Cartesian axes X, Y and Z be u_x, u_y and u_z. Then if the total number of molecules is N, the number dN of molecules having components of velocities between u_x and $u_x + du_x$, between u_y and $u_y + du_y$ and between $u_z + du_z$ can be written as

$$N\, f(u_x)\, f(u_y)\, f(u_z) \tag{5.19}$$

where f is some function to be determined. Since the axes are arbitrary, dN can depend only on the molecular speed u, where

$$u^2 = u_x^2 + u_y^2 + u_z^2. \tag{5.20}$$

The distribution must therefore satisfy the relation

$$f(u_x)f(u_y)f(u_z) = \phi(u_x^2 + u_y^2 + u_z^2), \tag{5.21}$$

where ϕ is some function. The solution of this is an exponential, and Maxwell showed that the resolved components of velocity in a given direction, e.g. u_x, have a distribution identical in form to the 'normal distribution' deduced by Laplace in his theory of errors:

$$dN_x = \frac{N}{\alpha^3 \sqrt{\pi}}\, u^2\, e^{-u^2/\alpha^2} du, \tag{5.22}$$

where α is a constant.

In his 1860 paper Maxwell also applied his distribution function to diffusion and heat conduction, but his treatment contained some computational errors and also some errors of principle; in particular he neglected the variations of pressure and density in setting up his distribution function. The errors were pointed out by Clausius who in 1862 proposed a modified theory which was also unsatisfactory since it used average velocities instead of a distribution. By 1864 Maxwell had developed a correct treatment, but being dissatisfied with some details did not publish it until three years later. In this 1867 paper Maxwell also gave an improved derivation of the velocity-distribution law, an explanation of Gay-Lussac's law of equivalent volumes and a treatment of the equilibrium in a column of gas under the influence of gravity.

In his 1860 paper Maxwell also considered the troublesome problem of the specific heats of gases, a matter that had been raised by Clausius in his 1857 paper. Clausius had pointed out that the ratio of specific heats should be less than 1.67 for molecules containing more than one atom, but he could not estimate the values being unconvinced of the need for equipartition of energy. Maxwell on the other hand concluded on the basis of classical mechanics, which he had studied very profoundly, that there must be equipartition of energy between the differenc modes of motion, and as a result was led to the conclusion that there were serious problems with the kinetic theory of gases. He agreed with Clausius that the ratio of specific heats must be 1.67 if the only motion that can occur is translational motion. If, however, molecules are spheroids there are three degrees of rotational freedom, each making a contribution of $\frac{1}{2}R$ to the specific heat; the ratio of specific heats is thus $(3R + R)/3R = 4/3 = 1.33$. However, the measured ratios for gases such as oxygen, hydrogen and nitrogen were about 1.4. Maxwell was discouraged by this discrepancy, and ended his paper by saying that he had proved that the kinetic theory 'could not possibly satisfy the known relation between the two specific heats of a gas'. In a paper that he gave in 1860 at a meeting in Oxford of the British Association his comment on the specific heat difficulty was:

This result of the dynamical theory, being at variance with experiment, overturns the whole hypothesis, however satisfactory the other results may be.

In subsequent publications, such as his 1875 article in the *Encyclopaedia Britannica*, Maxwell grappled further with this problem, but was never able to solve it. When he considered the spectral evidence he realized that there seemed to be an even greater difficulty, since the spectra showed that even for atoms there must be a number of additional types of motion each one of which, if the principle of equipartition of energy were valid, should make its contribution to the specific heat. Difficulties of this kind became particularly acute with the publication in 1875 of an important paper by August Adolph Kundt (1839–1894) and Emil Gabriel Warburg ((1846–1931), both of whom were professors of physics at the University of Strassburg. They measured the specific heat of mercury vapour and found C_p to be $(3/2)R$, so that the ratio C_p/C_v is 1.67, a value that Maxwell considered to be too large to be explicable in terms of kinetic theory, since the spectrum of mercury implies other modes of motion.

For diatomic molecules the specific heat results implied only two degrees of rotational freedom, and none of vibration. In 1877 a suggestion to explain this was made independently by Robert Holford Macdowall Bosanquet (1841–1912), a Fellow of St John's College, Oxford, and Ludwig Boltzmann. They pointed out that, for reasons then not clear, the rotation about the axis of symmetry does not contribute to the specific heat, so that only two degrees of rotational freedom are to be counted. We now know that this idea is essentially correct, the justification for the neglect of the rotation round the axis of a linear molecule being provided by quantum theory (Section 10.1).

In 1860 Maxwell moved from Aberdeen to King's College, London, where he designed an instrument for measuring gas viscosities. With it he carried out experiments in collaboration with his wife, the former Katharine Mary Dewar, who was the daughter of the Principal of Marischal College. The results confirmed Maxwell's conclusion that the viscosity is independent of the pressure, and he reported these results to the Royal Society in his Bakerian Lecture of 1866. One of his misgivings about the kinetic theory had therefore been resolved, but he was never able to reconcile the theory with the specific heats.

The statistical basis of the second law

Today we take it for granted that the second law of thermodynamics has a statistical basis, in that it applies because of the large numbers of molecules with which we are usually concerned. We tend to think that this is an obvious conclusion, but at first both Clausius and Boltzmann thought that the second law could be explained in terms of classical mechanics. Clausius continued to hold this opinion, but Boltzmann soon realized, after studying Maxwell's approach to the problem, that a statistical treatment is necessary, and he then made extremely important contributions from this point of view.

Scientific ideas usually come to light in formal scientific papers, but Maxwell's proposal was an exception. It took the form of an imaginary supernatural being, later called Maxwell's demon, which was born in a letter that Maxwell wrote to Peter Guthrie Tait on 11 December 1867. Maxwell and Tait were old friends, the two having been together at school, at the University of Edinburgh, and at Cambridge. Throughout their adult lives they carried on a lively and amusing correspondence—sometimes, since they were thrifty Scots, by means of post-cards which required only a halfpenny stamp rather than the penny stamp needed for a letter. Tait had written a book, *A Sketch of Thermodynamics*, which appeared in 1868, and before sending the manuscript to his publisher he asked Maxwell to read it and let him have comments. As was his habit (see Section 4.3), Tait had made a number of incorrect historical attributions in his desire to give credit to British (preferably Scottish) scientists. In his reply Maxwell remarked that he had not looked into the history and would not comment on Tait's priority allegations; one wonders if perhaps he was suspicious of them but did not want to get into an argument. Maxwell said that what he was prepared to do was 'in the way of altering the point of view here and there for clearness or variety, and picking holes here and there to ensure strength and stability'. The most important hole he picked was in relation to the significance of the second law, by showing how, in principle, it could be violated.

Maxwell considered a vessel divided into two compartments A and B, separated by a partition which had a hole in it that could be opened or closed by 'a slide without mass'. The gas in A was at a higher temperature than the gas in B, and Maxwell realized from his work on velocity distributions that each gas would contain molecules having velocities covering a range of values. 'Now', wrote Maxwell, 'conceive a finite being who knows the paths and velocities of all the molecules by simple inspection but who can do no work except open a hole in the diaphragm by means of a slide without mass'. This creature would open the hole for an approaching molecule in A when its velocity was less than the root mean square velocity of the molecules in B. He or she would allow a molecule from B to pass through the hole into A only when its velocity exceeded the root mean square velocity of the A molecules. These procedures were to be followed so that the number of molecules would remain the same in both A and B. As a result of this process, said Maxwell, 'the energy in A is increased and that in B is diminished; that is, the hot system has got hotter and the cold colder and yet no work has been done, only the intelligence of a very observant and neat-fingered being has been employed'. If we could do this, the second law would be violated but, Maxwell added, 'only we can't, not being clever enough'. It is interesting to note that in his comment about the employment of 'a very observant ... being', Maxwell was foreshadowing modern information theory. Later Maxwell emphasized that his intention had been 'to show that the 2nd law of thermodynamics only has a statistical certainty'. In a letter to J. W. Strutt (later Lord Rayleigh), written in December, 1870, Maxwell commented that

The 2nd law of thermodynamics has the same degree of truth as the statement that if you throw a tumblerful of water into the sea, you cannot get the same tumblerful out again.

Maxwell's idea of a doorkeeper who could bring about a violation of the second law was soon taken up by others. It was Kelvin who first referred to the creature as a demon, and he later playfully endowed it 'with arms and legs—two hands and ten fingers suffice'. The word 'demon' was at that time understood in its classical sense of a being, perhaps an angel, that was intermediate between a god and a man; we now think of a demon as an evil spirit, but there was no suggestion that Maxwell's demon was of this type!

Maxwell himself wrote no major paper on thermodynamics. His ideas on the subject are mainly known to us through his correspondence with Tait and others, through his lectures at Cambridge, and through his *Theory of Heat*, the first edition of which appeared in 1871—the year in which Maxwell somewhat reluctantly became the first Cavendish professor at Cambridge. This book was one of a series described by the publisher as 'textbooks of science adapted to the use of artisans and of students in public and science schools'. One wonders what many of these intended readers made of Maxwell's expositions of some of the latest contributions to thermodynamics, including Willard Gibbs's treatments. In the various editions of this book Maxwell explained his ideas on the statistical basis of the second law, and he made it clear that some processes that are consistent with the laws of mechanics are nevertheless impossible, as expressed by the restrictions of the second law.

It is interesting to note that the earlier discussions of the second law made no reference to a process that today is often used as a good example of the application of the law, namely the mixing of gases at constant temperature. Maxwell's demon could reverse such a process; air, for example, could be separated into oxygen and nitrogen by the use of a demon who would allow only oxygen molecules to pass in one direction and only nitrogen molecules to pass in the other. The earlier workers thought of the second law as relating only to the passage of heat from a higher temperature to a lower one, and an isothermal mixing process is not obviously related to such transfers of heat. It was not until

Ludwig Boltzmann

(1844–1906)

Boltzmann was born in Vienna and attended the University of Vienna where his teachers included Josef Stefan (1835–1893) and Josef Loschmidt (1821–1895). He received his doctorate in 1867, and became professor of theoretical physics at the University of Graz in 1869. Boltzmann had a rather unhappy and discontented disposition, and he made a number of moves during his career. In 1873 he became professor of mathematics at the University of Vienna, but in 1876 returned to the University of Graz as professor of experimental physics—rather inappropriately, as his work was entirely theoretical. From 1889 to 1893 he was professor of theoretical physics at the University of Munich. In 1894 he returned to the University of Vienna as professor of theoretical physics but in 1900, being unhappy at the University of Vienna, he accepted Ostwald's invitation to teach at the University of Leipzig. Soon, however, he decided that he preferred Vienna after all and returned there in 1902 to succeed himself as professor of theoretical physics. At Vienna he also lectured on the philosophy of science, replacing Ernst Mach who had retired for reasons of health; since Mach did not believe in the reality of atoms and took strong exception to Boltzmann's entire approach, this replacement brought about a profound change in the teaching of the philosophy of science at that University!

Boltzmann's early work on the second law of thermodynamics is mentioned in Section 4.3, and his great contributions to the distribution of energy and to molecular statistics are described in Section 5.3. In 1866, while still a student, he attempted to explain the second law on the basis of mechanics, but two years later, having read Maxwell's papers, he realized that this approach was incorrect and that a statistical treatment was necessary. In an important paper which appeared in 1868 he extended Maxwell's treatment of the distribution of molecular speeds and showed that the probability that a system is in a state having energy E is proportional to $\exp(-E/k_BT)$, where k_B is now called the Boltzmann constant. In his mathematical

Boltzmann had related an entropy increase to an increase in probability that mixing became recognized as governed by the second law.

The conclusions that Maxwell had reached about the statistical nature of the second law were in marked contrast to the opinions of Clausius, and of Boltzmann in his earlier work. The Austrian physicist Ludwig Boltzmann (1844–1906) began to make important contributions to kinetic theory while still a student at the University of Vienna, his interest having been inspired by two of his teachers, Josef Stefan (1835–1893) and Johann Josef Loschmidt (1821–

procedures he somewhat anticipated quantum theory by regarding the energy as made up of small packets, which at the end of his derivation he took to be of zero size.

In an important paper of 1872 Boltzmann developed the theory of the approach of systems to equilibrium, and of transport processes in gases. His theorem on the approach to equilibrium has become known as Boltzmann's H theorem. In later papers he dealt with the so-called reversibility and recurrence paradoxes, and with the ergodic hypothesis (Section 5.3). In an important paper of 1871 Boltzmann showed how all the properties of a system can be calculated from his distribution function, and this can be regarded as the birth of the science to which Willard Gibbs later gave the name statistical mechanics. In 1877 he gave his famous relationship (eqn 5.26) between entropy and probability.

Boltzmann's papers on thermodynamics and statistics, for which he is now chiefly remembered, form only about half of his publications. He also made important contributions to a wide range of topics in mathematics, physics, chemistry and philosophy. He had a particular interest in Maxwell's electromagnetic theory.

Although Boltzmann and Ostwald were on good terms personally, their philosophies of science were completely different. Ostwald's views were similar to those of Mach; he considered Boltzmann's statistical treatment based on kinetic theory to be of no value at all, even as a hypothesis, and he gave them no support in his influential books. Boltzmann found that he was completely deserted by his Continental colleagues, his principal supporters being in England. This realization led to his increasing despondency, several unsuccessful suicide attempts, and his final successful suicide in 1906, when he was on holiday at Duino, near Trieste.

References: S. G. Brush, DSB, 2, 260–268 (1970); Brush (1964, 1967, 1976).

1895), and later by Maxwell's publications. The first paper of any significance that Boltzmann wrote, in 1866 when he was 22 and a year before he obtained his doctorate, was entitled 'On the mechanical meaning of the second law of thermodynamics'. He started by commenting that the first law had been known for a long time, and for a man of 22 about 15 years can seem a long time. He noted that the second law could only be established by 'roundabout and uncertain methods', and said that his object was to give a 'purely analytical, completely general, proof of the second law of thermodynamics, as well as to discover the theorem in mechanics that corresponds to it'. His attempts to do so were by no means satisfactory, as he himself soon realized.

Clausius also made many efforts to explain the law as a necessary consequence of the principles of mechanics. In 1862 he introduced a concept which he called 'disgregation', defined as 'the degree to which the molecules of a body are dispersed'. In 1870 and 1871 he also published papers in which he claimed to have provided a mechanical explanation of the second law, the title of his second paper being 'On the reduction of the second law of thermodynamics to general mechanical principles'. This paper was at once followed by one from Boltzmann, then at the University of Graz, pointing out that his 1866 paper included essentially the same treatment as that given by Clausius, and concluding with the rather sardonic comment:

I can only express pleasure that an authority with Herr Clausius's reputation is helping to spread the knowledge of my work in thermodynamics.

In a paper of 1872 Clausius graciously conceded Boltzmann's claim, explaining that 'extraordinary demands' on his time, due to moves from Zürich to Würzburg and then to Bonn, had made it difficult for him to follow the scientific literature. In any case these attempts were unsatisfactory, as shown by Maxwell and others, including Boltzmann himself. Maxwell pointed out in particular that with the idea of disgregation Clausius was employing a concept that can have no experimental or theoretical significance.

Letters written by Maxwell to Tait after these mechanical theories had been put forward showed that Maxwell regarded them with some amusement, realizing that Clausius and Boltzmann—whom he referred to as 'these learned Germans'—had missed the point. For example, in a letter to Tait dated 1 December 1873, Maxwell wrote

It is rare sport to see these learned Germans contending for the priority of the discovery that the 2nd law of $\Theta\Delta$cs [thermodynamics] is the Hamiltonische Prinzip.

This letter was signed dp/dt, which was a private joke between Maxwell and Tait, who had used the equation $dp/dt = JCM$, the latter being Maxwell's initials; for a detailed explanation of this rather subtle and contrived joke see Klein's article on the Maxwell demon cited at the end of this chapter. Maxwell never attacked Clausius and Boltzmann directly, but made his own position clear in his *Theory of Heat*.

For reasons that are difficult to understand, Clausius never accepted Maxwell's statistical interpretation of the second law. Boltzmann's later attitude, however, was quite different. In 1868, two years after his unsuccessful attempt to explain the second law on purely mechanical grounds, he published a paper of great importance in which he extended Maxwell's theory and treated the distribution of energy. In his 1866 paper he had made no mention of Maxwell's work, but he was then learning English, particularly with a view to studying Maxwell's papers on the electromagnetic theory, and by the time he wrote his 1868 paper he had also become familiar with Maxwell's theory of the distribution of velocities. He extended it by considering the situation in which one of the particles in a system is at a potential energy V. By applying Maxwell's procedures he then deduced that the fraction $f(u)$ of molecules having speeds between u and $u + du$ is

$$f(u) = \text{constant} \times e^{-j(mu^2/2 + V)}. \tag{5.23}$$

The constant j could be related to the absolute temperature by comparing the theoretical pressure with that given by the experimental gas laws, and is found to be $1/k_BT$, where k_B, now called the *Boltzmann constant*, is the gas constant R divided by the Avogadro constant, i.e., is the gas constant per molecule. Equation 5.23 may therefore be written as

$$f(u) = \text{constant} \times e^{-(mu^2/2 + V)/k_BT} \tag{5.24}$$
$$= \text{constant} \times e^{-E/k_BT}, \tag{5.25}$$

where E, equal to the kinetic energy $\frac{1}{2}mu^2$ plus the potential energy V, is the total energy of the system. This general result, now referred to as the *Boltzmann distribution*, provides the basis of the subject of statistical mechanics.

In the same 1868 article Boltzmann derived the Maxwell distribution law in another way. He made the assumption that the total amount of energy is distributed among molecules in such a manner that all combinations of energies are equally probable. Entirely for mathematical convenience, he regarded the energy as composed of a large number of tiny packets of energy, and he applied the methods of combinatorial analysis. The expression he finally obtained reduced to Maxwell's velocity distribution expression if the packets of energy were taken to be infinitesimally small. This approach is interesting since it foreshadowed Planck's quantum theory, according to which there really are small packets or quanta of energy. In fact, Planck made good use of many of Boltzmann's mathematical procedures in developing his quantum theory (Section 10.1).

Another important problem dealt with by Boltzmann was the way in which systems approach a state of equilibrium, his major contribution on this subject appearing in 1872. Approach to equilibrium had been considered from two different points of view: in terms of the dissipation of energy, following Kelvin, or in terms of an increase of entropy, as considered by Clausius. Boltzmann's important contribution in this paper was to show in detail how an increase of entropy corresponds to increasing molecular randomness. His treatment of the problem involved the introduction of a function, now given the symbol H, which involves Maxwell's distribution function. Boltzmann showed that H is bound to decrease with time unless it has the Maxwellian form, in which case it has already reached its minimum value and remains constant. The importance of the function H is that it extends the definition of entropy to include non-equilibrium states which were not covered by the thermodynamic definition.

The theorem was originally referred to as 'Boltzmann's minimum theorem', but is now usually known as 'Boltzmann's H theorem'.

Soon afterwards Boltzmann dealt with another problem, the so-called 'reversibility paradox' which had been raised by Kelvin in a paper that appeared in 1874. Kelvin maintained that there is a contradiction between one of the basic premises of Boltzmann's H theorem, namely the reversibility of individual collisions, and the irreversibility predicted by the theorem for the approach to equilibrium of a system not initially at equilibrium. Kelvin argued that if one considered the passage of a system towards equilibrium, one could then imagine all the motions to be reversed; the process would then be occurring in the opposite direction, away from equilibrium, with equal probability. Boltzmann's response to this, in a paper that appeared in 1877, was that there is a vast number of possible equilibrium states, and that for only a tiny fraction of these would it be possible for the system to move away from equilibrium. He admitted that one could chose states that would move away from equilibrium with an increase in H and a decrease in entropy. However, such a situation would be highly unlikely; there are vastly more equilibrium states that would *not* move away from equilibrium.

A related problem, referred to as the 'recurrence paradox', arose from a theorem in mechanics first propounded in 1890 by the French physicist and philosopher Jules Henri Poincaré (1854–1912). According to this theorem, any system must eventually reach any specified configuration. Poincaré, and later the German mathematician and physicist Friedrich Ferdinand Zermelo (1871–1953)—whom Boltzmann often referred to as 'dieser Helunke' (that rascal)—argued that therefore the H theorem cannot always be valid, since a system at equilibrium can pass to any other state. Boltzmann's answer to this challenge, published in 1896, was similar to his answer to the reversibility paradox. On the molecular scale the equilibrium state consists of a vast number of possible configurations. It is true that in principle a particular chosen state can recur if one waits long enough, but one would have to wait an immensely long time. Thus, although the H theorem could in principle be violated, such a violation is highly unlikely.

Although Boltzmann had made great contributions to kinetic theory he became increasingly despondent, partly as a result of the strong attacks that were made towards the end of the nineteenth century on the real existence of atoms. These attacks, made particularly by Ernst Mach and Wilhelm Ostwald, struck at the roots of Boltzmann's entire life's work. He ably defended the atomic theory against its critics, but his distress became more acute and he made several unsuccessful suicide attempts. Finally, in 1906, on holiday with his wife and daughter and while they were out swimming, he hanged himself from the window of his hotel room. It is a sad irony that when soon afterwards Jean Perrin produced evidence for atoms that convinced even Ostwald (Section 9.1), his work was greatly influenced by Boltzmann's own contributions to the statistics of molecular motions.

5.4 Statistical mechanics

The expression 'statistical mechanics' was first used by Willard Gibbs in the title of his book *Elementary Principles in Statistical Mechanics Developed with Special Reference to the Rational Foundations of Thermodynamics*, which was published in 1902. However, the birth of statistical mechanics can reasonably be set at 1871, for in that year Boltzmann published a paper of great importance in which he showed how his distribution function allowed all of the properties of a system to be calculated. This is the problem with which statistical mechanics is mainly concerned. The methods Boltzmann used in that 1871 paper are essentially those commonly used today. Similar matters were treated in two papers Boltzmann published in 1877. In one of these he gave his famous relationship between entropy and probability, W:

$$S = k_B \ln W. \tag{5.26}$$

In this equation, which is engraved on Boltzmann's tombstone in Vienna, W is the number of possible molecular configurations corresponding to a given state of the system. These molecular configurations are now referred to as *microstates* and are defined with respect to finite cells in phase space. This equation allows an expression for the entropy to be obtained from the statistical distribution, and from this expression the other thermodynamic properties can be calculated.

Boltzmann based his treatment on the assumption that all microstates corresponding to the same total energy have the same a *priori* probability. In support of this assumption Boltzmann and Maxwell introduced what later came to be known as the *ergodic hypothesis*, according to which systems pass with equal probability into all states. There has been considerable confusion about what precisely Boltzmann and Maxwell meant by this hypothesis, and later writers have often used the expression in a sense different from what was originally intended; for a detailed discussion the reader is referred to an article by Brush (1967–8). The ergodic hypothesis led to considerable controversy, which came to a head in 1911 when Paul Ehrenfest (1880–1933) and his wife Tatyana (1876–1964) published an article in which they argued that ergodic systems are probably non-existent. They conceded, however, that 'quasi-ergodic' systems, which pass 'as close as one likes' to every possible state, might well be found. In any case, few physicists today are much concerned with the ergodic hypothesis, although ergodic theory has become an active branch of pure mathematics.

More recent treatments of systems at equilibrium by the methods of statistical mechanics often do not quite follow Boltzmann's original methods, but instead employ a function called the *Zustandsumme*, which was first used by Max Planck. This German word has been translated as 'sum over states' and more often and less appropriately as 'partition function'.

The idea of the partition function comes at once from Boltzmann's distribution law, according to which the number n_i of molecules having energy E_i is proportional to $\exp(-E_i/k_BT)$ where k_B is the Boltzmann constant:

$$n_i = A \exp(-E_i/k_BT), \qquad (5.27)$$

where A is a constant. If N is the total number of molecules in a system,

$$N = \Sigma\, n_i = A\, \Sigma\, \exp(-E_i/k_BT), \qquad (5.28)$$

and therefore

$$\frac{n_i}{N} = \frac{\exp(-E_i/k_BT)}{\Sigma\, \exp(-E_i/k_BT)}. \qquad (5.29)$$

The denominator in this expression is called the partition function, today usually given the symbol q when a single particle is concerned:

$$q = \Sigma\, \exp(-E_i/k_BT). \qquad (5.30)$$

When an assembly of particles is considered, the partition function is usually denoted as Q. If an expression has been obtained for Q as a function of volume and temperature, the thermodynamic properties can all be calculated.

An important refinement introduced by Gibbs in his 1902 book on statistical mechanics was the concept of a *statistical ensemble*. An ensemble is a large number of systems, and Gibbs introduced the concept with a view to illustrating fundamental principles; later workers have made good use of ensembles in computational techniques.

The advantage of using an ensemble is that the energies of individual atoms can be regarded as independent variables. Statistical averaging can be carried out by considering a range of energies from zero to infinity, the probability of a system having an energy E being proportional to $\exp(-E/k_BT)$. A collection of all of the systems was called by Gibbs a *canonical ensemble*. A collection of systems all having the same energy is called a *microcanonical ensemble*. Gibbs showed that a suitably chosen microcanonical ensemble leads to the same thermodynamical properties as a canonical ensemble.

Another important concept introduced by Gibbs is the *grand canonical ensemble*, in which the number of particles is allowed to vary as well as the energy. This concept has proved particularly useful in dealing with chemical equilibrium among several substances, since it allows the calculation of the chemical potential for each component of the system.

The most important of the early contributions to statistical mechanics were those of Boltzmann and Gibbs, but mention should also be made to some now largely forgotten work of Einstein. The third paper he published, in 1902 when he was 23 years old, was on the statistical basis of thermodynamics, and he later published a number of other papers on similar problems. He was much interested in Boltzmann's H theorem and the ergodic hypothesis. Important developments in statistical mechanics were also made by the American physicist Richard Chase Tolman (1881–1948) whose book *The Principles of Statistical*

Mechanics, published in 1938, was a classic that exerted a wide influence. Many applications of statistical mechanics were also made in *Statistical Mechanics* by the British physicist Ralph Howard Fowler (1889–1944) and in *Statistical Thermodynamics* by Fowler and the British physical chemist Edward Armand Guggenheim (1901–1970).

An important matter discussed by Tolman is the *principle of microscopic reversibility at equilibrium*. His statement of this principle was as follows:

In a system at equilibrium any molecular process, and the reverse of that process, occur on the average at the same rate.

A related principle, referred to as the *principle of detailed balance at equilibrium*, was put forward by Fowler in his book, and may be stated as follows:

In a system at equilibrium each collision has its exact counterpart in the reverse direction, so that at equilibrium the rate of every chemical process is exactly equal to the rate of the reverse process.

These principles are of great value in dealing with chemical reactions, but serious errors can arise if one applies them to systems that are not at equilibrium.

Suggestions for further reading

For accounts of the discovery of the gas laws see
I. Bernard Cohen, Newton, Hooke, and 'Boyle's Law' (discovered by Power and Towneley), *Nature*, **204**, 618–621 (1964).
J. B. Conant, Robert Boyle's experiments in pneumatics, in *Harvard Case Histories in Experimental Science* (Ed. J. B. Conant, Harvard University Press, 1957, pp. 1–63.
C. Webster, The discovery of Boyle's law, and the concept of the elasticity of air in the seventeenth century, *Arch. Hist. Exact Sci.*, **2**, 441–502 (1965).

For accounts of the emergence of atomic theories and their reception, see
W. H. Brock and D. M. Knight, The atomic debates; 'Memorable and interesting evenings in the life of the Chemical Society', *Isis*, **56**, 5–25 (1965).
W. H. Brock (Ed.), *The Atomic Debates. Brodie and the Rejection of Atomic Theory*, Leicester University Press, 1967.
E. H. Hiebert, The energetics controversy and the new thermodynamics, in *Perspectives in the History of Science and Technology* (Ed. D. H. D. Roller), University of Oklahoma Press, Norman, Oklahoma, 1971, pp. 67–86. This article discusses Ostwald's opposition to the atomic theory, and Boltzmann's attack on him at Lübeck.
R. H. Silliman, William Thomson: Smoke rings and nineteenth-century atomism, *Isis*, **54**, 461–474 (1963).
A. W. Thackray, The origins of Dalton's chemical atomic theory: Daltonian doubts resolved, *Isis*, **57**, 35–55 (1966).

A. W. Thackray, The emergence of Dalton's atomic theory, 1801–1808, *British J. Hist. Sci.*, **3**, 1–23 (1966).
A. W. Thackray, *Atoms and Powers*, Harvard University Press, 1970.

There are many publications dealing with the kinetic theory of gases, for example
S. G. Brush, *Kinetic Theory, Vol. 1: The Nature of Gases and of Heat. Vol. 2: Irreversible Processes*, Pergamon Press, Oxford, 1965.
S. G. Brush, *The Kind of Motion we Call Heat: A History of the Kinetic Theory of Gases in the 19th Century*, North-Holland Publishing Company, Amsterdam, 1976.
S. G. Brush, *Statistical Physics and the Atomic Theory of Matter: From Boyle and Newton to Landau and Onsager*, Princeton University Press, 1983.
Elizabeth W. Garber, Clausius's and Maxwell's kinetic theories of gases, *Hist. Stud. Phys. Sci.*, **2**, 299–319 (1972).
Elizabeth W. Garber, Aspects of the introduction of probability into physics, *Centaurus*, **17**, 11–39 (1972).
Elizabeth W. Garber, Molecular science in late nineteenth century Britain, *Hist. Stud. Phys. Sci.*, **9**, 265–297 (1978).
P. M. Harman, *Energy, Force and Matter*, Cambridge University Press, 1982.
Martin J. Klein, Maxwell, his demon, and the second law of thermodynamics, *American Scientist*, **58**, 84–97 (1970). This article explains, in an Appendix, the significance of dp/dt, which Maxwell sometimes used as his signature. By a somewhat contrived argument Maxwell and Tait used the equation $dp/dt = JCM$ as a way of expressing the second law of thermodynamics; t is temperature, not time.
Martin J. Klein, Mechanical explanation at the end of the nineteenth century, *Centaurus*, **17**, 58–82 (1972).
N. G. Parsonage, *The Gaseous State*, Pergamon Press, Oxford, 1966.
R. D. Present, *Kinetic Theory of Gases*, McGraw-Hill, New York, 1960.
C. Truesdell, *Essays on the History of Mechanics*, Springer-Verlag, Berlin, 1969, especially Chapter VI, Early kinetic theories of gases, pp. 278–394.

For accounts of the history of statistical mechanics see
S. G. Brush, Foundations of statistical mechanics, *Arch. Hist. Exact Sci.*, **4**, 145–183 (1967–8).
S. G. Brush, *The Kind of Motion we Call Heat* (see above), pp. 583–615.
J. Mehra, Einstein and the foundations of statistical mechanics, *Physica*, **79A**, 447–477 (1975).
L. Rosenfeld, On the foundations of statistical mechanics, *Acta Physica Polonica*, **14**, 3–39 (1955).

CHAPTER 6

Chemical spectroscopy

During the nineteenth century it became increasingly clear that spectroscopy has important chemical applications. Substances could often be identified by their spectra, and later in the century spectroscopy played an important role in the discovery of new elements. Spectra also provided valuable information about the nature of atoms and molecules, although sorting out the evidence was unexpectedly difficult—indeed impossible until the quantization of energy had been recognized. Some of the more important contributions to spectroscopy, especially in its chemical aspects, are listed in Table 6.1.

The coloured spectrum has been known from early times; it was discussed in

TABLE 6.1   Highlights in chemical spectroscopy

Author	Date	Contribution	Section
Melvill	1752	Flame spectroscopy	6.1
Lambert	1760	Absorption law	6.4
W. Herschel	1800	Infrared	Introduction
Ritter	1801	Ultraviolet	Introduction
Fraunhofer	1817	Spectral lines	6.1
J. Herschel; Draper	1840	Photography of spectra	6.1
Stokes	1852	Fluorescence	6.1
Beer	1852	Absorption by solutions	6.4
Kirchhoff	1859	Principles of spectroscopy	6.2
Bunsen and Kirchhoff	1860	First spectroscope	6.1
Balmer	1885	Spectral series	6.5
Stoney	1888	Orbiting electrons	6.6
Rydberg	1890	Spectral series	6.5
Zeeman	1897	Magnetic effects	6.6
Bohr	1913	Atomic theory and spectra	10.1
Raman and Krishnan	1928	Raman spectra	6.7
Herzberg; Mulliken	1928 ff.	Quantum mechanics of spectra	6.7
Townes	1954	Masers	6.7
Mössbauer	1958	Recoilless emission of γ rays	6.7
Maiman	1960	Lasers	6.7
Kasper and Pimentel	1965	Chemical lasers	6.7

about 63 AD by the Stoic philosopher Lucius Annaeus Seneca (*c*.4 BC–64 AD). Newton's great contribution, first reported in the *Philosophical Transactions* in 1672, was that a prism splits white light into its constituent colours and that a second prism reconstitutes it into white light. In 1800 the great astronomer William Herschel (1738–1822) used thermometers with blackened bulbs to measure the heating powers of various parts of the spectrum (Figure 6.1). He found a steady increase in heating power from the violet to the red, and also noticed that there was heating even when the bulb was placed in the dark region beyond the red. In this way he discovered the infrared spectrum. Herschel was a refugee from Germany who first earned his living in England as an organist, music teacher and composer; he became interested in astronomy, began to construct telescopes, and made important contributions to spectroscopy.

The ultraviolet region was discovered in 1801 by the German scientist Johann Wilhelm Ritter (1776–1810), who observed the blackening of silver chloride brought about by radiation beyond the violet end of the spectrum. A year later the English chemist William Hyde Wollaston (1766–1828) independently obtained evidence for the ultraviolet spectrum. He also observed something that Newton had unaccountably missed—or at any rate had not reported—dark lines in the spectrum of sunlight passed through a prism. Wollaston mistakenly regarded these lines as boundaries between the coloured bands, and he believed that there were four colours in the spectrum. In 1817 these dark lines were rediscovered and more satisfactorily interpreted by Joseph

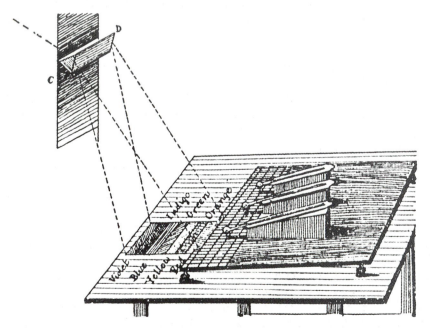

FIGURE 6.1 William Herschel's diagram of the arrangement he used to study the heating powers of various regions of the spectrum (Herschel, 1800).

von Fraunhofer (1787–1826), after whom they are now named. Fraunhofer was a Bavarian lens manufacturer who studied the refractive indices of glasses of various colours. In doing so he saw a pair of bright yellow lines in the spectrum of the flame he was using. He then passed sunlight through a prism and observed a number of dark lines. In later experiments he saw lines in the spectra of the moon, the planets and the stars. He observed that the lines found with light from various sources were not always in the same parts of the spectrum. This and other evidence convinced him that the lines were not due to his instrumentation but were characteristic of the light he used and of the substances through which it passed. To designate the more prominent spectral lines Fraunhofer suggested the use of capital letters from A to H, A being in the red and H in the violet, and this notation is often used today; the lines in the yellow region of the spectrum, now known to be due to sodium, are called D lines. Fraunhofer's fruitful investigations were interrupted by his early death.

6.1 Spectral analysis

Even before these fundamental studies had been made, a few chemists had used coloured lines in spectra for the identification of materials placed in flames. In 1752 the Scotsman Thomas Melvill (1726–1753), who had studied divinity at the University of Glasgow, studied the flame spectra of various salts, and is regarded as the founder of the technique of flame spectroscopy; his promising investigations were cut short by his early death in Geneva. Similar identifications were later made by two of the great pioneers of photography, Sir John Herschel (1792–1871) and Henry Fox Talbot (1800–1877).

An important contribution was made in 1834 by the Scottish physicist David Brewster (1781–1868) who did much work in optics and other branches of physics; in 1816 he achieved some popular fame with his invention of the kaleidoscope. He observed the Fraunhofer lines when white light is passed through various gases, and suggested for the first time a method of chemical analysis in which 'substances might be characterized by their action on different parts of the spectrum'. His studies were extended by John Frederic Daniell (1790–1845), who in 1831 had become the first professor of chemistry at King's College, London. Some of his spectroscopic work was done in collaboration with William Allen Miller (1817–1872) who succeeded him as professor at King's College.

In 1852 fluorescence was discovered by George Gabriel Stokes (1819–1903), who from 1849 until his death 54 years later was Lucasian professor at Cambridge—a position that Newton had held. He observed that a solution of quinine sulphate, when viewed by transmitted light, has a blue colour, and he showed that this is due to the absorption of ultraviolet radiation, the light emitted being of a lower frequency; this has come to be called Stokes's law of radiation. In his honour, lines of longer wavelength (lower frequency) than that of the incident radiation in modern Raman spectroscopy (Section 6.7) are known as Stokes lines; those of shorter wavelength are called anti-Stokes lines.

Robert Wilhelm Eberhard Bunsen

(1811–1899)

Bunsen was born in Göttingen, where his father was professor of modern languages, and received his doctorate from the university of Göttingen in 1830, his thesis being on a problem in physics. From 1830 to 1833 he travelled widely on the continent of Europe and met a number of prominent scientists, including Liebig in Giessen, and Gay-Lussac, in whose Paris laboratories he worked for a period. In 1833 he became *Privatdozent* at the University of Giessen, and in 1836 succeeded Friedrich Wöhler as professor at the Polytechnic School in Kassel. From 1838 to 1851 he was professor of chemistry at the University of Marburg. For some time in 1851 he was at the University of Breslau, where he became acquainted with Gustav Kirchhoff, with whom he later did important work in spectroscopy. In 1852 Bunsen was made professor at the University of Heidelberg, where he remained for the rest of his life, retiring from his professorship in 1889 at the age of 78.

Bunsen was an extremely conscientious teacher and became completely immersed in his research. He gave many lectures, and attracted to his research laboratories many who later became distinguished scientists; among them are Adolph von Baeyer, Dmitri Mendeleev, Lothar Meyer, Hermann Kolbe, Edward Frankland, William Hillebrand, and Henry Roscoe. Bunsen had very broad scientific interests, and paid much attention to geology, joining a number of geological expeditions and carrying out analyses of geological materials. He never married; there is a story that he once proposed to a young lady and was accepted, but then became so immersed in his research that he forgot about the matter and later proposed to her again, this time meeting with a refusal.

Bunsen's earliest research was in inorganic chemistry, and between 1837 and 1843 he worked on problems of organic chemistry, particularly the arsenic-containing cacodyl compounds. This work caused him to suffer from arsenic poisoning, and in 1843 an explosion of cacodyl cyanide caused him the loss of sight

The word 'fluorescence', which Stokes invented, has no etymological connection with light. It derives from the name of the fluorescent mineral fluorspar, the name of which comes from the Latin *fluor*, flowing, and the German *Spat*, which means a crystalline material or spar. Stokes developed a technique for studying ultraviolet light by causing it to produce fluorescence in a solution of a blue dye. This procedure was used for a time until it was replaced by photographic methods.

During the 1850s Stokes developed many of the physical principles of spectra

in his right eye. For the rest of his life his work was mainly in inorganic and physical chemistry.

Between 1838 and 1846 Bunsen carried out some rather practical work on industrial gases. From about 1840 he worked on electrochemical cells and brought about a number of improvements, particularly with the 'Bunsen battery' which used carbon as the negative electrode. He began to use electrochemical techniques to isolate pure metals, such as some of the rare-earth elements, and to examine their properties. He devised a sensitive ice calorimeter in connection with these investigations. During the 1850s he investigated, in collaboration with Henry Roscoe, the photochemical combination of hydrogen and chlorine (Section 8.3). At about the same time Bunsen began to study the combustion and explosion of gases, work that led to the famous burner that is named after him but which he perhaps did not invent.

In the 1860s Bunsen worked with Kirchhoff on spectroscopy, and invented a spectrometer that led to the discovery of new elements (Section 6.2).

Although he had a broad interest in many areas of science, Bunsen paid little attention to theories, his approach being largely intuitive and empirical. His students have reported that his lectures made no mention of Avogadro's hypothesis or the periodic system, even though two of his former students, Mendeleev and Lothar Meyer, had been greatly concerned with the periodic relationships. In this respect Bunsen differs from most of the leading workers in physical chemistry, but he nevertheless made many contributions that were important in the later theoretical development of the subject.

References: Susan G. Schacher, DSB, 2, 586–590 (1970); Oesper (1927, 1975); Roscoe (1900).

and of spectroscopic analysis, but much of this work remained unpublished. He did, however, communicate it to William Thomson (Kelvin), who presented it to his students at the University of Glasgow. Stokes claimed no priority for his work, always conceding to Bunsen and Kirchhoff the distinction of putting spectral analysis on a firm foundation.

Robert Bunsen (1811–1899) and Gustav Robert Kirchhoff (1824–1887) had worked together at the University of Breslau. In 1852 Bunsen became professor of chemistry at Heidelberg, where he began research on combustion

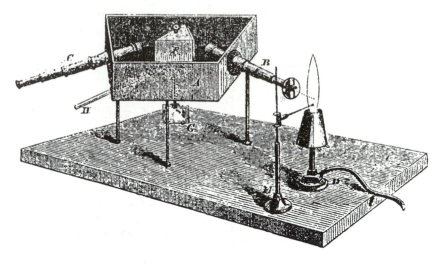

FIGURE 6.2   Bunsen and Kirchhoff's diagram of their spectroscope (Bunsen and Kirchhoff, 1860).

and flames, in the course of it using the famous laboratory burner which he did not perhaps invent but which now bears his name. In 1854 Kirchhoff joined Bunsen at Heidelberg as professor of physics, and suggested to him that his studies on the effects of salts on the colour of flames would be much improved by the use of a prism to produce the spectra. The two then collaborated on the construction of a spectroscope (Figure 6.2) which although simple was invaluable for its purpose.

Early developments in spectroscopy went hand in hand with the invention of photography and improvements in photographic techniques. Before it was possible to take photographs, spectroscopy had to be done entirely by visual observation, and this had serious disadvantages: there was available no permanent record which could be checked by others, and only the visible region of the spectrum could be studied. In the scientific journals the early spectra had to be presented as wood-cuts or etchings prepared from hand-drawn sketches, an example of which is shown in Figure 6.3. Other excellent examples of hand-drawn spectra were published in 1845 by William Allen Miller. Bunsen and Kirchhoff's first paper on their spectroscope, published in 1860, contains some tinted engravings of spectra.

The announcement in 1839 of two quite different photographic techniques was mentioned in Section 1.3. The technique of Daguerre in France had the advantage of giving rather sharp pictures, but the disadvantage that the photographs (daguerreotypes) were on a metal surface so that copies could not be obtained directly. The photographs obtained in England by the early methods devised by Henry Fox Talbot were not as well defined but were on paper, and

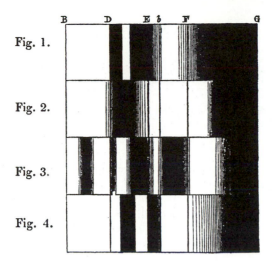

FIGURE 6.3 An example of a wood-cut prepared from hand-drawn sketches of spectra (Stokes, 1864). The spectra are of blood, under various conditions.

techniques were soon discovered which allowed any number of copies ('positives') to be obtained from a single 'negative'. The great advantages of photography in all aspects of life, including spectroscopy and other scientific fields, were at once recognized, and much effort went into improving the techniques. In France, for example, Fizeau succeeded by means of an electrochemical technique in converting a daguerreotype into an engraving, which could be used for the publication of spectra in scientific journals. This technique was not much used, however, since the alternative photographic and photomechanical techniques that evolved from Talbot's method eventually superseded Daguerre's technique.

In 1851, for example, Frederick Scott Archer (1813–1857) introduced a wet-plate process in which a film of collodion was attached to a plate of clear glass. Since there was now no paper base the images were much clearer, and this method soon became widely used both by scientists and by artistic photographers. Dry plates were introduced into photography in the 1870s, and were much more convenient than the wet plates. One manufacturer of dry plates was the American inventor George Eastman (1854–1932), who in about 1878 succeeding in attaching the light-sensitive emulsions to strips of paper, and later to celluloid, which could be rolled on spools. This new technique was also of great practical value to spectroscopists and indeed to all photographers.

A particular advantage of photography was that it allowed spectra to be studied outside the visible range of the spectrum; previously ultraviolet spectra had to be investigated by Stokes's technique of using a fluorescent solution or screen, and infrared spectra by their heating effects. Success in photographing

William Allen Miller

(1817–1870)

Miller was born in Ipswich, Suffolk, and received some of his education at a Quaker school, the influence of which was reflected in his conscientiousness and simplicity of character. At the age of 15 he was apprenticed for five years to his uncle, who was a surgeon at the Birmingham General Hospital. In 1837 he became a medical student at King's College, London, and was taught chemistry by J. F. Daniell. In 1870 Miller spent a few months in Justus Liebig's laboratory at Giessen, and on his return Daniell, who was much impressed by Miller's ability, appointed him demonstrator in chemistry. In 1841 Miller was appointed assistant lecturer to Daniell, and in the following year he became the first person from King's College to obtain the MD degree of the University of London. In collaboration with Daniell he carried out some work on electrolysis which was published in 1844. In their paper Daniell and Miller explained concentration changes in different parts of a solution in terms of different mobilities of cations and anions, a hypothesis that was later further developed by Hittorf (Section 7.2).

King's College, London, was controlled to a considerable extent by the Church of England, its Charter stipulating that 'no person who is not a member of the United Church of England and Ireland . . . shall be competent to fill any office in the College. . .'. When Daniell died in 1845 efforts were made to replace him by Liebig, but since he was a Lutheran the Archbishop of Canterbury and others refused their permission. Instead the appointment went to Miller, whose short stay with Liebig was apparently not considered to have tainted his ecclesiastical orthodoxy.

Miller's 25-year occupancy of the chair was a distinguished one. His experimental work, like that of Daniell, was carried out in an improvised laboratory established in a lumber room below the seats in the chemistry lecture theatre. Soon after taking up the professorship Miller began work on spectral analysis (Section 6.1), which became his main interest. In an important paper which appeared in 1845 he published for the first time detailed drawings of the spectra of a number of inorganic salts.

Miller was one of the pioneers in the application of photographic methods to spectroscopy. In the 1860s he published several papers which described photo-

in both the ultraviolet and infrared regions was achieved as early as 1840, the year following the first announcement of the techniques of photography. This important contribution was made by Sir John Frederick William Herschel, Bt. (1792–1871), who was the only child of Sir William Herschel; he was created a baronet in 1838. John Herschel was a scientist of great versatility

graphic techniques, involving quartz prisms and wet collodion plates, applied to the study of visible and ultraviolet spectra in the laboratory. In 1862 he entered into a collaboration with William (later Sir William) Huggins (1824–1910) who lived near him and had established a private observatory, at Tulse Hill, now in greater London but then in the country. Together they observed visually, and photographed, the spectra of the moon, Jupiter, Mars and a considerable number of stars, comparing them with spectra obtained in the laboratory. They concluded that chemically all of the stars have much in common with the sun. Their photographs of spectra were not very satisfactory, because of the inadequacy of the clockwork siderostat that they were using to keep the telescope pointing at the stars during the long exposures; by 1880, after Miller's death, Huggins had obtained a superior siderostat and was able to take excellent astronomical photographs.

Miller's important and influential book, *Elements of Chemistry: Theoretical and Practical*, in three parts, began to appear in 1855, the first volume having the subtitle *Chemical Physics*. During the first years of his professorship Miller continued to use Daniell's *Introduction to the Study of Chemical Philosophy*, and was reluctant to prepare his own book because of loyalty to his predecessor. In his book he said that in many places he had left untouched 'the work of his late master'.

Miller was one of the founders of the Chemical Society of London, and was twice its President. He was deeply religious, of exemplary character and of great sensitivity. On one occasion he was appearing as an expert witness in a court of law, and when the Judge made a remark that appeared to cast doubt on his veracity he fainted and had to be carried out of the court. In 1870 the President of the British Association for the Advancement of Science was Thomas Henry Huxley (1825–1895), an avowed agnostic (a word he coined himself), and Miller planned to attend the Liverpool meeting of the Association in order to combat Huxley's views. However, he was taken ill on the journey, and was unable to attend the meetings, dying of apoplexy in Liverpool two weeks later.

Reference: J. D. North, DSB, 9, 391–392 (1974); 'C. T.' (1871); Hey (1955).

and originality, now chiefly remembered for his work in astronomy, but he did important work in other fields and in particular was a highly competent chemist; in 1815, in fact, he had missed by one vote being elected professor of chemistry at Cambridge. He had made some earlier studies of photography, mentioned in Section 8.4, and on March 5, 1840, he presented to the Royal Society a pioneering paper entitled 'On the chemical action of the rays of the solar spectrum on preparations of silver and other substances, both metallic and non-metallic; and on some photographic processes'. This paper described a very detailed investigation of the photographic effects of different spectral regions, including the ultraviolet and infrared. A five-page summary of the work appeared in the *Proceedings of the Royal Society*, and a detailed 60-page account in the *Philosophical Transactions* of 1840. Herschel reported further work along the same lines in subsequent publications. In an important paper that appeared in 1842 Herschel reported success with colour photography, and coloured photographs he took of the solar spectrum are held by the Royal Society.

John William Draper (1811–1882) was also successful in photographing the solar spectrum in about 1842. He was born in England but was then teaching chemistry and physiology at New York University, at the same time carrying out many important investigations particularly relating to optics. The technique he used for photographing spectra was daguerreotypy, and a drawing of a daguerreotype of the solar spectrum appears in a paper he published in the *Philosophical Magazine* in 1843. He was one of the first to show that Fraunhofer lines exist outside the visible region of the spectrum. One of the monochrome plates he made at the time, which came into the possession of Sir John Herschel, is now in the National Museum of Photography in Bradford, England. In this 1843 paper he also reported having taken a daguerreotype of a spectrum taken using a diffraction grating, which had been ruled for him by the United States Mint. This was probably the first photograph taken of a diffraction spectrum.

Another to have been highly successful in photographing spectra was Alexandre Edmond Becquerel (1820–1891), who obtained his doctorate from the Université de Paris in 1840 and had been interested in Daguerre's invention from the time of its first announcement in 1839, when he was only 19 years of age. By 1843 he had made an excellent daguerreotype of the solar spectrum, and an engraving made from it was later published in his book *La Lumière, ses causes et ses effets* (Light, its causes and effects), which appeared in 1867. This engraving is reproduced as Figure 6.4; the visible region includes prominent Fraunhofer lines designated A to H, while lines L to R and many others are in the ultra-violet. By 1845 Becquerel had recorded coloured solar spectra on daguerreo-type plates; examples still survive at the Conservatoire des Arts et Métiers in Paris and at the Science Museum in London, and hand-tinted engravings made by Becquerel from his daguerreotype plates are also reproduced in *La Lumière*. To produce the coloured daguerreotypes he built up layers of various chlorides and other chemicals on the plates, his technique being described in a paper in the *Comptes rendus*. Becquerel later became professor of physics at the Con-servatoire and then became Director of the Musée d'Histoire Naturelle, succeeding his father Antoine César Becquerel (1788–1878) who had done important early work in electrochemistry; since Antoine remained active to a ripe old age and died only 13 years before his son Edmond he was able to enjoy his son's successes. Edmond's own son, Antoine Henri Becquerel (1852–

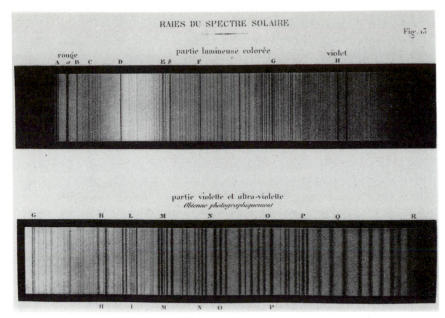

RAIES DU SPECTRE SOLAIRE

Fig. 15

rouge
A *a* B C   D   E *b*   F   G   H

partie lumineuse colorée

violet

partie violette et ultra-violette
Obtenue photographiquement

G   H   L   M   N   O   P   Q   R

H   I   M   N   O   P

FIGURE 6.4 An engraving drawn from a daguerreotype of the solar spectrum made in 1843 by Edmond Becquerel and shown in his 1867 book *La lumière, ses causes et ses effets*. Many Fraunhofer lines are to be seen in the ultraviolet.

1908), shared with the Curies the 1903 Nobel Prize for physics, for his discovery of radioactivity.

Photography of ultraviolet spectra was also carried out later by William Allen Miller at King's College, London; in 1862 he published some excellent photographs of the ultraviolet spectra of a number of metallic salts.

Another important development in spectroscopy was the introduction of diffraction gratings, which give better resolved spectra than prisms. As early as 1823 Fraunhofer had ruled some gratings, but they were not very effective. Mention has already been made of Draper's use of a grating in 1843. The main pioneer of gratings was the American physicist Henry Augustus Rowland (1844–1901), who towards the end of the 1870s ruled gratings of high quality with lines much closer together than previously possible; he achieved nearly 15 000 lines per inch. His gratings brought about a considerable improvement in the quality of spectra.

One troublesome matter in early spectroscopy was the chemical purity of the materials used. There was particular interest in, and for a time much confusion about, the two closely-spaced yellow D lines now known to be due to sodium. They were found not only in the flame spectra of sodium salts, but also with many other salts and with substances such as wood. Since water was a common ingredient in such substances it was understandably thought that the D lines

John William Draper

(1811–1882)

Draper was born in St Helens, Lancashire, the son of a Methodist preacher. In 1829 he began premedical studies at the University of London (later called University College) on Gower Street. At an early age Draper became interested in the chemical effects of light, which was his main interest throughout his life. At the time Oxford and Cambridge held the monopoly of granting degrees in England, and Draper was granted a 'certificate of honours in chemistry' instead of a degree.

In 1832 Draper left England for Virginia, accompanied by his new wife, his mother and his three sisters. In the farmhouse in which they resided he established a primitive laboratory in which he carried out a number of significant investigations, including an unsuccessful attempt to photograph the solar spectrum. At the same time he undertook medical studies at the University of Pennsylvania, obtaining the MD degree in 1836 on the basis of a thesis on glandular action which involved some consideration of the phenomenon of osmosis. From 1836 to 1839 he taught natural philosophy at the Hampden-Sidney College in Virginia.

In 1838 Draper was appointed professor of chemistry and physiology at New York University, where he remained until his death. Three years later he was one of the founders of the New York University School of Medicine, which had a some-what loose connection with the University itself; he later became President of the School.

Draper was one of the pioneers of photography, the discovery of which was announced by Daguerre and Talbot in 1839 (Sections 6.1 and 8.4). Previously Draper had been successful in taking temporary photographs by coating surfaces with potassium bromide and other materials. When the details of Daguerre's process were made available in September, 1839, Draper at once began to make use of the technique, and by December had been successful in obtaining an excellent portrait of his sister Dorothy Catharine, with an exposure of 65 seconds. He announced this achievement in the *Philosophical Magazine* of March, 1840; the portrait still survives, and is probably the oldest photographic portrait. At about the same time Draper took a daguerreotype of the moon, the first astronomical photo-graph to be taken.

In 1841 Draper announced the principle that for photochemical action light rays have to be absorbed (Section 8.4). Unknown to him. this principle had already been formulated by von Grotthuss in 1817, and it is now known as the Grotthuss–

Draper law. In 1843 he devised what he called a *tithonometer* to measure the intensity of light, basing it on the extent of photochemical reaction between hydrogen and chlorine. It was the prototype of the more accurate device, the *actinometer*, later designed by Bunsen and Roscoe.

Other important contributions of Draper, made in the 1840s, include photography using a diffraction grating, and photography in the infrared and ultraviolet (Section 6.1); he was one of the first to demonstrate the existence of Fraunhofer lines outside the visible spectrum. In 1847 he formulated principles of radiation that to some extent anticipated the more complete and mathematical principles later enunciated by Kirchhoff (Section 6.2).

Draper was elected the first President of the American Chemical Society in 1876, and in the following year was elected a member of the National Academy of Sciences. After about 1855 he became interested in aspects of philosophy and religion, and published a number of articles and books on those topics, including *History of the Conflict Between Religion and Science* (1874), which was a defence of science against attacks based on religious arguments. A paper he wrote for the 1860 meeting at Oxford of the British Association for the Advancement of Science was one factor leading to the famous confrontation between T. H. Huxley and Bishop Wilberforce.

Draper's son Henry Draper (1837–1882) followed very much in his father's footsteps. He was a medical student at New York University, obtaining his MD degree in 1858. From 1860 until his death he held various professorships at New York University, succeeding his father in the joint professorship of physics and chemistry. Henry Draper became one of the leading astronomers of his time. He and his wife, who assisted him in much of his research, established an observatory in their house on Madison Avenue, and they were particularly successful in taking astronomical photographs. Draper obtained satisfactory photographs of stellar spectra in 1877, shortly after this was first achieved by William Huggins. Henry Draper's career was cut short at the age of 45 when he succumbed to double pleurisy, contracted on a hunting trip in the Rocky Mountains when he was exposed overnight to low temperatures with no shelter.

Reference: D. Fleming, DSB, 4, 181–183 (1971).

were due to water. Later it was found that they are due to sodium, which is present in materials such as wood, and is an impurity in many salts unless they are carefully purified. After this had been realized spectroscopists took much more care with the chemical purification of the materials they examined.

6.2 Kirchhoff's principles of spectroscopy

In 1847 a contribution of importance was made by John Draper at New York University. He showed that all solids become incandescent, emitting a continuous spectrum, and that with increasing temperature there is a shift in the spectrum emitted. Another investigation along the same lines was reported in 1848 by Balfour Stewart (1828–1887), who at the time was at the University of Edinburgh and soon afterwards became Director of the Kew Observatory. Stewart had investigated the abilities of materials to emit and absorb heat radiation of various wavelengths, and he found a correlation between absorptive and emissive powers. Two years later Kirchhoff came to similar conclusions on the basis of experiments on optical spectra. Kirchhoff's work was more extensive than that of Draper and Stewart, and was expressed in a more precise and mathematical form. Also, since it was presented in a way that was more applicable to chemistry and astronomy, it became better known and exerted a wider influence. Later, however, Draper and Stewart and their supporters laid claims to their priority.

Kirchhoff made some important observations in the autumn of 1858. It had been realized that the dark D lines noted first by Fraunhofer in the solar spectrum coincided with the yellow lines emitted by flames containing a sodium salt. It was also known that the dark D lines in the solar spectrum could be made much darker by interposing a sodium flame. Kirchhoff concluded that a substance emitting a spectral line also has absorptive power for the same line, and that dark lines in the solar spectrum can be ascribed to absorption by the solar atmosphere. The way was then open for ascertaining the chemical composition of the sun and other stars from a study of their spectra. Kirchhoff was at once able to identify nine elements as present above the surface of the sun. This was the first application of spectral analysis to astronomy.

Shortly afterwards Kirchhoff formulated a number of important principles of spectroscopy. He based his ideas on the fact that a sufficiently hot gas will produces a continuous spectrum, that a gas at a lower temperature produces a spectrum consisting of bright lines or bands, and that if light giving a continuous spectrum is passed through a gas the spectrum may show dark Fraunhofer lines. He concluded that if there is radiative equilibrium at a given temperature, the ratio of the absorptive and emissive powers at a given wavelength is independent of the substance employed and is therefore a universal function of wavelength and temperature. In 1862 he introduced the concept of a black body (*schwarze Körper*) which completely absorbs all radiation incident upon it. If the absorptive power of a black body is defined as unity at all wavelengths, its emissive power is expressed by a universal function which Kirchhoff obtained empirically and expressed as a mathematical equation. This function thus expresses the spectral distribution of radiation in equilibrium with a black body at a given temperature. By means of a device having properties approximating those of a black body it is thus possible to determine experimentally this universal function, by measuring emissive power over a range of temperature. Kirchhoff's law was thus fundamental to the thermodynamics of radiation, and in the hands of Planck and Einstein it later proved to be the key to the new world of quantum theory (Section 10.1).

Kirchhoff used an arbitrary scale to express his spectral lines, but in 1869 the Swedish physicist Anders Jonas Ångstrom (1814–1874) introduced a great improvement. In his book *Recherches sur le spectre solaire*, which contains an atlas of the solar spectrum and gives the wavelengths of over a thousand lines obtained by the use of diffraction gratings, Ångstrom expressed his wavelengths in units of 10^{-10} m. This Ångstrom unit, although not strictly an SI unit, is approved by the International Union of Pure and Applied Chemistry 'for temporary use with SI units.'

6.3 Spectroscopic discovery of elements

Kirchhoff's work provided chemists and astronomers with a powerful analytical tool, and also a valuable aid in the discovery of new elements. In 1861 Kirchhoff and Bunsen announced the discovery of caesium which they found in the salts of mineral waters and identified from its blue spectral lines. The name they gave to this new element is derived from the Latin word *caesius* which means bluish grey and refers to the colour of the sky. In the same year they discovered rubidium in the mineral lepidolite, identifying it by two red lines in the spectrum. The name they gave to it is derived from the Latin *rubidus* which means dark red, or ruby red.

Bunsen and Kirchhoff tried to reserve this area of research to themselves, but others naturally felt entitled to enter the field. In 1861 William Crookes (1832–1919) discovered thallium by means of the bright green line in its spectrum, and gave it its name derived from the Greek θαλλος (thallos) which means a young shoot or twig.

Spectroscopy also played an important role in the discovery of the rare earths and the noble gases. The story of the discovery of helium is an interesting but involved one, and in brief is as follows. On 18 August 1868, a total eclipse of the sun was visible in India, and a number of scientists went there to make observations of the solar prominences. One who examined photographs of the spectra was Joseph Norman Lockyer (1836–1920) who although a civil servant at the War Office had already in his spare time done valuable work in astronomical spectroscopy. Having no previous experience of laboratory spectroscopy, and realizing that this was important for the interpretation of astronomical spectra, he turned to Edward Frankland (1825–1899) who had recently become professor at the Royal College of Chemistry which was then in London's Oxford Street. Frankland was already a distinguished organic chemist, and he allowed Lockyer the use of a room in his laboratories and part-time use of some research assistants.

Lockyer was particularly interested in a so-called D_3 line in the yellow region of solar spectra that had been obtained during the eclipse in India. It was known that the well-known sodium D line was in fact two lines close together, called the D_1 and D_2 lines. The D_3 line could not be obtained from any substance available in the laboratory, and Lockyer boldly suggested that it was caused by a new element, found in the sun but apparently not on earth. He gave this new element the name helium, from the Greek ἥλιος (helios), the sun.

Frankland, who had shown great interest in Lockyer's work, did not accept this suggestion, and was caused some embarrassment in 1871 when Kelvin, then President of the British Association, announced that Frankland and Lockyer had discovered a new element. Frankland did not think it appropriate to make a public disclaimer, but he wrote to Lockyer—the letter still exists—making it clear that he did not believe in the new element. Partly because he respected Frankland's opinion Lockyer himself said little in public about his proposed new element, and even his book *Chemistry of the Sun* (1887) did not mention it.

Much later, in 1895, William Ramsay (1852–1916), who with Lord Rayleigh had already discovered argon, began to investigate the gas produced by the mineral cleveite. On examining its spectrum he found a line which he remembered, from a lecture by Lockyer that he had attended many years previously, to have been called the D_3 line and to have been identified as relating to a new element. He sent a sample to Lockyer, who was then Director of the Solar Physics Laboratory in South Kensington, and who turned over his whole laboratory to the study of the material. The existence of helium was confirmed and soon afterwards, in 1897, Lockyer was knighted; Ramsay was knighted in 1902 and in 1904 received the Nobel Prize for chemistry 'in recognition of his discovery of the indifferent gaseous basic materials in the atmosphere and determining their place in the periodic table'.

Lockyer continued to make important contributions to spectroscopy, but perhaps his most important achievement was his founding in 1869 of the journal *Nature*, which he edited for the first 50 years of its distinguished existence. Lockyer's personality was a complex one, in that he was highly opinionated and engaged in acrimonious quarrels with many of his colleagues. Clerk Maxwell, an enthusiastic versifier, composed the following ditty about him:

> And Lockyer, and Lockyer,
> Gets cockier and cockier,
> For he thinks he's the owner
> Of the solar corona.

6.4 The Lambert–Beer law

Besides being important in the discovery of new elements and in qualitative analysis, spectroscopy began to be used for quantitative estimations. Such work is based on what has come to be called the Lambert–Beer law, which is a combination of relationships put forward by Lambert in 1760 and by Beer in 1852.

Johann Heinrich Lambert (1728–1777), who was born in Alsace, was for the most part a mathematician but he carried out some experiments in photometry and in 1770 published an important book on the subject. In it he showed that the intensity of light falls exponentially as it passes through a medium of uniform transparency, with the result that the intensity I of light emerging from a medium of thickness l is related to the original intensity I_0 by

$$I = I_0 \, e^{-al}, \tag{6.1}$$

where a is a constant which is characteristic of the medium. Unknown to Lambert this result had been obtained earlier by the French mathematician Pierre Bouguer (1698–1758) who did important work on the measurement of luminosities and is considered the founder of the science of photometry.

Equation 6.1 is equivalent to

$$\ln (I_0/I) = al \tag{6.2}$$

and in modern work it is usual to use common logarithms:

$$\log_{10} \frac{I_0}{I} = A = bl \tag{6.3}$$

where $b \, (= 2.303 \, a)$ is a constant and A is now known as the *absorbance*; it was formerly called the extinction.

The contribution of August Beer (1825–1863) was to show that for a solution containing an absorbing solute the coefficient b, and therefore the absorbance A, is proportional to its concentration c. One can thus write

$$\log_{10} \frac{I_0}{I} = A = \varepsilon cl \tag{6.4}$$

where ε is now called the *absorption coefficient*; it was formerly called the extinction coefficient. This equation is sometimes called Beer's law; however, Beer did not make use of the exponential relationship 6.1, and it is more appropriate to call it the Lambert–Beer law.

The Lambert–Beer law is fundamental to the measurement of concentrations by colorimetry, in which the colour intensities of two solutions are compared, often visually. In the 1940s colorimeters became largely superseded by spectrophotometers, in which a narrow range of wavelengths is used and in which light absorption is measured by means of photocells. With these instruments determinations can also be made in the infrared and ultraviolet regions of the spectrum.

6.5 Spectral series

The invention of the spectroscope, by Bunsen and Kirchhoff in 1859, led to a vast amount of information about the spectral lines obtained from various substances. Attempts were made to devise empirical relationships involving the frequencies or wavelengths of the lines, and to provide some theoretical explanations.

One of the first to try to explain spectral lines was George Johnstone Stoney (1826–1911), who for some years was Secretary of Queen's University of Ireland, in Dublin. Today Stoney is chiefly remembered for having first suggested the word 'electron' and his work on spectroscopy is now hardly ever mentioned. In its time, however, it exerted much influence even though some of it proved to be wrong.

In papers that appeared in 1868 and 1871 Stoney developed the mathematical theory of a vibrating molecule, expressing the motion according to Fourier's theorem as a series of sines and cosines. He pointed out that if the first term in the series corresponds to simple harmonic motion of period T, the remaining ones represent harmonics having periods of $T/2$, $T/3$, etc. Thus, he argued, one periodic motion may be the source of a whole series of lines, and additional series will result if there are other fundamental modes of vibration. He mentioned, for example, three lines in the hydrogen spectrum, designated, H, F and C, with the following wavelengths:

$$H : 4102.37 \times 10^{-12} \text{ m}$$
$$F : 4862.11 \times 10^{-12} \text{ m}$$
$$C : 6563.93 \times 10^{-12} \text{ m}$$

These lines, argued Stoney, can be explained if there is a fundamental vibration whose wavelength is $131\,277.14 \times 10^{-10}$ m, the three lines corresponding to the 32nd, 27th and 20th harmonics.

About ten years after Stoney proposed this idea serious doubts were cast upon it by Arthur Schuster (1851–1934), who was then working with Clerk Maxwell at the Cavendish laboratory in Cambridge; later, when he was professor of physics at Owen's College, Manchester, he was recognized as one of the world's leading spectroscopists. Schuster's argument was that if a hundred or more harmonics are considered, it is possible to arrive entirely by chance at any set of spectral lines, within the experimental error. He demonstrated that this was probably the explanation of the agreement obtained by Stoney, and most people were convinced that this was the case.

Whereas Stoney's treatment of spectral lines was based on theory, the next contribution to the problem, that of Balmer, was purely empirical, and a theoretical explanation could not be provided until much later. Johann Jakob

Balmer (1825–1895) was from 1859 until his death a teacher of geometry at a girls' secondary school in Basel, Switzerland. In 1871 Ångström had measured the wavelengths of four lines in the visible spectrum of hydrogen, and Balmer showed in 1885 that they could be accurately represented by the formula

$$\lambda = \frac{m^2 h}{m^2 - n^2} \tag{6.5}$$

where $h = 3645.6 \times 10^{-10}$ m and $n = 2$, The wavelengths of the four lines were obtained by putting m equal to 3, 4, 5 and 6. Balmer also predicted that with $m = 7$ there should be a fifth line, with $\lambda = 3969.65 \times 10^{-10}$ m, which is also within the visible range of the spectrum. He later learnt that there is indeed a line of this wavelength, and that there were lines in the ultraviolet spectrum corresponding to $m = 8$, 9, 10 and 11.

For some years after it appeared, Balmer's formula created much puzzlement, as there seemed to be no possible theoretical explanation. His relationship was of a completely different form from that of Stoney, which although apparently unsatisfactory did make some sense. An explanation for the Balmer series was eventually provided in 1913 by Bohr's atomic theory (Section 10.1).

In 1890 the Balmer formula was generalized by Johannes Robert Rydberg (1854–1919) of the University of Lund in Sweden. He introduced the concept of the wavenumber $\bar{\nu}$, the reciprocal of the wavelength, and first proposed the empirical formula

$$\bar{\nu} = \frac{1}{\lambda} = n_0 - \frac{N}{(m + \mu)^2} \tag{6.6}$$

where n_0 and μ are constants peculiar to the spectral series, N is a constant common to all series and all chemical elements, and m is a positive number that can have any integral value. Later Rydberg realized that there is a connection between three series of spectral lines that had been classified by the Cambridge chemists George Downing Liveing (1827–1924) and James Dewar (1842–1923) as 'principal', 'diffuse' and 'sharp' lines. Rydberg then proposed for all atomic lines a general formula that can be written as

$$\bar{\nu} = \frac{1}{\lambda} = R \left(\frac{1}{n_1^2} - \frac{1}{n_2^2} \right) \tag{6.7}$$

where n_1 and n_2 are integers and R is now known as the Rydberg constant. In 1908 the Swiss spectroscopist and theoretical physicist Walter Ritz (1878–1909) suggested that the frequency of every spectral line can be expressed as the difference between two 'spectral terms', and this became known as the *Ritz combination principle*.

Until 1908 there was no evidence for series other than the Balmer series for which $n_1 = 2$. In that year the German physicist Friedrich Paschen (1865–1940) discovered in the far infrared spectrum of hydrogen two lines corresponding to $n_1 = 3$, and lines in the ultraviolet were discovered at Harvard by

Theodore Lyman (1874–1954). Lines corresponding to $n_1 = 4$ (the Brackett series) and $n_1 = 5$ (the Pfund series) were later discovered at Johns Hopkins University.

6.6 Nineteenth-century theories of spectra

A number of efforts were made in the nineteenth century to explain the details of spectra, but with the benefit of hindsight we can see that until the quantum theory had been formulated such efforts could not get to the root of the problem. It is nevertheless instructive to consider some of the ideas put forward, since they played some role in pointing out the difficulties and in leading to their eventual solution.

Mention was made in Section 5.2 of the various theories of the nature of atoms and molecules that were being considered during the nineteenth century. Special reference was made to the discussions of Clerk Maxwell in various publications, particularly in the 1870s. It was noted that the fundamental difficulty he encountered was in reconciling the evidence from specific heats with that from spectra. To explain spectra it seemed necessary to take account of molecular vibrations, but such vibrations did not appear to make any contribution to the specific heats.

At about the same time that Maxwell was struggling with this problem, one important clarification was achieved. It had long been known that a given substance can produce different spectra under different experimental conditions. In the high temperatures of flames line spectra, consisting of sharp lines rather than bands, are observed, while at lower temperatures, band spectra are obtained. These differences caused considerable confusion until it was realized that band spectra are caused by molecules, and line spectra by atoms formed as a result of dissociation at higher temperatures. Some credit for that conclusion is due to Lockyer, but unfortunately he confused the situation and brought about considerable controversy, finally being shown to be wrong as to the details. Lockyer went so far as to say that he had proved experimentally that at high temperatures pure heavier elements are decomposed into hydrogen and other lighter elements. A number of chemists, notably James Dewar, were highly critical of this conclusion, and were able to show that the elements used by Lockyer contained traces of hydrogen and other elements as impurities.

More reliable evidence and conclusions regarding atomic and molecular spectra were obtained by Pierre Gabriel Georges Salet (1844–1894), who worked in the Paris laboratories of the organic chemist Charles Adolphe Wurtz (1817–1884). In an important series of articles, beginning in 1875, Salet was able to make a clear distinction between the band spectra produced by the molecules and the line spectra obtained when atoms are formed as decomposition products at higher temperatures.

During the 1880s there came an important change in the attitude to spectra with the acceptance of the electromagnetic theory of radiation. This theory was developed by Clerk Maxwell in a series of publications beginning in 1855, and it was essentially completed in 1873 with the appearance of his book *Treatise on Electricity and Magnetism*. In 1888 the German physicist Heinrich Rudolf Hertz (1857–1894), of the University of Kiel, was the first to give an experimental demonstration that supported the theory. He produced radiation by means of an electric spark, in this way providing the basis for radio broadcasting and related applications that are so common today.

By this time various lines of evidence, including Faraday's work on electrolysis (Section 7.1) and Crookes's production of cathode rays, had led to the conclusion that atoms are electrical in nature. A particularly significant contribution was made in 1888 by Johnstone Stoney, who for the first time suggested the idea of orbital electrons in atoms. He tried to develop this idea into detailed explanations of spectra, but in the absence of quantum theory these were inevitably in disagreement with experiment. As a result his suggestion of orbital electrons did not receive much attention. Gradually, however, Stoney's idea that spectra have an electrical origin began to be accepted. Thus at the 1893 meeting of the British Association Oliver (later Sir Oliver) Lodge (1851–1940) expressed the opinion that

radiation is due to the motion of electrified parts of molecules, not to the molecules as a whole.

Further support for the electromagnetic theory of radiation was provided by the discovery in 1897 by Pieter Zeeman (1865–1943) of the effect that is now known by his name. Zeeman was at the time working at the University of Leiden under the direction of the distinguished physicist Hendrick Anton Lorentz (1853–1925). Little of his previous work had been in spectroscopy, but at Lorentz's suggestion he investigated whether an electric field affected the light from a flame. In the course of this work he found that spectral lines can be split in a magnetic field. In his own discussion of this result he made no reference to electrons or other electrical charges, but his work was soon provided with such an interpretation. The Zeeman effect was in fact important in the formulation of atomic theories, such as Bohr's theory and its extensions (Section 10.1).

6.7 Modern chemical spectroscopy

Until the advent of the quantum theory in 1900, spectra were impossible to interpret. The first discussion of spectra in the light of quantum theory was given in 1911 by Neils Bjerrum, and is considered briefly in Section 10.1. That section also discusses the interpretation of atomic spectra that was given by Neils Bohr on the basis of his theory of hydrogen-like atoms.

The development of spectroscopy subsequent to the formulation of quantum mechanics (Section 10.2) has been very extensive, and can only be touched upon here, with special reference to chemical applications.

Gerhard Herzberg

(b. 1904)

Herzberg was born on Christmas Day in 1904 in Hamburg, Germany, where he received his early education. In 1924 he became a student at the Technische Hochschule in Darmstadt, obtaining his Ph.D. degree in 1928 for a thesis in the field of spectroscopy. From 1928 to 1929 he was a postdoctoral fellow at the University of Göttingen and did some work with Walter Heitler, who was then *privatdozent* and assistant to Max Born, the professor of theoretical physics. It was during this period that Herzberg published his suggestion that chemical bonding can be understood in terms of bonding and antibonding electrons. He then went to the University of Bristol for another postdoctoral fellowship, at the invitation of J. E. Lennard-Jones, and carried out further research in spectroscopy and molecular-orbital theory. He returned to Darmstadt in 1930 as *privatdozent*, and continued his work in spectroscopy.

In 1935, finding it inadvisable to remain in Nazi Germany, Herzberg accepted an appointment at the University of Saskatchewan, at the invitation of John W. T. Spinks, the professor of chemistry, who was later President of the University. There, besides teaching and carrying out research, Herzberg completed the German and English versions of his *Atomic Spectra and Atomic Structure*, and wrote his books on *Diatomic Molecules*. Some of his research in Saskatoon was concerned with the spectra of free radicals and free-radical ions, found in planets and also produced in the laboratory.

From 1945 to 1948 Herzberg was professor of spectroscopy at the Yerkes Observatory of the University of Chicago, where he continued his spectroscopic research and also did a little teaching. In 1949 he returned to Canada as Director of the division of physics at the National Research Council in Ottawa, retiring from the directorship in 1969 with the title of Distinguished Research Scientist. In 1975 the National Research Council created the Herzberg Institute of Astrophysics, in which Herzberg has continued to carry out research and revise his many monographs in order for them to keep pace with developments in the field.

Herzberg has made many pioneering contributions in spectroscopy and molecular theory, and most of them are of particular interest to chemists; it is therefore appropriate that although he is a physicist his 1971 Nobel Prize was for chemistry.

Reference: Herzberg (1985).

No contribution to the application of quantum mechanics to spectroscopy has been more important than that of Gerhard Herzberg (b. 1904). Besides his many articles, Herzberg has written a number of books which are often said to comprise the 'bible' of spectroscopy, since they present the whole subject in a comprehensive and authoritative way. His first book, *Atomic Spectra and Atomic Structure*, appeared in German in 1936 and a year later in English. He later wrote three books on molecular spectra, and has written amusingly on how as the subject developed his originally planned single book kept splitting in two, like Richard Wagner's *Ring Cycle*, which ended up as four operas. His proposed single book on molecular spectra first split into *Spectra of Diatomic Molecules*, which appeared in 1939, and *Spectra of Polyatomic Molecules*, which split into a book on *Infrared and Raman Spectra* and one on electronic spectra. Many of the procedures used for deriving bond distances and other molecular properties from spectra were first described in these books.

Herzberg's research has also made an outstanding contribution to the development of spectroscopy. Much of it is of considerable interest to chemists, especially his studies of the spectra of free radicals. His work has been done mainly in the laboratories of the National Research Council in Ottawa, and he won the 1971 Nobel Prize for chemistry.

Raman spectra

When a beam of monochromatic light is scattered on its passage through a medium most of the scattered light has the same wavelength as the incident radiation. Some of the scattered light may, however, have a longer wavelength than the incident light, and some may have a shorter wavelength. This effect, now known as the Raman effect, was discovered in 1928 at the University of Calcutta by Chandrasekhara Venkata Raman (1888–1970) and Kariamanikkam Srinivasa Krishnan (1898–1961). Many honours were bestowed upon Raman soon after this discovery; he was knighted in 1929 and received the 1930 Nobel Prize for physics. So confident was Raman of receiving this prize that he made boat reservations for his wife and himself before the award was announced. Krishnan was ignored in the first round of awards, but was knighted in 1946 at the last investiture held by the British administration in India.

An important advantage of Raman spectroscopy is that one can choose a convenient region of the spectrum in which to work. With ordinary spectroscopy it is necessary to work in the infrared and microwave regions in order to study vibrational and rotational transitions, and this has technical disadvantages; with Raman spectroscopy one can choose a more convenient region. Also, certain lines appear in Raman spectra that do not appear in ordinary spectra. There are therefore advantages in using both techniques.

Laser spectroscopy

Since the 1950s spectroscopic techniques have been greatly enhanced by the use of lasers. This field originated in 1954 with the introduction by the

American physicist Charles Hard Townes (b. 1915), at Columbia University, of the *maser*; the word is an acronym for microwave amplification by stimulated emission of radiation. In 1960 Theodore Harold Maiman (b. 1927), of the Hughes Research Laboratories in Miami, constructed a similar device which operates in the visible and ultraviolet regions of the spectrum. Such devices are known as *lasers*, this word being an acronym for light amplification by stimulated emission of radiation. Maiman's device produced a laser pulse; the first continuously operating (c.w., or continuous wave) laser was constructed in 1961 by A. Javan and his co-workers at the Bell Telephone Laboratories.

The principle of a maser or laser is that energy is supplied to a medium in such a way that a population inversion is produced; there are more molecules in excited states than correspond to the normal Boltzmann distribution. As a result it is possible to produce a beam that is of high intensity and is also highly monochromatic and coherent. Lasers therefore have advantages for spectroscopic and photochemical investigations.

Success has also been achieved with the construction of chemical lasers, in which the population inversion is produced by means of a chemical reaction. This development originated in theoretical work done at the University of Toronto by John Charles Polanyi (b. 1929), who in 1959 and 1961 discussed reactions in which such an inversion could be brought about. He considered as a good possibility the reaction

$$H + Cl_2 \rightarrow HCl + Cl,$$

and suggested that a laser could be constructed on the basis of the population inversion of vibrational states that is produced in this reaction. In 1965 J. V. V. Kasper and George Claude Pimentel (1922–1989), at the University of California, announced that they had followed up this suggestion and had succeeded in constructing a practicable laser based on this reaction.

Some applications of lasers in photochemistry will be discussed in Section 8.4.

Magnetic resonance spectroscopy

In magnetic resonance spectroscopy an energy level is split into two levels by the application of a magnetic field, and there can be absorption of radiation corresponding to the energy difference between the two levels. In *electron spin resonance spectroscopy* the energy difference between levels is created by applying a magnetic field to molecules containing unpaired electrons. There are two quantum numbers for electron spin and therefore two ways in which the electrons can be aligned in the magnetic field; there are therefore two energies. The radiation required for electron spin resonance is usually in the microwave region of the spectrum; that is, the frequencies are in the gigahertz region (1 $GHz = 10^9 \ s^{-1}$).

Electron spin resonance was first observed in 1944 by the Russian physicist Eugeny Konstantinovich Zavoisky (1907–1976) at the Kazan State University.

Later Brebis Bleaney (b. 1915) and his co-workers at Oxford developed the field very extensively, and the technique is now widely used. The instrumentation is sensitive enough to detect minute quantities of paramagnetic substances, such as free radicals present as intermediates in chemical and biological processes.

Whereas electron spin resonance is accessible only to molecules showing paramagnetism, which means that they must have an unpaired electron, *nuclear magnetic resonance* can be used more widely since it involves nuclear spin, which many nuclei have. The interaction between a nucleus and a magnetic field is some three orders of magnitude smaller than that for an electron, and the energy splitting is correspondingly smaller. The frequencies required for resonance to occur are in the megahertz region (1 MHz = 10^6 s^{-1}); these are short-wave radio frequencies.

Nuclear magnetic resonance (NMR) was discovered independently and almost simultaneously in 1946 by two groups of workers, Edward M. Purcell (b. 1912) and his colleagues at the Massachusetts Institute of Technology Radiation Laboratory, and Felix Bloch (1905–1983) and colleagues who did their work partly at the Radio Research Laboratory at Harvard, and partly at Stanford University. It is interesting that for a time the two groups were working within 4 kilometres of each other, but partly because of wartime secrecy were unaware of each other's investigations.

Today MNR spectroscopy is employed very widely, instruments being found in most chemical research laboratories. Important modern developments, referred to in Section 8.5, have been the application of NMR techniques to the study of rapid processes.

Mössbauer spectroscopy

In 1958, in Munich, the German physicist Rudolf Ludwig Mössbauer (b. 1929) discovered what has come to be called the Mössbauer effect. It is concerned with the emission of γ (gamma) rays, which under ordinary circumstances have a considerable spread of wavelength. In Mössbauer's technique the emitted nuclei are part of a crystal, the recoil of which is very small, and as a result the wavelength spread is much narrower, sometimes as small as one part in 10^{13}.

For this achievement, which has useful applications in spectroscopy, Mössbauer was awarded a Nobel Prize in physics in 1961. He was working at the California Institute of Technology, but in 1964 he returned to Munich as professor of physics at the Technische Hochschule.

Mass spectrometry

Is is convenient to discuss mass spectrometry here, even though it is of an essentially different character from the other types of spectroscopy dealt with in this section. We have been concerned so far with electromagnetic radiation, whereas mass spectrometry relates to beams of charged particles.

The origins of mass spectrometry lie in the many nineteenth-century investigations of the properties of electric discharges. Among the early studies of discharges are those of Humphry Davy in 1822 and of William Grove in the 1850s. In 1852 Grove described in some detail the influence of the pressure of the gas on the nature of the discharge. At first discharges were thought to involve only electromagnetic radiation. In the 1870s William Crookes (1832–1919), one of the most colourful figures in the history of science, began a detailed investigation of electric discharges, his work being an offshoot of his invention in 1875 of the 'light-mill' or radiometer, which later became a popular toy. He showed that the beam emitted by the cathode in a discharge tube is deflected in a magnetic field (Figure 6.5), and in this way established that it carries a negative electric charge. In 1892 Heinrich Hertz showed that the radiation could penetrate thin metal films.

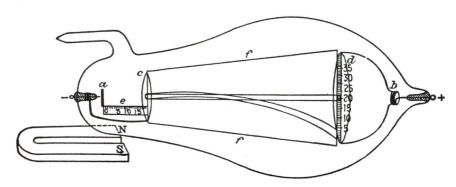

FIGURE 6.5   Crookes's diagram of the cathode-ray tube he used to demonstrate the deflection of the beam by a magnetic field (Crookes, 1879).

Beginning in about 1890 Joseph John Thomson (1856–1940) carried out a series of detailed investigations on the cathode rays. If they are in fact negatively charged they should be deflected in an electric as well as a magnetic field, but initial attempts with electric fields proved unsuccessful. This difficulty was overcome by Thomson, once he had recognized the reason for it. In his own words (in a rather long sentence!):

We must remember that the cathode rays, when they pass through a gas, make it a conductor, so that the gas acting like a conductor screens off the electric force from the charged particle, and when the plates are immersed in the gas, and a definite potential difference established between the plates, the conductivity of the gas close to the cathode rays is probably enormously greater than the average conductivity of the gas between the plates, and the potential gradient on the cathode rays is therefore very small compared with the average potential gradient.

He realized that he must obtain a better vacuum so as to reduce the screening effect of the gas. At the time the techniques for producing high vacua were nothing like as satisfactory as they later became, but by evacuating his apparatus for a number of days Thomson achieved sufficiently low pressures to be able to measure the deflections in an electric as well as a magnetic field.

The way was now open for Thomson's famous experiments in which he measured the ratio of the charge to the mass of the cathode-ray particles, which he always called by the old-fashioned name of corpuscles; this word had often been applied by Boyle, Newton and other early investigators to atoms and molecules. Thomson balanced the effects of electric and magnetic fields, finding a value for the charge/mass ratio of approximately -2×10^8 coulombs per gram (C g^{-1}). This result was of great significance, since in 1874 Johnstone Stoney had deduced from the value of the Faraday constant and other data that the corresponding ratio relating to a hydrogen ion is about 10^5 C g^{-1}. The two values differ by a factor of about 2000, which implies one of three possibilities: (1) that the charge on the cathode-ray particles is larger than that on a hydrogen ion, (2) that the mass is smaller, or (3) that both factors are involved. By this time experiments had been begun which indicated the charge on the 'corpuscle' to be the same as that on the hydrogen ion. As a result Thomson, at a Friday evening discourse at the Royal Institution on 30 April, 1897, made the momentous announcement of the existence of charged particles considerably smaller than atoms.

The first investigations on the charge of the electron were reported in 1897 by John Sealy Edward Townsend (1868–1957), a graduate of Trinity College, Dublin, who in 1895 had become a research student under Thomson. The technique he used was to cause the electrons to become attached to water droplets and to observe their rate of fall under gravity so that their radii could be estimated using Stokes's law; from the total charge and the number of droplets the charge on each droplet could be calculated. The value he obtained for the unit of charge was 5×10^{-10} esu, which is 1.57×10^{-19} C; the accepted modern value is 1.602×10^{-19} C, so that Townsend was in error by only 2%. From 1900 and for over 40 years Townsend was Wykeham professor of physics at Oxford.

In 1903 Townsend's procedure was improved by H. A. Wilson who measured the rate of movement of the droplets under the combined action of gravity and an electrostatic field. These early investigations led to the more detailed work at the University of Chicago by the American physicist Robert Andrews Millikan (1868–1953), who in 1906 attached the electrons to drops of water, and then in 1911 used oil drops to reduce evaporation. The value he obtained for the charge on the electron is close to the one accepted today.

These investigations on the electron all relate to mass spectrometry, but of even greater relevance were the parallel investigations of positively charged particles. In 1886, during discharge-tube experiments in which the cathode was perforated, Eugene Goldstein (1859–1930), of the University of Berlin, observed faint streamers of light behind the cathode. He referred to them as *Kanalstrahlen* (canal rays), and in 1898 Wilhelm Wien (1864–1928) showed that they were deflected by a magnetic field. These rays were very carefully investigated by Thomson during the course of his investigations of the cathode rays, and in 1899 he commented:

The results of these experiments show a remarkable difference between the properties of positive and negative electrification, for the positive, instead of being associated with a constant mass $\approx 1/1000$ of that of the hydrogen atom, is found to be always connected with a mass which is of the same order as that of an ordinary molecule, and which, moreover, varies with the nature of the gas in which the electrification is found.

The first thorough investigation of the positive rays was made by Thomson in 1907, and the apparatus he later used, illustrated in Figure 6.6, may be said to be the first mass spectrometer.

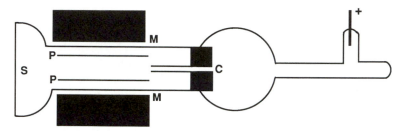

FIGURE 6.6 Apparatus used by J. J. Thomson in 1910 to investigate positive rays. The beam is deflected by the poles MM of an electromagnet, and by the parallel plates PP which were connected to an electric battery. This instrument is the prototype of a mass spectrometer.

Thomson was well aware of the potential usefulness of the positive ray machine that he had devised. In 1913 his book *Rays of Positive Electricity and their Application to Chemical Analysis* was published, and he remarked that one of the main reasons for writing the book was to induce chemists to try this method of analysis. He added:

I feel sure that there are many problems in chemistry which could be solved with far greater ease by this than by any other method. The method is surprisingly sensitive— more so even than the method of spectrum analysis, requiring an infinitesimal amount of material, and does not require this to be specially purified.

This prediction has been amply justified, in that in modern laboratories mass spectrometry is a widely used technique for analysis in research and in technology. Thomson won the 1908 Nobel Prize in physics, particularly for his work on the electron, and later a number of his research assistants received the same honour for work that stemmed from his pioneering investigations

Aside from the simple instrument described in Thomson's 1907 paper, the first mass spectrometers were built independently and at almost the same time by Dempster and by Aston. Arthur Jeffrey Dempster (1886–1950) was born and educated in Canada but did most of his work at the University of Chicago, where he was professor of physics. Francis William Aston (1877–1945) began his career as a chemist and later worked under Thomson at Cambridge, where he developed a mass spectrometer that was a great improvement over the simple instrument used by Thomson. Dempster's first paper on his mass spectrometer appeared in 1918, Aston's in 1919, and both continued to make modifications for a number of years. They both succeeded, by somewhat different techniques, in improving the focussing of the beams, and in this way increasing the accuracy of their instruments. Aston subsequently made a number of important discoveries about isotopes and the nature of atomic nuclei, and he was awarded the 1922 Nobel Prize for chemistry. Dempster discovered the isotope uranium-235 which has played an essential role in connection with nuclear energy.

Suggestions for further reading

For a general account of early spectroscopy see
W. McGucken, *Nineteenth Century Spectroscopy*, Johns Hopkins University Press, Baltimore, 1969.

Various special aspects of spectroscopy are treated in
W. H. Brock, Lockyer and the chemists; the first dissociation hypothesis, *Ambix*, **16**, 81–89 (1969).
A. J. Meadows, *Science and Controversy: A Biography of Sir Norman Lockyer*, Macmillan, London, 1972. Besides being an excellent biography, this book gives a valuable account of spectroscopy in the second half of the nineteenth century.
T. H. Pearson and A. J. Ihde, Chemistry and the spectrum before Bunsen and Kirchhoff, *J. Chem. Education*, **28**, 267–271 (1951).
H. G. Pfeiffer and H. A. Liebhafsky, The origins of Beer's law, *J. Chem. Education*, **28**, 123–125 (1951).

For an account of the development of photography that includes some spectroscopic and other scientific applications see
Janet E. Buerger, *French Daguerreotypes*, University of Chicago Press, 1989. This book includes some coloured reproductions of early spectra obtained by Edmond Becquerel.
Larry J. Schaaf, *Out of the Shadows: Herschel, Talbot, and the Invention of Photography*, Yale University Press, 1992.

I know of no general history of spectroscopy that covers post-quantum aspects. The following publications deal with special topics:
M. J. Beezley, *Lasers and their Applications*, Taylor and Francis, London, 1971; Chapter 1 (pp. 1–19) gives a good historical account of the subject, with full references.

J. C. D. Brand, The discovery of the Raman effect, *Notes & Records of the Royal Society*, **43**, 1–23 (1989).

G. Herzberg, Molecular spectroscopy: A personal history, *Ann. Rev. Phys. Chem.*, **36**, 1–30 (1985).

F. A. Miller and G. B. Kauffman, C. V. Raman and the discovery of the Raman effect, *J. Chem. Education*, **66**, 795–801 (1989).

J. H. van Vleck, A third of a century of paramagnetic relaxation and resonance, in *Magnetic Resonance* (Eds C. K. Coogan, N. S. Ham, S. N. Stuart, J. R. Pilbrow and C. V. H. Wilson), Plenum Press. New York, 1970.

For an account of early work on the electron see

B. A. Morrow, On the discovery of the electron, *J. Chem. Education*, **46**, 584–588 (1969).

Mass spectrometry is treated in

J. H. Beynon and R. P. Morgan. The development of mass spectrometry: An historical account, *Int. J. Mass Spectrometry and Ion Physics*, **27**. 1–30 (1978).

CHAPTER 7

Electrochemistry

Electrochemistry, concerned with the relationship between electricity and chemical properties, is a particularly important branch of physical chemistry, having a special significance for the rise of physical chemistry. The work in electrochemistry that was done in the 1880s by three men who were often known, sometimes derisively, as *die Ioner*—the Ionists—played an especially important role in causing physical chemistry to become recognized as a separate and important branch of chemistry. The three men, van't Hoff, Ostwald and Arrhenius, made important contributions also in other aspects of physical chemistry, particularly chemical kinetics, and were actively involved in the launching of the first journal devoted exclusively to physical chemistry, the *Zeitschrift für physikalische Chemie*. The major achievements in electrochemistry are summarized in Table 7.1.

The existence of what came to be called electricity was known to the ancient Greeks. Thales of Miletus (*c.*636–*c.*546 BC), statesman, mathematician, philosopher and engineer, made many important contributions to knowledge, and discovered that amber, when rubbed, acquires the property of attracting small pieces of pith or cork. Many centuries later William Gilbert (1540–1603) extended these observations, showing that other substances, such as glass and sulphur, show the same attracting properties when rubbed. To explain this behaviour he coined the word 'electric' from the Greek ηλεκτρον (electron), meaning amber. Gilbert, who became physician-in-ordinary to Queen Elizabeth and to King James I, also made pioneering investigations on magnetism, described in his book *De magnete, magneticisque corporibus, et de magno magnete tellure* (On magnetism, the magnetism of bodies, and the powerful magnetism of the earth), which was published in 1600 and exerted a wide influence, being much admired by Francis Bacon and Galileo among others. The word 'electricity' was first used in 1646 by the writer and physician Sir Thomas Browne (1605–1685) in his *Pseudodoxia epidemica: or, Enquiries into very many received Tenets and commonly presumed Truths*.

Electricity was at first regarded as little more than a curious and unimportant phenomenon, although a few significant investigations were made. In 1729 Stephen Gray (1666–1736) distinguished between conductors and non-conductors (insulators) of electricity; his work appeared in the *Philosophical Transactions* in 1732. At about the same time Charles-François de Cisternay Dufay (1666–1736) concluded that there are two kinds of electricity. The electricity

TABLE 7.1   Highlights in electrochemistry

Author	Year	Contribution	Section
Nicholson and Carlisle	1800	Electrolysis of water	7.1
von Grotthuss	1805	Mechanism of conductance	7.1
Faraday	1834	Laws of electrolysis	7.1
Daniell	1836	Chemical cells	7.4
Grove	1839	Fuel cell	7.4
Hittorf	1853	Migration of ions	7.2
Clausius	1857	Existence of free ions	7.2
Kohlrausch	1869	Conductance-concentration relationships	7.2
Helmholtz	1871	Fixed double layer	7.5
Planté	1879	Storage battery	7.4
Arrhenius	1887	Electrolytic dissociation	7.2
Ostwald	1888	Dilution law	7.2
van't Hoff	1890	Osmotic pressures of electrolyte solutions	7.2
Tafel	1905	Overvoltage	7.5
Bjerrum	1909	Interionic forces in solution	7.3
Gouy	1910	Diffuse double layer	7.5
Chapman	1913	Theory of diffuse double layer	7.5
Debye and Hückel	1923	Theory of strong electrolytes	7.2
Stern	1924	Fixed and diffuse double layers	7.5

produced by rubbing a glass rod with dry silk he called *vitreous* electricity, while that on resin or ebonite rubbed with flannel he called *resinous* electricity. He showed that there is repulsion when two objects bearing the same kind of electricity are brought together; when one object bears vitreous electricity and the other resinous electricity there is attraction between them.

These observations led to the formulation of the two-fluid theory of electricity. Bodies that were not electrified were supposed to contain equal quantities of vitreous and resinous fluids. An electrified body was supposed to have gained an additional quantity of one fluid and to have lost an equal amount of the other, the total amount of electrical fluid remaining the same. In those times the idea of an imponderable fluid was a popular one; heat, light, magnetism and electricity were all commonly regarded as invisible and weightless fluids which could pass with ease from one body to another.

The two-fluid theory of electricity was not, however, accepted by the American statesman and scientist Benjamin Franklin (1706–1790). His investigations led him to reject the two-fluid theory in favour of a one-fluid theory of electricity, according to which there is a single electrical fluid which non-electrified bodies were considered to possess in a certain normal amount. A body having more than this amount was said by Franklin to be positively charged, while one deficient in the fluid was negatively charged. This turned out to be an unfortunate choice; a body having an excess of electrons is in fact negatively charged, and what was later described as a flow of current through a wire in one direction is a flow of electrons in the opposite direction. Franklin's ideas about electricity were expounded in his major publication, *Experiments and Observations on Electricity, made at Philadelphia in America*, which was published in London in 1751.

An important contribution was made by the French military engineer Charles Augustin de Coulomb (1736–1806). In 1784 he invented a torsion balance, consisting of a bar hung at its centre by a long fine wire, and with it measured forces of magnetic and electrical attraction. In this way he showed that the forces are proportional to the product of the electrical charges and inversely proportional to the square of the distance between them. This law, now known as Coulomb's law, had been deduced ten years earlier by Henry Cavendish (1731–1810), who showed experimentally that there is no charge on the interior of a spherical charged conductor, and Newton had proved mathematically that this can only arise if the inverse square law applies.

In 1776 Joseph Priestley (1733–1804) summarized all that was then known about electricity in a book entitled *The History and Present State of Electricity, with Original Experiments*. When this book appeared the really important work on electricity was still to come, the subject being completely transformed by the discovery of the electric current. The first work which led to investigations of electric currents was done in the late eighteenth century by Luigi Aloisio Galvani (1737–1798), who described his findings in his book *De viribus electricitates in motu musculari commentarius* (Commentary on the forces of electricity in muscular motion), which appeared in 1791. Galvani found that when metals are inserted into frogs' legs, muscular contractions occurred, and he attributed the effects to 'animal electricity'. Others had made similar observations, and Galvani's interpretations of his results were incorrect, but his name has been preserved in a number of expressions. A flow of electricity is sometimes called 'galvanic electricity', and an electrochemical cell producing electricity is often called a 'galvanic cell'. Iron which has been treated electrically so that it is protected by a layer of zinc is called 'galvanized iron', and the expression is used even when electricity is not used for a similar type of protection. An instrument used for detecting and measuring current is called a galvanometer, and a person may be galvanized into action, electrically or otherwise.

More important contributions were made by the Italian physicist Alessandro Volta (1745–1827) of the University of Padua. He recognized that in Galvani's experiments the function of the frog's leg was to act as a detector of electricity, and realized that the electricity was being produced from the metals and solutions present. In 1800 Volta invented the pile known by his name, consisting of a series of discs of two different metals, such as silver and zinc, separated

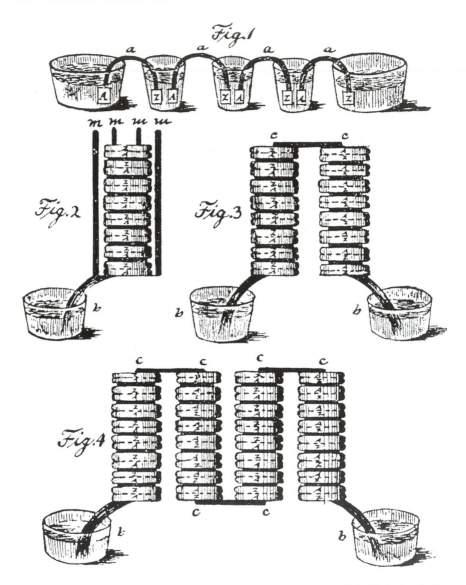

FIGURE 7.1 Diagrams given by Volta in his first account of his 'Voltaic piles'.

by paper moistened with brine. He found that a current of electricity was produced and announced this discovery in a letter, dated 20 March 1800, and written in French, to Sir Joseph Banks (1743–1820), the President of the Royal Society of which Volta was a Fellow. Banks arranged for the letter to be presented to the Society on 26 June 1800, and later in the year it appeared in the *Philosophical Transactions*. Figure 7.1 shows diagrams that accompanied Volta's communication.

7.1 Electrolysis

Even before Volta's letter appeared in print another paper was published which described experiments making use of the information presented by Volta, and using piles of the same kind. This paper showed that the electricity generated in the piles was capable of bringing about the production of hydrogen and oxygen gases from water. The author of this paper was William Nicholson (1753–1815) and the experiments had been done in collaboration with his friend Anthony Carlisle (1768–1840). The reason that this paper appeared before Volta's is an interesting one. After receiving the letter from Volta, Sir Joseph Banks had shown it to a number of acquaintances, including Carlisle, a fashionable London surgeon who was later knighted and became President of the College of Surgeons. Carlisle at once passed the information on to Nicholson, a competent amateur scientist who had founded the journal that was usually known as *Nicholson's Journal*. Without delay the two constructed Voltaic piles of their own, often using half-crowns as the silver discs; one of their piles, for example, consisted of '17 halfcrowns, with a like number of pieces of zinc, and of pasteboard, soaked in salt water. . .'. They then inserted wires from the two ends of their pile into a dish of water. When they used copper wires from the pile, hydrogen gas was evolved at one wire, and the other was oxidized. With either platinum or gold wires they found that hydrogen gas was evolved at one wire and oxygen gas at the other. Nicholson immediately announced the results in his journal, the paper appearing in July, 1800, in advance of Volta's own paper.

Nicholson and Carlisle's discovery that electricity can cause water to produce hydrogen and oxygen created as great a stir as any scientific discovery ever made. The surprise was not so much with the *fact* that the gases were produced, but with *where* they were produced. It seemed that if the gases were produced at all, it should be in one place. The puzzle was why the gases were produced separately at the two wires; in Nicholson's own words,

It was with no little surprise that we found the hydrogen extricated at the contact with one wire, while the oxigen (*sic*) fixed itself in combination with the other wire at the distance of almost two inches.

We can look at the matter as follows: imagine the water at the wire at which hydrogen is evolved. Why is not the oxygen, also presumably formed from the decomposition of water, evolved at the same place? Why and how does it apparently burrow its way through the solution and appear only at the other wire?

When Nicholson and Carlisle's experiments were done the great 'chemical revolution' had only just occurred. Lavoisier had displaced the old phlogiston theory by the idea that hydrogen and oxygen are chemical elements, and that water is a compound of them, but this was by no means universally accepted. Dalton's atomic theory was still to come, and ideas about electricity were still controversial. In view of this situation it is not surprising that Nicholson and Carlisle's results were puzzling and that alternative explanations were put forward.

Their experiments were at once repeated by many investigators. One of these was Johann Ritter (1776–1810), of the University of Jena, who used his results as the basis of a powerful attack on Lavoisier's theories. As a German he took pride in the phlogiston theory of his fellow-countrymen Johann Becker (1635–1682) and Georg Ernst Stahl (1660–1734). Also, on philosophical grounds he disliked Lavoisier's assumption that chemistry should be a quantitative science, based on the principles of physics. Some of Ritter's experiments were done with several V-shaped tubes connected electrically in series, and he still observed the evolution of gases at the wires that were connected to the Voltaic pile. From such results he concluded, in a paper that appeared in 1801, that it was not possible for the gases to be produced from the decomposition of water, since there was no way they could travel through the solution, and through the wires connecting the V-tubes, to the wires that were attached to the Voltaic pile. The truth, he argued—and to emphasize it he set it in a single line of his article— must be that

Water is an element.

In his view, oxygen was water plus positive electricity, while hydrogen was water plus negative electricity.

Ritter's views at once attracted considerable attention, sometimes favourable and sometimes accompanied by indignation. The Danish physicist Hans Christian Oersted (1777–1851), later famous for his work on electricity and magnetism, agreed enthusiastically with Ritter. Many British workers also accepted his conclusion, and between 1803 and 1809 *Nicholson's Journal* published a series of reports on Ritter's work, reports which Faraday later took very seriously. The French were naturally unhappy with the attack on Lavoisier, and the Institut de France set up a committee to look into the matter. Its Chairman was Georges Cuvier (1769–1837), famous as a zoologist and paleontologist but also highly versed in the physical sciences. Cuvier was very fair-minded, and in his report which appeared later in 1801 he accepted Ritter's theory as a not unreasonable one. He also suggested two other possibilities. One was that there actually was some unknown mechanism by which gases and other materials can travel invisibly through a solution. The other was that water does not have a constant composition, a view that had been put forward by France's leading chemist, Claude-Louis Berthollet (1748–1822). By this view 'galvanic action' could remove hydrogen and oxygen from water at different parts of the liquid, not decomposing it but merely modifying its composition. This argument was not, however, very convincing, since careful work done by Lavoisier and others had always found water to have a constant composition.

A theory that proposed a mechanism for the transport of materials through solutions was suggested in 1806 by a young German nobleman, Theodor

Christian Johann Diedrich von Grotthuss (1785–1822). While still a student in Italy he postulated that 'galvanic action' polarizes water to such a degree that the elements can pass through the solution in opposite directions, always remaining bound to a partner until released at the poles. The process, represented in Figure 7.2, is something like what happens in a square dance in which each person passes along a chain of people, continually taking the hands of those moving in the opposite direction. We now know that von Grotthuss's mechanism does not apply in general, the current instead being carried by ions

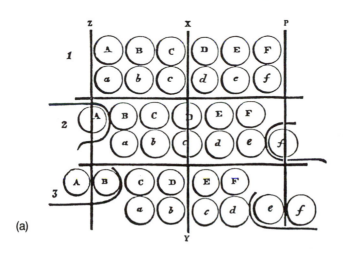

(a)

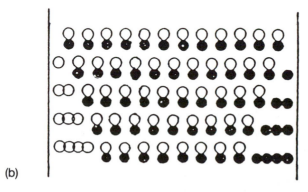

(b)

FIGURE 7.2 Diagrams illustrating ideas about the mechanism of conductivity: (a) A diagram given by Daniell and W. A. Miller (1844). (b) Part of a diagram given by Hittorf (1853). The species represented by open circles are moving to the left, at each jump becoming attached to another partner.

which can move without always being attached to a partner. However, a mechanism similar to that of von Grotthuss is believed to apply to the movement of a hydrogen or hydroxide ion through a solution. In the case of a hydrogen ion, a proton is transferred from H_3O^+ along a stream of water molecules, and such a mechanism is today called a Grotthuss mechanism.

The work of Nicholson and Carlisle inspired much further experimental work on what Faraday later called electrolysis. Humphry Davy (1778–1829), at the Royal Institution in London, put forward a theory that was similar to that of von Grotthuss, but was more general since he considered the electrolysis of salts in solution and also of fused salts. Davy made no reference to von Grotthuss, and his work may have been done quite independently. Davy made good use of electrolysis in the isolation of elements. For example, he passed the current from a powerful battery through fused potash and soda, and isolated for the first time the elements potassium and sodium. Later, by a similar procedure, he prepared barium, magnesium, calcium and strontium. On the basis of work of this kind Davy advanced the hypothesis that electrical attractions are responsible for the formation of chemical compounds. A more extreme view was taken by the great Swedish chemist Jöns Jacob Berzelius (1779–1848), who considered that every chemical compound is formed by the union of two oppositely electrified atoms or groups of atoms.

Until the 1830s all investigators of electrolysis assumed that the wires coming from a voltaic pile and inserted into a solution—'poles' as they were generally called—exerted an 'action at a distance', following the inverse square law. It was natural to assume this since in 1784 Coulomb had shown that this law applies to bodies charged with static electricity. A different conclusion was, however, drawn by Michael Faraday, who began to study electrolysis in the 1830s at the Royal Institution where he had succeeded Davy.

Faraday's investigations on electricity, magnetism and electrolysis were of the greatest importance, One of his earliest contributions, made in 1833, was to show that electrolysis can be brought about by electricities produced in a variety of ways, such as from electrostatic generators, voltaic cells, and electric fish. In particular, he showed that electrolysis can occur if an electric discharge is passed through a solution, without any wires being introduced into it. Experiments reported in 1834 convinced him that if electrolysis is brought about by inserting wires into a solution, and passing a current, the effect cannot be explained in terms of action at a distance. He performed one experiment in which two solutions were separated from one another by a 70-foot string soaked in brine. Gases were evolved at the two wires, and Faraday thought it impossible that the effect would extend over such a length if the inverse square law applied. He carried out another experiment in which a solution was placed near a source of static electricity which produced an intense electric field. No electrolysis occurred, and Faraday concluded that it is necessary for a discharge to take place or for a current to flow. He also demonstrated that the effects of electrolysis do not follow straight lines, which they would do if there were action at a distance. Also, Faraday argued, if a substance were attracted to a wire by the inverse square law, would it not remain bound to the wire rather than being released from it?

Faraday's conclusion that action at a distance does not apply led him to dislike the word 'pole' that had come to be used in connection with electrolysis, since he felt that it implied a centre of force. In its place, at the suggestion of his friend and personal physician Whitlock Nicholl (1786–1838) he introduced the word *electrode*, from the Greek ηλεκτρον (electron) and ὁδος (hodos), meaning 'way'. Nicholl also suggested the word *electrolysis*, from λυσις (lysis), meaning 'splitting', and also the word electrolyte, to mean something that undergoes electrolysis.

As Faraday proceeded with his investigations into electrolysis he came to view the process as follows. He considered that the electrical effect is transmitted from molecule to molecule, and that the affinities of the components of water, or of a dissolved salt, are weakened, so that each component would be able to leave its partner and jump to another close by. In this respect his theory was similar to that of von Grotthuss and Davy, who, however, had assumed action at a distance.

For other help with words Faraday approached William Whewell of Trinity College, Cambridge, who had shown much interest in Faraday's research. In a letter dated 24 April 1834, Faraday asked Whewell for words that would refer to the regions of solution present at the surfaces of the positive and negative electrodes. He pointed out that if we imagine the north and south magnetic poles of the earth to be due to an electric current passing round lines of latitude, the current would travel like the sun from east to west. He said that he had considered the words 'eastode' and 'westode', but realized that 'these are words which a scholar . . . could not suffer for a moment'. In his letter of reply dated the next day Whewell suggested the words *anode* and *cathode*, and explained why he favoured them. 'Anode' comes from the Greek ανα (ana), upwards, and ὁδος (hodos), a way, and therefore suggests the rising of the sun, in the east. 'Cathode' comes from κατα (kata), downwards, and is related to the setting of the sun, in the west. In a letter dated 5 May 1834, Whewell suggested three other words that are used today, *anion*, *cation* and *ion*, the latter being a general term for either an anion or a cation. The Greek word ιον (ion) is the neuter present participle of the word ιεναι (ienai), which means 'to go'; an ion is therefore something that moves. Whewell explained to Faraday why we must say cathode and not catode, but cation and not cathion; the word ὁδος (hodos) has an aspirate h, indicated by the accent over the o, whereas ιον (ion) does not.

Faraday made good use of the words cation and anion, but commented to Whewell that he thought he would find little use for the word ion. Now that it is known—which Faraday did not know—that ions do exist in solution, the word has found a wide use. The words cathode and anode have undergone a significant and undesirable change of meaning since Faraday's time. Faraday intended the words to apply to the electrode *surfaces*, or to regions of the

Michael Faraday

(1791–1867)

Faraday was born near London—in Newington, Surrey—into a family of very limited means, and he received little education. At the age of 14 he was apprenticed to a kindly newsagent and bookbinder, G. Ribeau, who encouraged him to read the books he was binding. His interest in science was particularly aroused by Jane Marcet's *Conversations on Chemistry*, which first appeared in 1809 and passed through many editions. This book, inspired by Humphry Davy's lectures at the Royal Institution, was of sparkling interest and had much effect on Faraday.

Through a fortunate chance Faraday became a laboratory assistant to Davy, and during 1813 and 1814 he accompanied Davy and his wife on a visit to the Continent. Although for some of the time he had to act as Davy's valet, he gained much from the experience of coming into contact with many of the leading scientists of the time. In 1825 he succeeded Davy as Director of the Royal Institution laboratories, and in 1833 he also became Fullerian Professor of Chemistry, a post that was created and endowed for him by John Fuller, a Member of Parliament. Fuller had attended many lectures at the Royal Institution, and had often slept through them; his endowment was 'in gratitude for the peaceful hours thus snatched from an otherwise restless life'.

Faraday's first work was in analytical chemistry, for which he soon established a reputation. After 1821 much of his work was on electricity and magnetism, and in 1831 he discovered the phenomenon of electromagnetic induction. In the following year he embarked on investigations designed to see whether the electricities produced in various ways—from electrostatic generators, voltaic cells and electric fish—were one and the same. He established their identity in part by his work on electrolysis, and he was led in 1834 to his two laws of electrolysis (Section 7.1).

Faraday was ignorant of mathematics beyond simple arithmetic, but he developed

electrolyte close to the surface, *in an electrolytic cell*. However, by a perhaps inevitable extension of meaning they were soon applied to the electrodes themselves. This is not so bad, but a further extension of meaning has been most unfortunate. It is now common to apply the words to the electrodes of a battery or other device that generates electricity. This means that whereas in an electrolytic cell the anode is the positive electrode, in a battery it is the negative electrode. This confusing extension of meaning was never intended by Faraday, who would certainly not have approved of it. At the present time there is no

qualitative theories, particularly as to the nature of electricity. He rejected the idea that in electrolysis there is action at a distance; instead an electric strain is set up in the solution. In Faraday's view an electric current is due to a buildup and breakdown of strain within the particles. Beginning in about 1845 Faraday developed his ideas about electricity and magnetism in a direction that led to the establishment of modern field theory. In the 1850s and 1860s Clerk Maxwell developed Faraday's ideas into his mathematical theory of electromagnetic radiation (Section 1.4).

Faraday was a man of great kindness and modesty, and deeply religious, being a loyal member of the Sandemanian Church. He declined all honours, and avoided scientific controversy. Since he preferred to work alone and never had an assistant he founded no school of research. However, he disseminated science in a variety of other ways, for example through his Friday evening discourses which he founded at the Royal Institution, and of which he gave over a hundred himself. He published four hundred and fifty scientific papers, and much other work of significance has been found among the seventeen thousand experiments described in his laboratory notebooks.

Faraday's mental faculties declined seriously after the mid-1850s, but by 1860 he had recovered sufficiently to give his famous Christmas lectures on the 'Chemical History of a Candle'. In 1862 he resigned from the Royal Institution, and until his death in 1867 resided in a house provided for him by Queen Victoria. He is buried in Highgate Cemetery, but his grave there is not visited as frequently as that of Karl Marx.

References: L. Pierce Williams, DSB, 4, 527–540 (1971); Macdonald (1964); Porter (1981); Tyndall (1869); Thompson (1898); Williams (1987).

need for the words anode and cathode; the expressions 'positive electrode' and 'negative electrode' are more helpful and are quite unambiguous.

Faraday's laws of electrolysis

In a paper that appeared in 1834 Faraday suggested two fundamental laws of electrolysis. It is often thought that these two laws were discovered empirically, but this is not the case; they were deduced by Faraday on the basis of his ideas about how electricity interacts with a solution, and were later confirmed by him.

If electrolysis had proceeded by action at a distance, as had been thought previously, the amount of material deposited or evolved would have depended on the electric potential and on the size of the vessel used. According to Faraday's view, however, the amount would depend only on the product of the current and the time, i.e., on the quantity of electricity that passes through the solution, and Faraday confirmed that this is the case; this was his first law. And since, in Faraday's view, the electricity was concerned with weakening the affinities between the constituents of matter, the amount should depend on the equivalent weight of the substance evolved or deposited. Faraday never expressed the second law in the explicit form in which it is used today, but his experiment showed it to be obeyed. It is ironic that he should have deduced his laws of electrolysis on the basis of ideas that we now know to be incorrect in one respect. The effect of the current is not to bring about dissociation, as Faraday thought; instead the ions are already present and the electric field brings about their movement towards the electrodes, where they are neutralized. The relationship between the quantity of electricity and the amount of material deposited or evolved arises not because the electricity is concerned with dissociation, but because it is concerned with neutralization.

Electricity and matter

Faraday's laws of electrolysis had a far-reaching significance, since they suggested that electricity itself is not continuous, and that fundamental particles of electricity are in some way associated with atoms. Faraday himself came close to reaching this conclusion, for he wrote

... if we adopt the atomic theory or phraseology, then the atoms of bodies which are equivalent to each other in their ordinary chemical action, have equal quantities of electricity naturally associated with them.

On the whole, however, Faraday tended to adhere to the idea that electricity is a fluid.

It was not for some years that more definite statements were made about the relationship between electricity and matter. In the 1873 edition of his *A Treatise on Electricity and Magnetism* Clerk Maxwell wrote

If we ... assume that the molecules or the ions within the electrolyte are actually charged with *definite quantities* of electricity, positive and negative, so that the electrolytic current is simply a current of convention, we find that this tempting hypothesis leads us into very difficult ground. Suppose we leap over this difficulty by simply arresting the constant value for the molecular charge, and that we call this ... *one molecule of electricity*.

A similar idea was advanced in 1874 by Johnstone Stoney at a meeting of the British Association, but the paper was not published until 1881. In that year Helmholtz presented the Faraday Memorial Lecture to the Chemical Society and discussed the atomic nature of electricity in the following terms:

Now the most startling result of Faraday's law is perhaps this. If we accept the hypothesis that the elementary substances are composed of atoms, we cannot avoid concluding that electricity also, positive as well as negative, is divided into portions, which behave like atoms of electricity.

In 1891 Johnstone Stoney suggested for the first time the word electron for the unit of negative electricity. Up to this time the unit of electricity was not associated with any physically significant particle, but in 1897 Joseph John Thomson (1856–1940), the Cavendish professor at Cambridge, performed his classic experiments on the particles produced in a cathode ray tube, measuring the ratio of charge to mass (Section 6.7). This work clearly established the real existence of the electron—always referred to by Thomson as a 'corpuscle'. This work played an important role in convincing even the sceptical Ostwald that atoms really exist (Section 5.2).

7.2 Electrolytic conductivity and dissociation

The way in which electricity passes through a solution of an electrolyte remained obscure for many years, and was not properly understood until 1887 when Arrhenius proposed that substances like sodium chloride actually exist in solution as independent ions. Previously it had been assumed that sodium chloride molecules are present intact in solution, and electrolysis was explained in terms of the dissociation of salt molecules by the action of the electric current. When Faraday spoke of the movement of cations and anions through a solution he was thinking of them as formed only when the current flowed. Modifications to this view were put forward in 1857 by Clausius. His suggestion was that as a result of collisions between solvent and solute molecules a small number of ions are always present, some of the collisions being sufficiently energetic to cause dissociation into ions. The movement of these ions would provide the transport of charge when a current was passed.

In 1853 Johann Wilhelm Hittorf (1824–1914) began work on conductivity at the Royal Academy of Münster. At first he followed Faraday's ideas as to the nature of the process, but after the appearance of Clausius's 1857 paper he accepted the idea that a small number of free ions are always present. He made measurements of the concentration changes in the neighbourhood of the two electrodes when the current is passed. He realized that these changes are occurring because the cations and anions are moving at different speeds. He introduced the idea of the *transport number*, which is the fraction of the current carried by a particular ion. Thus if the speed with which the cation moves under unit potential gradient (known as the *mobility* of the ion) is u_+, and the mobility of the anion is u_-, the transport numbers of the cation and anion are

$$t_+ = \frac{u_+}{u_+ + u_-} \qquad t_- = \frac{u_-}{u_+ + u_-}. \qquad (7.1)$$

He showed that the ratio of the decrease of concentration in the neighbourhood of the anode to the decrease in the neighbourhood of the cathode is equal to u_+/u_-. He devised experimental techniques for measuring the concentration changes near to the positive and negative electrodes, and determined transport numbers for a number of salt solutions, and hence the relative velocities of the ions under a given potential gradient. Figure 7.2 shows diagrams given by Daniell and Miller, and by Hittorf, illustrating early ideas about the mechanism of electrolytic conduction.

Further significant experimental advances were made in the 1860s and 1870s by Friedrich Wilhelm Georg Kohlrausch (1840–1910), and these were of particular importance in providing a background for Arrhenius's theory of electrolytic dissociation; before proposing it, in fact, Arrhenius had spent some time in Kohlrausch's laboratories in Würzburg. Kohlrausch was the first to measure satisfactorily the conductivities of solutions of electrolytes. Direct currents are unsuitable because of polarization effects, and Kohlrausch used alternating currents. Also, in an 1874 paper Kohlrausch and Otto Natalis August Grotrian (1847–1921), described work with spongy electrodes of large area which reduced the surface density of any deposit. A telephone receiver was used to respond to alternating currents, and it was found that when polarization was eliminated, Ohm's law was obeyed; i.e, the current was proportional to the electromotive force. Kohlrausch realized that this result presented difficulties for the idea that the electric current has to perform the work of splitting molecules into ions, a point that had been made in 1857 by Clausius on the basis of less convincing evidence. However, he was not prepared to take the bold step of concluding that ions are present in appreciable amounts even when a current is not passing through the solution.

One important contribution made by Kohlrausch was his use, in a paper that appeared in 1878, of the concept of *molar conductivity*, which had been suggested the previous year by Robert Lenz (1833–1903) of the University of St Petersburg. The specific conductivity is not a useful quantity for comparing the behaviour of different solutions, since a solution of higher concentration tends to have a higher conductivity simply because it contains more conducting material. There is a need for a property in which there is a compensation for differences in concentration, and the molar conductivity—which Kohlrausch called the 'molecular conductivity'—has this advantage. It can be visualized as the conductivity between two parallel plates a fixed distance (e.g. 1 cm) apart, of such an area that one mole of the solute is present between them. What we now call the molar conductivity was for a period called the 'equivalent conductivity', but the term 'equivalent' is no longer recommended by IUPAC. The concept, however, is still retained; we speak, for example of the molar conductivity of $\frac{1}{2}CuSO_4$ which is the same as the equivalent conductivity of $CuSO_4$.

In the same paper by Kohlrausch and Grotrian that appeared in 1874, for a number of solutions they plotted molar conductivity against concentration, and by extrapolating to zero concentration obtained a value Λ_0. They observed, for example, for corresponding sodium and potassium salts, that the differences between the Λ_0 values are independent of the nature of the anion. Similar results were obtained for a variety of pairs of salts with common cations or anions, in both aqueous and non-aqueous solutions. Kohlrausch explained this behaviour in terms of his *law of independent migration of ions*. Each ion is supposed to make its own contribution to the limiting molar conductivity Λ_0, irrespective of the nature of the ion with which it is associated. In other words, Λ_0 can be split into λ_{+0}, the contribution of the cation, and λ_{-0}, the contribution of the anion:

$$\Lambda_0 = \lambda_{+0} + \lambda_{-0} \qquad (7.2)$$

Kohlrausch further pointed out that the molar conductivity gives a measure of the sum of the ionic velocities in the two directions. He combined his measured values of Λ_0 with Hittorf's transport numbers, and so obtained the individual λ_{+0} and λ_{-0} values and hence the velocities with which various ions move under a unit potential gradient. For example, under a potential gradient of 1 volt per centimetre he found that hydrogen ions have a velocity of about 0.003 cm s^{-1} at ordinary temperatures, while the ions of neutral salts move at only about one-fifth of that speed. His value for the hydrogen ion was confirmed in 1886 by Oliver Joseph Lodge (1851–1940), who followed the ions as they moved through a gel coloured by an indicator sensitive to acid. This so-called *moving-boundary method* was also used by William Cecil Dampier Whetham (later known as Sir William Dampier, 1867–1952), who studied the movement of neutral ions by observing their motion in coloured salts or by the formation of precipitates.

Arrhenius's theory of electrolytic dissociation

Great clarification of the problem of electrolytic conductivity came with the work of Svante August Arrhenius (1859–1927) in the 1880s. In 1884 Arrhenius submitted to the University of Uppsala a Ph.D. dissertation which included a number of measurements of conductivities and an interpretation of them. Arrhenius did not, in this dissertation, suggest that dissociation into ions occurred in solution; instead he postulated that some unspecified 'active parts' of the molecules were formed and were responsible for carrying the electric current. Arrhenius's dissertation, in French, was not clearly written and he had incurred the hostility of his professors, the chemist Per Theodor Cleve (1840–1905) and the physicist Thobias Robert Thalén (1827–1905). Both these men were highly competent in their own fields, but were unable to appreciate Arrhenius's conclusions. As a result, the dissertation was placed only in the fourth class (*non sine laude approbatur*, approved not without praise). His defence of the dissertation fared a little better, being approved with praise (*cum laude approbatur*) rather than not without praise. In the ordinary way these grades would have kept Arrhenius out of any academic appointment.

Somewhat discouraged, Arrhenius sent copies of his dissertation to several

Svante August Arrhenius

(1859-1927)

Arrhenius was born in Vik (Wijk), near Uppsala, Sweden. He attended the University of Uppsala but in 1878, not wishing to carry out research under any professor at Uppsala, moved to Stockholm to work under the physicist Erik Edlund at the Swedish Academy of Sciences. He presented his thesis, on the conductivities of electrolytic solutions, to the University of Uppsala in 1884, but it achieved only a poor rating from the examiners (Section 7.2). For the next five years Arrhenius travelled extensively and visited a number of research centres in Europe. He then held some teaching positions, and in 1895 was appointed professor of physics at the newly formed University of Stockholm, serving as its Rector from 1897 to 1902. From 1905 until his death he was Director of physical chemistry at the Nobel Institute in Stockholm. He received many invitations from universities outside Sweden, including Giessen and Berlin, but he preferred to remain in his own country.

Arrhenius's Ph.D. thesis did not suggest the dissociation of electrolytes into ions, but his discussions with Ostwald, van't Hoff and others led him to postulate his theory of electrolytic dissociation in 1887. He continued for many years to work on electrolytic solutions, but to the end of his life never accepted the fact that strong electrolytes are completely dissociated, and that their behaviour is strongly influenced by interionic forces.

In 1889 he discussed the temperature dependence of reaction rates on the basis of an equation that had been given earlier by van't Hoff, and this equation is now usually known as the Arrhenius equation (Section 8.2).

Arrhenius worked in a variety of fields, and made important contributions to immunochemistry, cosmology, the origin of life, and the causes of the ice age. It was he who first discussed the 'greenhouse effect', calling it by that name. He published many papers and a number of books, including *Lehrbuch der kosmischen Physik* (1903), *Immunochemistry* (1907), and *Quantitative Laws in Biological Chemistry* (1915).

He received the Nobel Prize in chemistry in 1903, the citation making reference to the 'special value of his theory of electrolytic dissociation in the development of chemistry'.

References: H. A. M. Snelders, DSB, 1, 296–302 (1970); Hiebert (1982); Laidler (1983, 1984, 1986); Root-Bernstein (1980); Servos (1990); Walker (1928).

prominent scientists, including Clausius, Lothar Meyer, van't Hoff and Ostwald. Neither Clausius nor Meyer replied, but van't Hoff responded favourably. Ostwald's reaction was an immediate and practical one. He had himself been making measurements of electrical conductivities of solutions of a number of salts, and had also made measurements of their catalytic activities (Section 8.5), finding a correlation between the two sets of results. He at once prepared a short paper, which appeared in 1884, describing his own results and making full acknowledgement of the importance of Arrhenius's results. Ostwald also travelled to Uppsala to discuss with Arrhenius the significance of their two sets of results—a remarkable action on the part of an already distinguished professor towards a young man who had almost failed to obtain his degree. Ostwald also offered Arrhenius an appointment at the University of Riga, but for personal reasons Arrhenius was unable to accept it. On hearing of this offer the authorities at the University of Uppsala naturally wondered if they had been mistaken in their low assessment of Arrhenius, and in 1885 offered him a travelling fellowship that would keep him out of the country, and in little position to cause them further embarrassment or annoyance, for the next five years.

Arrhenius made a most profitable use of his travelling fellowship, spending time with Ostwald in Riga, with Kohlrausch in Würzburg, with Boltzmann in Graz, with van't Hoff in Amsterdam, and finally again with Ostwald who had moved in 1887 to Leipzig. By that year Arrhenius had been able to formulate his theory of electrolytic dissociation, his paper on it appearing in the June, 1887, issue of the *Zeitschrift für physikalische Chemie* which Ostwald and van't Hoff had founded in that year. According to the theory, there exists an equilibrium in solution between undissociated molecules AB and the ions A^+ and B^-:

$$AB \rightleftharpoons A^+ + B^-$$

This proposal was qualitatively different from the previous idea that ions are formed only when the current flows through the solution, and quantitatively different from the suggestion of Clausius that trace amounts of ions are present at all times. Arrhenius proposed that the ions are always present, sometimes in substantial amounts, in equilibrium with the undissociated molecules, and that the effect of an applied emf is not to produce them but to cause them to move through the solution.

In 1888 Ostwald expressed the behaviour quantitatively in terms of his so-called dilution law. If α is the degree of dissociation of AB the concentration of undissociated electrolyte is $c(1 - \alpha)$, where c is the concentration, and the ionic concentrations are αc. The equilibrium equation is thus

$$\frac{c^2\alpha^2}{c(1 - \alpha)} = \frac{c\alpha^2}{1 - \alpha} = K, \tag{7.3}$$

Friedrich Wilhelm Ostwald (1853–1932)

Ostwald, of German ancestry, was born in Riga, Latvia, which was then one of the provinces of the Russian Empire. He became a student at the University of Dorpat (now Tartu, Estonia) where, in spite of devoting much time to literature, music and painting, he managed to obtain his doctoral degree in chemistry in 1878. After a period as a schoolmaster he became professor of chemistry at the Riga Polytechnic Institute in 1881. In 1887 he was appointed professor of physical chemistry at the University of Leipzig where he remained until 1906.

Ostwald was a prodigious worker on a wide range of topics. Besides his work in physical chemistry he was much concerned with colour theory, the history and philosophy of science, a world language, and world peace. He exerted a wide influence and did much to establish the field of physical chemistry; many who later distinguished themselves in physical chemistry had spent some time in his laboratories. In the 1880s he helped Arrhenius to formulate the theory of electrolytic dissociation, and he derived the 'Ostwald dilution law' (Section 7.2). He made important contributions to kinetics, especially with his work on catalysis (Section 8.6), and the 1909 Nobel Prize for chemistry was awarded to him 'in recognition of his work on catalysis and for his fundamental investigations in the matter of chemical equilibrium and reaction velocities'. He was the first to realize that a catalyst acts without altering the energy relations of the reaction, and that it usually speeds up a reaction by lowering the activation energy. In his textbooks, particularly his *Lehrbuch der allgemeinen Chemie*, he lucidly expounded the principles of thermodynamics and kinetics.

where K is the dissociation constant. The fraction α is the ratio of Λ at any concentration to Λ_0 at infinite dilution, and thus

$$\frac{c(\Lambda/\Lambda_0)^2}{1 - \Lambda/\Lambda_0} = K. \tag{7.4}$$

Ostwald and others found that this equation interpreted the variation of Λ with c for some electrolytes, although there were sometimes serious deviations, to be mentioned later.

Support for Arrhenius's theory was also provided by van't Hoff's conclusion that the osmotic pressures of solutions of electrolytes are abnormally high (Section 4.4.), his equation being

$$\pi = icRT, \tag{7.5}$$

Ostwald's opinions were often controversial and sometimes wrong. Although he had supported the idea of electrolytic dissociation he denied the real existence of atoms, ions and molecules for many years, believing them to be no more than convenient fictions; he thought that everything could be explained in terms of energy. He was only finally convinced, in the first few years of the twentieth century, by the work of J. J. Thomson on cathode rays, and of Perrin on Brownian movement.

When he was in his late forties Ostwald became disillusioned with chemistry and the academic life. He requested the authorities at the University of Leipzig to relieve him of his lecturing duties, but this was denied. There was also trouble about the funeral of Johannes Wislicenus (1835–1902), whose family wished it to take place in the chapel of the University of Leipzig. Since Wislicenus was a professed agnostic this was refused, but Ostwald arranged for the funeral to take place in the physical chemistry lecture room, to the profound displeasure of the university administration. Because of friction of this kind Ostwald moved in 1906 to his country estate, named Energie, at Groszbothen in Saxony. There he continued to work actively on a wide variety of topics, mostly outside the field of science. After 1918 he devoted himself particularly to colour theory.

References: E. N. Hiebert and A. G. Körber, DSB, Supplement, 15, 455–469 (1978); Bancroft (1933); Donnan (1933); Hiebert (1971, 1982); Root-Bernstein (1980); Servos (1990).

where i is a factor that is greater than unity. Suppose that one molecule of an electrolyte would produce n ions if there were complete dissociation. If the degree of dissociation is α, the number actually produced is αn, and the number of undissociated molecules is $1 - \alpha$, so that the total number of particles produced from one molecule is

$$i = 1 - \alpha + \alpha n. \tag{7.6}$$

Therefore

$$\alpha = \frac{i - 1}{n - 1}. \tag{7.7}$$

These relationships were given by Arrhenius in his 1887 paper. It is thus possible to calculate the degree of dissociation α from the van't Hoff i obtained from the osmotic pressure data. For weak electrolytes the values so obtained were in reasonable agreement with Λ/Λ_0, and this provided support for Arrhenius's theory.

Objections to electrolytic dissociation

Today the idea of dissociation into ions is universally accepted, but for a number of years after Arrhenius made his proposal there were a few investigators who strongly disagreed. A particularly vigorous opponent was Henry Edward Armstrong (1848–1937), who from 1884 to 1911 was professor of chemistry at the Central Technical College, one of the many colleges which eventually became absorbed into the Imperial College of Science and Technology which is part of the University of London. Armstrong, a large man who always expressed himself forcefully, was a distinguished organic chemist who also did a considerable amount of work in electrochemistry. He was particularly outspoken on the subject of the Arrhenius theory and even lampooned the idea of ionic dissociation in two fairy tales he wrote, *A Dream of Fair Hydrone*, and *The Thirst of Salt Water*. Armstrong was aptly described by an obituarist, Sir F. Keeble, as having 'all the spare parts of genius but not the long patience to put them together'.

Another who took strong exception to the theory was Percival Spencer Umfreville Pickering (1858–1920), also an eccentric and controversial individual. Pickering had private means and after a brief appointment at Bedford College, London, he established a private laboratory in London in 1887, the year that Arrhenius's theory was published. From then until 1896 he carried out extensive investigations on solutions of electrolytes, publishing some 70 papers in which he tried to show that there was no dissociation into ions. His experimental work was carefully carried out and provided detailed information on the densities, freezing points and conductivities of solutions. His conclusions, however, were not well received, and he was much discouraged. In 1896 he turned to an entirely different line of investigation, scientific fruit farming, which he also approached with his characteristic disregard for conventional ideas, but at which he was more successful than with his electrochemical work.

The objections raised by Armstrong, Pickering and a few others were based on the fact that the Arrhenius theory seemed to be regarding water as playing only a passive role, providing, in Armstrong's expressive phrase, merely 'a dance floor for ions'. On the contrary, wrote Armstrong, 'water is of all substances known to us not only the most active and useful but the most marvellous'. As an alternative to the theory of dissociation into ions, it was suggested that hydrates are formed in solution, involving the undissociated electrolytes, and that these hydrates have varying capacities for carrying an electric current. The variations of conductivity with concentration that had been observed by Kohlrausch and Arrhenius were attributed to shifts in equilibrium for these various processes of hydrate formation. Pickering in particular produced evidence, from his measurements of the densities of solutions, for the existence of many hydrates in solutions.

These criticisms of the Arrhenius theory in its original simple form did indeed have some substance, but the remedy was not to overthrow the theory but to modify it. Considerable enlightenment came in 1894 when Drude and Nernst developed a theory of the electrostriction, or binding, of ions to surrounding water molecules (Section 7.4). This treatment at last gave water and other solvents a more active role in solutions, and greatly diminished the force of the objections that had been raised to the existence of ions in solution.

By the turn of the century the theory of electrolytic dissociation had been generally accepted, but a few remained unconvinced. Even as late as the 1930s Louis Kahlenberg (1870–1941) was still giving, at the University of Wisconsin, a course on electrochemistry that took no account of the existence of ions in solution. It is also of interest to recall that although Ostwald played such an important role in the development of the theory of electrolytic dissociation, he believed for most of his career that atoms, molecules and ions were no more that a convenient fiction and did not really exist.

Strong electrolytes

Even in the early days of Arrhenius's theory of electrolytic dissociation there were indications that his assumption that the decrease in Λ with increasing concentration is due to a decrease in the degree of dissociation does not apply to strong electrolytes. By strong electrolytes is meant all ordinary salts and some acids and bases, and it is now known that their dissociation into ions is complete and that the undissociated molecules do not exist. There is thus, for strong electrolytes, no shifting equilibrium between ions and undissociated molecules, as postulated by Arrhenius and assumed in Ostwald's dilution law. Another explanation is needed for the decrease in Λ which does occur for strong electrolytes as the concentration increases.

One type of experimental evidence for the new ideas about strong electrolytes was that the Ostwald dilution law was not satisfactorily obeyed by many strong electrolytes, in that the values of K calculated from the Ostwald equation were not true constants but varied, sometimes by many powers of ten, with the concentration. This fact had been pointed out to Arrhenius by van't Hoff and Nernst even before the Ostwald dilution law appeared in print, but Arrhenius was able to put them off by pointing out that weak electrolytes provide a more satisfactory test of the theory because of the greater range of molar conductivities! It is somewhat amusing that in the first edition of his *Theoretische Chemie*, which appeared in 1891, Nernst wrote that 'the Ostwald formula does not exactly fit the observed values in the case of highly dissociated acids and salts'. In view of the fact that the Ostwald 'constants' were known to vary by many powers of ten as the concentration was varied, this comment must be the understatement of the century. Arrhenius was held in very high esteem at the time, and as Wolfenden has pointed out in his review of the 'anomaly of strong electrolytes' (1972), there was something of a conspiracy of silence among the Ionists with regard to these embarrassing anomalies. The Ionists were convinced that Arrhenius was right in saying that there was extensive dissociation of electrolytes, and were not going to be put off by quantitative discrepancies, however large.

Another important piece of evidence was that values of the conductivity ratio α $(= \Lambda/\Lambda_0)$ sometimes differed significantly from those obtained from the van't Hoff factor (eqn 7.5). There was also a difficulty with the heats of neutralization of strong acids and bases, which did not vary with concentration in the way they would have done if the equilibrium shifted. A variety of investigations in the 1890s had uncovered evidence of this kind, and had led to the opinion that some modification to Arrhenius's theory was needed.

Several suggestions were made in order to provide a solution to the difficulty, by the Australian physicist William Sutherland (1859–1911) in 1902 and later, by G. N. Lewis in 1908–1912, and by the British physicist Samuel Roslington Milner (1875–1958) in 1912 and later. These suggestions involved the idea that electrostatic attractions and repulsions play an important role in solutions of strong electrolytes. Milner's work was of particular significance in that it provided a satisfactory and detailed treatment of these interionic forces, with special reference to the part they play in electrolytic conductance. It seems likely that had it not been for a severe illness which interrupted his work Milner would have arrived at the same kind of treatment that was later given by Debye and Hückel.

A particularly clear-cut piece of experimental evidence that did not support the idea of incomplete dissociation of strong electrolytes was presented by Niels Janniksen Bjerrum (1879–1958), who was then at the University of Copenhagen. In 1909 Bjerrum read a short paper to the Seventh International Conference on Applied Chemistry, held in London. In it he presented spectroscopic evidence that certain salts are completely dissociated in solution, and he attributed conductivity changes to interionic forces and not to varying degrees of dissociation. Arrhenius, who was present at the meeting, took Bjerrum's proposal as a personal affront, and refused to discuss the matter. In 1916 Bjerrum, who had become professor at the Royal Veterinary and Agricultural College in Copenhagen, presented further and more convincing evidence to the Sixteenth Scandinavian Scientific Meeting in Oslo. The Chairman was Arrhenius, who again refused to consider this modification of his theory. In 1919, at a meeting of the Faraday Society, a paper by Arrhenius, presented *in absentia*, discussed the matter in some detail, concluding that Bjerrum's idea 'seems not to agree very well with experiment'. We now know that Bjerrum was quite correct, but to the end of his life in 1927 Arrhenius never accepted the new ideas about strong electrolytes, even though by then the evidence had become overwhelming.

A complete and detailed treatment of strong electrolytes in solution was put forward in 1923 by Peter Joseph Wilhelm Debye (1884–1966) and Erich Hückel (1896–1980), both of whom were then at the University of Zürich.

Their treatment was based on the ideas that Milner had formulated, and it gave a mathematical treatment of the conductivity of a solution of a strong electrolyte in terms of the attractions and repulsions between the ions, which become more pronounced as the concentration is increased. Because of these attractive and repulsive forces the distribution of ions in a solution is not completely random. In the neighbourhood of any positive ion there tend to be more negative than positive ions, whereas near to a negative ion there are more positive than negative ions. As a result there is a negatively charged ionic 'atmosphere' round each positive ion, and the treatment showed that its effect on a given ion is the same as if an equal but opposite charge were distributed on the surface of a sphere which has the ion at its centre. The radius of this imaginary sphere is called the *radius of the ionic atmosphere*. As the concentration is increased this radius becomes smaller and the ionic atmosphere has a greater effect on the central ion.

The Debye–Hückel treatment showed that there are two ways in which the ionic atmosphere affects the conductivity. When there is a potential gradient across an ionic solution, each positive ion moves towards the negative electrode and must drag along with it its ionic atmosphere, which is attracted in the opposite direction by the electric field. The ionic atmosphere thus becomes asymmetric rather than spherical, the charge density behind the moving ion being greater than that in front, so that the motion of the ion is retarded. This effect is referred to as the *relaxation* or *asymmetry effect*.

A second factor results from the tendency of the ionic atmosphere to bind water molecules, because of ion-dipole interactions. There is thus a flow of solvent molecules in the direction opposite to the movement of each ion, and as a result there is an additional retarding effect, known as the *electrophoretic effect*. This word derives from the Greek word φερειν (pherein), meaning to carry or to bear.

Debye and Hückel arrived at an equation which represents the molar conductivity Λ as a function of concentration, and it was found to agree well with the experimental results at low concentrations. In their treatment they assumed the ions to travel in straight lines, neglecting the zigzag Brownian motion brought about by the collisions of the surrounding solvent molecules. Their theory was improved in 1926 by Lars Onsager (1903–1976), who took Brownian movement into account. The resulting Debye–Hückel–Onsager equation is of the form

$$\Lambda = \Lambda_0 - (P + Q\Lambda_0)\sqrt{c}, \tag{7.8}$$

where P and Q can be expressed in terms of various fundamental constants and properties of the system.

The equation is obeyed quite satisfactorily at low concentrations—up to about 10^{-3} M for a uni-univalent electrolyte—but serious deviations are obtained at higher concentrations. Gronwall, Victor K. La Mer (1895–1966) and co-workers at Columbia University in New York improved the derivations by including additional terms in the expansions of exponential functions. E. Wicke and Manfred Eigen (b. 1927) took into account the finite sizes of ions, a matter neglected in the previous formulations. These improved treatments lead to complicated expressions for the conductivity that are difficult to apply to the experimental results. Another factor that must be taken into account is ion association. Although covalent molecules such as NaCl do not exist, it is possible for ions to become associated in solution with the formation of ion pairs Na^+Cl^-. In the case of ions of higher charge, ion-pair formation is even more important. A treatment of ion association was given in 1926 by Bjerrum.

Thermodynamics of ions

Even before the Debye–Hückel theory was formulated some consideration had been given to the thermodynamic properties of ions. In 1901 G. N. Lewis introduced the concepts of activities and activity coefficients (Section 4.4). For a reaction

$$aA + bB \rightleftharpoons yY + zZ$$

the equilibrium constant can be written as

$$K = \frac{[Y]^y[Z]^z}{[A]^a[B]^b} \frac{\gamma_Y^y \gamma_Z^z}{\gamma_A^a \gamma_B^b} = \frac{a_Y^y a_Z^z}{a_A^a a_B^b}, \tag{7.9}$$

where the γs, the activity coefficients, multiply the concentrations so as to give activities, a. For solutions, activity coefficients can be defined in a number of ways, for example with reference to a solution of zero concentration, for which they are taken to be unity. There is then the important question of how the activity of an ion will change as the concentration of the solution is changed.

In 1921 G. N. Lewis and Merle Randall introduced the important concept of the *ionic strength* of a solution. This quantity, usually given the symbol I, is defined as

$$I = \tfrac{1}{2} \Sigma z_J^2 c_J, \tag{7.10}$$

where c_J is the concentration of an ion J and z_J is its charge number (i.e., +1 for Na^+, −2 for SO_4^{2-}). The summation was taken over all the ions in the solution, and the factor of one-half was arbitrarily introduced so that the ionic strength of a uni-univalent electrolyte would be the same as its concentration. Lewis had realized from various lines of evidence that the effect of a dissolved salt on thermodynamic properties, such as solubility, is determined not by the concentration of a salt but by its ionic strength. For example, a 0.1 M solution of Na_2SO_4, having an ionic strength of 0.3 M, has much the same effect on thermodynamic properties as a 0.3 M solution of NaCl which also has an ionic strength of 0.3 M.

When the Debye–Hückel theory was formulated it was realized that it involves the ionic strength, and that the radius of the ionic atmosphere is inversely proportional to the square root of the ionic strength. A year after the

Debye–Hückel theory appeared the Danish physical chemist Johannes Nicolaus Brönsted (1879–1947), in collaboration with La Mer, derived on its basis a relationship between the activity coefficient γ_i of an ion i and the ionic strength:

$$\log_{10} \gamma_i = -z_i^2 B \sqrt{I}, \qquad (7.11)$$

where z_i is the charge number of the ion i and B can be expressed using the Debye–Hückel theory in terms of certain properties of the solution. According to this equation the activity coefficient γ_i is unity at zero ionic strength, and decreases towards zero as the ionic strength increases. Equation 7.11, referred to as the *Debye–Hückel limiting law*, is obeyed quite satisfactorily at very low ionic strengths, but deviations from it occur at higher ones.

7.3 Electrochemical cells

During the years immediately following Volta's invention in 1800 of his pile, numerous attempts were made to devise other cells that would generate electricity. Such cells have been called voltaic cells, galvanic cells, electrochemical cells and voltaic batteries. Strictly speaking the word 'battery' should only be used when two or more unit cells are connected together.

Several cells were devised by Daniell at King's College, London. One of his more useful cells consisted of an outer cylinder containing copper sulphate solution, inside which was an ox-gullet bag containing a rod of amalgamated zinc in dilute sulphuric acid. It produced an electromotive force (emf) of about 1.1 V. In a later modification the ox-gullet was replaced by a porous pot which contained the acid and the zinc electrode. Daniell was a close friend of Faraday, and he first described this cell in letters to him, and later in the *Philosophical Transactions* in 1836. In 1839 he put together a battery of 70 of his cells and produced a brilliant electric arc which produced 'sunburn' and some eye injury to himself and others.

In 1839 Grove devised a cell consisting of zinc in sulphuric acid and platinum in nitric acid, the two liquids being separated by means of a porous pot. This cell produced an emf of 1.8–2.0 V—greater than given by Daniell's cells—and cells of this type began to be used by Faraday in his demonstrations at the Royal Institution. Later, in 1842, Daniell alleged that his own cells had inspired the Grove cell, but Grove denied this and in 1842 and 1843 there was a sharp exchange of published letters between the two men.

Grove also devised what he called a gas voltaic battery, consisting of platinum electrodes immersed in acid solution, with hydrogen bubbled over one electrode and oxygen over the other. The source of the emf is the energy released by the combination of the two gases. Later Grove used other combinations of gases, such as hydrogen and chlorine, and oxygen and carbon monoxide. Cells of this type are today called fuel cells, and they are still the subject of much research to make them more effective for commercial use. Grove's first description of his gas battery was a brief account in a paper in 1839; further details appeared from 1842 to 1845.

An important innovation came later with the introduction of dry cells which are easier to transport than ones containing solutions or strong acids. As early as 1802 a dry cell had been constructed by Johann Ritter, but the development of dry cells sprang from the work of the French engineer Georges Leclanché (1839–1882) who, however, did not construct a dry cell himself. In 1868 he devised a cell in which rods of zinc and carbon were dipped into a solution of ammonium chloride. Wet cells of this type were used well into the twentieth century for door-bells and other purposes where electricity is required only occasionally and for short periods. The familiar dry cells now used very widely are a development of the Leclanché cell.

The cells mentioned so far are called primary cells. They cannot be regenerated by electrical means but only by replacing the solutions or electrodes. A storage cell or accumulator, which could be recharged by the passage of a current, was first devised by Ritter in 1803, but it was the work of the French physicist Gaston Planté (1834–1889) that led to the great development of storage batteries, which are used so widely today. Planté began his work in the 1860s and first demonstrated his batteries in 1879; they came into commercial use about two years later. The Planté cell consists of lead plates immersed in sulphuric acid. When the cell is charged the positive plate becomes plated with lead peroxide, and as the cell is discharged this process occurs in reverse.

Thermodynamics of electrochemical cells

The first attempt to relate the emf of a cell to the thermodynamics of the chemical reactions occurring in the cell was made by Joule in 1840 in connection with his investigations of the mechanical equivalent of heat. In that year an account of his work was presented to the Royal Society by Faraday, and Joule's own papers appeared in 1841 and 1843. His careful experiments led to an important conclusion about the heat produced when a current passes along a wire; in his own words

The calorific effects of equal quantities of transmitted electricity are proportional to the resistance opposed to its passage, whatever may be the length, thickness, shape, or kind of metal which closes the circuit; and also that *caeteris paribus*, these effects are in the duplicate ratio of the quantities of transmitted electricity, and, consequently, also in the duplicate ratio of the quantity of electricity.

By 'quantity of transmitted electricity' Joule meant the current; by 'duplicate ratio' he meant the square. His conclusion was therefore that the heat produced is proportional to the square I^2 of the current and to the resistance R. Since it is also proportional to the time t, Joule had shown that the heat is proportional to I^2Rt. By Ohm's law the resistance R is the potential drop V divided by the current I, and the heat is therefore proportional to IVt, or to VQ, where Q, equal to It, is the quantity of electricity. In modern (SI) units the product of emf (unit: volt, V) and quantity of electricity (unit: coulomb, C) is heat (unit: joule, J).

In a paper in 1851, however, Joule went wrong in supposing that the heat of reaction is equal to the electrical work; in fact at constant pressure it is the decrease in Gibbs energy that equals the work done. Such an error is understandable, since the second law of thermodynamics was still not fully understood. The same error was made by Helmholtz in his *Über die Erhaltung der Kraft* (1847) and by Kelvin in a paper published in 1851. In that paper Kelvin calculated the emf of the standard Daniell cell from the heat of reaction, and obtained 1.074 V, which is close to the measured value. However, it happens that in that cell the entropy change is small, so that the heat evolved is close to the work done (i.e., $\Delta H \approx \Delta G$).

That the work done by an electrochemical cell is equal to the decrease in what came to be called the Gibbs energy was first pointed out by Willard Gibbs in his 1878 paper. This idea was later extended by Helmholtz in a paper that appeared in 1882. If the cell is balanced by an external emf so that it does no work, the decrease in Gibbs energy is then equal to the heat produced in the external circuit:

$$\Delta G = -QV = -QE \qquad (7.12)$$

where E, the emf of the cell, is the value of the electrical potential V when the cell is doing no work. For a process

$$aA + bB \rightleftharpoons yY + zZ$$

it follows from Faraday's laws of electrolysis that if the process is accompanied by the passage of z mol of electrons, the current Q that passes is zF, where F is the Faraday constant. Helmholtz was thus led to the equation

$$\Delta G = -zFE \qquad (7.13)$$

which is the fundamental thermodynamic equation for an electrochemical cell.

7.4 Nernst's electrochemistry

In the 1880s and 1890s Walther Nernst carried out a number of investigations of great importance on electrochemical cells and the properties of ions in solution. During this period he almost single-handedly transformed electrochemistry from a subject in which there was little understanding of anything beyond the basic principles into one in which all the thermodynamic aspects were well understood.

Nernst obtained his Ph.D. degree in 1887 at the University of Würzburg where Kohlrausch was professor, but as was the custom at the time he studied at three other universities during the previous four years. After a period at the University of Zürich he attended the thermodynamics lectures of Helmholtz in Berlin. At the University of Graz he studied under Boltzmann whose interpretations of

Walther Nernst

(1864–1941)

Nernst was born in Brieson, West Prussia (now Wabrzezno, Poland), and he studied physics and chemistry at the Universities of Zürich, Berlin, Graz and Würzburg, obtaining his doctorate from Würzburg in 1887. In that same year he joined Ostwald in Leipzig and began lecturing there in 1889. In the following year he moved to the University of Göttingen, where in 1891 he became professor of physical chemistry, at once setting about to establish a distinguished institute of physical chemistry. In 1905 he became professor of physical chemistry at the University of Berlin, and after 1922 he was at the same time President of the Physikalisch-technische Reichsanstalt of Berlin-Charlottenburg.

Nernst ranks with van't Hoff, Ostwald and Arrhenius as one of the great pioneers of physical chemistry, and his work covered an even wider range than that of the other three. His Nobel Prize for chemistry, awarded to him in 1920, was particularly 'in recognition of his thermochemical work', and it is his work on his heat theorem, sometimes called the third law of thermodynamics, that has received the most recognition. His previous work in electrochemistry, described in Section 7.4, has, however, had an even greater impact on the development of physical chemistry and perhaps on the physical sciences in general. In his first work in Ostwald's laboratory at Leipzig, between 1887 and 1889, he developed fundamental relationships relating to electrochemical cells. Later, particularly during the time he was at Göttingen, he worked on a number of other problems in electrochemistry, including the theory of the electrostriction of ions (1894), the theory of the solubility product (1901), and what has come to be called the Nernst diffusion layer. In 1897 he suggested a new way of determining dielectric constants, involving alternating currents.

By the time Nernst had taken up his appointment in Berlin his interests had shifted in the direction of the behaviour of solids at low temperatures. His famous heat theorem (Section 4.5) was announced in 1906, and for many years his laboratory was devoted to the study of the thermal properties of substances. This

physical processes in terms of atomic and statistical concepts made a deep impression on him. While he was in Würzburg he became acquainted with Arrhenius who had just completed his theory of electrolytic dissociation. Later Arrhenius introduced Nernst to Ostwald who in 1887 had become professor of physical chemistry at the University of Leipzig. Ostwald appointed Nernst his research assistant, a position he held until 1891 when he moved to Göttingen.

In Ostwald's laboratory the emphasis was on electrolyte theory, thermo-

work inspired much later work, particularly that carried out in the Clarendon Laboratory at Oxford. Nernst also made one important contribution to chemical kinetics, his interpretation in 1918 of the photochemical combination of hydrogen and chlorine in terms of a chain reaction (Section 8.4).

Nernst exerted a wide influence in many ways. His book *Theoretische Chemie*, which first appeared in 1893 and went through many editions and translations, was the leading book in physical chemistry until the 1920s; in its preface Nernst said that his object was to present 'all that the physicist must know of chemistry, and all that the chemist must know of physics'. Many who later did important work in physics and chemistry gained much from their early association with Nernst; among them may be mentioned F. A. Lindemann (later Lord Cherwell), G. N. Lewis and Irving Langmuir.

Nernst had an unusual personality, and Einstein, himself hardly conventional, said that 'he was so original that I have never met anyone who resembled him in any essential way'. Nernst's character showed many contradictions: he could be extremely generous and helpful, and sometimes be devastatingly sarcastic. In many ways he was fairminded, but was by no means noted for his modesty; his students have reported that in his lectures he presented physical chemistry in a way that left the impression that he had done all the work himself. The author remembers vividly a lecture he gave in Oxford in 1937. He ended it, as apparently he ended many of his lectures, by commenting that it had taken three people to formulate the first law of thermodynamics, two for the second law, but that he had been obliged to do the third law all by himself. He added that it followed by extrapolation that there could never be a fourth law.

References: E. N. Hiebert, DSB, 10, 432–453 (1974); Hiebert (1978, 1982, 1983); Mendelssohn (1973); Servos (1990).

dynamics and the colligative properties. Several others who later did distinguished work in physical chemistry were also in Ostwald's laboratory at the time: for example James Walker (1863–1935), later professor at Dundee and Edinburgh and author of one of the earliest textbooks on physical chemistry; Wilhelm Meyerhofer, who did pioneering work on the phase rule (Section 4.4); and Ernst Otto Beckmann, well known for his work on the colligative properties (Section 4.4). Although Nernst did not agree with Ostwald's rejection of the existence of atoms he profited greatly by their association and always retained a great respect for him.

Nernst's first electrochemical paper from Ostwald's laboratory appeared in 1888 and was concerned with the diffusion of electrolytes, with the potential differences established at an interface between two solutions, and with the emf of a concentration cell. He began with the diffusion equation that had been given by the German physiologist Adolf Eugen Fick (1829–1901). According to what is known as *Fick's first law of diffusion*, the rate of diffusion of a solute across an area A is given by

$$\frac{dn}{dt} = -DA\frac{dc}{dx} \qquad (7.14)$$

where dc/dx is the concentration gradient and the proportionality constant D is known as the diffusion constant. The negative sign takes care of the fact that if there is a positive concentration gradient from left to right, the direction of diffusive flow is from right to left.

Nernst first gave a brief discussion of the relatively simple problem of the application of Fick's law to a non-electrolyte. He then considered the diffusion of a 1:1 electrolyte in which the two ions have different mobilities u_+ and u_-, realizing that the diffusion coefficient for the pair of ions will depend on some kind of mean of the two mobilities. He showed that this mean mobility is in fact

$$u_+ = \frac{2u_+u_-}{u_+ + u_-} \qquad (7.15)$$

He then obtained a relationship between the diffusion coefficient D_+ for a 1:1 electrolyte and the mobilities of the individual ions. In modern notation this relationship is

$$D_\pm = \frac{RT}{F}\frac{2u_+u_-}{u_+ + u_-}. \qquad (7.16)$$

Nernst did not actually express the relationship in this form, giving a numerical value in place of the factor RT/F. Since the equation in the form of eq 7.16 has a connection with an equation derived in 1905 by Einstein relating the diffusion coefficient to the frictional coefficient in a diffusion process, it is often now referred to as the *Nernst–Einstein equation*. However, Nernst himself certainly had the idea of the equation in his 1888 paper even though he did not express it in that form.

In the same 1888 paper Nernst treated the liquid junction potential that is established at the boundary between two solutions of a strong electrolyte having different concentration. He considered in particular the boundary between two solutions of hydrogen chloride of concentrations c_1 and c_2. The H^+ ions move faster than the Cl^- ions from the more concentrated to the less concentrated solution, but this produces a *Doppelschicht* (electrical double layer) in which the electric field retards the H^+ ions and speeds up the Cl^- ions so that in the steady state both move at the same velocities. His expression for the liquid junction potential is equivalent to

$$E_{lj} = \frac{RT}{F} \frac{u_+ - u_-}{u_+ + u_-} \ln \frac{c_1}{c_2} = \frac{RT}{F} (t_+ - t_-) \ln \frac{c_1}{c_2}, \tag{7.17}$$

where t_+ and t_- are the transport numbers; he did not, however, write down the RT/F term but gave a numerical factor. Nernst called attention to the fact that the liquid junction potentials depend only on the ratio of the concentrations and not on their absolute values. Thus if the two solutions are replaced by another pair at higher concentrations in which the ratio is the same, the liquid junction potential will remain unchanged. Nernst referred to this as the *principle of superposition*, and mentioned it in many of his discussions of electrochemical cells.

In his second electrochemical paper, which appeared in 1889, Nernst treated the case of a concentration cell, in which two solutions of concentrations c_1 and c_2 are separated by a porous membrane, with an electrode of the same material in each solution; an example of such a cell is

$$\text{Ag} \mid \text{AgNO}_3 (c_1) : \text{AgNO}_3 (c_2) \mid \text{Ag}$$

The driving force which produces the emf is the tendency of the two concentrations to become equal, and Nernst deduced that the emf is proportional to the logarithmic ratio of the osmotic pressures due to the ions in the two solutions. The osmotic pressures are proportional to the ionic concentrations, and for a cell in which the liquid junction potential is eliminated Nernst obtained for the emf an expression equivalent to

$$E = \frac{RT}{F} \ln \frac{c_2}{c_1}. \tag{7.18}$$

Again, the expression he gave had a numerical factor instead of RT/F. If the liquid junction potential (eqn 7.17) is included the cell potential is

$$E = \frac{RT}{F} [(1 - (t_+ - t_-)] \ln \frac{c_2}{c_1} = \frac{RT}{F} 2t_- \ln \frac{c_2}{c_1}, \tag{7.19}$$

since $t_+ + t_- = 1$. Nernst carried out experimental studies of cells of this and other types for many years. For strong electrolytes he and his co-workers observed discrepancies between c_1/c_2 values found using this equation and c_1/c_2 values deduced from conductivity measurements. This provided evidence that the Arrhenius theory of a shifting equilibrium does not apply to strong electrolytes, which are completely dissociated at all concentrations.

Nernst's third electrochemical paper, which appeared in 1889, is over 50 pages long and comprehensive since it was his *Habilitationsschrift*, presented to the University of Leipzig in order to secure a lectureship. This paper reviews some of the material in the previous two papers and also deals with a number of other problems. For example, he arrived at the expression for the emf of a cell other than a concentration cell, such as

$$Zn \mid ZnCl_2(c) \mid Hg_2Cl_2(s) \mid Hg$$

Here there is no liquid junction, and the source of the emf is a chemical reaction rather than a concentration difference. The cell reaction for this particular cell is

$$Zn + Hg_2Cl_2 \rightleftharpoons Zn^{2+} + 2Hg + 2Cl^-$$

and when this occurs two electrons are transferred from one electrode to the other. In general the cell reaction may be written as

$$aA + bB \rightleftharpoons yY + zZ$$

and the number of electrons denoted as z. If E_0 is the emf of a standard cell, having unit concentrations, the standard Gibbs energy change for the cell reaction is [compare eqn 7.13]

$$\Delta G^0 = -zFE^0 \tag{7.20}$$

Insertion of equations 4.14 and 4.15 into the thermodynamic relationship between ΔG^0 and ΔG then led Nernst to a relationship equivalent to

$$\Delta G = \Delta G^0 + \frac{RT}{zF} \ln \frac{[Y]^y[Z]^z}{[A]^a[B]^b} \tag{7.21}$$

Equations 7.19 and 7.21 are now generally known as *Nernst equations*.

After carrying out these investigations Nernst continued to make many further theoretical and experimental studies on electrochemical cells. In 1891 he left Leipzig to become professor of physical chemistry at Göttingen, and in 1894 he published a number of very important papers. In one of them he pointed out that as a result of the high dielectric constant of a solvent such as water the tendency of dissociation to occur will be increased. A paper with Paul Karl Ludwig Drude (1863–1906) treated the binding of solvent molecules to ions, the word *electrostriction* being coined to refer to such binding. In their quantitative treatment they regarded the solvent as a continuous medium. This paper was important with respect to the controversy that still existed as to whether dissociation actually occurs in solution. Arrhenius's opponents emphasized the importance of solvation, which they thought could explain the conductivity results without the need to postulate dissociation. Drude and Nernst's work showed that solvation effects are indeed important in the case of the ions themselves. In 1920 Max Born presented another treatment of the solvation of ions, again regarding the solvent as a continuous dielectric. Such theories are useful as a first approximation; treatments in terms of the molecular structures of solvents have been carried out more recently, but they are much more complicated and do not bring about a great improvement.

In 1893 appeared the first edition of Nernst's *Theoretische Chemie*, devoted mainly to thermodynamics, kinetic theory and chemical kinetics. In this first edition electrochemistry was treated as part of thermodynamics and not at great length, but by the second edition, which appeared in 1898, the electrochemistry chapter was greatly expanded. The book differed markedly from those of Ostwald in its acceptance of the atomic theory. There were many editions of the book, the fifteenth edition being published in 1926, and until the 1920s it was the foremost textbook of physical chemistry.

A matter dealt with by Nernst in 1901 was the theory of the *solubility product*, of particular importance in analytical chemistry. In 1904 he published an important paper on the theory of the layer that exists at an electrode, now known as the *Nernst diffusion layer*. During his later career Nernst's main interests gradually shifted from 'wet' to 'dry' chemistry and physics, particularly to his work on the heat theorem described in Section 4.5, but he continued to make significant contributions to solution chemistry.

7.5 Electrode processes

After the thermodynamics of ions and of cells had been elucidated, some attention began to be devoted to the nature of the processes that occur at electrodes during an electrolysis or during the operation of an electrochemical cell. The first significant work along these lines was done by Julius Tafel (1862–1918) at the University of Würzburg. Tafel was primarily an organic chemist; he had been an assistant to Emil Fischer (1852–1919) and was active in the field of preparative electro-organic chemistry, which was a popular field in the late nineteenth century.

In the early 1900s Tafel investigated the electrolysis of solutions in which hydrogen gas is evolved at the negative electrode. He found that in order for electrolysis to proceed at an appreciable rate the voltage that must be applied must be greater than the equilibrium voltage; the additional voltage is called the *overvoltage*. Tafel obtained an empirical expression for the relationship between the overvoltage η and the current density i (the current per unit area of electrode):

$$\eta = a + b \ln i \qquad (7.22)$$

This equation, fundamental to the study of the kinetics of electrode processes, has come to be called the *Tafel equation*.

It took a number of years for the significance of this equation to be understood, in terms of theories of overvoltage. In 1930 T. Erdey-Gruz and Max Volmer (1885–1965) gave a kinetic, as opposed to a thermodynamic, derivation of the Nernst equation 7.21. They gave expressions for the rates of the electrode processes occurring in forward and reverse directions: when the cell is operating reversibly these rates are equal, but when a current flows the rate in one direction is not exactly balanced by the rate in the other, and there is then an overvoltage.

The logarithmic relationship between the overvoltage and the current density, as expressed by the Tafel equation 7.22, can be understood from this point of view. According to the Arrhenius equation for reaction rates (Section 8.2) there is a linear relationship between the logarithm of the rate of a reaction and the height E of the energy barrier for the reaction (the activation energy). The current density i is proportional to the rate of the electrode reaction that controls the flow of current, and the overvoltage applied has the effect of lowering the energy barrier to reaction.

In the early 1930s Ronald Wilfred Gurney (1898–1953), then at the University of Manchester, published an important theoretical treatment of the electrolytic evolution of hydrogen gas from an acidified solution. In 1932 R. H. Fowler, with whom Gurney had worked at Cambridge, gave a similar theoretical discussion. These treatments dealt with the transfer of a proton from H_3O^+ in solution to the surface of the metal electrode, taking into account hydration effects but not considering the possible adsorption of hydrogen atoms at the surface. Account was taken of the distribution of energy of the electrons in the metal (in modern terms, of the *Fermi levels*) and of the energy levels in the H_3O^+ ion. An important feature of Gurney's treatment is that he suggested, for the first time for any process, that there might be *quantum-mechanical tunnelling* through the energy barrier, His idea was that the rate is controlled by the transfer of electrons and that these did not have to surmount the barrier but could tunnel through it, a possibility not permitted in classical mechanics. Quantum-mechanical tunnelling has later been established for many chemical reactions but careful studies by Brian E. Conway and others at the University of Ottawa have not yet revealed any such evidence for an electrode process.

The Gurney-Fowler treatment of electrode processes has been extended in various ways. In 1934 Bawn and Ogden, at the University of Manchester, presented a quantum-mechanical treatment of the transfer of a proton at the surface of an electrode, and two years later John Alfred Valentine Butler (1899–1977), then at the University College of Swansea, gave a treatment in which account was also taken of adsorption at the surface of the electrode. When adsorption was neglected, as in Gurney's treatment, the predicted activation energies for the discharge of a proton were much higher than the experimental values and were independent of the nature of the surface. Butler was able to account for the substantial differences in activation energy for nickel and mercury surfaces. Soon afterwards, in 1939, Henry Eyring, Samuel Glasstone and the present author at Princeton University gave an interpretation of electrode processes on the basis of potential-energy surfaces and transition-state theory which had been developed in 1935 (Section 8.3). Much additional work has been done in this important field; the matter is complicated, different mechanisms being important under different conditions.

The electrical double layer

An important matter relating to electrode processes is the electrical nature of the surface between two phases, such as a solid and a solution. It was suggested

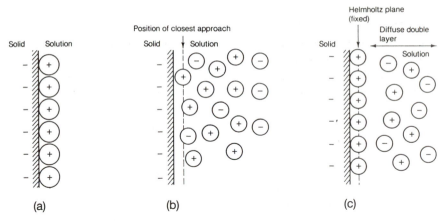

FIGURE 7.3 Three models for the structure of the electrical double layer: (a) The Helmholtz fixed double layer (1879). (b) The Gouy–Chapman diffuse double layer (1910–1913). (c) The Stern double layer (1924), a combination of a and b.

in 1879 by Helmholtz that such a surface can be regarded as bearing positive or negative charges, and that a layer of opposite charges will be attracted to them. If the charges on the surface are negative, as in Figure 7.3(a), there will be a layer of positive charges. This *fixed double layer* is now often referred to as the *Helmholtz double layer*.

This idea was modified in 1910 by the French physicist Georges Gouy (1854–1926) and in 1913 by the British chemist David Leonard Chapman (1869–1959). They pointed out that the Helmholtz theory is deficient in neglecting the Boltzmann distribution of the ions. They suggested that on the solution side of the interface there is not a simple layer of ions but an ionic distribution that extends some distance from the surface; in other words, there is a *diffuse double layer* as shown in Figure 7.3(b). In the example shown in the diagram, the surface is negative, so that there are more positive ions than negative ions in close proximity to the surface. The idea behind the Gouy–Chapman theory is very similar to that in the later Debye–Hückel theory of the ionic atmosphere surrounding an ion (Section 7.2). Indeed, Chapman's mathematical treatment in 1913 of the distribution of ions at a surface is essentially a two-dimensional version of the Debye–Hückel theory which appeared ten years later.

The Gouy–Chapman theory did not prove entirely satisfactory, and in 1924 a further modification was made by the German physicist Otto Stern (1888–1969), whose model is represented in Figure 7.3(c). Stern combined the fixed double-layer model of Helmholtz with the diffuse double-layer model of Gouy and Chapman; there is a fixed layer at the surface and also a diffuse layer. To explain certain results, however, it has been found necessary to develop even more elaborate models.

Polarography

An important analytical technique that involves electrodes at which there is an overvoltage is polarography, which was developed by the Czech physical chemist Jaroslav Heyrovsky (1890–1967), at first at University College, London, and then at Charles University in Prague. It depends on the use of a dropping mercury electrode, in which drops of mercury constantly fall through a solution. Dropping mercury electrodes had been used much earlier, for example by Ostwald in 1887, and Nernst later worked out the theory of them. The surface of the mercury in the dropping electrode is being continuously renewed. If a steadily increasing potential is passed through a cell involving a dropping mercury electrode, the current varies according to a pattern which depends on the nature of the ions present in the solution.

Beginning in about 1918 Heyrovsky developed the dropping mercury electrode into a highly effective and delicate device for chemical analysis. His first paper on the subject appeared in 1922, and he coined the word 'polarography' in 1925. In 1950 the Polarographic Institute was founded in Prague, and Heyrovsky was appointed its first Director. He was awarded the 1959 Nobel prize for chemistry for this work. Polarography is now used widely as an analytical tool, and by the time of Heyrovsky's death in 1967 over 20 000 papers had been written on the technique and its applications.

Suggestions for further reading

General accounts of the history of electrochemistry are to be found in

J. R. Partington, *A History of Chemistry*, Macmillan, London, 1964: Vol. 4, Chapter 21, 'Electrochemistry'.

G. Dubpernel and J. H. Westbrook (Eds.), *Selected Topics in the History of Electrochemistry*, The Electrochemical Society, Princeton, 1978.

J. T. Stock and M. V. Orna, *Electrochemistry, Past and Present*, American Chemical Society, Washington, DC, 1989. The introductory paper, by J. T. Stock, 'Electrochemistry in retrospect: an overview', is a useful review of the subject.

For accounts of Faraday's work on electrolysis see

F. A. J. L. James, Michael Faraday's first law of electrochemistry, in *Electrochemistry, Past and Present*, op. cit., pp. 32–49.

L. Pearce Williams, *Michael Faraday: A Biography*, Chapman and Hall, London, 1965.

For further details about Faraday's electrochemical terms see

S. Ross, Faraday consults the scholars: The origins of the terms of electrochemistry, *Notes & Records of the Royal Society*, **16**, 187–220 (1961).

The development of Arrhenius's theory of electrolytic dissociation is discussed in

R. B. Root-Bernstein, The ionists: founding physical chemistry, 1872–1890, Ph.D. Thesis, Princeton University, 1980.

K. J. Laidler, Chemical kinetics and the origins of physical chemistry, *Arch. Hist. Exact Sci.*, **32**, 43–75 (1985).

For accounts of the investigations leading to the understanding of strong electrolytes see

K. S. Pitzer, Gilbert N. Lewis and the thermodynamics of strong electrolytes, *J. Chem. Education*, **61**, 104–107 (1984).

J. H. Wolfenden, The anomaly of strong electrolytes, *Ambix*, **19**, 175–196 (1972).

Various other aspects of solutions are treated in the following publications:

R. G. A. Dolby, Debates over the theory of solutions, *Hist. Stud. Phys. Sci.*, **7**, 297–404 (1976).

K. J. Laidler, A century of solution chemistry, *Pure & Appl. Chem.*, **62**, 2221–2226 (1990).

Lars Onsager, The motions of ions: principles and concepts, *Science*, **166**. 1359–1364 (1969).

For accounts of Nernst's work on electrochemistry see

Mary D. Archer, Genesis of the Nernst equation, in *Electrochemistry, Past and Present*, op. cit., pp. 115–126.

E. N. Hiebert, Nernst and electrochemistry, in *Selected Topics in the History of Electrochemistry*, op. cit., pp. 180–200. The Archer and Hiebert articles nicely complement each other; Archer gives more of the scientific details, Hiebert more of the background.

For accounts of the kinetics of electrode processes see

B. E. Conway, Historical development of the understanding of charge-transfer processes in electrochemistry, in *Electrochemistry, Past and Present*, op. cit., pp. 152–164.

The development of polarography is discussed in

M. Heyrovsky, L. Novotny and I Smoler, Past and future of the dropping electrode, in *Electrochemistry, Past and Present* (op. cit.), pp. 370–379).

L. R. Sherman, Jaroslav Heyrovsky (1890–1967), *Chemistry in Britain*, **26**, 1165–1168 (1990).

P. Zuman, With the drop of mercury to the Nobel Prize, in *Electrochemistry, Past and Present* (op. cit.), pp. 339–369.

CHAPTER 8

Chemical kinetics

Today the study of the rates of chemical reactions, and of the factors on which they depend, is usually referred to as 'chemical kinetics'. The expression 'chemical dynamics' is now applied to an important branch of chemical kinetics that is concerned with the details of the individual chemical events that occur during the course of a chemical reaction. There has been a change of meaning, since in the last century what we now call chemical kinetics was usually called chemical dynamics, which referred also to the study of matter at equilibrium. Ostwald, for example, in his 1904 Faraday Lecture, defined chemical dynamics as 'the theory of the progress of chemical reactions and the theory of chemical equilibrium', and a similar definition was given by van't Hoff in his *Lectures on Theoretical and Physical Chemistry* (1899–1900).

Hardly any quantitative studies of the rates of chemical reactions were made until the latter part of the eighteenth century. In 1777 Karl Friedrich Wenzel (1740–1793), the Director of the foundries at Freiburg, described some measurements of the rates of solution of metals in acids but gave no details, merely saying that the rate increased with increasing concentration of acid. In 1818 the French chemist Louis Jacques Thénard (1777–1857) described some studies of the rate of decomposition of hydrogen peroxide, a substance he had just discovered. Thénard was a chemist of great distinction but like most other chemists of his time his main interest was in the isolation of new elements and compounds, and to him rate measurements were not of great interest or significance.

Some of the more important contributions to chemical kinetics are listed in Table 8.1. The birth of the subject is often taken to have occurred in 1850, when the German chemist Ludwig Ferdinand Wilhelmy (1812–1864) studied the rate of inversion of sucrose. This pioneering work is of great significance, since it was the first in which a quantitative approach was taken to reaction rates. Wilhelmy, a *Privatdozent* at Heidelberg from 1849 to 1854, used a polarimeter to follow the reaction at various concentrations of acid, and observed that the instantaneous rate of change of the sugar concentration was proportional to the concentration of both the sugar and the acid. He set up, for the first time, a differential equation expressing the rate as proportional to the sugar concentration, integrated it, and showed that the resulting equation was consistent with his experimental results. He also examined the influence of temperature on the reaction rate, and proposed an empirical equation to

TABLE 8.1   Highlights in chemical kinetics

Date	Author	Contribution	Section
1850	Wilhelmy	Rate-concentration dependence	8.1
1865	Harcourt and Esson	Time course of reactions	8.1
1884	van't Hoff	Differential method; temperature dependence	8.1, 8.2
1889	Arrhenius	'Arrhenius equation'	8.2
1891	Ostwald	Theory of catalysis	8.6
1899	Chapman	Theory of detonation	8.5
1913	Chapman	Steady-state treatment	8.4
1914	Marcelin	Potential-energy surfaces	8.3
1917	Trautz; W. C. McC. Lewis	Collision theory	8.3
1918	Nernst	Atomic chain mechanism	8.4
1921	Langmuir	Surface reactions	8.6
1921–1922	Lindemann; Christiansen	Unimolecular reactions	8.5
1927–1928	Semenov; Hinshelwood	Branching chains	8.5
1931	Eyring and M. Polanyi	Potential-energy surface for $H + H_2$	8.3
1934	Rice and Herzfeld	Organic chain mechanisms	8.5
1935	Eyring; Evans and M. Polanyi	Transition-state theory	8.3
1949	Porter and Norrish	Flash photolysis	8.4
1954	Eigen	Relaxation methods	8.5
1980	J. C. Polanyi	Spectroscopy of transition species	8.7

interpret the behaviour, Wilhelmy's scientific career was a short one; after teaching for five years at Heidelberg he retired to private life in 1854 at the age of 42. Little notice was taken of his important work until 1884, when Ostwald called attention to it, and in 1891 Ostwald reprinted Wilhelmy's paper in his *Klassiker*.

Wilhelmy's work was followed by some investigations by the French chemists Pierre Eugène Marcellin Berthelot (1827–1907) and Léon Péan de Saint-Gilles (1832–1863) who in 1862 began to publish the results of a number of studies of the reaction between ethanol and acetic acid to give ethyl acetate and water.

Augustus George Vernon Harcourt

(1834–1919)

Harcourt, who was born in London, was a member of a family that achieved distinction in a variety of fields. His paternal grandfather was Archbishop of York, and his uncle, the Rev. William Vernon Harcourt, who had a strong interest in science, particularly chemistry, was one of the founders of the British Association for the Advancement of Science, becoming its President in 1839. The latter's son, Sir William Harcourt, was twice Chancellor of the Exchequer.

Vernon Harcourt was educated at Balliol College, Oxford, and while still an undergraduate became an assistant to Sir Benjamin Brodie (1817–1880), the professor of chemistry, who for lack of more suitable accommodation had been carrying out his research and teaching in an improvised laboratory in a Balliol cellar. In 1858 Harcourt helped Brodie to transfer his equipment from this cellar to the not quite completed chemistry laboratory which was part of the University Museum.

In the following year Harcourt was appointed Dr Lee's Reader in Chemistry at Oxford, a post that had been created by Christ Church and which he retained until his retirement in 1902. He was also appointed a 'Student' of Christ Church, the title given by that college to what in any other college would be called a 'Fellow'. Although he was probably the most distiguished Oxford chemist of his time Harcourt never became a professor, perhaps because he preferred his college appointment which did not carry the responsibilities and distractions of a professorship.

Their interest was mainly in the equilibrium that became established, but they also obtained some results on the rate of combination of ethanol and acetic acid, finding it to be proportional to the product of the reactant concentrations.

At about the same time the Norwegians Guldberg and Waage carried out their work on the 'law of mass action'. As noted in Section 4.4, they arrived at equilibrium equations on the basis of assumed kinetic equations, a procedure which is not satisfactory, and their investigations did not really contribute to the understanding of reaction rates.

8.1 The course of chemical change

The beginning of a new phase in chemical kinetics was the result of a collaboration between a chemist and a mathematician. Beginning in 1865, Augustus

Christ Church placed at Harcourt's disposal a building that since 1767 had been used mainly as an anatomy laboratory, but partly for the teaching of chemistry. Harcourt converted it into a laboratory for teaching and research in chemistry, and it was used for that purpose until 1941. Beginning in about 1865 Harcourt embarked on a programme of research in chemical kinetics, paying particular attention to the reaction between hydrogen peroxide and hydrogen iodide, and that between potassium permanganate and oxalic acid (Section 8.1). His very precise rate measurements were analysed by the mathematician William Esson, since Harcourt had only a limited ability in mathematics. Their collaboration lasted for over half a century, their latest work being on the temperature dependence of reaction rates (Section 8.2). Harcourt also carried out some work of a very practical kind; he designed the pentane lamp as a standard of brightness, and he developed a chloroform inhaler that was used for many years.

Harcourt was an excellent teacher of chemistry, and he exerted a considerable influence on the development of physical chemistry. Several of his students achieved distinction, notably H. B. Dixon, D. L. Chapman and N. V. Sidgwick. A significant fraction of British physical or physical-organic chemists are his chemical descendants.

References: E. E. Daub, DSB, 6, 109–110 (1972); Dixon (1920); King (1981, 1982, 1983, 1984); Laidler (1985, 1988); Shorter (1980).

George Vernon Harcourt (1834–1919), a Student [Fellow] of Christ Church, Oxford, carried out some important experimental studies of the kinetics of the reaction between hydrogen peroxide and hydrogen iodide, and of that between potassium permanganate and oxalic acid. Recognizing that his mathematical skills were insecure Harcourt enlisted the collaboration of a Scottish colleague, William Esson (1838–1916), who was a Fellow of Merton College and tutor in mathematics. Esson devoted his full attention to kinetic problems, carrying out no other research, and the collaboration lasted for over half a century. Starting with the differential equations for several types of chemical reactions, Esson integrated them and obtained expressions for the amounts of product formed as a function of time. He dealt in this way with what are now called 'first-order' reactions, in which the rate is proportional to the concentration of a single reacting substance, with 'second-order' reactions in which the rate is proportional to the product of two concentrations (or to the square of one), and with consecutive reactions in which the product of one first-order reaction under-goes another first-order reaction. The resulting equations were then used to analyse the experimental results that Harcourt obtained in his laboratory at Christ Church.

The next event of great importance in the history of chemical kinetics was the publication, in 1884, of van't Hoff's *Études de dynamique chimique*, which as noted in Section 4.4 was also significant for the development of chemical thermodynamics. Particularly during the period that van't Hoff was at the University of Amsterdam, from 1878 to 1896, he and his students carried out investigations of great importance on chemical equilibrium and chemical kinetics. Van't Hoff had a genius for reaching important generalizations from the results of a few well-planned experiments. During the first few years that he was at Amsterdam he published little but was carrying out crucial experiments which were described and interpreted in the *Études*, a book that is packed with important and novel material. Subsequent to its publication in 1884 van't Hoff continued work on chemical equilibrium and kinetics until switching his in-terest to the phase rule after he went to Berlin in 1896 (Section 4.4).

In his book van't Hoff extended and generalized the mathematical analysis of the course of chemical change that had been given by Wilhelmy and by Harcourt and Esson. One important innovation was that he recognized the significance of what we now call the 'order' of a reaction, although he did not use that term; it was introduced in 1887 by Ostwald. Reactions were referred to by van't Hoff as 'monomolecular' if their rates were proportional to the first power of the concentration of a reactant. He called them 'bimolecular' if the rates were proportional to the square of a reactant concentration or to the product of two reactant concentrations. Today we follow Ostwald's terminology in referring to these classes of reactions as 'first order' and 'second order'. These expressions relate to the empirical equations representing the depend-ence of rate on concentration. We now use the expressions 'unimolecular', 'bimolecular' and 'trimolecular' to indicate the number of molecules entering into an individual chemical act, as deduced from the kinetic evidence; the order does not always correspond to the molecularity. Although he used a different terminology van't Hoff was quite aware of this distinction, and he referred not only to the 'molecularity' [order] but to the 'number of molecules entering into reaction'. To avoid confusion, in what follows the modern terminology will always be used.

In the sections of the *Études* dealing with kinetics van't Hoff discussed a number of examples, giving many previously unpublished experimental results that had been obtained by himself and his assistants in the Amsterdam labor-atories. As an example of a first-order reaction he considered the decomposi-tion of arsine in the gas phase, a reaction he found to occur to an appreciable extent on the surface of the reaction vessel:

$$2AsH_3 \rightarrow 2As + 3H_2.$$

He wrote down the differential equation for this reaction,

$$-\frac{dc}{dt} = kc \tag{8.1}$$

where c is the concentration of the reactant at time t and k is a constant, known as the rate constant. He integrated the equation, obtaining an expression for c as a function of t, and obtained from his data the constant k. He recognized that for a first-order reaction the fraction of the substance being consumed in a given time is independent of the initial amount.

As an example of a second-order reaction van't Hoff considered the hydrolysis of ethyl acetate by caustic soda, a reaction he represented as

$$NaOH + CH_3COOC_2H_5 = CH_3COONa + C_2H_5OH$$

This reaction had been investigated experimentally by van't Hoff's assistant Lodewijk Theodorus Reicher (1857–after 1933), who published his results in 1885–1887. Again, van't Hoff showed that the integrated rate equation was in agreement with the experimental results. Van't Hoff also derived the integrated rate equation for a reaction of order n. He could not, however, find examples for which n is greater than two, and these are indeed rare; a few reactions with $n = 3$ have been found, but there are still no known reactions with n greater than three.

In the *Études* van't Hoff introduced a general method, still useful today, for analysing kinetic results and determining the order of a chemical reaction. This is the *differential method*, which involves measuring rates, v, at various concentrations c of the reactant. Then, if the relationship between the two is of the form

$$v = kc^n, \tag{8.2}$$

the value of n is the slope of a plot of $\log v$ against $\log c$. He applied this method, for example, to some results that had been obtained by Reicher on the action of bromine on fumaric acid.

In 1896 a second edition of the *Études* appeared, now in German, and in the same year there was an English translation. The German edition was prepared in collaboration with Ernst Julius Cohen (1869–1944) who was then van't Hoff's assistant in Amsterdam and who succeeded him as professor there in 1896, finally becoming a victim of the Nazis in the concentration camp at Auschwitz. The English translation was prepared by Thomas Ewen (1868–1955) who had been an assistant to van't Hoff in Amsterdam and was then a demonstrator in chemistry at the Yorkshire College of Science in Leeds (which in 1904 developed into the University of Leeds). These 1896 editions are quite similar to the first edition, but contain some additional experimental data. Also, Arrhenius's theory of electrolytic dissociation (Section 7.2), published in 1887, allowed a more satisfactory interpretation of some of the work on the kinetics of reactions in solution.

Another book that had an important influence on the development of chemical kinetics was Ostwald's *Lehrbuch der allgemeinen Chemie*, the first volume of which appeared in 1885, the second in 1887. The writing of this book was done while Ostwald was professor of chemistry at the Riga Polytechnic Institute, and after William Allen Miller's *Chemical Physics* (1855) it was the first book to deal in a comprehensive way with the new field of physical chemistry. In 1886 Ostwald studied van't Hoff's *Études*, and the second volume of his *Lehrbuch* pays generous tribute to it and shows much evidence of its influence. It was in this second volume that Ostwald introduced the expression 'order of reaction' to replace van't Hoff's 'molecularity'. The second edition of the *Lehrbuch* came out in parts from 1891 to 1902, and was never completed. In it Ostwald dealt with other kinetic problems of fundamental importance. In Volume 2 , Part 1, which appeared in 1893, he considered successive reactions, in which the product of one reaction undergoes a second reaction, the product of which may undergo further reactions. In that same volume he introduced his 'method of isolation'; if the rate of a reaction depends on the concentration of two reacting substances, the concentration of one of them can be held at a high value so that it does not change appreciably during the process. The order with respect to the other reactant can then be obtained.

In various publications Ostwald put forward a number of other kinetic principles and procedures. In 1888 he introduced the idea of the *half life* of a reactant, which is the time it takes for half of a reactant to be consumed. He showed how, by finding out how half lives vary with the initial concentrations, one can determine the order of a reaction. In a paper in 1900 he discussed coupled reactions, such as two reactions having a common reactant occurring in the same system:

$$A + B \rightarrow Y,$$
$$A + C \rightarrow Z.$$

He and others had observed that in some systems one reaction is sometimes accelerated by the other, but he was unable to find any satisfactory explanation for this effect. This was inevitable, because at the time little was known of the mechanism of reactions; we now realize that an intermediate in one reaction can affect the rate of another.

By the turn of the century a fair amount was understood about the time course of reactions of various types, but almost nothing was known about the mechanisms of reactions. An understanding of such mechanisms had to await the development of new experimental techniques, for example techniques for the detection of reaction intermediates such as free radicals.

8.2 Temperature dependence of reaction rates

Studies of the temperature dependence of the rates of chemical reactions played a particularly important role in chemical kinetics. The problem proved to be a difficult one, and for a period of about 60 years, from about 1850 to 1910, in spite of considerable effort, there was much uncertainty and controversy. An important monograph on chemical kinetics by Joseph William Mellor (1869–1938), published in 1904, quoted Ostwald as saying that temperature dependence 'is one of the darkest chapters in chemical mechanics'. A similar comment was made by van't Hoff in his *Vorlesungen* which appeared in 1898; he presented a number of possible equations for temperature dependence and concluded that 'it is so far impossible to choose between the above equations'.

In Wilhelmy's pioneering publication of 1850 he suggested a temperature dependence equation, but it does not satisfy the necessary condition that rate equations for a reaction in forward and reverse directions must show a temperature dependence that is consistent with that for an equilibrium constant. In 1862 Berthelot proposed the equation

$$k = Ae^{DT}, \tag{8.3}$$

where A and D are constants. This equation, which does satisfy the condition, was taken seriously for about 50 years.

The next significant contribution, the most important of all, was made in 1884 by van't Hoff in his *Études*. He began with the equation for the temperature dependence of an equilibrium constant (eqn 4.32), which in modern notation can be written as

$$\frac{d \ln K_c}{dT} = \frac{\Delta U^\circ}{RT^2}. \tag{8.4}$$

Here K_c is the equilibrium constant expressed in terms of concentrations, and ΔU° the standard change in internal energy. Suppose that a reaction

$$A + B \underset{k_{-1}}{\overset{k_1}{\rightleftharpoons}} Y + Z$$

is such that the rate from left to right is $k_1[A][B]$ while that from right to left is $k_{-1}[Y][Z]$. At equilibrium the two rates are equal, and K_c is the ratio of the rate constants,

$$\left(\frac{[Y][Z]}{[A][B]}\right)_{eq} = K_c = \frac{k_1}{k_{-1}}. \tag{8.5}$$

Equation 8.5 may thus be written as

$$\frac{d \ln k_1}{dT} - \frac{d \ln k_{-1}}{dT} = \frac{\Delta U^\circ}{RT^2}. \tag{8.6}$$

It was then argued by van't Hoff that the rate constants will be influenced by two different energy terms E_1 and E_{-1}, and he therefore split eq. 8.6 into the two equations

$$\frac{d \ln k_1}{dT} = \frac{E_1}{RT^2} \qquad (8.7)$$

and

$$\frac{d \ln k_{-1}}{dT} = \frac{E_{-1}}{RT^2}. \qquad (8.8)$$

The two energies must be such that $E_1 - E_{-1} = \Delta U°$.

Van't Hoff was aware that $\Delta U°$, which appears in the equilibrium equation, is not necessarily temperature independent, and he therefore realized that the energies E_1 and E_{-1} which appear in equations 8.7 and 8.8 may also be temperature dependent. In the *Études* he considered two possibilities. One is that E_1 is independent of temperature, in which case eq. 8.7 can be integrated to give (with the subscript dropped)

$$\ln k = \text{constant} - \frac{E}{RT} \quad \text{or} \quad k = A \, e^{-E/RT}. \qquad (8.9)$$

The factor A was long known as the frequency factor, and is now known as the *pre-exponential factor*. If this equation applies, a plot of ln k against $1/T$ gives a straight line, and this is the case for many reactions. From the slope, equal to $-E/R$, the energy E, known as the *activation energy*, can be calculated.

The second possibility considered by van't Hoff in the *Études* is that E has the form

$$E = B + DT^2 \qquad (8.10)$$

where B and D are temperature independent. Equation 8.9 then becomes

$$\frac{d \ln k}{dT} = \frac{B + DT^2}{RT^2} \qquad (8.11)$$

and integration gives

$$\ln k = \text{constant} - \frac{B - DT^2}{RT} \qquad (8.12)$$

or

$$k = A \exp\left[-(B - DT^2)/RT\right] \qquad (8.13)$$

Most of the data can be fitted satisfactorily to this equation.

In 1889 Arrhenius published an important contribution to the problem, taking van't Hoff's treatment as his starting point. He noted that the magnitude of the temperature effect on the rates of chemical processes is generally much too large to be explained on the basis of how temperature affects the molecular translational energies, or in terms of how temperature affects the viscosity of the medium. He suggested that an equilibrium is established between normal re-actant molecules and active ones, which are able to form products without the addition of further energy. The equilibrium between normal and active mole-cules will shift with temperature in accordance with van't Hoff's equation, so that the rate of reaction will vary with temperature in the same way. He did not consider a temperature dependence of the energy E, and therefore adopted van't Hoff's simpler equation 8.9. As a result of Arrhenius's interpretation this equation has come to be called the Arrhenius equation, although it was certainly first given by van't Hoff. With characteristic modesty van't Hoff himself tended to leave the impression that Arrhenius had given the equation first!

Another variant of van't Hoff's equation was suggested in 1893 by his student D. M. Kooij:

$$\ln k = A' - \frac{B}{T} + m \ln T \quad \text{or} \quad k = AT^m e^{-E/RT}, \tag{8.14}$$

where A', B and m are temperature independent constants and $E = BR$. From both the theoretical and empirical points of view this equation is now regarded as the most satisfactory. Data sometimes show significant deviations from the simple Arrhenius equation 8.9, but usually they can be fitted well to eqn 8.14.

In their Bakerian Lecture to the Royal Society in 1895 Harcourt and Esson presented yet another empirical equation for the temperature dependence of rate constants:

$$\ln k = \text{constant} + m \ln T \quad \text{or} \quad k = AT^m. \tag{8.15}$$

They argued that this equation fits the data better than any other, and in 1912 they presented additional data in support of their claim.

In 1898 van't Hoff pointed out that most of the equations that had been proposed for temperature dependence were special cases of the equation

$$\ln k = A' - \frac{B}{T} + m \ln T \quad \text{or} \quad k = AT^m \exp[-B - DT^2)/T]. \tag{8.16}$$

This can be called a three-parameter equation, since the temperature depend-ence is determined by the three constants B, m and D. The three one-parameter equations obtained by dropping two of these parameters are listed in Table 8.2, as is also the two-parameter equation of Kooij.

Table 8.2 indicates the type of plot that will give a straight line for each of the equations. It may seem surprising at first sight that the same data usually give equally good straight lines with each of the plots, so that it is difficult to discriminate between the equations on purely empirical grounds. The reason is that over the narrow temperature ranges usually employed in kinetic studies T, $1/T$ and $\ln T$ are more or less linearly related to each other. In measuring rates one is usually restricted to fairly narrow ranges because of the strong

TABLE 8.2   Temperature-dependence equations

$k = Ae^{DT}$	Berthelot, 1862	ln k vs. T
$k = Ae^{-B/T}$	van't Hoff, 1884	ln k vs. $1/T$
	Arrhenius, 1889	
$k = AT^C$	Harcourt and Esson, 1895	ln k vs. ln T
$k = AT^C e^{-B/T}$	Kooij, 1893	$\ln(k/T^C)$ vs. $1/T$

temperature dependence observed; if a rate is very low or very high it cannot easily be measured.

The activation energy and pre-exponential factor

The final decision as to the most satisfactory equation was therefore based on theoretical arguments. The Arrhenius equation 8.9, and the equation 8.14 given by Kooij, can be interpreted in terms of an energy barrier to reaction, as was done by van't Hoff and Arrhenius; if E is the height of the barrier, the fraction of molecules having energy in excess of E is equal to $\exp(-E/RT)$. The equation of Harcourt and Esson, on the other hand, was theoretically sterile; no physical interpretation could be given to the parameters in the equation. For that reason equations 8.9 and 8.14 are the ones that have been used almost universally since about 1910.

Today the activation energy is usually given the symbol E_a, and it is regarded as an experimental quantity defined on the basis of the Arrhenius equation, eqn 8.9. The definition is

$$E_a = -R\left(\frac{\partial \ln k}{\partial (1/T)}\right)_p,$$
(8.17)

the subscript p indicating constant pressure. Only for an elementary reaction, occurring in one step, is the activation energy necessarily related to the height of an energy barrier; for a composite mechanism, the reaction occurring in more than one step, the activation energy is a composite function of different energies.

The pre-exponential factor A is defined in terms of equations 8.9 and 8.17. Once the activation energy has been obtained by the use of eqn 8.17, A can be calculated at a given temperature from k at that temperature, by use of eqn 8.9.

8.3 Theories of reaction rates

Until about 1900 chemical kinetics was almost entirely an empirical subject, little being understood about how the rates of reactions can be interpreted in terms of molecular motions. Notable exceptions to this neglect of theory were the publications in 1867 and 1872 of Leopold Pfaundler (1839–1920), who was then professor of physics at the University of Innsbruck. Pfaundler based his discussion on the idea of dynamic equilibrium, which had been suggested in the 1850s by Williamson and by Clausius. Prior to their work it had been assumed that chemical equilibrium is reached when the forces on a system are equal and opposite, and that all reaction has then ceased; this had been the point of view of Guldberg and Waage. It had been emphasized by Williamson and Clausius, on the other hand, that when a chemical system reaches equilibrium, chemical change is still occurring, at equal rates in forward and reverse directions. This was the point of view adopted by van't Hoff, as well as by Pfaundler earlier.

Pfaundler's first paper, which appeared in 1867, was mainly concerned with dissociation of gases at high temperatures. Although Maxwell's first paper on the distribution of molecular speeds had appeared in 1860, Pfaundler made no reference to it, basing his treatment on Clausius's qualitative discussion of velocity distributions. He suggested, for the first time, that only those molecules having more than a critical amount of energy are capable of undergoing reaction. Pfaundler developed this idea more fully in his second paper, which appeared in 1872, still making no reference to the work of Maxwell. Pfaundler's work is now largely forgotten, but van't Hoff paid tribute to it in his *Études*, and later mentioned that it had greatly helped him in the development of his ideas.

Aside from this suggestion of Pfaundler, and its development by van't Hoff and Arrhenius, little understanding of the theory of reaction rates was gained until well into the twentieth century. As far as elementary reactions were concerned, progress then occurred along three paths: on the basis of thermodynamics, of the kinetic theory of gases, and of molecular statistics. All three elements are combined in the modern treatments.

Thermodynamic aspects

On the thermodynamic side, the first important contribution was made by the French chemist René Marcelin (1885–1914) whose promising career at the École Polytechnique in Paris was cut short by his death in action in the early days of World War I. In a paper presented in 1910 to the Académie des Sciences he proposed rate equations based on the concept of affinity, a quantity used by earlier workers in different senses. Marcelin's affinity A can be interpreted as the negative of the standard molar Gibbs energy change ΔG° in going from the initial state to an activated state which can at once give products, although the introduction of the idea of an activated state did not come until a few years later. Marcelin gave an equation which can be written as

$$v = \text{constant} \ [\exp(A_1^{\ddagger}/RT) - \exp(A_{-1}^{\ddagger}/RT)], \tag{8.18}$$

or, in terms of Gibbs energy changes, as

$$v = \text{constant} \ [\exp(-\Delta^{\ddagger}G_1^{\circ}/RT) - \exp(-\Delta^{\ddagger}G_{-1}^{\circ}/RT)] \tag{8.19}$$

where the subscript 1 refers to the forward reaction and −1 to the reverse reaction. In 1911, at the University of Amsterdam, the Dutch chemists Philip Abraham Kohnstamm (1875–1951) and Frans Eppo Cornelis Scheffer

(1883–1954) published similar and rather more explicit ideas. In the 1920s these ideas were further developed in a number of publications by Scheffer, who in 1917 had moved to the Technische Hogeschool at Delft, in collaboration with his student Wiedold Frans Brandsma (1892–1964). In modern notation their main argument may be summarized as follows. The equilibrium constant K_c for a reaction may be expressed in terms of the standard Gibbs energy change as

$$\ln K_c = -\frac{\Delta G^\circ}{RT}. \tag{8.20}$$

As was done by van't Hoff for d ln K_c/dT (eqn 8.6), they split this equation into

$$\ln k_1 - \ln k_{-1} = -\frac{\Delta G^\circ}{RT}. \tag{8.21}$$

They then suggested that the expressions for the individual rate constants should be

$$\ln k_1 = -\frac{\Delta^{\ddagger} G_1^\circ}{RT} + \text{constant} \tag{8.22}$$

$$\ln k_{-1} = \frac{\Delta^{\ddagger} G_{-1}^\circ}{RT} + \text{constant} \tag{8.23}$$

For any reaction one can therefore write

$$k = \nu \exp(-\Delta^{\ddagger} G^\circ / RT) \tag{8.24}$$

where ν is a factor which is the same for all reactions, and $\Delta^{\ddagger} G^\circ$ is now known as the Gibbs energy of activation. This equation at once leads to

$$k = \nu \exp(\Delta^{\ddagger} S^\circ / R) \exp(-\Delta^{\ddagger} H^\circ / RT) \tag{8.25}$$

In this way the Dutch workers introduced the important concepts of the standard entropy of activation, $\Delta^{\ddagger} S^\circ$, and the standard enthalpy of activation $\Delta^{\ddagger} H^\circ$. They were not, however, able to give an explicit expression for the multiplying factor ν.

Collision theory

Entirely different attacks on the problem of the pre-exponential factor A (eqn 8.9) were made in 1916–1918 by Max Trautz (1879–1960), then at Heidelberg, and by William Cudmore McCullagh Lewis (1885–1945) of the University of Liverpool. Because of World War I Trautz and Lewis were unaware of each other's work. Their treatments were developed within the framework of the radiation hypothesis, now known to be inapplicable, according to which chemical reaction is brought about by the absorption of infrared radiation from the walls of the reaction vessel. An important reason for the popularity of this hypothesis for some time was that unimolecular gas reactions (to be considered briefly in Section 8.5) seemed to be incapable of explanation in terms of molecular collisions. The radiation hypothesis appeared to offer the only explanation, and by some investigators was assumed to apply to reactions of all types. By the middle of the 1920s the hypothesis had been largely discredited, mainly through the efforts of the American physical chemist Irving Langmuir (1881–1957) and of the British physicist Frederick Alexander Lindemann, later Viscount Cherwell (1886–1957). Even after the radiation hypothesis had been abandoned the theories of Trautz and Lewis remained useful, since although they had embraced the theory that radiation is responsible for the activation of molecules, they regarded the rate of reaction as determined by the frequency of molecular collisions.

Lewis's treatment was more explicit and comprehensible than that of Trautz, and in brief was as follows. According to the simple collision theory developed by Clausius and others, for a gas containing molecules A and B regarded as hard spheres, the total number of collisions per unit volume per unit time is

$$Z_{AB} = N_A N_B d^2_{AB}\left(8\pi k_B T \frac{m_A + m_B}{m_A m_B}\right)^{1/2}. \tag{8.26}$$

Here N_A and N_B are the numbers of molecules per unit volume, d_{AB} is the sum of their radii (the distance between their centres on collision), m_A and m_B are their masses, k_B the Boltzmann constant, and T the absolute temperature. According to the ideas of Trautz and Lewis, this collision density Z_{AB} multiplied by the fraction $\exp(-E/RT)$, the fraction of collisions for which there is sufficient energy E, is the rate of reaction. This gives, for the rate constant k,

$$k = L d^2_{AB}\left(8\pi k_B T \frac{m_A + m_B}{m_A m_B}\right)^{1/2} e^{-E/RT}, \tag{8.27}$$

where L, the Avogadro constant, has converted molecular units to molar units. Lewis applied this formula to the reaction $2HI \rightarrow H_2 + I_2$, and obtained remarkably good agreement with experiment. However, it was later realized that there are many reactions, particularly some involving reactant molecules of greater complexity, for which there are large discrepancies, sometimes of several powers of ten, between calculated and observed values. Another difficulty with this approach is that if the treatment is applied to forward and reverse reactions one does not obtain the correct expression for the equilibrium constant, which should explicitly involve entropy and enthalpy changes for the overall reaction.

In spite of these weaknesses, the simple collision theory did play a role of great importance in the development of chemical kinetics. During the 1920s and 1930s in particular a great deal of important experimental work was done in kinetics, and its interpretation in terms of collision theory led to conclusions of considerable significance. These interpretations were easily modified to take account of the more sophisticated theories.

Statistical treatments

The third line of development of theories of reaction rates, in terms of molecular statistics, began with Pfaundler's qualitative discussion of 1867, and in the end was the most fruitful. Following Pfaundler's work little was done along those lines until 1912, when the French physical chemist A. Berthoud used the Maxwell–Boltzmann distribution law to obtain an expression for the rate constant. Two years later Marcelin made a very important contribution by representing a chemical reaction by the motion of a point in phase space and, by applying Gibbs's statistical–mechanical procedures, obtained an expression for the rate. These ideas were later extended by the British physicist James Rice (1874–1936) who suggested a more precise formulation for the activated state in a chemical reaction. According to these treatments the rate of a reaction is proportional to the concentration of activated complexes.

In 1919 a significant approach to reaction rates was made by the Austrian (later American) physicist Karl Ferdinand Herzfeld (1892–1978). He considered the dissociation of a diatomic molecule,

$$AB \rightleftharpoons A + B$$

He applied statistical mechanics to the equilibrium constant and kinetic theory to the reverse reaction, in this way obtaining for the rate constant of the dissociation reaction the expression

$$k = \frac{k_B T}{h} [1 - \exp(-h\nu/k_B T)] \, e^{-E/RT}, \tag{8.28}$$

where ν is the vibrational frequency of the bond and h is the Planck constant (Section 10.1). This expression is of special interest since it is the first time that the expression $k_B T/h$, a feature of the later transition-state theory, appeared in a rate equation.

In 1920 Tolman developed, on the basis of statistical, mechanics a relationship that gives significance to the experimental activation energy E_a defined by eqn 8.17. According to his treatment, as later improved by Fowler and Guggenheim, the activation energy per molecule ($\varepsilon_a = E_a/L$) is given by

$$\varepsilon_a = \text{(average energy of all molecules undergoing reaction)} - \text{(average energy of all reactant molecules).} \tag{8.29}$$

Potential-energy surfaces and transition-state theory

A development of great importance for rate theory was the construction by Henry Eyring (1901–1981) and Michael Polanyi (1891–1976) of a potential-energy surface for a chemical reaction. This work. published in 1931, was done at the Kaiser Wilhelm Institut in Berlin where Polanyi was professor of chemistry and Eyring a US National Research Fellow, having previously been at the University of Wisconsin. They dealt with the reaction $H + H_2 \rightarrow H_2 + H$, the

Henry Eyring

(1901–1981)

Eyring, of German and English descent, was born in Colonia Juarez, Mexico, becoming a naturalized American citizen in 1935. He first studied mining engineering at the University of Arizona, and obtained a master's degree in metallurgy in 1924. He then went to the University of California at Berkeley, obtaining a Ph.D. degree in physical chemistry in 1927. After teaching at the University of Wisconsin, where he collaborated on some experimental research with Farrington Daniels, Eyring spent a year with Michael Polanyi at the Kaiser Wilhelm Institüt in Berlin. In 1931 he became professor of chemistry at Princeton, and in 1946 moved to the University of Utah as Dean of Graduate Studies and professor of chemistry, remaining there until the end of his life.

Eyring's research covered a wide range. Much of it was in chemical kinetics, but he also worked on the kinetics of physical processes such as diffusion and on the structures of solids and liquids. He carried out little experimental work, but his theoretical approaches were always closely related to experimental results.

One of his more important contributions was his construction in 1931, with Michael Polanyi, of a potential-energy surface for the $H + H_2$ reaction (Section 8.3). Perhaps of even greater significance was his formulation, in 1935, of transition-state theory (Section 8.3). Subsequently Eyring applied this theory to a wide variety of chemical and physical processes, including processes occurring in biological systems, and he also extended the theory in various ways. He wrote a number of books, including *The Theory of Rate Processes*, coauthored in 1941 with S. Glasstone and K. J. Laidler, which presented a systematic account of transition-state theory and many of its applications, and *The Theory of Rate Processes in Biology and Medicine*, written in 1984 with F. H. Johnston and Betsy J. Stover.

Eyring was a devout and active member of the Mormon Church, on which he published a number of articles. His and G. N. Lewis's failure to obtain Nobel Prizes has been a matter of surprise among many physical chemists.

References: K. J. Laidler, DSB, Supplement 2, 17, 279–284 (1990); Heath (1980, 1985); Hirschfelder (1982); Urry (1982).

prototype of a large class of abstraction reactions. They used semi-empirical procedures, based on quantum-mechanical principles but including experimental data on vibrational frequencies and energies of dissociation. They constructed a three-dimensional diagram showing potential energy as a function of the two shorter H–H distances in a linear $H \cdot \cdot H \cdot \cdot H$ complex. Later

Eyring and his students at Princeton carried out dynamical calculations on surfaces of this kind.

In 1932, a year after Eyring and Polanyi constructed their surface, an important contribution was made by H. Pelzer and the Hungarian (later American) physicist Eugene Paul Wigner (b. 1902); this work was done at Göttingen shortly before Wigner moved to Princeton, to which Eyring also moved in 1931. Pelzer and Wigner treated the passage of a system over a potential-energy surface, and their work is of importance in being the first to focus attention on the col, or saddle point, of a potential-energy surface. They realized that the motion of the system through the col controls the rate of reaction, and they obtained an expression for the rate in this way. In a second paper published in 1932 Wigner discussed the possibility of quantum-mechanical tunnelling through the barrier of a potential-energy surface. Previously Gurney had discussed the possibility of tunnelling in electrode processes at surfaces (Section 7.5), and Wigner's paper was the first to treat tunnelling in an ordinary chemical reaction. For such reactions there is now much experimental evidence for tunnelling, particularly for processes at low temperatures involving the transfer of hydrogen atoms.

The paper of Pelzer and Wigner contributed significantly to the formulation in 1935 of what has come to be called *transition-state theory*. This theory, which provides a valuable framework for the understanding of all types of chemical and physical processes, was developed independently by Henry Eyring at Princeton and by Michael Polanyi and Meredith Gwynne Evans (1904–1952) at the University of Manchester. The essence of the theory is that species represented by points near to the col of a potential-energy surface, i.e., the activated complexes, are considered to be in a state of quasi-equilibrium with the reactants, so that their concentration can be calculated. The rate of reaction is then the concentration of these complexes multiplied by the frequency with which they are converted into products. According to the theory this frequency is $k_B T/h$, where k_B is the Boltzmann constant, h the Planck constant and T the absolute temperature. In terms of partition functions the rate equation for a bimolecular reaction turns out to be

$$k = \frac{k_B T}{h} \frac{q_{\ddagger}}{q_A \, q_B} \exp(-E_0/RT) \tag{8.30}$$

where q_A and q_B are the partition functions for the reactants A and B, and $q_{\ddagger}$ is the partition function for the activated complex. The energy E_0 is the hypothetical activation energy at the absolute zero, and is the difference between the zero-point energy of the activated complex and that of the reactants.

In terms of thermodynamic functions, the rate expression according to transition state theory is

$$k = \frac{k_B T}{h} \exp(\Delta^{\ddagger} S^\circ/R) \exp(-\Delta^{\ddagger} H^\circ/RT) \tag{8.31}$$

The enthalpy of activation $\Delta^\ddagger H°$ is slightly different from the energy E_a defined by eqn 8.17, and from the energy E_o that appears in eqn 8.30.

Although it is still difficult to make reliable rate calculations on the basis of transition-state theory, or indeed of any theory, transition-state theory has proved of great value in providing a working tool for those who are not so much concerned with making accurate calculations as with gaining some insight into how chemical and physical processes occur. The theory leads to a useful qualitative insight, unmatched by any other theory, into such matters as solvent effects on reaction rates, relative rates of homologous processes, kinetic-isotope effects, and pressure influences.

Composite mechanisms

So far attention has been confined to elementary reactions, in which the process occurs in one stage, the reactants passing through an activated state on their way to becoming products. However, as the investigation of reaction mechanisms has proceeded, it has become clear that the majority of the chemical reactions with which we are familiar are not elementary but occur in more that one step. The terms 'composite', 'complex' and 'stepwise' are used to refer to this situation. Some of the earlier workers in kinetics were aware of this, but lacking the appropriate experimental techniques were unable to investigate the mechanisms. The first paper published by Harcourt and Esson did treat certain types of consecutive reactions, in which the product of one reaction undergoes a further reaction, and the later textbooks of van't Hoff and Ostwald included similar discussions. These treatments were all rather formal, however, with little said about the actual reactions that were occurring.

After certain experimental techniques had been developed it became possible to probe the actual mechanisms of composite reactions. The main problem was that many of the reaction intermediates that exist during the course of chemical reactions are unstable and therefore have very short lifetimes. Atoms and free radicals, such as OH and CH_3, have unpaired electrons and are present for only a fraction of a second and can only be detected by means of the high-speed techniques that are mentioned in Sections 8.4 and 8.5. The study of photochemical reactions has played a particularly important role in the elucidation of reaction mechanisms and deserves a section to itself.

8.4 Photochemical reactions

The subject of photochemistry comprises the study of chemical reactions in which electromagnetic radiation plays some role, either by being absorbed or emitted. When a reaction is initiated by the absorption of radiation it is appropriate to speak of *photolysis* (Greek φωτος (photos), light; λυσις (lysis), loosening), since the effect of the light is to loosen chemical bonds and often to break them. When radiation is emitted in a chemical reaction we speak of *chemiluminescence*.

Elizabeth Fulhame

Almost nothing is known about Mrs Fulhame beyond the fact that she wrote an important book which was published in 1794. She was apparently married to Dr Thomas Fulhame, a physician who had been a student of Joseph Black at Edinburgh, where he obtained his MD degree. In her book Mrs Fulhame said that she had begun her scientific work in about 1780, so that perhaps she was born about the middle of that century. In 1810 she was reported to be an honorary member of the Philadelphia Chemical Society, but that is the last we hear of her.

Her remarkable book, published privately by herself and printed by a London printer, was entitled *An Essay on Combustion, with a View to a New Art of Dying and Painting, wherein the Phlogistic and Antiphlogistic Hypotheses are Proved Erronious*. Her name is given in the book as Mrs Fulhame, but we know her first name to be Elizabeth from the registration of the book at Stationers Hall in London. The book contains two pioneering contributions. She was the first to achieve photoimaging, in that she obtained patterns by impregnating cloth with gold salts and other materials and exposing them to light (Section 8.4). Also, she demonstrated that water acts as an intermediary in certain oxidation processes, and she proposed for the first time specific reaction mechanisms (Section 8.6).

The main purpose of her extensive experiments was to provide support for the new ideas of Lavoisier—who unhappily never saw the book since he died under the guillotine just six months before it was published. For a time the book attracted a good deal of attention. In 1799 rather sharply-worded objections to it were raised by the somewhat irascible and eccentric Irish chemist William Higgins (1763–1825), who claimed that she had stolen the ideas from his *A Comparative View of the Phlogistic and Antiphlogistic Theories*, published in 1789. Higgins had suggested that

---

Reactions brought about by the action of electromagnetic radiation have been studied since the eighteenth century, but their existence was recognized much earlier. Alexander the Great (356–323 BC), for example, used cloth strips, impregnated with photosensitive dyes, to coordinate his military attacks at a set time after dawn.

Photochemical imaging

In the eighteenth century it was discovered that silver salts are blackened under the influence of light. This finding was extended in the nineteenth century into the technique of photography which plays an enormously important role in our

water is essential to the rusting of iron, but made no mention of its function in other types of reactions. However in a book he published in 1799 Higgins wrote

Had this fair author read my book, and indeed I suppose she did not, (having quoted every other treatise upon the subject) no doubt she would have been candid enough to do me the justice of excepting me from the rest of my cooperators in science, when she told them they erred for having overlooked this modification of their doctrine, and also when she adduced it as an original idea of her own.

Higgins was a difficult man, and his criticism was unjustified, as also was his accusation of plagiarism against John Dalton (Section 5.2).

In 1800 Joseph Priestley, by that time residing in Pennsylvania, published his book *The Doctrine of Phlogiston Established, and that of the Composition of Water Refuted.* In it he mentioned that he had met Mrs Fulhame in London in 1793. Although he by no means agreed with her conclusions she apparently found their discussions encouraging, and it may be that this meeting caused her to publish her book in the following year. Priestley was a member of the Philadelphia Chemical Society, and perhaps it was he who proposed her for membership. Her book was reprinted in Philadelphia in 1810.

One of the reviews of Mrs Fulhame's book, by C. Coindet, treated it very favourably and gave an extended summary of it. A German translation of it was published in 1798. The book therefore received a good deal of recognition for a time, but was soon forgotten.

References: Davenport & Ireland (1989); Laidler (1986); Schaaf (1990); Wheeler and Partington (1960).

everyday lives and also in scientific research. Photography and its modifications and extensions, such as photocopying and television, have been referred to as *photochemical imaging*, or *photoimaging*.

The first work in photochemical imaging was done in the last two decades of the eighteenth century by Elizabeth Fulhame, about whom almost nothing is known. She was the wife of an English or perhaps Irish doctor, Thomas Fulhame, and lived from about the middle of the eighteenth century until some time in the nineteenth century. In 1794 she published privately a remarkable book *An Essay on Combustion* in which she made two pioneering contributions: one was concerned with photochemical imaging; the other dealt with what came to be called catalysis, to be considered in Section 8.6. She was the first to be successful in creating rudimentary images on dyed materials.

During the next few decades more work of this kind was done, and it all contributed to the final success of photographic techniques. In 1798 Count Rumford (Benjamin Thompson) published an account of experiments along the same lines as those of 'the ingenious and lively Mrs Fulhame', as he called her, saying that his results agreed entirely with hers, and that her work had inspired his own. At about the same time Thomas Wedgwood (1771–1805), a son of the famous potter Josiah Wedgwood (1730–1795), carried out similar investigations, the results of which were included in a paper that Humphry Davy published in 1802 on his own studies of photochemical imaging. This paper made no mention of Mrs Fulhame, but presumably Wedgwood and Davy were aware of her work.

The investigations mentioned so far created images by placing flat objects in contact with sensitized materials. Thus, although contributing to the development of photography, they cannot be classified as photography proper since they made no use of cameras. The first to have used a camera to secure an image was Joseph Nicéphore Niépce (1765–1833), who achieved success in about 1816. In about 1837 Louis Jacques Mandé Daguerre (1787–1851), a painter of theatrical scenery and a showman, and the inventor of the diorama, made his great achievement, the *daguerreotype* as it came to be called. A daguerreotype is created from a silver-plated sheet of copper, and even the earliest daguerreotypes were remarkably clear. Their main disadvantage was that it is not possible to make direct copies.

In Britain in the meantime investigations with the same objective, but with the use of quite different techniques, had been pursued by William Henry Fox Talbot (1800–1877). In about 1833 he deposited silver chloride on paper, and obtained images by placing an object such as a leaf in contact with it and exposing it to sunlight. He then made the paper fairly insensitive to further exposure to light by washing with a strong salt solution. He was also able to prepare positives from negatives. Later he was successful in making negatives and then positives by the use of small box cameras. His 'photogenic drawings', as he called them, were much less sharply defined than those obtained by Daguerre, but ultimately his method was more useful since it led, after a number of refinements, to the techniques in use today. Since a negative was involved, any number of copies could be made. Talbot first announced his method on 25 January 1839, at one of the regular Friday evening meetings of the Royal Institution, and with Michael Faraday's cooperation; he exhibited samples of his work, including a photograph of his own house taken in 1835. In a paper presented on 20 February 1839, sufficient details were given to enable others to repeat the experiments, and a full account of this paper appeared in the *Proceedings of the Royal Society*. Although Daguerre has priority in announcing his technique, Talbot has priority in giving a full scientific account of his method of taking a photograph by means of a camera.

At about the same time that Daguerre and Talbot announced their techniques, Sir John Herschel was carrying out similar studies. On hearing, in January 1839, of the successes achieved by Daguerre and Talbot he at once began a detailed investigation of a number of related problems, being particularly concerned with trying to improve Talbot's techniques. On 14 March 1839. he communicated to the Royal Society a paper in which he introduced the procedure, still used today, of 'fixing' negatives with a solution of sodium thiosulphate, which he called 'hyposulphite', a name preserved in the word 'hypo', still employed by photographers. At this meeting Herschel exhibited a number of samples of the photographs he had taken, one being of the famous telescope that his father, Sir William Herschel, had built at Slough.

Reference has already been made, in Section 6.1, to Herschel's success in 1840 in taking photographs in the ultraviolet and infrared. Some of his later papers are remarkable for the fact that Herschel, like Edmond Becquerel in France, was at that early date pursuing the idea of colour photography. It was Herschel who first used the word 'photography', instead of 'photogenic drawing' which Talbot had used, and to refer to 'negatives' and 'positives'. In 1842 Herschel describe a procedure for producing a 'cyanotype'—or as we should say blueprint—and the making of blueprints is the oldest modern photographic technique still used more or less in its original form. All of Herschel's work was done in the spirit of pure scientific enquiry. He never took out a patent or profited in any way from his work, and he rarely took a photograph for pictorial purposes; most of his experimental images were contact copies of engravings, so that he was able to study the chemical aspects and avoid any complications from optical effects.

It was the method initiated by Talbot that ultimately became the photography that is practised today. In 1840 Talbot introduced the *calotype*, a refinement of his photogenic drawings, and patented the process on 8 February 1841, describing it to the Royal Society on 10 June. As early as March 1840, Herschel had described to the Royal Society a method of precipitating on glass a coating of photosensitive material, in this way making it possible to improve the quality of the positive prints that could be made. Reference was made in Section 6.2 to the introduction by Scott Archer of the important technique of coating glass with collodion, and to other modifications that made photography more convenient and eventually easily accessible to the general public.

Principles of photochemistry

The fact that radiation can only induce chemical reaction if it is first absorbed was demonstrated in 1819 by von Grotthuss, whose important work on electrolysis was mentioned in Section 7.1. His conclusion was rediscovered in 1841 by John William Draper (1811–1882) and is now often referred to as the Grotthuss-Draper law. The way in which radiation induces chemical reaction was not, however, understood at all until the idea of the quantization of energy and radiation had evolved. Previously photochemical reactions had been 'understood' in terms of a mysterious property called 'tithonicity', a word that derives from Tithonus, a legendary Trojan who was loved by the dawn goddess Eos (Aurora) and was eventually changed into a grasshopper.

In the nineteenth century and the early years of the twentieth, much effort went into the study of the reaction between hydrogen and chlorine gases with the formation of hydrogen chloride,

$$H_2 + Cl_2 \rightarrow 2HCl.$$

This reaction was first carefully studied in the 1860s at Heidelberg by Robert Bunsen (1811-1899) and his English student Henry Roscoe (1833–1915). One important observation they made was that a small amount of light can bring about the conversion of a great deal of material; they calculated from their results that the light emitted by the sun in one minute and falling on a hydrogen–chlorine mixture would bring about the conversion of 25×10^{12} cubic miles of it into hydrogen chloride. The result is significant in connection with the high quantum yields found for the reaction.

The reaction has many complicating features, and it continued to be investigated for many decades, notably by the German kineticist Max Bodenstein (1871–1942), who worked at several universities including the University of Hannover, and by D. L. Chapman, who did most of his work in the laboratories of Jesus College, Oxford. Both Bodenstein and Chapman were skilful and careful investigators, but in spite of their extensive studies the elucidation of the reaction proved difficult, mainly because of an exasperating lack of reproducibility in the experimental results. It is Chapman who deserves much of the credit for overcoming the experimental difficulties. He found, for example, that the reaction is extremely sensitive to the nature of the surface of the vessel, and to the presence of minute traces of impurities. As a result, in the earlier work different investigators using different vessels and having different amounts of trace impurities in their gases obtained widely different results, and the same investigator often obtained different results at different times. Also, the reaction showed induction periods, of variable lengths, during which nothing happened. Chapman was finally able to show how reproducible results could be obtained by rigorous control of the nature of the surface of the vessel and of the purification of the gases.

Photochemical equivalence

It was not until the advent of the quantum theory in 1900 (Section 10.1) that photochemical reactions could be at all understood. In 1905 Einstein proposed that radiation under certain conditions behaves as a beam of particles, later called photons, and this suggestion was the key to the whole problem of photochemistry. In 1908 Johannes Stark (1874–1957), and later Einstein in 1912, proposed that one quantum of radiation is absorbed by one molecule. This relationship is now almost always called *Einstein's law of photochemical equivalence*; no doubt Stark's thoroughly unpleasant character, and particularly

David Leonard Chapman

(1869–1958)

Chapman was born in Wells, Norfolk, and became an undergraduate at Christ Church, Oxford, where he was a student of Vernon Harcourt. He taught for a period at Manchester Grammar School, and in 1897 was appointed by H. B. Dixon to be a lecturer at Owens College, Manchester (which later became part of the University of Manchester). In 1907 he was elected a Fellow of Jesus College, Oxford, with the special responsibility of directing the College chemistry laboratories which had just been founded in honour of Sir Leoline Jenkins, a former benefactor of the College. His leadership of the laboratories was a distinguished one, and he retained his appointment until 1944.

Chapman was a man of great modesty, and his accomplishments have not received the recognition they deserve. He had a proficiency in mathematics and physics that was unusual among chemists of his generation, and this was reflected in his excellent teaching and research. His first research was on detonation in gases, and in 1899 he presented the first reliable theory of it (Section 8.5), basing his treatment partly on the experimental evidence obtained by Dixon. Some of his equations were arrived at independently by Émile Jouguet, and the region immediately behind a detonation wave is still referred to as the 'Chapman-Jouguet layer'. In 1913 Chapman presented an important treatment of capillarity, suggesting the existence of what has come to be called the 'Gouy–Chapman layer' (Section 7.5).

Much of Chapman's research was on the chemical kinetics of gas reactions, and in it he was ably assisted by his wife, Muriel C. C. Chapman. In 1910 he used the surface/volume ratio method to determine the relative importance of surface reaction in the ozone decomposition (Section 8.6). The painstaking studies by Chapman and his wife on the photochemical hydrogen–chlorine reaction led to great clarification of the experimental aspects of that highly elusive reaction, and made possible the understanding of its mechanism (Section 8.4).

In 1913 Chapman first proposed and applied the steady-state treatment (Section 8.4), which is usually attributed to Bodenstein—who did make much use of it and ably defended it against its critics. In 1924 Mrs Chapman made a careful investigation of the effect of light intensity on the hydrogen–chlorine reaction, and her results led Chapman to develop the rotating-sector technique for measuring the mean lives of chain carriers. This procedure has subsequently been used widely for various types of reactions, particularly polymerizations.

References: Bowen (1958); Hammick (1959); Laidler (1988).

his later association with the Nazis and his pathologically vicious attacks on Jewish and other scientists, has something to do with this preference for giving the credit to Einstein. The energy of a photon of radiation is equal to the product of the frequency v and the Planck constant h, and in a chemical reaction the product hv is usually used to denote a photon.

The law of photochemical equivalence has proved essential to the interpretation of the results of photochemical experiments, since it enables the rates of formation of active species to be calculated from the results of optical experiments. A mole of photons is now called an *einstein*, the name having been proposed in 1929 by Max Bodenstein and Carl Wagner. Modern work with lasers has shown that because of the high concentration of photons, a molecule may absorb more than one photon in the primary photochemical act, and this is referred to as *multiphoton absorption*. For ordinary radiation, however, the law of photochemical equivalence is always accepted as applying to the formation of active species in primary processes. For many years the situation was confused, and even at the Faraday Society Discussion of 1926 there was much dissension as to whether the law of photochemical equivalence is valid. This was due in part to failure to realize that an excited species may emit radiation before it can give rise to species such as atoms and free radicals which play an active part in the chemical reaction. A second complication is that the active species produced in the initial photochemical act may undergo subsequent reactions in which more than one molecule of reactant is transformed into product. The ratio of the number of reactant molecules transformed to the number of photons absorbed is known as the *quantum yield* or *photon yield*.

For example, in the photochemical decomposition of hydrogen iodide into hydrogen and iodine, Emil Warburg (1845–1931) found in 1918 that two molecules of hydrogen iodide are transformed when one photon of radiation is absorbed. The mechanism he proposed to explain this is

$$HI + hv \rightarrow H + I$$
$$H + HI \rightarrow H_2 + I$$
$$I + I \rightarrow I_2$$

so that the overall reaction, obtained by adding these equations, is $2HI + hv \rightarrow H_2 + I_2$. Although one photon interacts with one molecule of hydrogen iodide, in accordance with the law of photochemical equivalence, two molecules are transformed.

Chain reactions

An important contribution made by Bodenstein was to suggest the idea of a chain reaction. In 1913 he observed that the hydrogen–chlorine reaction deviates considerably from the law of photochemical equivalence, one photon of light sometimes bringing about the conversion of millions of molecules of hydrogen and chlorine into hydrogen chloride. To explain this result he made a clear distinction between primary and secondary photochemical processes. The primary act is the one in which the light is absorbed and produces active species, and he suggested that in this primary process the law of photochemical equivalence is obeyed, one photon interacting with one molecule. To explain the fact that one photon brings about the conversion of many molecules, Bodenstein suggested that secondary processes occur in which one active species can bring this about. Walter Dux (1889–1987), who was Bodenstein's research assistant at the time, recalled that on one occasion Bodenstein removed his gold watch chain and discussed the transmission of an influence along the chain. During World War I Bodenstein patriotically contributed this historically important gold chain to the German war effort, replacing it with a stainless steel chain. Dux later had a replica gold chain made and presented it to the University of Hannover, where it remains.

Although Bodenstein had the right idea about chain reactions, it appears that the expression was not used until 1921 when it appeared (as *Koederreaktion* in Danish) in the Ph.D. thesis of J. A. Christiansen. Also, Bodenstein did not correctly identify the chain reaction. In 1913 he and Dux suggested that the active species are Cl_2^+ ions, produced by the interaction of a photon with a Cl_2 molecule. Bodenstein later realized that there was not enough energy to form ions, and in 1916 he suggested that excited chlorine molecules were formed. In 1931, after having accepted Nernst's suggestion that it is atoms that are involved, Bodenstein commented:

For the hydrogen-chlorine reaction I assumed—in 1913—that the primary act is an ionization of the chlorine molecule

$$Cl_2 + E = Cl_2^+ + \theta$$

I ought to have known even then that this was impossible—I did not. Later, in 1916, I suggested an excited molecule of chlorine, $Cl_2 + E = Cl_2^*$.

Nernst's mechanism for $H_2 + Cl_2$

It sometimes happens that the solution to a scientific problem is reached, not by those who have worked long and hard on the problem, but by someone who has had little interest in it. This was so here, for it was neither Bodenstein nor Chapman, but Nernst, who obtained the solution. Nernst had made pioneering contributions to electrochemistry and thermodynamics, and had shown much interest in quantum theory, but had paid little attention to kinetics. He suggested in 1918 that the primary process in the hydrogen–chlorine reaction is

$$Cl_2 + h\nu \rightarrow 2\ Cl$$

and that each Cl atom produced can bring about a pair of reactions in which two HCl molecules are formed without any loss of a chlorine atom:

$$Cl + H_2 \rightarrow HCl + H$$
$$H + Cl_2 \rightarrow HCl + Cl$$

Since a chlorine atom is regenerated when this pair of *chain-propagating* reactions occurs, a single chlorine atom can bring about the conversion of many hydrogen and chlorine molecules into hydrogen chloride.

Paradoxically, the reason that Chapman and Bodenstein did not suggest this mechanism was that—unlike Nernst—they knew too much about the experimental details of the reaction! As originally formulated, Nernst's mechanism did not satisfactorily explain the kinetic behaviour. The dependence of the rate of the reaction on the reactant concentrations and on the light intensity was inconsistent with the original Nernst mechanism, which could also offer no interpretation of the effects of trace impurities such as oxygen, or of the induction periods. As a result, a good deal of further work was necessary, and it finally emerged that Nernst's mechanism was along the right lines, but that some additional reactions had to be added in order for it to be completely satisfactory.

It turned out that another photochemical reaction, that between hydrogen and bromine, would have been a better one for the testing of Nernst's type of chain mechanism. This reaction was studied by Bodenstein and Lütkemeyer in 1924, but in the meantime the non-photochemical (thermal) reaction had been investigated by Bodenstein and the American physical chemist Samuel Colville Lind (1879–1968) in 1907. They found that the thermal reaction is not particularly sensitive to the nature of the surface or to added impurities, but is inhibited by the product hydrogen bromide. The measured rates of formation of hydrogen bromide were reproducible and could be fitted to the expression

$$v = \frac{k[H_2][Br_2]^{1/2}}{1 + [HBr]/m[Br_2]}, \tag{8.32}$$

where k and m are constants. In 1919 and 1920 the following mechanism for the thermal reaction was proposed independently by Christiansen, Herzfeld and Michael Polanyi:

$$
\begin{array}{rl}
(1) & Br_2 \rightarrow 2Br \\
(2) & Br + H_2 \rightarrow HBr + H \\
(3) & H + Br_2 \rightarrow HBr + Br \\
(-2) & H + HBr \rightarrow H_2 + Br \\
(-1) & 2Br \rightarrow Br_2
\end{array}
$$

The fourth reaction, which is the reverse of the second, is responsible for the inhibition by hydrogen bromide; the degree of inhibition is reduced by the addition of bromine, and this is explained by the competition between reactions 3 and −2. The first reaction occurs thermally, bromine molecules being dissociated to some extent as a result of collisions. When the photochemical reaction was later studied by Bodenstein and Lütkemeyer it was realized that its mechanism is very similar to that of the thermal reaction, the only difference being that the initial dissociation of the bromine molecule is brought about by a photon instead of thermally. Comparison of the thermal and photochemical reactions led to further understanding of the mechanistic details.

Steady-state hypothesis

In dealing with reactions of this kind, in which intermediates are present at very low concentrations, it is convenient to make use of the so-called *steady-state* hypothesis, or *stationary-state* hypothesis, in order to obtain an expression for the rate of the overall reaction in terms of the concentrations of reactants. According to this hypothesis, the concentration of an atom or radical, which is small in comparison with that of the reactants, remains more or less constant throughout the main course of the reaction. Thus for an atom or radical X one may write

$$\frac{d[X]}{dt} = 0. \tag{8.33}$$

The use of this hypothesis greatly simplifies the solution of the kinetic equations for composite mechanisms, and usually leads in a simple way to an overall rate equation that is correct to a good approximation. The hypothesis is usually attributed to Bodenstein, but six months before his paper appeared in 1913 the hypothesis had been clearly stated by Chapman and Underhill who applied it to the hydrogen–chlorine reaction, even though they did not know its mechanism. At first the hypothesis was a matter of some controversy. The German kineticist Anton Skrabal (1877–1957), of the University of Graz, criticized it in 1927, but his criticisms were at once effectively answered by Bodenstein, and in 1940 a treatment of the theory was given by the Russian scientist D. A. Frank-Kamenetsky. It is impossible to give a general proof of the hypothesis, valid for all types of mechanisms, but there seems no doubt that it is true to a good approximation provided that it is not applied to intermediates at concentrations that are too high. Initially there is a period, referred to as the *pre-steady-state* or *induction* period, or sometimes as the *transient phase*, in which the concentrations of reactants are rising from zero to their steady-state values.

Chapman made another important contribution to the study of short-lived intermediates in reaction systems by his introduction of the rotating-sector technique. The idea of employing this procedure came to him as a result of some investigations by his wife, Muriel C. C. Chapman, who showed in 1924 that the rate of the hydrogen–chlorine reaction depends sometimes on the first power of the light intensity and sometimes on the intensity to a lower power. These and other observations led Chapman to think of using a rotating slotted disk in order to produce alternating periods of light and dark; the excited species would be generated in the light and would decay in the periods of darkness. Chapman realized that by varying the speed of rotation it would be possible to measure the mean lifetimes of the active species. Results from this technique were described in a paper by Chapman, Briers and Walters, in which the mathematical theory of the sector effect was worked out. A later paper with Briers described an application of the technique to the photochemical reaction between hydrogen and bromine. This rotating-sector technique has been used

George Porter, Lord Porter of Luddenham

(b. 1920)

Porter was born in Stainforth, Yorkshire. He was a student at Leeds University until 1941 when he enlisted in the Royal Naval Volunteer Reserve and served until 1945 as a radar officer. From 1945 to 1948 he was a research student at Emmanuel College, Cambridge, and then for three years was a university demonstrator in chemistry. During this period he carried out research mainly in flash photolysis, partly in collaboration with R. G. W. Norrish, the professor of physical chemistry. In 1952, after a brief appointment with the British Rayon Research Association, he became professor of physical chemistry at the University of Sheffield.

In 1966 he became Director of the Royal Institution and at the same time Fullerian professor of chemistry and director of the Faraday Research Laboratory. He retired from the Royal Institution in 1987 to become research professor at Imperial College, London. From 1985 to 1990 he also served as President of the Royal Society.

Most of Porter's research has been related to his invention, in the late 1940s, of a remarkably effective flash photolysis technique (Section 8.4). Within a year or two the duration of the flash had been reduced to about a microsecond, and Porter developed an efficient 'pump and probe' method which allowed spectroscopic studies to be made a short time after the primary flash. During the next few years Porter and his co-workers used this technique to investigate the structures of a number of transient species, such as those in triplet excited states, and also the kinetics of rapid reactions particularly in solution.

very widely, particularly in the study of photochemical polymerizations. In 1956 Shepp developed a more general theory of the method.

Flash photolysis

Since about 1950 photochemistry has entered a new phase with the development of flash photolysis. In this technique a flash of high intensity but short duration brings about the formation of species such as atoms and free radicals, the structure and reactions of which can be studied by spectroscopic methods. This technique was first developed at Cambridge University by George (now Lord) Porter (b. 1920) and Ronald George Wreyford Norrish (1897–1978), their first paper on the subject appearing in 1949, with details in 1950. For this work they were awarded Nobel Prizes in 1967, together with Manfred Eigen (b. 1927) who had developed the relaxation techniques (Section 8.5) which are

By the late 1960s Porter and his colleagues had been successful in developing laser flash systems capable of dealing with processes occurring in the nanosecond (10^{-9} s) range. It was then possible to study, for example, the radiative decay of singlet-excited species, both in the gas phase and in solution. During the 1970s further developments in laser technology, in the form of mode-locked lasers, allow the production of a pulse of only a picosecond (10^{-12} s) duration, and Porter and his colleagues then made many investigations of physical processes such as vibrational redistribution and transfer of energy. In particular, with the use of these techniques Porter has been able to study many of the elementary processes involved in photosynthesis. In the early 1990s, at Imperial College, he began studies of the reaction centres of chloroplasts in the femtosecond (10^{-15} s) time range.

Porter's 21-year leadership of the Royal Institution brought that distinguished organization into an entirely new era. Like his predecessor Michael Faraday he gave many public lectures which have increased public awarenes of science, and he was concerned in the organization of conferences on matters of current interest and importance. In 1967 Porter shared the Nobel Prize in chemistry with Norrish and Eigen. He was knighted in 1972, created a life peer in 1990, and has been awarded the Order of Merit, the most exclusive af all British honours.

Reference: Laidler (1993).

also important for the investigation of rapid reactions. Shortly after the work of Porter and Norrish was done, flash photolysis equipment was built and used by Gerhard Herzberg (b. 1904) and Donald Allan Ramsay (b. 1922) at the National Research Council in Ottawa, and by Norman Ralph Davidson (b. 1916) at the California Institute of Technology in Pasadena.

In Porter's earliest experiments at Cambridge, in 1948, the duration of his flashes was about a millisecond, 10^{-3} s, and it is a remarkable fact that during the next four decades the duration of the flash was reduced by eleven powers of ten, to about 10^{-14} s, or 10 femtoseconds. As a result, techniques are now available for studying the kinetics of the fastest chemical and physical processes.

By 1950 flashes of a microsecond (10^{-6} s) duration had been produced, and were used for the study of the structures of free radicals in the gas phase, for following the kinetics of a number of fast reactions such as the combinations of atoms and free radicals, and for the study of the decay of species in triplet excited states. By 1966 Porter, who in that year became Director of the Royal Institution, had developed a highly efficient laser flash system capable of dealing with processes occurring in the nanosecond (10^{-9} s) range. This allowed the study of many additional processes, such as the decay of species in singlet excited states.

A flash of a nanosecond duration is adequate for the study of almost any purely chemical process, by which is meant a process in which there is a change in chemical identity, Accompanying chemical processes, however, there are always processes of a purely physical nature, such as energy redistribution and solvent reorganization, and these processes commonly occur in the picosecond (10^{-12} s) range. Flashes of these very short durations were achieved during the 1970s, notably by Charles Vernon Shank (b. 1943) and Peter M. Rentzepis (b. 1934) at the AT & T Bell Laboratories in New Jersey, where great advances in laser technology have been made. Many processes have now been studied with flashes of a picosecond duration. For example, in 1976 Porter and his colleagues at the Royal Institution made a study of the reaction between the purple dye thionine and ferrous ions in aqueous solution.

During the 1980s, largely due again to the efforts of Shank and his colleagues, it has become possible to achieve flashes of even shorter duration, of only a few femtoseconds (10^{-15} s). The uncertainty principle imposes a limit on what can be done with flashes of very short duration, and no further information would be revealed if shorter flashes were achieved; the limit has thus been attained. Much has already been done with the use of flashes of a few femtoseconds duration. For example transition species, which are intermediate between the reactants and products of an elementary reaction, have been investigated, particularly by Philip Russell Brooks (b. 1938) at Rice University and by Ahmed H. Zewail (b. 1946) at the California Institute of Technology. Geraldine Anne Kenney-Wallace (b. 1943) at the University of Toronto has used these short flashes to study dynamic processes in the liquid phase, such as the vibrations that develop when a reaction occurs in solution. Processes involved in photosynthesis, such as energy transfer and electron transfer in the chloroplast reaction centres, have been extensively investigated by Porter and his collaborators, some of the later work being done with flashes in the femtosecond time range.

8.5 Thermal reactions

Reactions that are not brought about by the action of light are referred to as thermal reactions, since the energy that allows them to occur is provided by the thermal energy in the reactant molecules. Many of the features of thermal reactions are the same as those of photochemical reactions, but a number of features required special attention.

After Trautz and Lewis had put forward their kinetic theory of collisions

(Section 8.3) many tests were made to see if the experimentally measured pre-exponential factors were properly explained by the treatment. As already noted, there was excellent agreement for the reaction $2HI \rightarrow H_2 + I_2$, but for other reactions, in the gas phase and in solution, there were serious discrepancies. It was realized that treating reactive molecular collisions as if the molecules were hard spheres was hardly satisfactory, and various devices were used to bring about improvement. With the advent of transition-state theory in 1935 collisions could be handled in a more satisfactory way. More recently the theoretical methods of molecular dynamics (Section 8.8) have been applied, but there are formidable difficulties.

Solvent effects presented a particular problem for the simple kinetic theory of collisions, and again transition-state theory proved more satisfactory in leading to a general understanding, although not in allowing absolute predictions to be made.

Trimolecular reactions in the gas phase also presented a difficulty for the kinetic theory of collisions, since it was difficult to treat a collision between three molecules. There are only a few known reactions of this type; one of them is

$$2NO + O_2 \rightarrow 2NO_2.$$

This reaction has the unusual feature of a negative temperature coefficient. This was satisfactorily interpreted in 1935 by Harold Gershinowitz and Eyring's application of transition-state theory, which leads to the conclusion that the activation energy is zero and that the pre-exponential factor is proportional to T^{-3}.

Unimolecular gas reactions

The early history of unimolecular gas reactions was one of considerable confusion, on both the experimental and theoretical sides. Many first-order gas reactions, such as the decomposition of ethane into ethylene and hydrogen, turned out to be chain reactions, and only a few are known to be elementary; an example is the isomerization of cyclopropane into propylene.

In 1919 the French physicist Jean Baptiste Perrin (1870–1942) argued that unimolecular gas reactions could be understood only if the energy required for reaction to occur is provided by radiation from the walls of the reaction vessel. This idea, known as the *radiation hypothesis*, had previously been discussed by a number of people, including Trautz and W. C. McC. Lewis, and it was believed by many that it provided an explanation for reactions of all types. In 1919, when Perrin applied the hypothesis to unimolecular gas reactions, no example of such a reaction had been investigated, and he took it for granted that they would remain first order at all pressures. His argument was that one could imagine expanding the reaction vessel to an infinite volume, when the molecules would be so far apart that no collisions could take place. First-order kinetics implies that the probability that a given molecule reacts is independent of the pressure, and if this probability remains the same as the pressure approaches zero, collisions can play no part in the reaction. Expressed in another way, the frequency of collisions depends on the square of the pressure; dependence of the rate on the first power therefore seemed to be impossible if collisions are responsible for reaction. The radiation hypothesis seemed to Perrin and others to offer the only alternative explanation.

The fallacy in the argument is that Perrin was assuming experimental behaviour that is not in fact obeyed. Unimolecular reactions do not remain first-order at low pressures.

In 1921 the Faraday Society held a discussion on 'The Radiation Theory of Chemical Action'. Perrin presented a lengthy paper, and the whole problem was thoroughly discussed, including the problem of unimolecular reactions. One who expressed strong opposition to the radiation hypothesis was F. A. Lindemann, who suggested an alternative explanation of unimolecular reactions in a short communication that was published in 1922. Lindemann's oral presentation was on 28 September 1921, and a few days later Christiansen published his Ph.D. thesis at the University of Copenhagen in which the same treatment was included, in a more detailed and explicit form than appeared in Lindemann's written submission.

The Lindemann–Christiansen hypothesis is the basis of modern theories of unimolecular gas reactions, although a number of modifications had to be made to it. The basic idea is that collisions between two molecules A lead to an energized species A^*:

$$A + A \rightleftharpoons A^* + A.$$

The species A^* can either undergo inactivation if it collides with another molecule A, or it can give rise to the products of reaction:

$$A^* \rightarrow Y + Z.$$

If the pressure is high enough the energized species A^* are at equilibrium and their concentration is proportional to [A]; the kinetics are therefore first-order. At low concentrations, on the other hand, the de-energization is much less rapid than the rate with which A^* is converted into products, and the rate of product formation is equal to the rate at which A^* molecules are formed; the kinetics are therefore second order.

It was later confirmed experimentally that unimolecular reactions do become second order at sufficiently low pressures. In its original form, however, the Lindemann–Christiansen hypothesis did not interpret the results quantitatively, and a number of refinements had to be made. An important modification was made in 1927 by Hinshelwood, who pointed out that the energy in an energized molecule A^* is distributed among a number of normal modes of vibration of the molecule. This idea led him to deal more satisfactorily with the rate with which molecules become energized on collision. Later theories of unimolecular reactions are based on, or are consistent with, Hinshelwood's formulation.

In 1927 a further extension of the theory of unimolecular reactions was made at the University of California by Oscar Knefler Rice (b. 1903) and Herman Carl Ramsperger (1896–1932). Their treatment took into account the fact that the rate with which an energized molecule A* is converted into products depends on the energy it contains, instead of being constant as Hinshelwood had assumed. Shortly after their paper appeared a similar treatment was given by Louis S. Kassel, of the University of Chicago. Kassel later developed his treatment in his book *The Kinetics of Homogeneous Gas Reactions*, which appeared in 1932 by which time he had moved to the US Bureau of Mines Experiment Station at Pittsburgh.

A further extension of the treatment of unimolecular reactions has been made by Rudolf A. Marcus (b. 1923), whose first publication on the problem appeared in 1952. His treatment is a quantum-mechanical one which takes into account the zero-point levels for the vibrations, and does not assume all vibrational frequencies to be the same.

Organic free radicals

It became increasingly clear that many thermal reactions, like many photo-chemical reactions, occur in more than one stage, often involving intermediates of very short lifetimes, such as atoms and free radicals. The participation of atoms as intermediates was recognized by Nernst in his mechanism for the photochemical hydrogen–chlorine reaction, and in 1925 Hugh Stott Taylor (1890–1974) suggested the possibility that organic free radicals might be involved in reactions. He discussed among other processes the mercury-sensitized reaction between hydrogen and ethylene, the quantum yield of which was greater than unity, suggesting the existence of chains. Taylor suggested that the hydrogen atoms produced in the photosensitized process bring about the following reactions:

$$H + C_2H_4 \rightarrow C_2H_5,$$

$$C_2H_5 + H_2 \rightarrow C_2H_6 + H.$$

At a Faraday Society Discussion in 1925 Taylor discussed this and other reactions, which he suggested were chain reactions involving free radicals as well as atoms.

These suggestions proved to be correct, but for a few years the evidence was indirect. An important development came in 1929 when Friedrich Adolf Paneth (1887–1958), then at the University of Berlin, developed with some collaborators a valuable technique for detecting free radicals in the gas phase. By decomposing a substance such as lead tetramethyl, $Pb(CH_3)_4$, he deposited a thin film (or 'mirror') of lead on the inner surface of a tube, and then found that if certain substances were passed through the tube the mirror was removed. He attributed this removal to the reaction of free radicals with the metal, with the formation of a volatile product; for example, if methyl radicals are present in a reaction system they react with lead to form lead tetramethyl, which is volatile and is removed:

$$4CH_3 + Pb \rightarrow Pb(CH_3)_4.$$

This technique for the detection of organic free radicals by metallic mirrors was greatly improved and extended by Francis Owen Rice (1890–1989) and his students at Johns Hopkins University. They used a number of metals in addition to lead, and were able to identify specific radicals by the products they formed with the metals. By means of this technique they were able to confirm that when substances like ethane, acetaldehyde and the ethers are undergoing thermal decomposition, there are free radicals present in the reaction system.

In 1934 Rice and the physicist Karl Herzfeld, who was then also at Johns Hopkins University, proposed detailed free-radical mechanisms for a number of organic decompositions, and showed that these mechanisms interpreted the kinetic behaviour found experimentally. For example, acetaldehyde when heated decomposes into methane and carbon monoxide,

$$CH_3CHO \rightarrow CH_4 + CO.$$

The kinetic order is 1.5, showing that the reaction cannot occur in one stage, for in that case the order would be unity. Rice and Herzfeld proposed a mechanism involving the participation of CH_3CO and CH_3 radicals, and showed that it leads to an order of 1.5. The thermal decomposition of ethane has an order of unity, and for some time was regarded as occurring in one stage. However, Rice's use of the mirror technique showed radicals to be present, and Rice and Herzfeld proposed a mechanism which led to the correct behaviour. Later evidence has confirmed their belief that these reactions occur entirely by chain reactions involving free radicals, although it has been found necessary to modify some of the mechanisms they originally proposed.

Gaseous explosions

It has long been known that combustion reactions involve some curious and interesting features. Soon after phosphorus was discovered in about 1669 Robert Boyle began to investigate the emission of light that occurs from it. He thought he had shown, but admitted that his results were inconclusive, that there was no glow in the complete absence of oxygen, or in pure oxygen at atmospheric pressure, a result that is consistent with what we now know about pressure limits for certain combustions and explosions. Later, in 1788, the French chemist Antoine Franois Fourcroy (1755–1809) found that phosphorus burns more rapidly in air than in pure oxygen. In 1817 Houton de Labillardière found that a mixture of phosphine (PH_3) and oxygen would explode if the pressure was decreased. Later van't Hoff found, and reported in the second

(1896) edition of his *Études*, that there are two explosion limits for the phosphine–oxygen reaction; at higher and lower pressures the mixture does not explode, but between certain limits there is an explosion.

During the nineteenth century a number of investigations were made with the object of discovering the nature of explosions and the conditions under which a slow combustion reaction could become an explosion. Several such studies were carried out in the 1850s by Robert Bunsen, some of whose conclusions were, however, misleading. For example, he claimed to have shown that the amounts of reaction products formed varied *per saltum*—by a jump, or discontinuously—as the amounts of the reactants were varied continuously. Bunsen also measured the rates with which explosion waves travel, and reported that they did not exceed a few metres a second. Much higher rates were measured subsequently.

More reliable and comprehensive work on gaseous explosions was later carried out by Harold Baily Dixon (1852–1930), who was a student of Vernon Harcourt at Christ Church, Oxford, and who carried out his first work on explosions in Harcourt's laboratories. This line of investigation was suggested to him by Harcourt, but Dixon's interest in explosions, which lasted to the end of his life, may have been influenced by the fact that in 1874 his father's house in Regent's Park, London, was seriously damaged by an explosion of gunpowder on the nearby Regent's Canal. Dixon first made careful measurements of the amounts of products formed and reactants remaining, and was one of the first to measure low pressures by means of a McLeod gauge designed in the 1870s by Herbert McLeod (1841–1923), who was professor of chemistry at the Royal Indian Civil-Engineering College at Coopers Hill. Dixon soon found that Bunsen had made errors in his chemical analyses, and that there were no discontinuities in the amounts of products formed. This work was done in the 1870s and reported by Dixon at the British Association meeting in 1880, a detailed account appearing in the *Philosophical Transactions* in 1884. At about this time the word *detonation* began to be applied to a more violent type of explosion in which a shock wave, associated with a sharp change of pressure, travels at speeds similar to that of the speed of sound. Originally the word 'detonation', as the name implies, referred to the *sound* of an explosion, but came to be used for the explosion itself.

In 1880 a violent explosion in a gas main near Tottenham Court Road, London, involving the loss of two lives and much property damage, provided strong evidence that flame speeds were much greater than Bunsen had thought. Vernon Harcourt was called in to investigate the accident, and he suggested to Dixon the desirability of making a systematic investigation of the rates of propagation of explosions in gases. By that time Dixon had become a Fellow of Trinity College, Oxford, and to make measurements on the rates with which explosion waves travel he set up long metal pipes under the Dining Hall of neighbouring Balliol College. To measure the rates he made use of an electric chronograph that had been designed by Charles John Francis Yule (1848–

1905) for some physiological experiments. In 1887 Dixon became professor of chemistry at Manchester, where he continued these experiments, now making use of an instrument similar to a so-called tram-chronograph that had been designed by the Revd Frederick John Smith (1848–1911), also a Fellow of Trinity College. Smith, who later amended his name to Jervis-Smith, also made some measurements of rates of detonation. The results of all of Dixon's investigations on explosions were described in three long papers in the *Philosophical Transactions*, appearing in 1884, 1893 and 1903. These three papers constitute a valuable monograph on explosions in gases.

Dixon also attempted to develop a theory of explosions. Previously the French chemists Pierre Eugène Marcellin Berthelot (1827–1907) and P. Vieille had developed such a theory, which was described in a book that Berthelot published in 1883. Their treatment, however, did not give satisfactory agreement with experiment as far as rates of detonation are concerned. Dixon was able to make some small improvements to the details of their theory, and these did lead to some improvement between calculated and observed rates. However, in 1897 D. L. Chapman became a member of Dixon's department at Manchester, and at Dixon's suggestion began to work on the problem. Also a former student of Harcourt at Christ Church, Chapman was highly competent in physics and mathematics as well as chemistry, and in 1899 he published a satisfactory treatment that gave excellent agreement with experiment. A similar treatment was later given by Émile Jouguet, and the Chapman–Jouguet treatment is still accepted as the standard theory of detonation waves.

In the course of his investigations on the reaction between carbon monoxide and oxygen Dixon obtained the surprising result that no explosion would occur if the gases were completely dried by means of phosphorus pentoxide. He performed a demonstration of this at the 1880 meeting in Swansea of the British Association. There was surprise and scepticism at first, but this particular effect was later confirmed and appears to be genuine. Later one of Dixon's students, Herbert Brereton Baker (1862–1935), devoted much of his career to investigating the effect of moisture on chemical change. His work was done first at Oxford and later at Imperial College, London, and he became known as 'Dry Reaction Baker' or 'Dry Baker'. He obtained many results that were controversial, particularly as he went so far as to maintain that all chemical reactions require the presence of water—a conclusion that Mrs Fulhame had implied nearly a century earlier (Section 8.6). Much work has been done on this matter, and it is now clear that water is required only in a few reactions, such as the oxidation of carbon monoxide which Dixon had investigated.

Branching chains

The Chapman–Jouguet treatment was concerned with the more physical aspects of detonation waves in gases, and there still remained the question of the detailed mechanisms of the chemical reactions that were occurring. This was a problem that could only be settled after more was known of reaction mechanisms, in particular of chain mechanisms.

In an important paper which appeared in 1923 Christiansen and the Dutch physicist Hendrick Anthony Kramers (1894–1952) reviewed the whole problem of chain reactions, and suggested the possibility of *branching chains*. A chain mechanism may involve propagating steps in which one chain carrier (atom or free radical) gives rise to more than one; an example is

$$H + O_2 \rightarrow OH + O.$$

Here the atom H has produced another atom O and also a free radical OH. When such chain branching occurs, the number of chain carriers can increase extremely rapidly, and an explosion may occur.

The first clear-cut evidence for chain branching was obtained in 1927 in Leningrad (formerly, and again later, called St Petersburg) by Nicolai Nicolaevich Semenov (1896–1986). He studied the reaction between phosphorus vapour and oxygen, and showed that there is a critical pressure above which explosion occurs but below which there is steady reaction. He related this result to the removal of chain carriers at the walls of the reaction vessel. At sufficiently low pressures the rate of removal of the chain carriers is large enough to counteract the effect of the chain branching. As the pressure is raised, however, it becomes harder for the chain carriers to reach the walls, and explosion occurs.

In 1927 Cyril Norman Hinshelwood (1897–1967), working in a converted lavatory at Trinity College, Oxford, began a similar line of investigation on the reaction between hydrogen and oxygen. His papers on this subject with his student Harold Warris Thompson (1908–1983) showed that in addition to a lower explosion limit there is also an upper limit, above which reaction is again slow. The upper limit was attributed to the removal of chain carriers in the gas phase rather than at the surface of the vessel. These contributions of Semenov and Hinshelwood played an important part in their being awarded jointly the 1956 Nobel Prize for chemistry.

Techniques for fast reactions

By the end of the 1930s most of the general principles relating to reaction mechanisms had become established. A kineticist at that time thought of the mechanisms of reactions in much the same way as does a kineticist today. What has been done more recently has been the filling in of many details. The later developments have become possible because of the invention and refinement of experimental techniques. For example, today free radicals are rarely studied using the mirror techniques, since spectroscopic and mass-spectrometric techniques have become sufficiently improved as to allow this to be done more effectively.

Of particular importance to kinetics has been the development of special

Sir Cyril Norman Hinshelwood (1897–1967)

Hinshelwood was born in London and won a scholarship to Balliol College, Oxford. Because of the First World War he was not able to go into residence at once, but worked from 1916 to 1918 on explosions at the Royal Ordnance Factory at Queensferry. Here his remarkable ability caused him to be referred to as the 'boy wonder', and he was promoted to be deputy chief chemist at the age of 21. After the war he took a shortened course at Oxford, and on graduating with distinction in 1920 was elected a research fellow of Balliol. In the following year he became Fellow and Tutor of Trinity College, Oxford, and from then until 1941 carried out research in kinetics in makeshift laboratories that had been established by that college. In 1927 he was appointed University Lecturer in Chemical Dynamics, an interesting title that was created for him. In 1931 he became head of the Balliol-Trinity laboratories for physical chemistry, effectively becoming at an early age a professor without a chair. In 1937 he was appointed Dr Lee's professor of chemistry at Oxford, a position he held until his retirement in 1964.

Hinshelwood's most productive years were his early ones, before he had the responsibility of administering a large physical chemistry laboratory which was completed in 1941. Like van't Hoff, he excelled in planning crucial experiments and in carrying them out with the simplest of equipment. From about 1920 to 1929 he investigated surface reactions, and on the basis of Langmuir's theory of adsorption he contributed greatly to the understanding of the mechanisms of such reactions (Section 8.6). In about 1924 he became interested in unimolecular gas reactions, and made an important contribution to the theory of such reactions in 1927, besides investigating a number of them experimentally (Section 8.5). From about 1927 to 1935 he investigated gaseous explosions, and almost simultaneously

techniques for the study of very rapid reactions; the ranges of the more important methods are indicated in Figure 8.1. The methods fall into two classes: *flow* methods, and *pulse* methods. The flow methods can be used for reactions having half lives of not less than about a millisecond (10^{-3} s). A flow method was first used in 1897 for gas reactions by Ernest Rutherford (1871–1937). For reactions in solution the flow techniques were improved, with particular reference to biochemical systems, at Cambridge University—in 1923 by H. Hartridge and Francis John Worsley Roughton (1899–1972) and in 1936 by G. A. Millikin. In 1940 the biophysicist Britton Chance (b. 1913), at the University of Pennsylvania, introduced the very useful *stopped-flow method*, in

with Semenov explained in 1928 their mechanisms in terms of branching chains (Section 8.5). Much of this work was done in a converted college lavatory, the original fittings of which were visible. This building was visited by Bodenstein, who remarked to Hinshelwood that 'Ihr Institut ist doch schrecklich'. Hinshelwood was not surprised at the adjective, but was amused at hearing the building referred to as an Institute. When he was investigating the hydrogen-oxygen reaction the humour of the fact that he was making water in a converted lavatory was not lost on him or his colleagues.

In 1929 Hinshelwood moved to a slightly more suitable building nearby that had previously been used as an engineering laboratory. During the 1930s he and his students did much work on the mechanisms of other chain reactions and of reactions in solution. In 1937 he became particularly interested in the kinetics of reactions occurring in bacterial cells, a field in which he worked until the end of his career and made important contributions.

Hinshelwood exerted a considerable influence in many ways. One of his books, *The Kinetics of Chemical Change in Gaseous Systems*, first published in 1926, made an important new approach to the subject; it underwent several revisions as the subject advanced. He was President of the Royal Society from 1955 to 1960, and President of the Classical Association in 1959. He was knighted in 1948, received a Nobel Prize in 1956, and was appointed to the Order of Merit in 1960.

References: E. G. Spittler, DSB, 6, 404–405 (1972); Bowen (1967); Laidler (1988); Thompson (1973).

which solutions are rapidly mixed and the flow then stopped, after which the reaction is followed photometrically. Chance has applied his technique to a wide variety of biological reactions.

The usefulness of the flow methods is limited by the speed with which gases or solutions can be mixed, which can never be done in less than about a millisecond. For half lives less than about a millisecond a *pulse* method has to be used. One class of pulse methods comprises the relaxation techniques, in which the reaction system is initially at equilibrium; the equilibrium is then disturbed in some way, and the system relaxes to a new state of equilibrium. The flash photolysis techniques, already considered (Section 8.4), are also pulse methods,

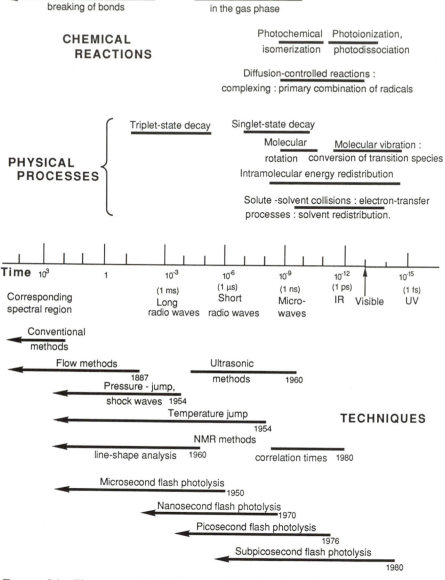

FIGURE 8.1   The range of half lives or relaxation times for some chemical and physical processes, and the ranges of times available to various experimental techniques. The approximate years in which the techniques were first used are indicated.

but are not necessarily relaxation methods since the system need not be at equilibrium initially. The flash photolysis methods are capable of investigating faster processes than is possible with any other technique.

An important group of relaxation methods was introduced in the 1950s by Manfred Eigen (b. 1927) at the University of Göttingen. A chemical system at equilibrium is disturbed in some way so as to cause the equilibrium to shift. For example, in the *temperature-jump* or *T-jump* method the temperature of a tiny cell containing a reaction mixture is raised by a few degrees, often in less than 10^{-7} s. As the system relaxes to its new state of equilibrium the process is followed by photometric or other techniques. By the use of this method, which is capable of following reactions with half lives down to about 10^{-9} s, Eigen and his co-workers have studied some of the fastest chemical reactions, such as the combination of hydrogen and hydroxide ions. For this contribution Eigen shared the 1967 Nobel Prize in chemistry with Norrish and Porter who had invented the flash-photolysis technique.

Ultrasonic techniques are being used more and more for the study of rapid reactions in solution. As early as 1910 Nernst realized that measurement of sound dispersion might provide kinetic information, and he tried unsuccessfully to determine the rate of dimerization of nitrogen dioxide, NO_2, in this way. Following work at Harvard by George Washington Pierce (1872–1956), Herzfeld and Rice in 1928 analyzed some of the data on sound dispersion and were able to interpret the results in terms of an exchange of energy between translational and internal modes. Important improvements in techniques during World War II, and the publication in 1959 of an authoritative book by Herzfeld and Theodore Aaron Litovitz (b. 1923), led to further applications of the ultrasonic method to processes such as proton transfer, conformational changes, and ion association. The technique is useful for processes occurring in the microsecond and nanosecond time range.

Nuclear magnetic resonance (NMR) spectroscopy, considered in Section 6.7, has also found application in high-speed kinetics. The analysis of linewidths in NMR spectra has provided valuable information about processes having relaxation times of 10^{-5} s or higher. A more recent technique, developed by P. Laszlo and others, involves correlation times, and allows one to study processes occurring in times of 10^{-8} s to 10^{-12} s. Similar investigations to those done with NMR can be done with electron paramagnetic resonance (Section 6.7), but this technique is limited to paramagnetic species such as free radicals.

8.6 Reactions in solution

Most of the reactions with which chemists are concerned occur in the liquid phase. The kinetic effects of solvents are therefore of great importance, and have been extensively investigated. Only a brief account can be given here.

An important early investigation of the problem was described in 1890 by the distinguished Russian physical chemist Nikolai Alexandrovich Menschutkin

(1842–1907) of the University of St Petersburg. He studied the rate of the reaction between ethyl iodide and triethylamine in a number of solvents:

$$(C_2H_5)_3N + C_2H_5I \rightarrow (C_2H_5)_4N^+I^-$$

He found, for example, that at 100°C the reaction in acetophenone proceeds over seven hundred times as fast as it does in hexane solution. The importance of this work was that it countered a conclusion that some people were drawing from van't Hoff's analysis of the osmotic pressures of solutions (Section 4.4), which were the same pressures that a solute would exert if it were in the gas phase at the same concentration. This result could be taken to suggest that the solvent did little more than fill up the space between solute molecules. Menschutkin's work showed that at least for some reactions and some solvents this conclusion was far from correct.

Van't Hoff was quick to appreciate the significance of Menschutkin's results, and within a few years he was demonstrating, in his lectures at the University of Berlin, a correlation between the rate of a Menschutkin reaction and the dielectric constant of the solvent; a version of these lectures was given in his *Vorlesungen über theoretische und physikalische Chemie* (Lectures on Theoretical and Physical Chemistry) which appeared, together with an English translation, at the turn of the century. Van't Hoff also pointed out that there is no correlation between the rate and the viscosity of the solvent, showing that this reaction is not controlled by the rate with which the reactant molecules diffuse together.

When Menschutkin published this work the Arrhenius equation (Section 8.2) was only a year old, and it was by no means generally accepted for another 25 years. Menschutkin therefore measured no activation energies, but in 1931 the German chemists Hans Georg Grimm (1887–1958), H. Ruf and H. Wolff reported a number of activation energies for reactions of the same type, and they also extended the range of solvents. Their temperature-dependence studies showed that there are variations in the pre-exponential factors; in other words, the variations in rates are not exactly correlated with the variations in activation energy. If the results are expressed in terms of enthalpies and entropies of activation there are certain correlations and compensations between the two parameters. The investigation of compensation and correlation effects has proved to be a fruitful one, and it is actively pursued today.

In 1930, a little before this work of Grimm, Ruf and Wolff was published, Henry Eyring and Farrington Daniels (1889–1972), at the University of Wisconsin, had reported detailed studies of another reaction, the decomposition of nitrogen pentoxide in various solvents. They found that some solvents, such as carbon tetrachloride, had little effect on the kinetics, whereas others, such as nitric acid, had a marked effect.

In the 1930s, work on reactions in solution was carried out in Hinshelwood's laboratories at Trinity College, Oxford. Hinshelwood had already done his important work on gaseous explosions (Section 8.5) and on reactions on surfaces (Section 8.7); his interest in solvent effects was stimulated by the arrival in his laboratory of Emyr Alun Moelwyn-Hughes (1905–1978), who himself was to have a distinguished career in that area of research. Hinshelwood and Moelwyn-Hughes attempted to study a Menschutkin reaction in the gas phase, in order to compare the behaviour with that in different solvents. This proved to be unexpectedly difficult since the reaction went much more readily on the surface of the vessel than in the gas phase. In carbon tetrachloride solution the reaction between triethylamine and ethyl iodide had a pre-exponential factor (A factor) that was smaller by many powers of ten than expected from simple collision theory, and the same appeared to be true of the reaction in the gas phase. The result was important in showing that the low A factors in a solvent are not just due to the solvent, but are also to some extent an intrinsic property of the reaction.

These conclusions led Hinshelwood and Moelwyn-Hughes to propose a modification of the interpretation that W. C. McC. Lewis had given to the Arrhenius equation. According to Lewis the pre-exponential factor A in the Arrhenius equation (8.9) is identical with the collision theory pre-exponential factor that appears in eqn 8.27:

$$A = L d_{AB}^2 \, 8\pi k_B T \left(\frac{m_A + m_B}{m_A m_B} \right)^{1/2} = z_{AB}. \tag{8.34}$$

The quantity z_{AB} is now called the collision frequency factor, or the collision density. The proposal of Hinshelwood and Moelwyn-Hughes was that the A factor should not be simply z_{AB}, but should be z_{AB} multiplied by a probability factor P:

$$A = P z_{AB} \quad \text{so that} \quad k = P z_{AB} e^{-E/RT}. \tag{8.35}$$

The factor P, which inevitably became known as the 'fudge factor', was intended to take care of the special conditions that had to be satisfied in order for suitably energized molecules to react on collision. One of these conditions was an orientation requirement, related to the need for certain parts of the colliding molecules to come together in order for there to be any possibility of reaction occurring.

The advent of transition-state theory in 1935 brought further clarification to this problem. According to that theory, the effect of a solvent on the rate of reaction depends on the difference between its effect on the initial reactants and the activated complex. A low pre-exponential factor, interpreted by Hinshelwood and Moelwyn-Hughes in terms of a low P factor, is interpreted by transition-state theory in terms of a negative entropy of activation. In the case of a Menschutkin reaction this is partly due to the complexity of the reactant molecules, as a result of which reaction can occur only if the reactant molecules come together with a suitable orientation. It is also partly due to the fact that the activated complexes have a considerable polar character; they are therefore substantially solvated and this lowers their entropy, giving a low A factor. Solvation also lowers their energy, which means that the activation energy is lowered. Thus solvents for Menschutkin reactions that are polar and therefore strongly solvating reduce the activation energy and increase the rate; there is, however, some compensation effect in that they also reduce the entropy of activation and therefore the A factor.

Many other important contributions have been made to the understanding of the mechanisms of reactions in solution, and particular mention should be made of the work of several people who were primarily organic chemists. One was Robert Robinson (1886–1975), who besides carrying out synthetic organic chemistry, particularly on the alkaloids, which led to his 1947 Nobel Prize, also developed a comprehensive treatment of organic reactvity, with special reference to the effects of substituents in organic compounds. His work was not concerned with kinetic aspects or with the effect of solvents. Important work on organic reactivity was also done by Christopher Kelk Ingold (1893–1970), much of whose work was done in collaboration with Edward David Hughes (1906–1963). Hughes and Ingold in particular made a study of solvent effects, and their work provided a conceptual framework by means of which a practical chemist can make qualitative predictions of the effects of different kinds of solvents on the rates of reactions.

Another important aspect of reactions in solution, particularly those involving organic molecules, is the matter of mechanisms and the effects of substituents. This is a vast subject which can only be referred to briefly here. Important contributions in this field were made by Hinshelwood, Moelwyn-Hughes, Ingold, and Hughes. Louis Plack Hammett (1894–1987) at Columbia University in New York dealt in particular with rate constants and equilibrium constants for series of homologous reactions involving an aromatic ring on which there are various substituents. He found linear relationships between the logarithms of equilibrium and rate constants, which means that there are linear relationships between Gibbs energies for overall reactions, and Gibbs energies of activation. Hammett demonstrated and discussed these relationships in his book *Physical Organic Chemistry*, which was published in 1940. Much further work has been done on these so-called 'free-energy relationships'.

Diffusion-controlled reactions

In his famous paper of 1889 on the temperature-dependence of reaction rates, Arrhenius emphasized that the rates he was concerned with were not dependent on the rates with which the reactant molecules diffused together. This is certainly the case for the reactions being studied at the time, but when extremely rapid reactions came to be studied it was realized that this is not always the case. If a bimolecular reaction in solution has a very low activation energy, the chemical process occurs at once when the reactant molecules come together; the observed rate is therefore the rate with which the molecules diffuse together.

There is thus a need for a theory of the rate with which solute molecules diffuse together. Such a theory was provided by Marian von Smoluchowski

(1872–1917), whose investigations on the theory of Brownian movement are described in Section 9.1. During the course of this work he obtained an expression for the rate with which two particles diffuse together, his paper on this problem appearing in 1917, the year of his untimely death from dysentery while serving as Rector of the Jagiellonian University in Cracow. In modern notation, and with some modifications, his result can be expressed in terms of the rate constant for the process as

$$k_D = 4\pi(D_A + D_B)d_{AB}, \tag{8.36}$$

where D_A and D_B are the diffusion constants of the species A and B, and d_{AB} is the final distance between their centres. This expression leads to a rate constant of approximately 7×10^9 dm^3 mol^{-1} s^{-1}. If the rate constant for the chemical process is substantially greater than this—as it is, for example, for the neutralization of ions—the reaction is completely diffusion controlled, and the rate constant will be approximately the value given above. If the rate constant is much less, as is often the case, there is no diffusion control. For intermediate situations there is partial diffusion control, the theory of which has been worked out.

Equation 8.36 applies only if the reacting species are uncharged. If they are ions of opposite signs, so that they attract one another, the diffusion rates will be enhanced; if they are of the same sign they will be reduced. The theory of diffusion control for ionic reactions was worked out in 1942 by Debye, and his treatment can be satisfactorily reconciled with the experimental results.

8.7 Catalysis

The investigation of catalysis is of particular significance in the development of physical chemistry, for two main reasons. One is that Ostwald, who was a great pioneer in physical chemistry, had a special interest in catalysis, and it was mainly for his work in this field that he was awarded the 1909 Nobel Prize in chemistry. The other reason is that catalysis is of great practical importance, many chemical and pharmaceutical industries relying heavily on the use of catalysts that allow products to be formed in reasonably short periods of time.

Although it was not until 1836 that catalysis was recognized as a distinct and general phenomenon, and that the name 'catalysis' was coined, there were several relevant earlier studies. Elizabeth Fulhame's 1794 book, mentioned in Section 8.4 as containing pioneering work on photochemical imaging, described extensive experiments which she interpreted as involving the participation of water which essentially acted as a catalyst. In the combustion of charcoal, for example, she suggested that

the carbone attracts the oxygen of the water, and forms carbonic acid, while the hydrogen of the water unites with oxygen of the vital air, and forms a new quantity of water equal to that decomposed.

Mrs Fulhame's idea that water is an intermediate in reactions was later developed extensively by a number of workers, particularly by H. B. Baker, as mentioned in Section 8.5.

In the early years of the nineteenth century a number of investigators discovered that metals and other solids enhanced the rates of various reactions. In 1813, for example, Thénard found that the rate of decomposition of ammonia was increased when certain metals were added to the system. In 1818 he discovered hydrogen peroxide and found that the rate of its decomposition was increased by the addition of metals and other substances. In 1817 Humphry Davy found that a platinum or palladium wire, previously heated to a temperature 'short of redness', glowed if it was inserted into a mixture of oxygen and coal gas. Many other similar effects were observed by Davy, some related to his invention by 1818 of the safety lamp. The German chemist Johann Wolfgang Döbereiner (1790–1849) designed a hydrogen–air lamp that was spontaneously ignited by means of spongy platinum; such lamps had some popularity until friction matches of the modern type became available later in the century. These investigations were greatly extended by Faraday, who in 1834 explained the action of solids in terms of what he called 'contact action'. He suggested that a solid exerts an attractive force on the molecules of a gas, which are therefore 'drawn into association'; gases which are 'simultaneously subjected to this attraction' are led to combine with each other. This explanation, which came to be called the *condensation theory*, was criticized as not explaining the specificity of surfaces.

Until 1836 the various examples of the enhancement of rates by added substances were treated as isolated examples, but in that year the Swedish chemist Jöns Jakob Berzelius (1779–1848) reviewed a number of examples, and concluded that

It is then proved that several simple and compound bodies ... have the property of exercising on other bodies an action very different from chemical affinity ... I do not believe that it is a force quite independent of the electrochemical affinities of matter ... but since we cannot see their connection and mutual dependence it will be more convenient to designate this force by a separate name. I ... will therefore call this force *catalytic force*. Similarly I will call the decomposition of bodies by this force *catalysis* ...'

Although Berzelius's idea of a catalytic force did not prove useful, his realization that a common type of effect is involved in various reactions, and his suggestion of the word 'catalysis', were important in inspiring much later work which led to an understanding of what occurred in different types of catalyzed processes. He coined the word 'catalysis' from the Greek λυσις (lysis), a loosening, and κατα (kata), which can mean 'down' and which also lends emphasis and can also be translated as 'wholly'. After the publication of Berzelius's paper the word 'catalysis' was used widely, but a substance that brought about catalysis tended to be called a 'contact substance', a term suggested in 1842 by the German chemist Eilhardt Mitscherlich (1794–1863). The word 'catalyst' was first proposed in 1885 by Henry Armstrong.

Catalysis by acids and bases

Important systematic investigations of catalyzed processes were made by Ostwald. In 1881, on being appointed professor of chemistry at the Riga Polytechnic Institute, he began to investigate the hydrolyses of various substances in the presence of acids. In papers that appeared from 1883 to 1888, with the general title 'Studien zur chemischen Dynamik', he reported that the rate of hydrolysis of a substance catalyzed by a mineral acid is independent of the nature of the acid. He also made measurements of the electrical conductivities of a number of weak acids, and found a relationship between these conductivities and the rates of the reactions they catalyze. Earlier he had found, in his work for his master's and doctor's degrees at the University of Dorpat, that the affinity between an acid and a base does not depend on the nature of the base.

Ostwald was at first unable to explain these results, particularly the fact that the acid-catalyzed rates were independent of the nature of the anion. In June, 1884, after receiving a copy of Arrhenius's Ph.D. dissertation, he at once realized that with a proper interpretation Arrhenius's results would throw light on the whole problem of catalysis by acids and bases. Ostwald immediately prepared a short paper, dated July, 1884, which described his own results on electrical conductivities and their relationship to reaction rates. This paper emphasized the importance of Arrhenius's results.

By 1887 Arrhenius, with help from his discussions with Ostwald and others, had formulated his theory of electrolytic dissociation (Section 7.2). The significance of Ostwald's correlation between rates and conductivities was now clear to Ostwald and Arrhenius. The conductivity of a solution of an acid provides an approximate measure of the hydrogen-ion concentration in the solution, so that Ostwald was observing a correlation between the rates of acid-catalyzed reactions and the concentrations of hydrogen ions.

In 1889 Arrhenius published an important paper in which he suggested that in acid catalysis a free hydrogen ion adds on to a molecule of 'substrate', the term later given to a substance undergoing a catalyzed reaction. The intermediate so formed, the protonated substrate, then undergoes further reactions, the hydrogen ion eventually being released. This theory, which became known as the *hydrion theory*, is along the right lines, but it did not at once lead to a quantitative interpretation of the kinetic behaviour. Arrhenius realized that complications arising from non-ideality had to be taken into account, but was not clear how this was to be done. Later, in 1899, he showed that the kinetics of acid-catalyzed reactions could be better understood if one arbitrarily used osmotic pressures instead of concentrations. At the time the reason for this was not clear but we can now understand it in terms of activity effects.

Activities and activity coefficients were introduced into thermodynamics in 1900–1907 by G. N. Lewis (Section 4.4), but the way they enter into rate equations was not satisfactorily solved for a number of years. Since equilibrium constants must be expressed in terms of activities rather than concentrations, it follows that activities must in some way enter into rate equations, since when a reaction has reached equilibrium the rates are equal in forward and reverse directions. Thus for a reaction

$$A + B \rightleftharpoons Z,$$

the equilibrium constant is

$$K = \frac{a_Z}{a_A a_B} = \frac{[Z]}{[A][B]} \frac{\gamma_Z}{\gamma_A \gamma_B}, \tag{8.37}$$

where the as are the thermodynamic activities and the γs the activity coefficients, which multiply concentrations to give activities. Rate equations must involve activity coefficients also, but in exactly what way was not at first clear.

One suggestion, made in 1908 by the Scottish chemist Arthur Lapworth (1872–1941), was that the rate of an elementary bimolecular reaction between A and B should be expressed as

$$v = k_o a_A a_B = k_o[A][B]\gamma_A \gamma_B. \tag{8.38}$$

This theory, which came to be called the *activity-rate theory*, received support from the American electrochemist Herbert Spencer Harned (1885–1969) and from W. C. McC. Lewis. It soon became clear, however, that this theory did not provide a satisfactory explanation of the effects of 'foreign salts' on reaction rates. A foreign salt is one that does not have an ion in common with any ion directly involved in the reaction, and for reactions between ions such foreign salts have a significant effect on the rate. For a reaction involving the interaction between two ions of the same sign, the addition of a small amount of a foreign salt always increases the rate; if the reaction is between ions of opposite signs, the effect of a foreign salt is to reduce the rate.

However, at low concentrations foreign salts always reduce the activity coefficients of ions, a result that is explained by the Debye–Hückel theory in terms of the effect of the ionic atmosphere. The activity-rate equation (8.35) therefore always predicts a decrease in rate when a foreign salt is added at low concentrations, but this is inconsistent with the actual behaviour when the reaction involves ions of the same sign.

The problem was essentially solved in the 1920s by three Danish physical chemists, Brönsted, Bjerrum and Christiansen. Their work was based on Lewis and Randall's introduction of the ionic strength (eqn 7.10) and on the Debye–Hückel theory. In 1922 Brönsted arrived at the rate equation

$$v = k_o[A][B] \frac{\gamma_A \gamma_B}{\gamma_X}, \tag{8.39}$$

where γ_X is the activity coefficient of some critical complex X that is formed from the reactants A and B. By applying eqn 7.11, which relates activity coefficients to the ionic strength I, to his rate equation 8.35 he arrived at the equation

$$\log_{10}k = \log_{10}k_o - B(z_A^2 + z_B^2 - z_X^2)\sqrt{I}, \qquad (8.40)$$

where k_o is the rate constant at zero ionic strength. Since the charge z_X on the critical complex is equal to $z_A + z_B$, this equation reduces to

$$\log_{10}k = \log_{10}k_o + 2Bz_Az_B\sqrt{I}. \qquad (8.41)$$

This relationship explains how an increase in ionic strength increases the rate when the reactant ions are of the same sign (z_Az_B is positive) and decreases the rate when z_Az_B is negative.

The reason that the activity-rate eqn 8.35 fails is that it leads not to eqn 8.36 but to an equation with the term z_X^2 missing:

$$\log_{10}k = \log_{10}k_o - B(z_A^2 + z_B^2)\sqrt{I}. \qquad (8.42)$$

Since $-B(z_A^2 + z_B^2)$ is bound to be negative, an increase in I must always decrease the rate, in disagreement with the results for ions of the same sign.

Christiansen's derivation of eq. 8.37 focussed attention on the frequency of collisions between reacting molecules, and took account of the fact that the probability of two ions being close to one another is affected by the electrostatic interactions. Little exception could be taken to his derivation, but there were difficulties with the simpler derivations given by Brönsted and Bjerrum. Brönsted's justification for his use of the kinetic activity factor $\gamma_A\gamma_B/\gamma_X$ in eqn 8.35 was by no means clear. Bjerrum considered the mechanism as

$$A + B \rightleftharpoons X \rightarrow \text{products}$$

and then assumed that the rate of breakdown of X is proportional to the concentration, rather than the activity of X. The rate is then

$$v = k'[X] = k\,[A][B]\,\frac{\gamma_A\gamma_B}{\gamma_X}. \qquad (8.43)$$

The difficulty is that, if X is an ordinary intermediate, in equilibrium with reactants, the rate of its breakdown would not be proportional simply to its concentration but would involve activity coefficients. However, as pointed out by R. P. Bell in 1941, if instead X is an activated complex, only in pseudo-equilibrium with A and B and bound to form products, the assumption that the rate is proportional to [X] is justified.

One important development in acid–base catalysis was the realization that catalysis can be brought about not only by hydrogen and hydroxide ions but also by general acids and bases as defined in 1923 by Brönsted and in the same year by the British chemist Thomas Martin Lowry (1874–1936). A species such as NH_4^+ is an acid since it can donate a proton, and the anion of an acid is a base since it can accept a proton. That such species can bring about catalysis was first established experimentally at the University of Leeds by Harry Medford

Dawson (1876–1939), in a series of investigations beginning in about 1912. Important relationships between catalytic activities and the strengths of acids and bases were proposed in 1928 by Brönsted.

During the course of his extensive investigations on catalysis Ostwald suggested several definitions of catalysis, all of them consistent with one another and more or less consistent with the definition used today. In Volume 2 of the uncompleted 1902 second edition of his *Lehrbuch*, for example, he suggested that

A catalyst is a substance that changes the velocity of a reaction without itself being changed in the process.'

Today we restrict the definition to a substance that *increases* the rate of a reaction. Formerly a substance that decreases the rate was often known as a 'negative catalyst' but this usage is now felt to be undesirable. So-called 'negative catalysts', better known as *inhibitors*, do not act in the same way as catalysts, but instead bring about their action either by destroying catalysts already present, or by removing active intermediates such as atoms and free radicals.

In a review 'Über Katalyse' that Ostwald published in 1901 he mentioned a number of different types of catalysis, including homogeneous catalysis such as acid–base catalysis, catalysis by enzymes which are the biological catalysts, and heterogeneous or surface catalysis.

Surface catalysis

Ostwald himself did not work on surface catalysis, the first systematic investigation of which was carried out by van't Hoff and described in his *Études*; it was concerned with the decomposition of arsine, AsH_3, into arsenic and hydrogen. By working with vessels of different shapes and sizes, in which the surface/volume ratio was varied, he found that the reaction proceeded more rapidly in a vessel having a larger proportion of surface. In 1910 the technique of varying the surface/volume ratio was applied more quantitatively by D. L. Chapman and H. E. Jones to the thermal decomposition of ozone, O_3, into oxygen.

Important advances in the field of surface catalysis were made by the American physical chemist Irving Langmuir (1881–1957). For most of his scientific career he was with the General Electric Company at Schenectady, New York, and much of his work, although very fundamental, was related to the practical problem of improving electric light bulbs. In careful studies of the interaction between gases and surfaces he reported in 1912 that hydrogen molecules are dissociated into atoms at hot metallic surfaces. Later he made important investigations on the adsorption of gases on surfaces, showing that it is of a chemical nature, i.e., is *chemisorption*, rather than physical or *van der Waals* adsorption.

Langmuir showed in 1921 that when surface catalysis occurs there is usually chemisorption, and he interpreted the kinetics in terms of what became known as the *Langmuir adsorption isotherm*, which relates the fraction of surface covered to the pressure of the gas (Section 9.3). He showed how the adsorption isotherm can be used to interpret the kinetics of a variety of surface reactions, and his work was greatly extended at Oxford by Hinshelwood in a series of investigations carried out in the 1920s. Other important advances were made at about the same time by Eric Keighley Rideal (1890–1974) at Cambridge University and by H. S. Taylor at Princeton. Rideal showed, for example, that a surface reaction between two substances sometimes involves the adsorption of only one of them, the other substance reacting directly with it from the gas phase. Taylor made two outstanding contributions to the understanding of catalysis at surfaces. In 1925 he showed that under ordinary conditions surfaces are of variable activity, and that chemical reactions occur mainly at certain active centres on the surface. In 1931 he showed that adsorption processes may involve an activation energy, particularly if dissociation (e.g., of H_2 into 2H) occurs during adsorption.

These pioneering and fundamental investigations on surface catalysis provided the groundwork for the very detailed and extensive studies that are carried out today as a result of the technical importance of catalysis. As a consequence of the development of new experimental methods, much more is now known of the detailed nature of surfaces and of adsorbed species.

8.8 Reaction dynamics

Reaction dynamics, or molecular dynamics, is an important and rapidly growing branch of science that is concerned with the fine details of what occurs during the course of chemical change. It deals with the intermolecular motions that occur, and with the relationships between the quantum states of the reactant and product molecules.

The investigations in this field had their origin in work carried out in the 1920s and 1930s by Michael Polanyi and his collaborators. This work was in part concerned with the radiation emitted when alkali metals react with halogens or certain halides in the gas phase. Some of the reactions were studied by the method of 'highly dilute flames', in which two reactants, such as sodium vapour and chlorine gas, were allowed to flow into an evacuated vessel from opposite ends. The gases mix by diffusion, and the product (e.g., sodium chloride) was deposited on the walls of the vessel. Measurements were also made of the rates of the reactions and of the radiation emitted. Polanyi concluded that in the reaction

$$Cl + Na_2 \rightarrow NaCl + Na,$$

the NaCl is formed in an excited vibrational state and can transfer its energy to a sodium atom which emits the characteristic yellow D radiation.

During more recent years this type of investigation has been greatly extended by Michael Polanyi's son John C. Polanyi (b. 1929), who has used infrared

John Charles Polanyi

(b. 1929)

John Polanyi was born in Berlin, where his father Michael Polanyi was professor of physical chemistry. When John was aged four his father became professor of chemistry at the University of Manchester, and it was in Manchester that John received his education. He was awarded the Ph.D. degree by the University of Manchester in 1952, and then spent two years as a postdoctoral fellow at the National Research Council of Canada in Ottawa. There he worked mainly in the kinetics laboratories of E. W. R. Steacie, but for a few months he was in the spectroscopic laboratories of Gerhard Herzberg. From 1954 to 1956 he was a research associate at Princeton.

In 1956 Polanyi was appointed to the University of Toronto, where he has remained in various positions ever since, being given the special title of University Professor in 1974. He at once embarked on a comprehensive programme of research in the rapidly developing field of reaction dynamics (Section 8.8) and has been particularly effective in combining both experimental and theoretical approaches. Some of his work has been concerned with the energy distributions among the products of an elementary reaction, and he introduced the technique of *infrared chemiluminescence* in which the products of reaction are studied by means of an infrared spectrometer. This work led to his suggestion that an infrared laser could be constructed using the reaction $H + Cl_2$ HCl + Cl, and this was done in 1965 by J.V.V. Kasper and G.C. Pimentel.

spectroscopy to study the vibrational states of reactants and products. His investigations have explored the relationship between the type of energy transfer that occurs when a reaction takes place and the shape of the potential-energy surface for the reaction.

Molecular beams

The experiments just described are referred to as *bulk* experiments. A different type of investigation, which leads to information of a similar kind, is carried out with molecular beams. In such experiments narrow beams of reactant molecules, in pre-selected states, are brought into contact with each other, often at an angle of 90°, and studies are made of the subsequent fate of the reactant and product molecules.

The first chemical experiment that can be described as a molecular-beam experiment appears to have been that of F. O. Rice, Harold Clayton Urey

Polanyi also dealt with the problem of the type of energy—translational, vibrational or rotational—that is most effective in leading to chemical reaction. He obtained information about the fate of energy that is in excess of that required to surmount an energy barrier. His work has provided valuable insights into the relationship between the shape of a potential-energy surface and the details of what happens in a chemical process.

Polanyi was the first to be successful in detecting *transition species*, which have configurations between those of the reactants and products of an elementary chemical reaction. More recently he has worked in *surface-aligned photochemistry*, which is concerned with the interaction of light with molecules adsorbed on surfaces and with exploring the details of the processes that occur.

Aside from his research, Polanyi has exerted a wide influence in many ways. He has eloquently explained to politicians the importance of both pure and applied science in the culture and economy of a country. He has a highly developed social conscience, and urges his fellow scientists to participate more in public affairs. He has devoted much attention to world peace, and has been a powerful advocate of nuclear disarmament. He has been untiring in pressing these views, in books and articles and in lectures he has given in many parts of the world. He shared the 1986 Nobel Prize in chemistry with Dudley Herschbach and Y. T. Lee.

Reference: Laidler (1993)

(1893–1981) and R. N. Washburn, who in 1928 studied the decomposition of nitrogen pentoxide in a narrow beam so as to reduce the effects of collisions. The first bimolecular reaction to be studied in a molecular beam was that between Cs and CCl_4, which was investigated at Birmingham University in 1954 by T. H. Bull and P. B. Moon. Shortly afterwards E. H. Taylor and S. Datz, and later in more detail E. F. Greene, R. W. Roberts and John Ross (b. 1926), investigated the reaction

$$K + HBr \rightarrow KBr + H,$$

and obtained information about collision yields and the energy of activation. Subsequent to these investigations there has been much activity in this field, and the techniques have been greatly improved. For example, P. R. Brooks and co-workers at Rice University have carried out molecular-beam investigations in which reactant molecules have been put into desired vibrational and rotational states by laser excitation, and the states of the products identified by their fluorescence. Other important molecular-beam studies have been carried out at Harvard by Dudley Robert Herschbach (b. 1932) and at the University of California at Berkeley by Yuan Tseh Lee (b. 1936). The 1986 Nobel Prize in chemistry was awarded jointly to Herschbach, Lee and Polanyi, for their pioneering work in molecular dynamics. Investigations in which reactants are put into known quantum states, and the quantum states of the products are determined, are referred to as 'state to state kinetics'.

The detection of transition species

An important recent development has been the spectroscopic detection of transition species, which are molecular entities having configurations intermediate between those of the reactants and products. The first successful detection was reported in 1980 by J. C. Polanyi and co-workers for the reaction

$$F + Na_2 \rightarrow F \cdots Na \cdots Na \rightarrow NaF + Na^*,$$

occurring in crossed molecular beams. The product Na^* is in an electronically excited state and emits the yellow D line. On both sides of this line there was 'wing' emission, and the evidence indicated that this was due to the transient species $F \cdots Na \cdots Na$. Similar success was more recently obtained for other reactions. As already mentioned (Section 6.4), transient species are now being investigated by the use of femtosecond flash photolysis.

Suggestions for further reading

General aspects of the history of chemical kinetics are treated in the following publications

M. H. Back and K. J. Laidler (Eds.), *Selected Readings in Chemical Kinetics*, Pergamon Press, Oxford, 1967.

E. Farber, Early studies concerning time in chemical reactions, *Chymia*, **7**, 135–148 (1941).

M. Christine King, Experiments with time: progress and problems with the development of chemical kinetics, *Ambix*, **28**, 70–82 (1981), 29, 49–61.

K. J. Laidler, Chemical kinetics and the origins of physical chemistry, *Arch. Hist. Exact Sci.*, **32**, 43–75 (1985).

J. R. Partington, *A History of Chemistry, Macmillan*, London, 1961, Vol. 4, Chapters 18 and 22.

The work of Vernon Harcourt is treated in

M. Christine King, The course of chemical change. The life and times of Augustus G. Vernon Harcourt (1834–1919), *Ambix*, **31**, 16–31 (1984).

J. Shorter, A. G. Vernon Harcourt. A founder of chemical kinetics and a friend of 'Lewis Carroll', *J. Chem. Education*, **57**, 411–416 (1980).

For historical accounts of temperature dependence in chemical kinetics see
K. J. Laidler, The development of the Arrhenius equation, *J. Chem. Education*, **61**, 494–498 (1984).
S. R. Logan, The origin and status of the Arrhenius equation, *J. Chem. Education*, **59**, 279–281 (1982).

The radiation hypothesis, and early work on unimolecular gas reactions, are discussed in
M. Christine King and K. J. Laidler, Chemical kinetics and the radiation hypotheses, *Arch. Hist. Exact Sci.*, **30**, 45–86 (1983).

Early theories of reaction rates, and transition-state theory, are reviewed in
K. J. Laidler and M. Christine King, The development of transition-state theory, *J. Phys. Chem.*, **87**, 2657–2664 (1983).

Interesting reminiscences about early work in chemical kinetics are to be found in
H. S. Taylor, Fifty years of chemical kineticists, *Ann. Rev. Phys. Chem.*, **13**, 1–18 (1962).
J. O. Hirschfelder, My adventures in theoretical chemistry, *Ann. Rev. Phys. Chem.*, **34**, 1–29 (1983).
J. O. Hirschfelder, My fifty years of theoretical chemistry. I. Chemical kinetics, *Ber. Bunsenges. phys. Chem.*, **86**, 349 (1982).

For reviews of photochemistry including flash photolysis see
P. R. Brooks, Spectroscopy of transition region species, *Chem. Revs.*, **88**, 407–428 (1988).
R. G. W. Norrish, Some fast reactions in gases studied by flash photolysis and kinetic spectroscopy, in *Les Prix Nobel en 1967*, Norstedt and Soner, Stockholm, 1968.
J. R. Partington, *A History of Chemistry*, Macmillan, London, 1961, Vol. 4, Chapter 22.
G. Porter, Flash photolysis and some of its applications, in *Les Prix Nobel en 1967*, Norstedt and Soner, Stockholm, 1968; reproduced in *Science*, **160**, 1299–1308 (1968).
G. Porter, 'Picosecond chemistry and biology', in *Picosecond Chemistry and Biology* (Eds T. A. M. Daoust and M. A. West), Science Reviews Ltd. (1983).
A. H. Zewail and R. B. Bernstein, Real time laser femtochemistry, *Chemical and Engineering News*, Nov. 7, 1988, pp. 24–43.
A. H. Zewail, The birth of molecules, *Scientific American*, December, 1990, pp. 76–82.

There are many reviews of the history of photography; a few that cover some of the scientific aspects are

Janet E. Buerger, *French Daguerreotypes*, University of Chicago Press, 1989. This discusses some of the early applications of photography to scientific problems, such as spectroscopy.

Helmut Gernsheim, *The Rise of Photography, 1850-1880: The Age of Collodion*, Thames & Hudson, London, 1988.

Beaumont Newhall, *The History of Photography*, The Museum of Modern Art, New York, 1982.

Michael Pritchard (Ed.), *Technology and Art: The Birth and Early Years of Photography*, Royal Photographic Society Historical Group, Bath, 1990. The article by Larry J. Schaaf discusses the early work of Mrs Fulhame and others on photoimaging.

Larry J. Schaaf, *Out of the Shadows: Herschel, Talbot, and the Invention of Photography*, Yale University Press, 1992.

A. V. Simcock, *Photography 150*, Museum of the History of Science, Oxford, 1989. This short pamphlet accompanying a small exhibition of early photographs at the Museum gives a very succinct and accurate account of the early work.

A. V. Simcock, Essay Review: 195 Years of Photochemical Imaging, *Annals of Science*, **48**, 69–86 (1991). This article begins with Mrs Fulhame's work of 1794.

For reviews of physical organic chemistry which cover kinetic aspects, such as solvent effects, see

K. J. Laidler, A century of solution chemistry, *Pure & Appl. Chem.*, **62**, 2221–2226 (1990).

M. D. Saltzmann, The Robinson–Ingold controversy: precedence in the electronic theory of organic reactions, *J. Chem. Education*, **57**, 484–488 (1980).

M. D. Saltzmann, The development of physical organic chemistry in the United States and the United Kingdom: 1919–1930, parallels and contrasts, *J. Chem. Education*, **63**, 588–593 (1986).

J. Shorter, Electronic theories of organic chemistry: Robinson and Ingold, *Natural Product Reports*, **4**, 61–66 (1987).

J. Shorter, The British school of physical organic chemistry, *Chemtech*, **2**, 252–256 (1985). The title of this article is misleading, having been changed without its author's approval from 'The contribution of British physical chemists to physical organic chemistry'.

J. Shorter, Hammett Memorial Lecture, *Prog. Phys. Org. Chem.*, **17**, 1–29 (1990).

W. A. Waters, Some comments on the development of free radical chemistry, *Notes & Records Roy. Soc.*, **39**, 105–124 (1984).

For accounts of the history of various aspects of catalysis see

P. Collins, Humphry Davy and heterogeneous catalysis, *Ambix*, **22**, 205–217 (1975).

P. Collins, Johann Wolfgang Döbereiner and heterogeneous catalysis, *Ambix*, **23**, 96–115 (1976).

K. J. Laidler, The development of theories of catalysis, *Arch. Hist. Exact Sci.*, **35**, 345–374 (1986).

K. J. Laidler, Chemical kinetics and the Oxford college laboratories, *Arch. Hist. Exact Sci.*, **38**, 197–283 (1988). Reactions on surfaces are discussed on pp. 260–263, acid–base catalysis on pp. 269–272.

For reviews of the work on intensive drying see

J. W. Smith, *The Effects of Moisture on Chemical and Physical Changes*, Longman, Green & Co., London, 1929.

C. N. Hinshelwood, School Science Review, No. 31, p. 169 (1927).

C. N. Hinshelwood, *Kinetics of Chemical Change in Gaseous Systems*, Clarendon Press, Oxford, 3rd edition, 1933, pp. 114, 242–247.

M. V. Farrar, The strange history of intensive drying, *Proc. Chem. Soc.*, May 1963, pp. 125–128.

K. J. Laidler, The development of theories of catalysis, loc. cit., pp. 354–358.

H. Goldwhite, Intensive drying: anomaly and the chemical community, *J. Chem. Education*, **67**, 657–659 (1987).

There are many reviews of reaction dynamics, but few of them indicate how the subject has developed. One that does so to some extent is

Keith J. Laidler, *Chemical Kinetics*, 3rd edition, Harper and Row, New York, 1987, pp. 449–490.

CHAPTER 9

Colloid and surface chemistry

By the beginning of the nineteenth century chemists had become aware that substances ordinarily considered insoluble, such as gold and sulphur, could sometimes exist in apparently homogeneous solution. The expression 'pseudo-solutions' came to be applied to such systems, which often had to be prepared in an indirect manner, and which were obviously of a different character from solutions of substances like salt and sugar. Much clarification of the problem was given by the Scottish chemist and physicist Thomas Graham (1805–1869), who is regarded as the founder of the science of colloid chemistry.

In 1829 Graham became professor of chemistry at the Andersonian College in Glasgow. There he carried out important work on the phosphates and arsenates, and also began to study gaseous diffusion. In 1829 he published the first of his papers on the subject of diffusion, and a paper that appeared in 1833

TABLE 9.1 Highlights in colloid and surface chemistry

Date	Investigator	Contribution	Section
1757	Franklin	Oil on stormy waters	9.2
1828	Brown	Motion of particles	9.1
1837	Graham	Identification of colloids	
1869	Tyndall	Light scattering	9.1
1871, 1899	Rayleigh	Theory of Tyndall effect	9.1
1891	Pockels	Liquid films on surfaces	9.2
1903	Zsigmondy	Ultramicroscope	9.1
1905	Einstein	Statistical theory of Brownian movement	9.1
1906	Smoluchowski	Collision theory of Brownian movement	9.1
1908	Perrin	Sedimentation of particles; Brownian movement	9.1
1916	Langmuir	Adsorption isotherms; theories of surface reactions	9.3
1917	Langmuir	Film balance	9.2
1923	Svedberg	Ultracentrifuge	9.1
1930	Tiselius	Electrophoresis	9.1

expressed the law of diffusion that is known by his name, stating that the rate of diffusion is inversely proportional to the square root of the density of a gas. He pointed out that a mixture of substances could be separated by diffusion, a principle that has found many applications; in the present century, for example, it was used for the separation of isotopes.

Graham later extended his diffusion studies to solutions, particularly after 1837 when he became professor of chemistry at University College, London, the first constituent college of the University of London. In work described in a comprehensive paper 'On the diffusion of liquids', published in 1850, he applied to liquids the same methods of investigation that 20 years earlier he had applied to gases. He also made studies of osmosis, and on the basis of that work he developed the technique of 'dialysis', a word he coined from the Greek δια (dia), through, and λυσις (lysis), loosening, or setting free. His dialyzers consisted of membranes through which solutions were caused to flow. By their means he distinguished between what he called *crystalloids*, which diffused rapidly through the membranes, and *colloids*, which diffused more slowly. He prepared many colloidal solutions and he distinguished between *sols*, such as colloidal solutions of gold, and *gels*, which are semi-rigid and have the liquid incorporated within them.

Much of the terminology and many of the concepts in the field of colloids originated with Graham. With regard to the word 'colloid' he wrote:

As gelatin appears to be its type, it is proposed to designate substances of the class as *colloids* [κολλα, glue], and to speak of their particular form of aggregation as the *colloidal condition of matter*. Opposed to the colloidal is the crystalline condition. Substances affecting the latter form will be classed as *crystalloids* ... Fluid colloids appear to have always a pectous [πηκτος, curdled] modification; and they often pass under the slightest influences from the first into the second condition... The colloidal is, in fact, a dynamical state of matter; the crystalloid being the statical condition.

As this passage indicates, he recognized that the essential difference between colloids and crystalloids is the state, and that the same substance can exist in either state according to conditions.

Colloids have been extensively investigated since Graham's time, but for several decades much of what was done was purely empirical, although a few useful generalizations were made. The classification of colloids proved to be somewhat controversial, and a number of suggestions were made. The most satisfactory proposals are probably the following. Colloidal dispersions of solids in liquids, known as sols, can be roughly classified as *lyophobic* or *lyophilic*. If the dispersion medium is water, lyophobic sols can be called *hydrophobic* sols. The word 'hydrophobic', from the Greek φοβος (phobos), fear of, was first used in 1905 by Perrin to denote a disperse phase, such as gold, which has a low affinity for water. Since there is a low affinity for the solvent, lyophobic sols are relatively unstable. Lyophilic sols are those in which there is affinity between the disperse phase and the molecules comprising the dispersion medium. These sols are much more like true solutions.

Important advances in the field of colloid chemistry came with advances in experimental techniques. One of these is chromatography, invented in 1906 by the Russian botanist Mikhail Semenovich Tswett (1872–1919); the word 'chromatography', coined by him, comes from the Greek χρωμα (chroma), meaning colour, and γραφειν (graphein), meaning to write. Other important techniques useful in colloid chemistry, are referred to in the following section. Developments in biological chemistry and industrial chemistry also contributed to further work on colloids. One who was personally responsible for much of the rise of interest in colloids was Wolfgang Ostwald (1883–1943), the son of Wilhelm Ostwald. Like his father, Wolfgang Ostwald was endowed with tremendous energy and enthusiasm. He had a vast knowledge of his subject, and being something of a showman he was a superb and popular lecturer. In 1913 and 1914, for example, he made a lecture tour in the United States, giving 56 lectures in two and a half months. In these lectures he eloquently stressed the wide distribution of colloids, as indicated by the following passage;

We need only to look at the sky, at the earth, or at ourselves to discover colloids or substances closely allied to them. We begin the day with a colloid practice—that of washing—and we may end it in one with a bedtime drink of colloid tea or coffee. Even if you make it beer, you still consume a colloid.

Ostwald's lectures later appeared in the form of a book which he wrote in part while serving in the trenches in World War I; an English translation of it, entitled *Introduction to Theoretical and Applied Colloid Chemistry*, appeared in 1917. Wolfgang Ostwald also edited an important journal, the *Kolloid Zeitschrift*, which he founded in 1906.

One whose interest in colloids was to some extent inspired by Wolfgang Ostwald was Wilder Bancroft, of Cornell University. In about 1910 he became attracted to the subject, as two decades earlier he had been attracted to the phase rule (Section 4.4), because of its non-mathematical character. Like Ostwald he worked extensively and enthusiastically on the subject, his papers appearing in the *Journal of Physical Chemistry*, which he had founded and edited for many years (Section 2.3); for a period about half of the papers in this journal were on colloid chemistry, a substantial proportion by Bancroft himself.

In spite of their great devotion to the subject of colloid chemistry, neither Wolfgang Ostwald nor Wilder Bancroft made a great success of their work. Both of them held opinions that were the subject of much controversy and criticism. They held to the position that colloids were all aggregates of smaller molecules and were not chemical compounds. It began to emerge, however, that some substances of biological importance, such as polysaccharides and proteins, were definite chemical compounds consisting of molecules of high molecular weight, i.e., macromolecules.

Bancroft's work on colloid chemistry led him into one field in which his work was harshly criticized. He began to study the effect of anaesthetics and of substances such as narcotic drugs on the colloidal state of the proteins of nerve cells. In the course of this work he concluded that sodium thiocyanate, which is also known as sodium rhodanate, is able to reverse the effects of drugs, and he embarked on a number of clinical trials to demonstrate this. His announcement of his conclusions led to much publicity, a number of newspapers and *Time* magazine announcing in 1931 that Bancroft had discovered a cure for alcoholism, insanity, and the abuse of drugs. In February, 1933, the New York section of the American Chemical Society voted to award Bancroft the Nichols medal for this work. This decision provoked strong criticism from many sources, including the American Medical Association, the opinion being that Bancroft's clinical tests had not been properly conducted. Worse still, a number of distinguished pharmacologists reported that they had been unable to repeat his results. Much embarrassed, the American Chemical Society asked Bancroft to accept the award for his work on the phase rule. This he refused to do, and the Society promptly announced that Bancroft had declined the Medal, and that there would be no award in that year.

9.1 Physical properties of colloidal systems

The literature of colloids is vast, and the present account can cover only the more physical aspects.

Light scattering

When light passes through a medium containing no particles larger than about 10^{-9} m in diameter the path of light cannot be detected. When larger particles are present, however, some of the light is scattered, and the incident beam passes through with weakened intensity. The first investigation of this was by the Irish-born physicist John Tyndall (1820–1893), a descendant of William Tyndale the famous translator of the Bible into English. In 1854 he became professor of natural philosophy at the Royal Institution where he was a colleague of Faraday whom he succeeded as Director in 1867. Tyndall reported in 1869 his investigation of what came to be called the Tyndall effect, a good example of which is found in a sunbeam, in which the light is scattered by dust particles. He suggested that the blue of the sky is due to the scattering of light by dust particles in the atmosphere. An alternative theory, due to James Dewar who had produced liquid oxygen and found it to be blue, was that the blue of the sky is due to the oxygen in it. Tyndall's theory was supported by the theoretical treatment of Lord Rayleigh.

John William Strutt (1842–1919), who became the third Baron Rayleigh in 1873, throughout his career had a particular interest in wave motion in all its aspects. His first investigation of light scattering by colloidal particles, carried out when he was a fellow of Trinity College, Cambridge, was published in 1871. If the intensity of the incident radiation is I_0, and l is the length of the path through the scattering medium, the intensity of the transmitted radiation is given by

$$I = I_0 \, e^{-\tau l} \tag{9.1}$$

where τ is known as the *turbidity*. Rayleigh deduced that for spherical particles having dimensions much smaller than the wavelength of the radiation, the turbidity is inversely proportional to the fourth power of the wavelength. The light at the blue end of the visible spectrum is therefore scattered more than the rest of the light, and this explains the blueness of the sky, as Tyndall had suggested. In 1879 Rayleigh succeeded Clerk Maxwell as Cavendish professor at Cambridge, and received the Nobel Prize for physics in 1904, the same year that William Ramsay received the Nobel Prize for chemistry.

When the particle dimensions are not small compared with the wavelength, and the particles are not spherical, the theory is much more complicated. Important contributions to the theory were made in 1908 at the University of Griefswald by Gustav Mie (1868–1957), and later in the United States by Bruno Hasbrouck Zimm (b. 1920) and by Paul Mead Doty (b. 1920).

Work on the Tyndall effect was extended by the Austrian-born chemist Richard Adolf Zsigmondy (1865–1929), who while employed at the Jena glassworks became interested in the scattering of light by particles of gold and other materials. He left the glassworks in 1900 and began to collaborate with H. Siedentopf, a physicist, on the construction of an *ultramicroscope*. In this instrument a light beam is passed through a colloidal system and observations made at right angles to the beam. Individual particles much too small to be visible directly can be seen as flashes of scattered light. Their paper announcing the construction of the ultramicroscope appeared in 1903. In 1908 Zsigmondy became professor at the University of Göttingen, and in 1925 he received the Nobel Prize for physics.

Brownian movement and sedimentation

In 1828 the Scottish botanist Robert Brown (1773–1858), who was then Botanical Keeper at the British Museum in London, reported that he had observed pollen grains under a microscope and that they were in constant and irregular motion. At first he thought the motion to be due to life within the grains, and believed that he had discovered the 'primitive molecule' of living matter. He later found that dye particles, undoubtedly non-living, showed the same motion, and had to modify his opinion.

The phenomenon was not satisfactorily explained for several decades, although many suggestions were made. Victor Regnault thought that the motion was caused by irregular heating by incident light, and others thought that electrical forces were in some way involved. In 1888 a careful investigation was made at the Université de Lyon by Georges Gouy, who showed that neither light nor a very strong electromagnetic field had any effect on the motion. He found that the motion is more lively the smaller the viscosity of the liquid. He concluded that the explanation lay in the thermal motions of the molecules in the liquid, and made some measurements of the velocities of the particles.

This point of view was developed considerably by Einstein in a paper that appeared in 1905 and in four other papers that were published during the next three years. When Einstein wrote his first paper on the subject he knew little of what had been observed on Brownian motion, being concerned with deducing relationships that could be used to test the atomic hypothesis and to determine atomic dimensions and the Avogadro constant. His 1905 paper, in fact, did not explicitly mention Brownian movement, but he began his second paper, of 1906, by mentioning that he had been informed that several physicists, including Gouy, had become 'convinced by direct observation that the so-called Brownian motion is caused by the irregular thermal movements of the liquid'. In later papers he referred specifically to experimental results on Brownian motion and related them to his treatment.

In these papers Einstein used the methods of statistical physics to give a comprehensive treatment of diffusion in liquids, and of the sedimentation of particles in a gravitational field. He considered suspended particles that were much larger than molecules, and which as a result of collisions with solute molecules acquired a random motion. He realized that even if such particles were large enough to be seen under a microscope, their speeds would be great and their directions would change very rapidly, with the result that it would not be possible to measure their velocities. What could be measured, however, was the average distance that a particle would move in a given time. Einstein's treatment led to the important result that the mean square distance $\overline{x^2}$ travelled in time t is

$$\overline{x^2} = 2Dt, \tag{9.1}$$

where D is the diffusion coefficient that appears in Fick's equation (7.14).

Einstein also showed that the frictional coefficient f for the movement of a particle through a liquid is related to the diffusion coefficient D by the equation

$$D = \frac{k_B T}{f}, \tag{9.2}$$

where k_B is the Boltzmann constant (the gas constant R per molecule) and T is the absolute temperature. In 1851 Stokes had deduced that for a particle of radius r in a liquid of viscosity η the frictional coefficient is given by

$$f = 6\pi r \eta. \tag{9.3}$$

The diffusion coefficient can thus be expressed as

$$D = \frac{k_B T}{6\pi r \eta}, \tag{9.4}$$

and this is now often known as the *Stokes–Einstein equation*.

Important work on Brownian movement was also done at about the same time by Marian Smoluchowski, some of whose work on diffusion was referred to in Section 8.6. Smoluchowski was professor at the University of Lvov from 1900 to 1913, and from about 1900 carried out theoretical studies on the Brownian movement but at first published none of it. When Einstein's 1905 paper on the subject appeared, Smoluchowski began to publish his work, his first paper appearing in 1906. His methods were rather different from those of Einstein, who had used statistical physics to obtain relationships that were not easily visualized; for example, Einstein did not deal with the collisions occurring between a particle and the surrounding solvent molecules. Smoluchowski did consider these collisions explicitly, and his final equations were very similar to those of Einstein. Later he proposed that Brownian movement of rotation could be studied experimentally by means of a tiny mirror supported in a fluid on a thin quartz fibre. This was later achieved, with results that agreed with Smoluchowski's treatment. Another interesting conclusion he reached was that density fluctuations in the atmosphere were capable of bringing about the light scattering that causes the sky to appear blue, which it would do even in the absence of dust particles. In 1913 Smoluchowski moved to the Jagiellonian University of Cracow, being elected its rector in 1917, but he died in that year at the age of 45.

Einstein's and Smoluchowski's treatments of the Brownian movement provided the basis for some of the experimental studies made by Jean Perrin (1870–1942) and his students at the Faculté des Sciences de Paris. His first investigations relating to the Brownian movement were, however, made in ignorance of their work. He realized that colloidally suspended particles would, as a result of collisions with the molecules of the fluid, distribute themselves vertically at equilibrium, and that an investigation of this distribution would be of value for various reasons. It would provide a test of the kinetic theory, which if confirmed would perhaps convince the sceptics, notably Ostwald, of the real existence of atoms. It would also provide a value for the Avogadro constant (then always called the Avogadro number).

According to the kinetic theory of gases, the molecules of a gas will distribute themselves in a gravitational field in such a manner that the pressure of the atmosphere diminishes as the distance from the earth increases. Perrin argued that the same would apply to colloidal particles suspended in a liquid, so that by making direct measurements of the vertical distribution it would be possible to test the theory. According to the theory, the numbers of particles n_1 and n_2, at heights h_1 and h_2 respectively, are given by

$$\ln \frac{n_1}{n_2} = \frac{LV(D - d)}{RT} (h_2 - h_1), \tag{9.5}$$

where V is the volume of a particle, D and d are the densities of the particles and the liquid, respectively, R is the gas constant, T the absolute temperature, and L the Avogadro constant.

Perrin worked with suspensions of the pigment gamboge in water, contained in a cell under the microscope, and counted the number of particles at successive depths differing by amounts of about 0.01 mm. At shallow depths the logarithmic law (eqn 9.5) was obeyed. The size of the particles could be obtained by direct measurement, so that everything in eqn 9.5 was known except L, which could therefore be calculated from the data on the distribution of particles. The value he obtained from his first experiments. published in 1908, was 7.05×10^{23} mol^{-1}; the value accepted today is 6.022×10^{23} mol^{-1}.

Only after Perrin had carried out these experiments did he learn of the theoretical work on the Brownian movement that had been done by Einstein and Smoluchowski. He then performed experiments to determine the mean square displacements of individual gamboge and mastic particles, in various periods of time. With the use of Einstein's equation 9.1 he could then determine the Boltzmann constant k_B and hence, since $k_B = R/L$, the Avogadro constant L. The values of L he obtained in this way, published in 1911, varied from about 6×10^{23} mol^{-1} to 9×10^{23} mol^{-1}, with an average value of about 7×10^{23} mol^{-1}. Perrin's results confirmed Einstein's conclusion that the displacement of a particle is proportional to the square root of the time, a conclusion that had come under attack. Perrin also investigated rotational Brownian movement, and again confirmed Einstein's conclusions.

The importance of Perrin's work was not so much in its estimation of the Avogadro constant as in its confirmation of the kinetic theory and therefore by implication of the real existence of atoms. It even convinced Ostwald, who as noted in Section 5.2 vigorously argued that the atomic theory was a 'mere hypothesis', useful perhaps in systematizing chemical knowledge but not at all convincing as far as the real existence of atoms was concerned. The various editions of his *Lehrbuch* and his *Grundriss* had developed the subject of physical chemistry without reference to atoms and molecules, which means, of course, that he had also ignored the kinetic theory and statistical mechanics. As a result of the work of J. J. Thomson (Section 10.3) and of Perrin he finally recanted and in 1909, in the preface to the fourth edition of his *Grundriss*, he wrote

I am now convinced that we have recently become possessed of experimental evidence of the discrete or grained nature of matter, which the atomic hypothesis sought for hundreds and thousands of years. The isolation and counting of gaseous ions, on the one hand, which have crowned with success the long and brilliant researches of J. J. Thomson, and, on the other, the agreement of the Brownian movement with the requirements of the kinetic hypothesis, established by many investigators and most conclusively by J. Perrin, justify the most cautious scientist in now speaking of the experimental proof of the atomic nature of matter.

He then added the comment that the laws of stoichiometry could be 'equally well, and perhaps better, represented without the aid of the atomic conception...' In other words, he was trying to justify his previous opinion that Dalton's evidence for the atomic theory was not compelling.

Perrin's experiments on sedimentation were limited by the fact that they were done in the earth's gravitational field, which means that there is no observable sedimentation of small particles such as proteins. To study smaller particles it is necessary to employ larger fields, which is done by means of an *ultracentrifuge*, in which solutions are rotated at high speeds. A rotational speed of 80 000 revolutions per minute produces a gravitational field of about 3×10^5 g. The development of such centrifugal techniques is due largely to the Swedish physical chemist Theodor Svedberg (1884–1971), of the University of Uppsala, whose work along these lines began in 1923 and who devoted much effort to the characterization of protein molecules and other macromolecules. For this work he was awarded the 1926 Nobel Prize for chemistry.

Electrical properties

The physical properties of colloidal systems depend to a great extent on their electrical nature. The matter is of great complexity, and even today it is necessary to rely to a great extent on empiricism.

It was noted in Sections 7.4 and 7.5 that potential differences become established at interfaces, and that electrical double layers are formed. In the case of small particles the surface to volume ratio is large, so that electrical effects play a particularly important role. The existence of electrical double layers gives rise to a number of *electrokinetic effects*, in which either an applied electric potential brings about motion, or motion produces an electric potential. The magnitude of an electrokinetic effect is determined by the *electrokinetic* or ζ *(zeta) potential*, which is defined as the difference in electrical potential between the fixed double layer, occurring at the distance of closest approach of the ions to the surface, and the bulk of the solution. An important electrokinetic property is *electrophoresis*, in which particles move in an electric potential, the direction and rate of movement depending on the zeta potential. The first experiments of any importance on electrophoresis were carried out in the 1930s by Arne Wilhelm Kaurin Tiselius (1902–1971), who studied under Svedberg at the University of Uppsala where he later taught for many years. He devised a special type of U-shaped apparatus and carried out extensive experiments on electrophoresis, particularly of blood proteins, some of his work being done in collaboration with Svedberg. He was awarded the 1948 Nobel Prize in chemistry.

9.2 Surface tension and surface films

The spreading of oil on the surface of water has attracted attention for many centuries. References to the stilling of waves by a layer of oil appear in the writings of Pliny the Elder (23–79 AD), who mentioned that divers made a practice of adding oil to the surface of water in order to see the bottom more clearly, the oil having the effect of smoothing the surface. In the Venerable Bede's *Ecclesiastical History*, which was written in 731 AD, he reported that St Aidan gave a young priest a cruise of oil to pour on the sea if it became stormy, and that this proved effective. Later, Scottish seal hunters made a practice of looking for smooth patches of water, since the oil from the seals calmed the waters immediately above.

Benjamin Franklin's observations

The first investigations of this effect were carried out by Benjamin Franklin (1706–1790). In June 1757 he was on a ship on its way from New York to England, and while close to Louisbourg, near the eastern tip of Cape Breton Island, Nova Scotia, he noticed something unusual; the wakes of two ships were remarkably smooth while all the others were 'ruffled by the wind'. His ship was one of a convoy, and his account in a paper in the *Philosophical Transactions* that appeared 17 years later is reproduced in Figure 9.1. Franklin saw at once that the captain's reply was a superficial one—in two senses of the word—and that some proper explanation should be sought.

(a)

XLIV. *Of the stilling of Waves by means of Oil. Extracted from sundry Letters between* Benjamin Franklin, *LL. D. F. R. S.* William Brownrigg, *M. D. F. R. S. and the Reverend Mr.* Farish.

(b)

In 1757, being at sea in a fleet of 96 sail bound against Louisbourg, I observed the wakes of two of the ships to be remarkably smooth, while all the others were ruffled by the wind, which blew fresh. Being puzzled with the differing appearance, I at last pointed it out to our captain, and asked him the meaning of it? "The cooks, says he, have, I suppose, been just emptying their greasy water through the scuppers, which has greased the sides of those ships a little;" and this answer he gave me with an air of some little contempt, as to a person ignorant of what every body else knew. In my own mind I at first slighted his solution, tho' I was not able to think of another. But recollecting what I had formerly read in PLINY, I resolved to make some experiment of the effect of oil on water, when I should have opportunity.

FIGURE 9.1 (a) The title of Benjamin Franklin's paper of 1774. (b) The passage in Franklin's paper in which he describes his observations, off Louisbourg in 1757, of the effect of oil on rough water.

Benjamin Franklin

(1706–1790)

Benjamin Franklin was born in Boston, Massachusetts, the son of a chandler and soap boiler who had emigrated from England in 1683. He received hardly any education, and at the age of ten was taken from school to assist his father who, noticing that he spent much time reading, decided that he should become a printer. At the age of 12 he was indentured to a printer, but within a few years he was able to break the indenture and settle in Philadelphia. From 1724 to 1726 he was in England where he became skilled in printing techniques, and on returning to Philadelphia he soon established his own business. He became highly successful and began to publish a number of periodicals, such as the *Philadelphia Gazette, Poor Richard: An Almanack*, and the *General Magazine*. He was active in the founding of the Academy of Philadelphia, which eventually became the University of Pennsylvania.

Franklin's scientific work, although of great importance, comprised only a small part of his activities. After 1748 he did not play an active role in the printing business, and in 1751 he was elected a member of the Pennsylvania Assembly and Alderman of Philadelphia; from 1753 to 1774 he served as Deputy Postmaster-General for the British colonies in North America. At the same time, over a period of years, he carried out a number of scientific investigations.

His work on electricity (Chapter 7) began in the 1740s. He performed a number of experiments with Leyden jars, elucidating the electrical action of pointed bodies and investigating electrostatic induction. His work led him to the conclusion that there is only one form of electricity; a body containing excess of it is positively charged, one deficient in it is negatively charged. His famous 'sentry-box' apparatus

Franklin did not let the matter rest, and on arriving in England he made a number of demonstrations and experiments. He was in England for the purpose of presenting a petition to King George II, but he also became active with the Royal Society, of which he had been elected a Fellow in the previous year for his important investigations on electricity (Chapter 7). He was a man of agreeable personality who had a number of good friends in England, some of whom lived near Clapham Common, about four miles from the centre of London where he resided. There he carried out a demonstration which he described in his 1774 paper as follows:

At length being at CLAPHAM where there is, on the common, a large pond, which I observed to be one day very rough with the wind, I fetched out a cruet of oil, and dropt a

to demonstrate the electrical nature of lightning was developed in 1752, and his kite experiment was done at about the same time. His work on electricity was described in papers in the *Philosophical Transactions* and a book *Experiments and Observations on Electricity, made at Philadephia in America* (1751). These contributions led to his being awarded the Copley medal of the Royal Society in 1753 and to his election to the Fellowship of the Society in 1756. He made a number of practical inventions, including a rocking chair, bifocal spectacles, the 'Franklin stove', and lightning rods.

Franklin's scientific achievements became widely known and led to his political and diplomatic successes. In 1757 he was sent by the Pennsylvania House of Assembly to petition King George II against repressive actions. On his way to England he became aware of the effect of oil on a rough sea (Section 9.2), and this observation led eventually to the practical trial offshore at Portsmouth in 1773. He made several subsequent trips to Europe and in 1773 was elected one of the eight foreign associates of the Académie des Sciences.

With Thomas Jefferson and John Adams he was one of the three authors of the American Declaration of Independence, and in 1776 he was one of three commissioners sent to negotiate a treaty with France. In 1781 he played an active role in negotiating the final peace with Great Britain. He was one of the very few statesmen whose reputation was first gained through his scientific contributions.

References: I. B. Cohen, DSB, 5, 129–139 (1972); Van Doren (1938).

little of it on the water . . . I then went to the windward side, where [the waves] began to form; and there the oil, though not more than a teaspoonful, produced an instant calm over a space several inches square, which spread amazingly, and extended itself gradually till it reached the lee side, making all that quarter of the pond, perhaps half an acre, as smooth as a looking glass.

The pond on which Franklin carried out this demonstration has been identified as Mount Pond, which is still known by that name. Franklin went on to say that he afterwards made a practice of taking with him, whenever he went into the country,

a little oil in the upper hollow joint of my bamboo cane, with which I might repeat the experiment as opportunity should offer; and I found it constantly to succeed.

Walking sticks with removable heads were not uncommon at the time, being used for example by physicians for transporting drugs. One of the places where Franklin demonstrated the effect of oil was Derwent Water in the Lake District. A clergyman named Farish became aware of these experiments, and a letter from him, mentioning additional examples of the effect of oil on water, is included in the paper that appeared in the *Philosophical Transactions* (Figure 9.1(a)). In 1773, a year before his paper was published, Franklin carried out a demonstration of wave-damping near Portsmouth, in the presence of some Fellows of the Royal Society.

It is interesting to use Franklin's rough data to make an estimate of the thickness of his oil films. He mentioned that a teaspoonful of oil produced a layer of half an acre, and if he had done the calculation, before the metric system had been introduced, it might have proceeded as follows. Half an acre is 2420 square yards, or 3 136 320 square inches. If the teaspoon had a capacity of a tenth of a cubic inch the thickness of the film is about 3.2×10^{-8} inches, which is 8×10^{-10} metres or 8 Ångstroms, which is reasonable; Rayleigh later estimated the thickness of a similar film to be 10 Å. It may at first sight seem surprising that Franklin did not make this simple calculation, but it would probably never have occurred to anyone at the time. A molecule was still something of an abstraction, perhaps believed in but not considered to have anything to do with one's experiments. When Franklin published his paper in 1774 a recent atomic theory, put forward in 1758, was that of Boscovich, who regarded atoms not as particles but as centres of force, of zero size (Section 5.2). Estimating the size of an atom might not then have seemed a reasonable thing to do.

Some years later, after Franklin had successfully negotiated peace with Great Britain, one of his Clapham friends, Christopher Baldwin, recalled his earlier experiments to him. In a letter dated 18 February 1783, Baldwin wrote

Tis you who have . . . again poured the oil of Peace on the troubled Wave, and stiled [*sic*] the mighty Storm!

A century after Franklin had carried out these experiments some remarkable and extensive trials of the wave-damping effect of oil were carried out in Scotland. They were organized by a Perth businessman, John Shields (1822–1890), a man of considerable ability and enthusiasm but with little knowledge of science. He was concerned about the many shipwrecks that were occurring at the time, well over a thousand each year with a considerable loss of life. In 1879 he patented a device for spreading oil from underground pipes at the entrances of harbours, and in 1882 carried out a test of the equipment in Peterhead harbour in Aberdeenshire. On the day of the test there was a very heavy sea, and the results of the test were considered to be very satisfactory. Little was said about this test in the scientific literature, but there was much in the press and the matter was discussed at some length in the House of Lords, mention being made of Franklin's earlier experiments. Recommendations were made for ships to carry oil for use on stormy days, and for harbours to be equipped with pipes for spreading oil. Little was done, however, perhaps because of concern for pollution of the oceans, or because the larger vessels used later were less subject to shipwreck.

In his 1774 paper Franklin had proposed an explanation for the effect of oil on a stormy sea. He suggested that the wind

in passing over the smooth surface of water, may rub, as it were, upon that surface, and raise it into wrinkles, which, if the wind continues, are the elements of future waves ... Now I imagine that the wind blowing over water thus covered with a film of oil, cannot easily *catch* upon it, so as to raise the first wrinkles, but slides over it.

This, however, was shown not to be correct by the Scottish meteorologist John Aitken (1837–1919), a man of independent means who followed no professional career but carried out a number of scientific investigations. He caused currents of air to pass over water surfaces, with and without layers of oil, and found little effect of the kind that Franklin had envisioned. Franklin had reported that oil produced no lowering of the waves, but prevented the waves from cresting and breaking, and Aitken noted that the same was true of the Peterhead tests. Aitken's careful experiments led him to the conclusion that the effect is due entirely to the change in surface tension brought about by the layer of oil. When a thin layer of oil is present new free surface cannot easily be formed; as a result small waves do not form on the surface of large ones, so that breaking of the crests does not occur so readily.

The work of Agnes Pockels and Lord Rayleigh

In 1891 a paper of great importance was published by Agnes Pockels (1862–1935), who had carried out her experiments using improvised equipment in the kitchen of her home in Brunswick, Lower Saxony. She became interested in such mundane matters as the behaviour of greasy washing-up water. She studied the relationship between the area occupied by a film and the surface tension, and found that the behaviour was different above and below a certain critical area.

Fräulein Pockels began her experiments in about 1882, when she was 20, but at first her work excited little interest. On 10 January 1891, she wrote a long letter in German to Lord Rayleigh, beginning in translation as follows:

MY LORD,—Will you kindly excuse my venturing to trouble you with a German letter on a scientific subject? Having heard of the fruitful researches carried out by you last year on the hitherto little known properties of water surfaces, I thought it might interest you to know of my own observations on the subject. For various reasons I am not in a position to publish them in scientific periodicals, and I therefore adopt this method of communicating to you the most important of them.

The letter then went on to describe in some detail a rectangular tin trough which Fräulein Pockels had designed and constructed, and by means of which

Agnes Pockels

(1862–1935)

Agnes Pockels was born in Venice, in the province of Venezia which at the time and for four more years was part of the Austrian Empire. When she was born her father, Theodor Pockels, was an officer in the Royal Austrian army. Because of ill health Captain Pockels took early retirement from the army, and after 1871 the family lived in a substantial house in Brunswick, Lower Saxony. Agnes was educated at the Municipal High School for Girls, and when she graduated higher education was not available to women. A few years later she could have been admitted to a university, but her parents asked her to remain at home. Both her mother and father suffered from ill health, and Agnes had the heavy responsibility of looking after them for many years. Some domestic and nursing help was available, but Agnes had to concern herself with the household management and some of the nursing.

In spite of these many distractions Agnes persistently carried out a number of important investigations on surface films which were to have far-reaching consequences. Her work was carried out for the most part at the kitchen sink, and later her sister-in-law, Frau E. Pockels, wrote

. . . what millions of women see every day without pleasure and are anxious to clean away, i.e., the greasy washing-up water, encouraged this girl to make observations and eventually [to enter into] scientific investigation.

From her diaries as well as her published papers it is possible to gain some detailed knowledge of the circumstances of her investigations of the properties of surface films (Section 9.2). Her experiments were begun in 1880, when she was 18, and in 1881 she measured surface tension by suspending small buttons, immersed

she had investigated films of various organic substances on water. Her letter concluded with:

I thought I ought not to withhold from you these facts which I have observed, although I am not a professional physicist; and again begging you to excuse my boldness, I remain, with sincere respect,

Yours faithfully.
AGNES POCKELS

Rayleigh, besides being a man of great prestige, had a remarkably kind, friendly and generous disposition, and was always helpful to others. He at once entered into a correspondence with Fräulein Pockels, but unfortunately his letters have

at the liquid surface, from a wooden beam balance. In 1882 she devised a rectangular trough with which, by means of a sliding strip of tin, she determined the relationship between surface area and surface tension. During the course of these investigations she discovered that the behaviour of a film was different above and below a certain critical area. Above this critical area the area varies to a considerable extent with the tension, while below it 'the displacement of the partition makes no impression on the tension'.

Fräulein Pockels found little support for her work from German scientists, but was fortunate to gain much help from her younger brother Friedrich (Fritz) (1865–1913) who became a professor of physics. It was Fritz who suggested that she should write to Lord Rayleigh, who arranged for her work to be published in *Nature*. During later years, thanks largely to Rayleigh's encouragment, she made a number of further investigations in which she covered the physical aspects of the problem in a comprehensive way. Her approach was more that of a physicist than of a chemist, and she made no deductions about molecular sizes or conformations.

Recognition of her contributions came rather late. In 1931 she shared the Laura Leonard Prize for her 'quantitative investigation of the properties of surface layers and surface films'. On the occasion of her seventieth birthday in 1932 the distinguished colloid chemist Wolfgang Ostwald, the son of Wilhelm Ostwald, published a tribute to her in the *Kolloid Zeitschrift*, and in the same year she was awarded an honorary doctorate by the Carolina-Wilhelmina University of Brunswick.

References: Giles & Forrester (1971); Derrick (1982).

not survived, the Pockels house having been destroyed by bombing in 1944. All of her letters were, however, preserved by Lord Rayleigh and still exist, and from them we can guess much of what he wrote to her. It appears that he first expressed some incredulity as to her sex, or perhaps was not certain whether the name 'Agnes' given to a German necessarily implied a woman, for her second letter included the passage

... with regard to your curiosity about my personal status, I am indeed a woman.

After some further correspondence in which certain technical details were clarified, Rayleigh offered to arrange for her work to be published in *Nature*, an offer she said would 'give me considerable pleasure'.

The editor of *Nature* at the time was still its founder, Sir Norman Lockyer, and Rayleigh wrote to him as follows on 2 March 1891:

I shall be obliged if you can find space for the accompanying translation of an interesting letter which I have received from a German lady, who with very homely appliances has arrived at valuable results respecting the behaviour of contaminated water surfaces. The earlier part of Miss Pockels' letter covers nearly the same ground as some of my own recent work, and in the main harmonises with it. The later sections seem to me very suggestive, raising, if they do not fully answer, many important questions. I hope soon to find opportunity for repeating some of Miss Pockels' experiments.

RAYLEIGH

The letter had been translated by Lady Rayleigh, and the translation with the introductory note by Rayleigh, appeared in the 12 March 1891, issue of *Nature*. The encouragement Pockels received from Rayleigh, and resulting from this publication, stimulated her to make further investigations; she published 13 papers from 1891 to 1918, and two later papers in 1926 and 1933.

From 1887 to 1905 Rayleigh held a professorship at the Royal Institution and the story is told that his interest in surface tension was aroused when he noticed that the woman who served tea at the Institution always put a little water in each saucer. When he asked her why she replied 'So that the cup don't slip.' (Whether this phenomenon is in fact a surface tension effect may be questioned; it might be due to the formation of a partial vacuum below the cup.)

Rayleigh began his work on surface films in about 1889, and published two papers on the subject in 1890. In a lecture he gave on 'Foam' at the Royal Institution in 1890 he described the use of a water-filled trough that was very similar to the one that Agnes Pockels had used. In later years Rayleigh made further investigations of surface films, and published an important paper on the subject in 1899. In this work he made use of the Pockels method of cleaning the water surface in the trough by moving a partition from one end to the other. From his results he deduced that the critical area discovered by Agnes Pockels is that at which the molecules are closely packed. He calculated the thickness of the film to be about 1.0×10^{-7} cm (10 Å) and remarked that there is 'a complete layer one molecule thick'. This is the first suggestion of the existence of a monomolecular layer on a surface.

Further work on surface films was carried out immediately prior to World War I by René Marcelin, whose important work in chemical kinetics was referred to in Section 8.3; his work on films, involving novel optical methods for the study of their growth, was done in Perrin's laboratories at the Faculté des Sciences de Paris. Marcelin was killed in action in September, 1914, before he had prepared any of his surface work for publication, but his brother André arranged for two papers to be published in 1918. André Marcelin, also of the Faculté des Sciences de Paris, was himself a distinguished surface chemist who published much on the subject between 1915 and 1968, including a book, *Solutions superficielles*, which appeared in 1931. Perrin also did some work on liquid films on surfaces, publishing papers on the subject in 1918 and 1919.

Langmuir's surface film balance

An important series of investigations on surface films was begun in 1917 by Irving Langmuir. He devised a *surface film balance* which had many of the same features as that used by Agnes Pockels, but also a number of technical improvements. His apparatus, shown in Figure 9.2, measures the surface pressure π of a film, which is defined as the force exerted on it divided by the length of the edge along which the force is exerted. It can be shown that for a film lying on a surface the surface pressure is equal to the decrease in surface tension brought about by the film.

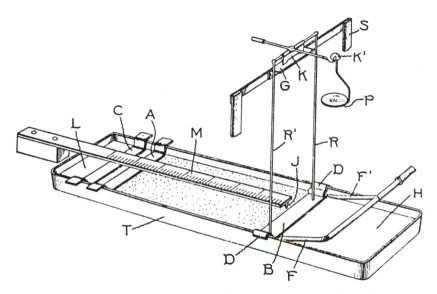

FIGURE 9.2 Langmuir's diagram of his original surface film balance.

Langmuir, and many subsequent investigators using a similar type of film balance, made numerous studies of the relationships between surface pressure and area, for a variety of films on the surface of water and sometimes of other liquids. Langmuir found that such films are of two general types. For some films, referred to as *coherent films*, there is a critical area, above which the pressure is small but below which there is little further decrease with increase in pressure. Films of long-chain fatty acids, such as stearic acid, behave in this way. The behaviour is analogous to that of a gas at a temperature below its critical temperature; at higher volumes and lower pressures the product of pressure and volume is roughly constant, but when the pressure becomes sufficiently high that the liquid is formed the compressibility is much smaller.

Irving Langmuir

(1881–1957)

Langmuir was born in Brooklyn, New York, and went to school there except for a period of three years while his father, an insurance executive, held an appointment in Paris. He attended a boarding school in the suburbs of Paris, afterwards attending high schools in Philadelphia and Brooklyn. After obtaining a degree in metallurgical engineering from Columbia University he worked at the University of Göttingen under Nernst, obtaining his Ph.D. degree in 1906 for a dissertation on the dissociation of gases at hot platinum surfaces. This investigation foreshadowed some of his later work.

For a short period Langmuir was an instructor in chemistry at the Stevens Institute of Technology in Hoboken, New Jersey. In 1909 he accepted an appointment in the industrial research laboratory of the General Electric Company at Schenectady, New York, where he remained until his retirement in 1950. His great successes there were made possible through the enlightened attitude of the research director, W. R. Whitney, who had been a student of Ostwald at Leipzig. Whitney encouraged his staff to tackle even the most practical problems by carrying out basic research of the highest quality.

Langmuir's first task was to investigate the thermal conduction of gases in the presence of hot metal filaments, with the object of improving the life of the electric light bulb, then very short. At that time such bulbs were evacuated, and it was thought that improvement would be obtained by improving the vacuum. Langmuir concluded that this was not the case, the short lives arising from the evaporation of the tungsten filaments. His investigations showed that the interaction between the

Langmuir showed that for films of fatty acids on water the critical area corresponds to the situation in which the long-chain molecules are standing up out of the liquid surface, with their polar groups in the water; they are packed closely together, so that further compression of the film is difficult. When the surface pressure reaches a very high value the area suddenly decreases, since the further compression has caused them to pile on top of one another.

Films of the second type, referred to as *non-coherent films*, are not as common. Such films obey, to a good approximation, the equation

$$\pi A = k_B T, \tag{9.6}$$

where A is the area per molecule forming the film; k_B is the Boltzmann constant, and T is the absolute temperature. This equation is the two-

filament and the gas in contact with it was of great importance, and that in some cases dissociation occurred at the surface (Section 9.3). Later, beginning in about 1916, Langmuir made further studies of the adsorption of gases at surfaces, developing the isotherm known by his name (Section 9.3). These ideas led to his formulation of a general treatment of the kinetics of reactions at surfaces (Section 8.6), presenting it to the Faraday Society in 1921.

In the meantime Langmuir had begun another line of investigation, concerned with the behaviour of films on water surfaces. In 1917 he described the construction of a surface film balance, based on the primitive device used by Agnes Pockels, but with a number of technical improvements. This instrument led to a deeper understanding of the nature of various types of films (Section 9.2).

In 1919 Langmuir created something of a stir by proposing extensions of G. N. Lewis's theory of valency (Section 10.3), and he widely publicized his ideas during the next few years. This led to a certain coolness between the two men, Lewis in particular not appreciating the fact that the theory was sometimes known as the Langmuir theory!

In 1932 Langmuir was awarded the Nobel Prize in chemistry for 'his discoveries and researches in the realm of surface chemistry.' During World War II he did some interesting research on the production of smoke screens and on the seeding of clouds with the object of inducing rain.

References: C. Süsskind, DSB, 8, 22–25 (1973); Schaffer (1958); Taylor (1958).

dimensional analogue of the ideal gas equation; non-coherent films thus behave like a two-dimensional gas, and are often called *gaseous films*. They are usually formed at higher temperatures, where the molecular energies tend to prevent the formation of condensed films, and are analogous to gases above their critical temperatures.

9.3 Adsorption on solid surfaces

Some of the work on the adsorption of gases on solids was referred to in Section 8.6 in connection with surface kinetics. Some of the earliest work in this field was done by Michael Faraday, who realized that reactions catalyzed at surfaces occur in adsorbed films, but thought that the main effect of the solid catalyst is to exert an attractive force on the gas molecules so that they are present at much higher concentrations on the surface than in the main body of the gas. That this idea needs to be modified is shown by the fact that in certain cases different surfaces give rise to different products of reaction; for example, ethanol decomposes mainly into ethylene and water on an alumina catalyst, and mainly into acetaldehyde and hydrogen on copper. Specific forces must therefore be sometimes involved when molecules become attached to surfaces.

As a result of the work of van der Waals on the equation of state for gases and on attractive forces in gases (Section 5.1), it came to be realized that molecules could become attached to surfaces by the same kind of forces, the van der Waals forces. This type of adsorption became known as *van der Waals adsorption*, or as *physical adsorption*. However, Irving Langmuir concluded in 1916 from very careful studies of gaseous adsorption that there is a more powerful type of adsorption, which he called *chemisorption*. In this type of adsorption the adsorbed molecules are held to the surface by covalent forces such as those occurring between bound atoms in molecules. Langmuir and others made measurements of the heats evolved per mole for this type of adsorption, and found them to be usually comparable to that involved in chemical bonding, namely 100 to 500 kJ mol^{-1}. By contrast the heats for van der Waals adsorption are usually less than 20 kJ mol^{-1}, so that there is a sharp distinction between the two types.

Langmuir realized that when chemisorption occurs a single layer will be formed after which the surface is saturated; any additional adsorption can occur only on the layer already present, and this is generally weak adsorption, occurring only to a limited extent. Langmuir thus emphasized that chemisorption involves the formation of a unimolecular layer, and many later investigations have confirmed that this is so.

Adsorption isotherms

Beginning in 1916 Langmuir developed equations relating the amount of gas adsorbed on a surface to the pressure of the gas; such equations are now referred to as *Langmuir adsorption isotherms*. He arrived at them by equating the rates of adsorption and desorption at equilibrium. For the simplest case, that of a gas adsorbed on a smooth surface without becoming dissociated, he found that the fraction Θ of the surface covered is related to the pressure p of the gas by

$$\Theta = \frac{Kp}{1 + Kp},\tag{9.7}$$

where K is an equilibrium constant for the adsorption process. He also obtained isotherms for adsorption with the gas dissociated on the surface (e.g., H_2 into 2H), and for the simultaneous adsorption of two gases on a surface. In 1932 Langmuir, who throughout his career was at the General Electric Research

Laboratory in Schenectady, New York, received the Nobel Prize in chemistry for 'his discoveries in the realm of surface chemistry'.

Much work on solid surfaces has been done in more recent years, particularly in view of the great industrial importance of catalysis. The conclusions drawn by H. S. Taylor, that adsorption may involve an activation energy, and that there is a variability of surface activity, were noted in Section 8.6, since they are of particular importance in connection with the kinetics of surface reactions. Modern work has been much concerned with the application of new techniques to the exploration of surfaces and surface layers.

Suggestions for further reading

For general accounts of colloid and surface chemistry see
J. R. Partington, *A History of Chemistry*, Macmillan, London, 1961, Volume 4, Chapter 23.

Historical aspects of colloid and surface chemistry are also included in
N. K. Adam, *The Physics and Chemistry of Surfaces*, Clarendon Press, Oxford, 1938.
G. M. Schwab, *Catalysis*, translated by H. S. Taylor and R. Spence, Van Nostrand, New York, 1937.
John W. Servos, *Physical Chemistry from Ostwald to Pauling*, Princeton University Press, 1990. This book includes, in interesting detail, an account of the work on colloids by Wolfgang Ostwald and Wilder Bancroft. It describes in some detail Bancroft's agglomeration theory of anaesthesia and the curious incident of the 1933 Nichols Medal of the American Chemical Society. (Watson: 'There was no 1933 Nichols Medal; Holmes: 'That was the curious incident'.)
A. Zangweil, *Physics at Surfaces*, Cambridge University Press, 1985. Pages 1–4 give a brief historical sketch, referring to some quite recent work.

For work on surface films see
C. H. Giles, Franklin's teaspoonful of oil (Part 1 of 'Studies in the early history of surface chemistry'), *Chemistry and Industry*, 1616–1624, 8 Nov., 1969.
C. H. Giles and S. D. Forrester, Wave damping: the Scottish contribution ('Part 2 of 'Studies in the early history of surface chemistry'), *Chemistry and Industry*, 80–87, 17 Jan., 1970.
C. H. Giles and S. D. Forrester, The origins of the surface film balance (Part 3 of 'Studies in the early history of surface chemistry'), *Chemistry and Industry*, 43–53 (2 Jan., 1971). This article contains portraits of Agnes Pockels, Rayleigh and others.
M. Elizabeth Derrick, Agnes Pockels, 1862–1935, *J. Chem. Education*, **59**, 1030–1031 (1982).

For Perrin's work on sedimentation and Brownian movement see

Mary Jo Nye, *Molecular Reality: A Perspective on the Scientific Work of Jean Perrin*, London and New York, 1972.

J. B. Perrin, *Brownian Movement and Molecular Reality*, translated by F. Soddy, Taylor and Francis, London, 1910.

J. B. Perrin, *Atoms*, translated by D. L. Hammick, Constable, London, 1916.

For a history of polymers see A. Morawetz, *Polymers*: *The Origins and Growth of a Science*, John Wiley, New York, 1985.

CHAPTER 10

Quantum chemistry

By the beginning of the twentieth century most scientists were satisfied that their basic problems were solved, all that remained being to fill in a few details. The principles of mechanics had been well worked out, heat was understood to be a mode of motion, and light to have wave properties. Some difficulties did, of course, remain, and were referred to in a lecture given by Lord Kelvin at the Royal Institution on 27 April 1900:

The beauty and clearness of the dynamical theory, which asserts heat and light to be modes of motion, is at present obscured by two clouds. I. The first ... involves the question, How could the earth move through an elastic solid, such as essentially is the luminiferous ether? II. The second is the Maxwell-Boltzmann doctrine of partition of energy.

In referring to the partition of energy Kelvin had in mind the problem of interpreting the specific heats of gases and of solids. Here the belief in the equipartition of energy, which Maxwell and Boltzmann seemed to have established beyond any doubt, was leading to the wrong answers (Section 5.3). Another difficulty, not mentioned here by Kelvin, was the interpretation of the spectra of atoms and molecules. The fact that spectral lines appear only at particular wavelengths had never been given an adequate explanation; some of the ideas put forward were mentioned in Section 6.5, and they were obviously inadequate.

In 1900 few if any scientists suspected that within a few years two revolutionary theories would require a complete revision of all scientific concepts. These were the quantum theory, born in 1900, and the theory of relativity, born in 1905. Einstein was the sole originator of the theory of relativity, and he also had much to do with the quantum theory, which has profound implications for chemistry. Until the quantum theory was introduced it had been taken for granted that energy, including radiation, is continuous; in other words, one could think of an atom or molecule as having any amount of energy. The essence of the quantum theory is that this is not the case; energy comes in small packets, or quanta (from the Latin *quantum*, how much?). The quanta are small even on the atomic scale, and on the macroscale they are negligible; in driving one's car one does not have to worry about quantum transitions. In view of this it is amusing to see that journalists, politicians and others are now fond of talking about 'quantum leaps', which they seem to think are very large.

After the quantum theory was born in 1900 little attention was paid to it for a number of years. Indeed its originator, Max Planck (1858–1947), remained sceptical about it and was only convinced of its truth by Einstein. Planck's main interest had always been in thermodynamics, and his quantum theory arose from his attack, along thermodynamic lines, on the problem of the spectral distribution of energy in black-body radiation. Planck's concentration on radiation was somewhat unfortunate for the acceptance of the theory, since black-body radiation was not a problem of great interest among physicists. X-rays had been discovered in 1895, radioactivity in 1896, the electron in 1897, and radium in 1898, and it was to such matters that the best minds in physics and chemistry were directing their attention. Not for a decade was it realized, except by Einstein and a few others, that the quantum theory is essential to the understanding of many problems besides radiation. Chemists rarely thought in terms of quanta until 1913 when Neils Bohr applied the quantum theory to the problems of atomic structure, which obviously had chemical implications. It is significant that Planck did not receive his Nobel Prize until 1919, after Bohr had applied the quantum theory to atomic structure.

Some of the highlights in quantum theory and in the later quantum mechanics are listed in Table 10.1.

10.1 The old quantum theory

As noted in Section 6.2, in 1860 Kirchhoff introduced the concept of a black body (*schwarze Korper*), which completely absorbs all of the radiation incident upon it. Subsequent to his work many other studies were made of black-body radiation. The Austrian physicist Josef Stefan (1835–1893), of the University of Vienna, concluded in 1879 that the total radiation emitted by a black body is proportional to the fourth power of the absolute temperature. This law received much experimental support and was derived theoretically by Boltzmann in 1884; it is now usually called the Stefan–Boltzmann law. In 1894 the German physicist Wilhelm Wien (1864–1928), who had been an assistant to Helmholtz, derived a relationship, now known as Wien's displacement law, which relates the spectral distribution at one temperature to that at another.

An important aspect of black-body radiation is the variation of the energy emitted with the frequency or the wavelength. Comprehensive investigations of this were carried out in the 1890s at the Physikalische-Technische Reichanstalt in Berlin by Otto Lummer (1860–1925) and Ernst Pringsheim (1858–1917). Their work was concerned with what is now called the spectral radiant energy density P_v, which is the energy at a frequency v, per unit volume and per unit frequency range. Until about 1900 measurements of the variation of P_v with the frequency were confined to the visible and near ultraviolet, and showed a decrease with increasing frequency. In 1896 Wien made an attempt to explain this behaviour, on the basis of the doubtful assumption that the intensity of the emitted radiation is a function of the velocity of the emitting molecules.

TABLE 10.1 Highlights in quantum chemistry

Author	Date	Contribution	Section
Planck	1900	Theory of black-body radiation	10.1
Einstein	1905	Quantization of radiation	10.1
Einstein	1907	Specific heats of solids	10.1
Bjerrum	1911–1914	Interpretation of spectra	10.3
Einstein	1912	Photochemical equivalence	10.1
Einstein and Stern	1913	Existence of zero-point energy	10.1
Bohr	1913	Atomic theory	10.1
Pauli	1923	Exclusion principle	10.1
Mulliken	1924	Confirmation of zero point energy	10.4
de Broglie	1924	Wave properties of particles	10.2
Born and Heisenberg	1925	Matrix mechanics	10.2
Dirac	1925	Quantum mechanics	10.2
Schrödinger	1926	Wave mechanics	10.2
Heisenberg	1927	Uncertainty principle	10.2
Heitler and London	1927	Valence-bond theory	10.4
Hund; Mulliken	1928	Molecular-orbital theory	10.4
Herzberg; Mulliken	1928 ff.	Quantum mechanics of spectra	10.3
Herzberg	1929	Bonding and antibonding electrons	10.4

By use of Maxwell's distribution function for velocities he obtained the expression

$$P_\nu = \alpha \nu^3 \, e^{-\beta \nu / T}, \tag{10.1}$$

where α and β are constants. The exponential factor explains the decrease in P_ν at high frequencies, and Wien's formula gave an adequate representation of all the data that were then available. The pre-exponential factor corresponds to an increase in P_ν with increasing frequency, and eqn 10.1 predicts that the function passes through a maximum, as shown in Figure 10.1. When Wien put forward his equation it was not known that there is a maximum, but experimental work at lower frequencies soon showed that this is the case.

Another theoretical treatment was given by Lord Rayleigh, in a two-page paper published in June, 1900. Commenting that Wien's assumptions were 'little more than a conjecture', Rayleigh applied the principle of equipartition of energy, and concluded that P_ν should be proportional to $\nu^2 T$. Five years later he

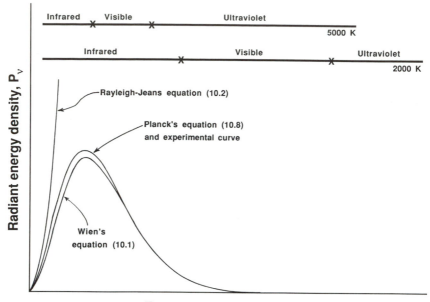

FIGURE 10.1   Plots of Wien's equation (10.1), the Rayleigh–Jeans equation (10.2), and Planck's equation (10.8). Since all three equations involve the ratio v/T of the frequency to the temperature, the radiant energy density is plotted against v/T. The approximate infrared, visible and ultraviolet ranges are shown at two temperatures.

elaborated his derivation, and after a correction had been made by James Hopwood (later Sir James) Jeans (1877–1946) the resulting expression was

$$P_v = \frac{8\pi v^2}{c^3} k_B T, \tag{10.2}$$

where k_B is the Boltzmann constant. This formula, which would be accurate if there were equipartition of energy, predicts a steady increase as v increases (Figure 10.1), and is completely at variance with the experimental results at the higher frequencies. This discrepancy, later referred to as the 'ultraviolet catastrophe', is a striking demonstration of the failure of the principle of equipartition of energy.

The particular inspiration for Planck's contribution of 1900 was some work that had just been done by two other workers at the Physikalische-Technische Reichanstalt in Berlin. Heinrich Rubens (1865–1922) and Ferdinand Kurl-baum (1857–1927) studied the spectral distribution over a much wider spectral range, including the infrared, and found that P_v does pass through a maximum, as shown in Figure 10.1. Their results were thus qualitatively consistent with Wien's formula, which applied very well at the higher frequencies. At the lower frequencies, however, Wien's formula predicted spectral densities that were too low, the discrepancies becoming greater as the frequency was lowered. The Rayleigh–Jeans formula, which was derived a little later, gives an excellent fit at very low frequencies, but does not predict the maximum and becomes seriously in error at higher frequencies.

These experimental results of Rubens and Kurlbaum were presented by Kurlbaum to the Berlin Academy on 25 October 1900, but a few days previously they had been communicated privately to Planck. In the discussion following the presentation of the paper Planck proposed the empirical equation

$$P_\nu = \frac{A\,\nu^3}{e^{\beta\nu/T} - 1}.$$ (10.3)

This expression differs from Wien's equation 10.1 only in the inclusion of unity in the denominator; it predicts a curve which always lies above the Wien curve (Figure 10.1). At higher frequencies, where unity in the denominator can be neglected, the equation reduces to

$$P_\nu = A\,\nu^3\,e^{-\beta\nu/T},$$ (10.4)

which is consistent with the results at higher frequencies and with Wien's equation 10.1. At lower frequencies $\exp(\beta\nu/T)$ reduces to $1 + \beta\nu/T$, and therefore

$$P_\nu = (A/B)\,\nu^2 T$$ (10.5)

(When a few years later the Rayleigh–Jeans equation 10.2 was developed it was seen to be consistent with eqn 10.5.) The day after the Berlin Academy meeting, Rubens, who had worked through the night, told Planck that his formula 10.3 fitted all of his data within the experimental error; more precise measurements that were subsequently made gave further confirmation of the equation.

Much encouraged, Planck worked hard during the next few weeks to find a theoretical basis for his empirical formula. As a student he had attended Kirchhoff's lectures at the University of Berlin, and had also been impressed by the work of Clausius and of Boltzmann, with whom he had had some discussions. Planck was aged 42 in 1900 and had done competent work, but nothing of great distinction. He based his theoretical treatment of radiation on the statistical interpretation that had been developed by Boltzmann. Purely as a mathematical convenience Boltzmann had often treated energy as coming in small packets, but at the end of his derivations he always required the energy units to become of zero size. Planck did not, however, carry out this final step, and he did not at first appreciate the great significance of this omission. He assigned to the oscillators in the emitting solid an energy E given by

$$E = nh\nu,$$ (10.6)

where ν is the fundamental frequency of the oscillator, n is a whole number and h a constant. In his treatment there are N_0 oscillators having zero energy, N_1

having energy $h\nu$, N_2 having energy $2h\nu$, and so on. He then applied the Boltzmann distribution to the oscillators, and finally obtained for the average energy of the oscillators the expression

$$\varepsilon = \frac{h\nu}{\exp(h\nu/k_BT) - 1},\qquad(10.7)$$

where k_B is the Boltzmann constant. To obtain P_ν this is multiplied by $8\pi\nu^2/c^3$, a factor derived by Planck in a paper of 1899; the result is

$$P_\nu = \frac{8\pi\nu^2}{c^3}\frac{h\nu}{\exp(h\nu/k_BT) - 1},\qquad(10.8)$$

which is of the same form as eqn 10.3. His paper giving this treatment appeared in December 1900; and 14 December 1900, when he presented the work to the German Physical Society, is often regarded as the birthday of the quantum theory.

From the standpoint of Planck's derivation it is easy to understand the failure of treatments based on the equipartition of energy. If there were equipartition, the energy in an oscillator of high frequency would be the same as that in an oscillator of low frequency, and ε, instead of being given by eqn 10.7, would be k_BT. This is what is assumed in the later Rayleigh–Jeans formula, and the energy density goes on increasing as the frequency increases. In a paper published in 1911 Paul Ehrenfest (1880–1933) referred to this deviation from experiment as the *ultraviolet catastrophe*. Planck's treatment avoided the catastrophe by reducing the amounts of energy that can go into the oscillators of high frequency.

An important contribution made by Planck in the same publication was to evaluate the constants h and k_B from the experimental data on the spectral distribution. His values in modern units, compared with the best modern values, are shown in Table 10.2. Surprisingly, Boltzmann had never emphasized the great importance of his constant k_B, and had never obtained a value for it; Planck was the first to do so. Noting that k_B is the gas constant per molecule, and therefore equal to the gas constant R divided by the Avogadro constant L, Planck also obtained a value for L, shown in Table 10.2. He then obtained a value for the elementary unit of charge, e, as the ratio of the Faraday constant to the Avogadro constant. All of his values are seen from Table 10.2 to be in reasonable agreement with the modern values. This contribution was of great importance, aside from the fact that it provided strong support to Planck's procedures.

In view of the present appreciation of Planck's quantum theory, his 1900 presentation would have been expected to attract immediate attention. Even if the theory had not been accepted at once, one would at least expect some discussion of it. In fact, for about five years hardly any notice at all was taken of Planck's paper. For example, the first edition of Jeans's comprehensive and authoritative *Dynamical Theory of Gases*, published in 1904, made no mention of

TABLE 10.2 Planck's and modern values of fundamental constants

Constant	Symbol	Values		SI Unit
		Planck	Modern[a]	
Planck constant	h	6.55×10^{-34}	6.626×10^{-34}	J s
Boltzmann constant	k_B	1.346×10^{-23}	1.381×10^{-23}	J K^{-1}
Avogadro constant	L	6.175×10^{23}	6.022×10^{23}	mol^{-1}
elementary charge[b]	e	1.56×10^{-19}	1.602×10^{-19}	C

[a] Taken from *Quantities, Units and Symbols in Physical Chemistry* (Ed. I. M. Mills), International Union of Pure and Applied Chemistry, Blackwell Scientific Publications, Oxford, 1988.
[b] Planck gave his value for e as 4.60×10^{-10} esu; 1 esu $= 3.336 \times 10^{-10}$ C.

Planck's work. The main reason for this is that the theory was presented only as applying to black-body radiation, and its wider implications were not recognized until later. For example, the application to specific heats, relevant to Jeans's book, did not come until 1907, when Einstein dealt with the problem, as discussed later.

When attention was finally paid to Planck's theory, it was at first usually unfavourable. Jeans, for example, in 1905 published a severe criticism of Planck's argument. His main objection was to Planck's failure to make the size of the quanta equal to zero, as Boltzmann had always done; in Jeans's words, 'statistical mechanics gives us the further information that the true value of h is $h = 0$'. The tone of Jeans's criticism was not that Planck had made a quantum hypothesis that could not be accepted; it was that he had made a mistake in his mathematics by forgetting the final stage of setting the constant h equal to zero.

Quanta of radiation

At first Planck's quantum theory, insofar as it was accepted at all, was regarded as applying only to the material oscillators—for example those in a solid, which could emit energy only in multiples of $h\nu$. Another important step forward was taken in 1905 by Einstein by his suggestion that radiation itself is quantized. At the time, and for four more years, Einstein was a junior official in the patent office at Berne, Switzerland, having become a Swiss citizen in 1901. The year 1905 was an *annus mirabilis* for him, since in the spring of that year he prepared three papers, each one of which was worthy of a Nobel Prize. The first of these was on the quantization of energy, which did actually get him his Nobel Prize in 1921. The second was on the Brownian movement (Section 9.1) and the third on the special theory of relativity. All three papers appeared in the *Annalen der Physik* in late 1905.

Modern textbooks often refer to Einstein's 1905 paper on quanta of radiation as his paper on the photoelectric effect, and say that it was based on Planck's 1900 paper on the quantization of oscillator energies. The first statement is inadequate and misleading, the second quite wrong. The paper consists of an introduction and nine sections, only one of which is concerned with the photo-electric effect. The first few sections present arguments, thermodynamic and otherwise, in favour of the idea that radiation in some experiments acts as a beam of particles, called quanta by Einstein and called photons by G. N. Lewis in 1926. The first six sections of the paper deal with general aspects of radiation, the last three with applications of the quantum concept to Stokes's ideas about fluorescence (Section 6.1), to the photoelectric effect, and to photoionization in gases.

A perusal of Einstein's paper shows that he made only a passing reference to Planck's work, which may have suggested the approach that Einstein made but certainly did not provide a basis for it. In fact, at this stage Einstein did not believe that oscillator energies are quantized, but only that radiation is quantized. His deduction that radiation is quantized was instead derived from Wien's treatment, and from a consideration of the experimental evidence regarding the absorption and emission of radiation, including the photoelectric effect. Soon after Einstein's paper appeared it was pointed out that his treatment led not to Planck's radiation law but to Wien's, which was inconsistent with the data on the spectral distribution. Only later, with reference to his work on specific heats in 1907, did Einstein accept Planck's point of view on the quantization of oscillator energies. Einstein's great contribution in his 1905 paper was his recognition that radiation itself is quantized.

A curious feature of Einstein's paper is that Planck's constant does not appear in it explicitly—another indication that at the time Einstein was not convinced by Planck's contribution. Einstein denoted by the symbol β what is equivalent to h/k_B, and he wrote R/N for k_B; in his paper the energy of a light quantum is not written as $h\nu$ but as $R\beta\nu/N$. Another curious feature is the title, which can be translated as 'On a heuristic viewpoint concerning the emission and transformation of light'. The word 'heuristic', meaning 'relating to dis-covery', hardly seems appropriate, since all research is related to discovery; by 'heuristic viewpoint' Einstein probably meant 'tentative hypothesis'. The American physicist Edward Uhler Condon (1902–1974) later commented that by the title Einstein perhaps meant 'that there is something in the paper, but he does not know quite what. At least that is what I mean when I say "heuristic" '.

It is perhaps true that the photoelectric effect does provide the most convincing evidence for Einstein's idea that radiation itself is quantized. The effect had been discovered in 1887 at the University of Kiel by the German physicist Heinrich Rudolf Hertz (1857–1894), who in his short lifetime of 37 years made such outstanding contributions, particularly in connection with electromagnetic radiation. Rather paradoxically, it was a series of experiments that convincingly confirmed Maxwell's electromagnetic theory of radiation that also produced some of the first evidence for the particle nature of radiation. Hertz discovered that the length of a spark induced in a secondary circuit was greatly reduced when the spark gap was shielded from the light of the spark in the primary circuit. He showed that ultraviolet light has the effect of increasing the length of the spark that can be produced. Similar effects, later realized to be due to ejection of electrons by the radiation, were reported in the same year, 1887, by Arthur Schuster at Manchester, and by Arrhenius who was on his travelling fellowship working with the electrochemist Kohlrausch in Würzburg.

Two years later, in 1889, detailed studies of the emission of electrons under the influence of light were reported by Johann Elster (1854–1920) and Hans Geitel (1855–1923). These men were often known as the Castor and Pollux of science, or as the Heavenly Twins, since they were inseparable both in their private lives and in their teaching and research. They had been school friends, and both became teachers at a gymnasium near Brunswick. When Elster married and moved into a house Geitel joined the couple, and the two friends carried out research in a laboratory they established in the house. They were often confused with one another, and a man who somewhat resembled Geitel was once addressed by a stranger as 'Herr Elster'. His reply was 'You are wrong on two counts: first, I am not Elster but Geitel, and secondly I am not Geitel.'

In 1898 J. J. Thomson, then the Cavendish professor at Cambridge, concluded that the particles ejected by the radiation are the same as those present in cathode-ray tubes (Section 6.7); these electrons he always referred to as 'corpuscles'. Later Phillip Lenard (1862–1946) showed that no emission occurs if the frequency of the radiation is less than a critical value ν_0, and that the energy of the ejected electron is equal to $h\nu - h\nu_0$ if radiation of frequency ν is employed. He found that the intensity of the radiation has no effect on whether or not electrons are emitted; it affects only the number of electrons emitted.

Einstein realized that these results are inexplicable in terms of the wave theory of radiation, according to which radiation of sufficiently high intensity, irrespective of frequency, would be able to eject electrons. With the particle theory, on the other hand, the explanation is straightforward; a particle of radiation is capable of ejecting an electron provided that its energy $h\nu$ is sufficiently large.

Like Planck's quantum theory, Einstein's theory of the quantization of radiation was slow to gain acceptance, since the wave theory of light was so strongly entrenched. For many decades the particle theory of radiation, advocated by Newton, Laplace and others, had been generally regarded as untenable (Section 3.3). Also, there appeared to be a logical difficulty with Einstein's theory: the energy of radiation is expressed in terms of its frequency, a concept that only has meaning in terms of the wave theory.

In his 1905 paper Einstein anticipated these objections and answered them. He began his paper by admitting that the wave theory of light was well established and was not to be displaced as far as properties such as diffraction and interference are concerned. But, he continued, such properties, in which light interacts with matter in bulk, relate to time averages. It is quite possible that the wave theory will prove inadequate whenever there are one-to-one effects such as the photoelectric effect and the emission and absorption of radiation. In other words, what is needed is a dual theory of radiation. For diffraction and interference the wave properties are relevant; for the photoelectric effect and for emission and absorption one must regard the radiation as a beam of particles.

Some idea of the scepticism with which Einstein's theory was regarded is provided by some of his personal experiences. In 1913 he was proposed for membership in the Prussian Academy of Science, as successor to van't Hoff, who had died in 1911. In their letter of recommendation four of the most eminent German physicists—Planck, Emil Warburg, Nernst and Rubens— spoke highly of Einstein's special theory of relativity and of his theory of specific heats. However, they added apologetically, 'That he may sometimes have missed the target in his speculations, as for example in his hypothesis of light quanta, cannot really be held too much against him. . . .' It is amusing to note that in the recommendation on behalf of Einstein for the Nobel Prize—in which Planck and Nernst were also involved—it was his work on light quanta that was emphasized; his work on relativity theory then seemed questionable. In fact, Einstein's work on quanta of radiation is probably of greater importance and wider significance—at any rate for chemists—than his better publicized work on relativity.

Planck himself had difficulty in accepting Einstein's theory of the quantization of radiation. To Planck the idea of quanta of energy was no more than an *ad hoc* hypothesis to explain black-body radiation; he was slow to realize that the quantum theory has a much wider significance. It was in his nature to be conservative, and having been brought up on classical physics he found it difficult to renounce the old ideas.

Specific heats of solids

In 1907 Einstein made another important contribution to quantum theory, with a paper entitled 'Planck's theory of radiation and the theory of specific heats'. This paper was of special significance, since it was the first to show that quantum theory was not only relevant to problems of radiation: it was able to explain the inconsistencies that had been plaguing the investigations of kinetic theory.

Reference has been made (Section 5.4) to difficulties that had been encountered, by Clerk Maxwell, Boltzmann and others, in explaining the specific heats of gases. There were similar problems with the specific heats of solids, and it was to these that Einstein directed his attention. In 1819 the French physicists Pierre Louis Dulong (1785-1838) and Alexis-Thérèse Petit (1791–1820) had stated their famous law:

Les atomes de tout les corps simples ont exactement la même capacité pour la chaleur. (The atoms of all simple bodies have exactly the same heat capacity.)

For half a century this statement had been accepted without much question. In 1840 and 1841 Regnault had reported extensive measurements of specific heats, the results all confirming the law to a good approximation.

After the kinetic theory was developed it appeared that the specific heat of a solid element should be $3R$, or approximately 6 cal K^{-1} mol^{-1} (≈ 25 J K^{-1} mol^{-1}), and measurements made at or above room temperature confirmed this prediction. However, when it became possible to make measurements at lower temperatures there were found to be serious discrepancies. The first suggestion that something was badly wrong was in 1872 when Wilhelm Edouard Weber (1804–1891) measured at Göttingen the specific heat of diamond at low temperatures. He found, for example, that at -50°C the value had fallen as low as 0.76 cal K^{-1} mol^{-1}. After 1896, when Karl Paul Gottfried Linde (1842–1934) in his ice-machine factory applied the Joule–Thomson principle to the liquefaction of air, it became possible to make more extensive low-temperature measurements, many of which showed anomalously low specific heats. In 1898 U. Behn of the University of Berlin made a number of such measurements, and concluded that the values seemed to be approaching zero as the temperature approached the absolute zero. Similar conclusions were drawn in 1903 by William Augustus Tilden (1842–1926), professor of chemistry at the Royal College of Science in London, and in 1905 by James Dewar, the professor of chemistry at Cambridge, who did most of his work at the Royal Institution where he held a joint appointment.

Einstein realized that this fall in specific heat with decreasing temperature can be explained in terms of the quantization of the energy in the oscillators. He assumed that all of the atoms in a solid vibrate with the same frequency ν. One mole of an element contains L atoms, and since the vibration occurs in three dimensions the total energy is (compare eqn 10.7)

$$E = \frac{3Lh\nu}{\exp(h\nu/k_B T) - 1}. \tag{10.9}$$

The specific heat per mole (the molar heat capacity) is thus

$$C_V = \frac{dE}{dT} = \frac{3R(h\nu/k_B T)^2 \exp(h\nu/k_B T)}{\{\exp(h\nu/k_B T) - 1\}^2}, \tag{10.10}$$

where R, the gas constant, is equal to Lk_B. At high enough temperatures the value of C_V is close to $3R$, but it decreases towards zero as the temperature is lowered.

Einstein's formula agreed reasonably well with experiment, as was shown, for example, by Nernst and Lindemann in 1911. The treatment was improved in 1912 by Debye, who developed a continuum theory of specific heats, the vibrational frequencies being allowed to have a range of values cut off at a maximum value. This gave improved agreement with the experimental results.

An important consequence of Einstein's work on specific heats was the organization of the first Solvay Congress, which was held in Brussels in 1911. This conference was mainly due to the initiative of Nernst, who had initially been unimpressed by the quantum theory. Later, however, after he had begun low-temperature measurements related to his heat theorem (Section 4.5), he realized that Einstein's equation gave a good interpretation of specific heats, and he became more enthusiastic. The conference was made possible by the generous support of the Belgian industrialist Ernest Solvay (1838–1922). Its title 'La théorie du rayonnement et les quanta' suggests that despite Nernst's own conviction that the theory had a broader significance, it was still generally thought to be mainly concerned with radiation. The Congress was attended by many of the world's scientific leaders, and played a powerful role in the development and extension of the quantum theory.

Specific heats of gases

Einstein's 1907 paper did not treat the specific heats of gases. In 1911, in a paper concerned with his heat theorem, Nernst pointed out that Einstein's approach should also be applicable to the energy of vibration of a gas molecule, and this idea was extended in some detail by Niels Bjerrum, then at the University of Copenhagen. In a series of papers that appeared from 1911 to 1914 Bjerrum applied the quantum theory to both spectra and the specific heats of gases, and discussed the relationship between the two properties. He pointed out that the specific heat contributions of the various types of motion depend in an important way on the spacing between the quantized energy levels. For translational energy the levels are so close together that transitions between them occur easily; translational energy therefore makes its full contribution to the specific heat except at extremely low temperatures. Vibrational energy levels, on the other hand, are so far apart that when the temperature is raised the system does not easily pass into a higher vibrational state, the theory being similar to that for a solid (eqn 10.10). Vibrations therefore make little contribution to the specific heat except at high temperatures. For rotational motion the quanta are intermediate in size; again the contribution to the specific heat falls as the temperature is lowered, but at room temperature it usually has its classical value of ½R for each degree of rotational freedom. Bjerrum had thus overcome the difficulty, which Clerk Maxwell had considered but could never solve, that motions revealed in the spectra are not revealed in the specific heats.

One particular matter that had been troublesome is that rotations of an atom, and rotation of a linear molecule about its axis, make no contribution to the specific heat. The reason is that the mass is almost entirely concentrated in the atomic nuclei, so that the moment of inertia corresponding to these rotations is very small. As a result, the quantum of rotational energy for an atom or for a linear molecule rotating about its axis is much larger than $k_B T$; such rotational transitions therefore do not occur to an appreciable extent.

In 1913 Einstein and his assistant Otto Stern (1888–1969) published a paper of great significance in which they suggested for the first time the existence of a residual energy that all oscillators have at the absolute zero. They called this residual energy the *Nullpunktsenergie*, which is usually rather unsatisfactorily translated as the *zero-point energy*. They carried out an analysis of the specific heat of hydrogen gas at low temperature, and concluded that the data are best represented if the vibrational energy is taken to have the form

$$\varepsilon = \frac{h\nu}{\exp(h\nu/k_{\mathrm{B}}T) - 1} + \frac{h\nu}{2}. \tag{10.11}$$

According to this, even at the absolute zero the energy has the value $\frac{1}{2}h\nu$. It may be mentioned that at the time Einstein, Ehrenfest and others considered the possibility that the vibrational energy has units of $\frac{1}{2}h\nu$ rather than $h\nu$. Ehrenfest continued to believe this for some time, but Einstein and Stern thought that the hydrogen data led to the conclusion that the energy levels corresponded to $(v + \frac{1}{2})h\nu$. where v is an integer, having a value of zero for the zero-point level.

Einstein and Stern were not entirely convinced that the zero-point energy was $\frac{1}{2}h\nu$. More direct and convincing evidence was obtained by Mulliken in 1924 (Section 10.3), and a theoretical interpretation came from quantum mechanics.

Photochemical equivalence

It was noted in Section 8.4 that Johannes Stark in 1907, and Einstein in 1912, put forward what is usually known as Einstein's law of photochemical equivalence, which provides a basis for all work in photochemistry.

Stark's enthusiasm for the quantum theory began in about 1907, and he began to apply it to a number of problems. He was an excellent experimentalist but a weak theoretician, and some of his ideas involved serious errors in physics; photochemical equivalence was one of his better ideas, but he did not explain it very clearly. By 1913 Stark had become an intemperate opponent of the quantum theory. As he grew older he became increasingly paranoid, and seemed to find it necessary to oppose any view that was becoming generally accepted. He quarrelled violently with all his colleagues, and when the Nazis came into power he adopted a vicious anti-Semitism in which he also attacked non-Jewish scientists—'white Jews' as he called them—who adopted a reasonable position. Planck, Sommerfeld, Heisenberg and many others came under his malevolent attack. He is now chiefly remembered for the 'Stark effect', discussed later, which he used as the basis of an attack on the quantum theory!

The Bohr atom

The next important event in the history of the quantum theory was its application to atomic theory. This was first done successfully by Niels Bohr in 1913, but mention should be made of an earlier attempt made by the Austrian physicist Arthur Erich Haas (1884–1941). This work is of interest as illustrating early failures to appreciate the broad significance of the quantum theory. In 1910, in support of his habilitation as *Privatdocent* at the University of Vienna, Haas submitted a treatment of the hydrogen atom that to a small extent anticipated the more comprehensive treatment given three years later by Bohr.

Niels Henrik David Bohr

(1885–1962)

Bohr was born in Copenhagen, and studied at the University of Copenhagen where he obtained his doctorate in 1911. He then worked for a short period at Cambridge with J. J. Thomson, but the relationship between them was an uncomfortable one and Bohr moved to the University of Manchester to work with Ernest Rutherford. This association was a happy and fruitful one, and Bohr began the development of his famous theory of hydrogen-like atoms (Section 10.1). After his return to Copenhagen in 1913 he completed this work, adding the application to atomic spectra, and his paper appeared in that year in the *Philosophical Magazine*. In 1914 Rutherford appointed Bohr to be Reader in theoretical physics at the University of Manchester, a position he held for a year.

Rutherford would have liked Bohr to have remained in England, but he preferred to return to his own country, and in 1916 he became professor of physics at the University of Copenhagen. Later he headed an institute in Copenhagen that was generously financed by the Carlsberg Breweries, and which rapidly became a world centre for theoretical physics. Bohr exerted a strong influence on the development of atomic and molecular physics, not only through his own researches but through his encouragement of others. Many who were later to achieve fame, including Pauli and Heisenberg, spent time in his laboratories.

Bohr received the Hughes Medal of the Royal Society in 1921, and was elected a Foreign Member of the Society in 1926. He was awarded the Nobel Prize for physics in 1922. He continued to work on a number of problems in nuclear physics, and in the 1930s proposed a model of the nucleus which involved strong coupling between its constituent particles.

Bohr visited the United States in 1939 to attend a scientific conference, and brought with him the news that nuclear fission had been accomplished. This played a large part in initiating the nuclear research carried out by the Allies during World War II, work that unlike the German effort was successful in producing an atomic bomb. In 1940 Denmark was occupied by the Germans but in 1943 Bohr was able to escape to Sweden at considerable risk. Later that year he was flown to England in a small plane, nearly losing his life for lack of oxygen. He then went to the United States where he worked on the atomic bomb project until 1945. Later he devoted much effort in the cause of peaceful uses of atomic energy, and he organized the first Atoms for Peace Conference in Geneva in 1955. In 1957 he received the first Atoms for Peace award.

References: L. Rosenfeld, DSB, 2, 239–254 (1970); Cockcroft (1963); Moore (1967); Pais (1991).

Haas's treatment involved the quantization of electronic orbitals, but it was rejected by the examiners and even somewhat ridiculed, one examiner describing as 'naive' Haas's introduction of quantum theory into atomic structure and spectroscopy; the theory was still generally regarded as concerned only with heat and radiation. Haas's application was rejected, and he only obtained his appointment two years later after he had submitted a treatment of a different topic. He eventually became professor of physics at Vienna, and for the last five years of his life was professor at the University of Notre Dame in Indiana.

Bohr's work on atomic theory was begun in Rutherford's laboratory at Manchester University. In 1912, at the age of 27, Bohr had gone to Cambridge to work with J. J. Thomson, the Cavendish professor, but the association was not a happy one. Sources differ as to details, but it appears that Bohr offered some well-meant but penetrating criticisms of the 'plum-cake' theory that Thomson had proposed in 1904, in which electrons were supposed to be imbedded in a positively-charged medium. Thomson resented the criticisms, and suggested to Bohr that he might do better elsewhere. Bohr then went to work with Rutherford, and the collaboration between the two became a very happy and fruitful one. Rutherford was himself not much of a theorist, and was initially somewhat wary of Bohr; however, partly because Bohr was a good soccer player, Rutherford soon accepted him and gave him much encouragement and cooperation. Bohr completed most of his theory of the atom in Manchester, but put important finishing touches to it, including his work on spectral lines, after he had returned to Copenhagen. Bohr's work was published in English, in the *Philosophical Magazine* in 1913.

Bohr's treatment is readily accessible in a number of places, but it should be noted that many modern textbooks give derivations that are rather different from that first given by Bohr. These more modern derivations, considered more satisfactory, involve the quantization of the angular momentum of the electron in its orbit. Bohr's original derivation, on the other hand, considered the quantizaton of the *energy* of the orbiting electron.

On the basis of purely classical mechanics Bohr first obtained an expression for the binding energy ε_b of an electron of charge e orbiting round a nucleus of charge Ze. This is the energy required to remove the electron to infinity, and is given by

$$\varepsilon_b = \frac{1}{2}mu^2 = \frac{Ze^2}{2r}, \tag{10.12}$$

where r is the radius of the orbit, m the mass of the electron, and u its velocity. An essential feature of Bohr's theory was that an electron could remain in an orbit without emission or absorption of radiation, which only occurs if there is a jump from one orbit to another.

Bohr first introduced quantization as follows. The frequency of rotation of an orbital electron is $u/2\pi r$, which may be denoted as ω. For a free electron the frequency is zero, so that the average frequency in the two states is $\omega/2$. Bohr then applied the quantum condition to the energy $h\omega/2$ corresponding to this average frequency:

$$\varepsilon_b = \tfrac{1}{2}nh\omega, \tag{10.13}$$

where n is the quantum number. This condition leads, after some rearrangement, to the conclusion that the binding energies for the permitted energy levels are given by

$$\varepsilon_b = \frac{2\pi^2 Z^2 e^4 m}{n^2 h^2}. \tag{10.14}$$

It must be admitted that Bohr's quantum condition 10.13, involving the average of the energies of the bound electron and the free electron, is rather contrived. Bohr showed later in his series of papers that his condition is equivalent to the quantization of the angular momentum. Thus from equations 10.12 and 10.13, with $\omega = u/2\pi r$ it follows that

$$\tfrac{1}{2}mu^2 = nh\omega/4\pi r \tag{10.15}$$

or

$$mur = n\frac{h}{2\pi}, \tag{10.16}$$

which is the quantum condition in terms of the angular momentum mur. Modern treatments usually use the condition 10.16 instead of 10.13 as being more plausible.

Today we think of Bohr's interpretation of atomic spectra as particularly convincing, and it is surprising to find that it was introduced as an afterthought. The quantization of the energy levels had been done before Bohr left Manchester. After his return to Copenhagen he was visited by the spectroscopist Hans Marius Hansen (1886–1956) who asked him if he had interpreted the spectral lines. Bohr replied that he had decided not to consider them as they would be impossibly complicated. Hansen, however, persisted, and drew Bohr's attention to the relatively simple spectral regularities which had been generalized in the Rydberg formula 6.7. Surprisingly, Bohr had not heard of these, and after looking into them realized that his expression 10.14 for the energy levels provided a complete explanation. For two quantum numbers n_1 and n_2 the energy difference is

$$E_1 - E_2 = \frac{2\pi^2 mZ^2 e^4}{h^2}\left[\frac{1}{n_1^2} - \frac{1}{n_2^2}\right], \tag{10.17}$$

and this is equal to the product of the frequency v of the spectral line and the Planck constant. The value of the Rydberg constant, in terms of wavenumbers $\bar{v}$ $(= v/c)$ is therefore

$$R = \frac{2\pi^2 mZ^2 e^4}{ch^3}. \tag{10.18}$$

On inserting the values of the constants Bohr obtained a value for the Rydberg constant for the hydrogen atom ($Z = 1$) that was in remarkably good agreement with the experimental value. He also obtained good agreement for the helium ion He$^+$ ($Z = 2$), for which the Rydberg constant is four times as large.

Particularly because of this good agreement, Bohr's theory was quickly accepted with some enthusiasm. Einstein and Jeans, for example, spoke highly of it. In September, 1913, at the 83rd meeting of the British Association for the Advancement of Science, the theory was discussed very favourably, and reports of it, as expounded at the meeting, appeared in *Nature* and in *The Times* of London. It is interesting that Bohr himself was somewhat less enthusiastic than others, regarding his theory as 'preliminary and hypothetical'.

Besides being of great significance for the interpretation of atomic structure and spectra, Bohr's theory provided further convincing evidence for the validity of the quantum theory, which was now recognized to be much more than an *ad hoc* hypothesis to explain black-body radiation. It related to specific heats, to the absorption of radiation, and to atomic structure, and evidently had wide implications.

At about the same time that Bohr put forward his atomic theory Johannes Stark, now at the Technische Hochschule in Aachen, made a discovery of great significance. He found that an electric field splits spectral lines, such as the lines of the Balmer series of hydrogen, into a number of component lines—just as Zeeman had earlier found a splitting brought about by a magnetic field. Stark's paper dealing with this appeared in November, 1913, just a few months after Bohr's paper was published. Shortly afterwards Emil Warburg interpreted the general features of the Stark effect on the basis of Bohr's theory, and a paper by Bohr that appeared in 1914 also showed that the theory provided some explanation of both the Stark and the Zeeman effects, although some of the details remained inexplicable. Stark himself, perversely and true to form, used his results as the basis of an attack on Bohr's theory and on the whole idea of quantization!

Einstein's support for the Bohr theory led him to propose, in 1916, an important theory of the absorption and emission of radiation. Consider two energy levels in an atom, a higher one E_m and a lower one E_n. Einstein's proposal was that transitions between these levels are governed by probability laws. During the time interval dt there is a probability, denoted by $A_m dt$, of a spontaneous transition, with emission of radiation, from the higher level to the lower one. In a radiation field there can also be an induced or stimulated transition, the probability of which is $B_m P dt$, where P is the energy density of the radiation. For the reverse process, the transition from the lower to the upper level, there is no spontaneous process, but there is a stimulated transition, of probability $B_n P dt$. Einstein proved that $B_m = B_n$, and he confirmed that his treatment was consistent with Bohr's fundamental relationship

$$E_m - E_n = h\nu. \tag{10.19}$$

Einstein also showed that Planck's distribution law, eqn 10.8, is obtained on the basis of his treatment.

It soon became clear that in spite of the considerable successes of the Bohr theory, it could by no means explain all the features of atomic structures and spectra. As early as 1891 the American physicist Albert Abraham Michelson (1852–1931) had discovered that the Balmer series was not composed of truly single lines, a result that could not be explained by Bohr's theory. The theory also could not give a complete explanation of the Zeeman and Stark effects. Further difficulties were encountered when attempts were made to apply the Bohr theory to atoms containing more than one electron, and to account for the building up of the periodic table.

Extensions of Bohr's theory

Some limited improvement was achieved by Arnold Johannes Wilhelm Sommerfeld (1868–1951), who had succeeded Boltzmann as professor at Munich. Bohr's theory had considered the possibility of elliptical as well as circular orbits, but with no corresponding quantization. In a paper that appeared in 1916 Sommerfeld suggested that there was a second quantum number, or *azimuthal quantum number, l,* associated with the eccentricity of an elliptical orbit. In a magnetic field an elliptical orbit would become oriented, and a third quantum number, the *magnetic quantum number, m,* was introduced. This treatment gave a partial explanation for the Zeeman splitting in a magnetic field, but there was more splitting than could be accounted for by the three orbital quantum numbers, and this was referred to as the *anomalous Zeeman effect.*

The explanation for the anomalous Zeeman effect eventually came with the introduction of the *spin quantum number, s,* relating to the spin of the electron on its own axis. The way that this spin quantum number came to light is a little complicated. Beginning in 1921, at the University of Frankfurt, Otto Stern and Walter Gerlach (1889–1979) carried out experiments in which they passed narrow beams of silver atoms through an inhomogeneous magnetic field. They observed that the beam was split into two beams, indicating that the silver atom had a magnetic moment, but at first there was uncertainty as to how this result should be explained; it was not known whether the moment was related to the spin of the nucleus or to the odd electron that a silver atom has. The explanation was finally provided in 1925 by Uhlenbeck and Goudsmit, but in the meantime the Austrian physicist Wolfgang Pauli (1900–1958) had suggested his *exclusion principle* to explain the structure of the periodic table. As originally expressed by Pauli in 1923, no more than two electrons in an atom can have the same set of values of the three orbital quantum numbers n, m and l. The idea of electron spin did not enter into the original formulation of the exclusion principle, but it was soon seen to be very relevant.

The suggestion of a fourth quantum number s, for the spin of the electron, was made in 1925 by the Dutch physicists George Eugene Uhlenbeck (1900–1988) and Samuel Abraham Goudsmit (1902–1978) who were then Ph.D.

the University of Leiden; they both obtained their degrees in 1927. Their proposal in their two-page paper was that there are two possible values for the spin quantum number. This idea at once provided a satisfactory explanation for the anomalous Zeeman effect, and also for the exclusion principle, which could now be changed from Pauli's original form to the statement that

in an atom no two electrons can have the same set of the four quantum numbers n, l, m and s.

The Bohr theory of the atom was certainly a remarkable achievement, since it provided a general interpretation of atomic structures and spectra, which previously had seemed quite incomprehensible. However, even with the improvements that were made to the theory, serious difficulties remained, as Bohr himself freely admitted. It did not, for example, seem possible to extend the theory in such a way as to give a complete treatment of atoms containing more than one electron, or even of the simplest of molecules. Aside from these difficulties, there was a feeling that it was arbitrary simply to add quantum restrictions to the old mechanics. What was needed instead was a new mechanics in which the quantum restrictions would emerge as a mathematical necessity.

10.2 Quantum mechanics

The first paper that outlined a new mechanics was by Max Born, with the title 'Zur Quantenmechanik'; it appeared in 1924, and this was the first time that the expression 'quantum mechanics' had been used. In 1925, especially in the second half of the year, there was a tremendous burst of activity in the field, particularly by Born, Heisenberg, Dirac and Schrödinger. It was Heisenberg's paper, completed in July of that year, that inspired those of Dirac and Schrödinger which appeared a few months later. Eventually all the formulations, although different in their approaches, turned out to be mathematically equivalent. The treatment of Schrödinger is usually preferred by chemists, as providing the greatest physical insight and as best lending itself to computations on complex systems.

Born and Heisenberg's matrix mechanics

In 1921 Max Born (1882–1970) became professor of physics at the University of Göttingen. His 1924 paper on quantum mechanics involved replacing certain differential operators in classical mechanics by other operators. In the spring of 1925 one of Born's research assistants was the 24-year-old Werner Heisenberg (1901–1976), who had obtained his doctorate two years earlier. Like some others who later had careers of great distinction he almost failed to obtain his degree, as he had neglected his laboratory work and was unable to answer any of the questions put to him by Wilhelm Wien (1864–1928), the professor of experimental physics. Wien wanted to fail him, but was persuaded by Sommerfeld to give him third-class standing in his oral examination. It is amusing to note that one of the topics that caused Heisenberg trouble was the resolving power of a microscope, and he still got this wrong when he proposed his uncertainty principle in 1927.

In the spring of 1925 Heisenberg was suffering from such a severe attack of hay fever that he obtained leave to stay for some time on Heligoland, a rocky island free from vegetation. While there he developed a modification of Born's treatment in which physical quantities were represented by time-dependent complex numbers which underwent a non-commutative multiplication rule (i.e., pq is not equal to qp). In July, 1925, Heisenberg sent his manuscript to Born who at once recognized its importance and submitted it for publication, in Heisenberg's name only, to the *Zeitschrift für Physik*; it appeared later in the year.

Born did not at once recognize the significance of Heisenberg's complex numbers, but 'after eight days of concentrated thinking and testing' realized that they had the properties of matrices. At the time matrices were known to some mathematicians, but were rarely used by physicists; Born had learnt about them in his student days 20 years earlier, and now recalled them. Heisenberg later admitted that he knew nothing of matrices; in developing his quantum mechanics he had reinvented matrix mechanics without realizing that he was doing so.

Born himself further developed Heisenberg's treatment in terms of matrices, partly in collaboration with Ernst Pasqual Jordan (1902–1980). This association came about in an interesting way. In the summer of 1925 Born travelled by train from Göttingen to Hannover and discussed with a colleague his struggles with matrix mechanics. Jordan, who was a student at Göttingen, overheard the conversation and told Born that he had made a special study of matrices, having studied under the mathematician Richard Courant. He offered his help to Born and this chance meeting led to important papers on quantum mechanics, the first joint paper being one in which Born and Jordan recast Heisenberg's treatment in the language of matrix mechanics.

Another contribution of Heisenberg, made shortly after his matrix mechanics, has proved of great importance. This was his *uncertainty principle*, or *principle of indeterminacy*, which may be expressed as

$$\Delta q \Delta p \approx \frac{h}{4\pi} \tag{10.20}$$

Here q is the position of a particle and p its momentum, and the equation indicates that the product of the uncertainties in the measurement of these quantities must be approximately the Planck constant h divided by 4π. Heisenberg arrived at the principle by carrying out a *gedanken* (thought) experiment in which a beam of light is used to determine the position and momentum of an electron. Because the light disturbs the electron there is a limitation on the accuracies of the measurements. The resolving power of a microscope is involved in this argument, and Heisenberg's treatment was incorrect; Born had to straighten the matter out for him.

Heisenberg won the 1932 Nobel Prize in physics for his quantum mechanics; it was actually awarded to him in 1933, the year in which Dirac and Schrödinger shared their Nobel Prize. It was always a matter of regret to Heisenberg that Born had no share in this honour. The award of the prize jointly to Born and Heisenberg would indeed have been more appropriate. Not only had Born initiated the idea of a quantum mechanics; he had shown the true significance of Heisenberg's obscure mathematics by relating it to matrix mechanics. A less generous man would have delayed the submission of Heisenberg's paper until a joint paper of broader significance could be prepared. Born did receive a Nobel Prize much later, in 1954, for his work on the quantum mechanics of collision processes.

Heisenberg, who had a wide popularity, remained in Germany and during World War II directed the German atomic energy programme, which proved unsuccessful. He by no means approved of Nazi policies and for this he suffered from the vicious attacks of Stark and Lenard, who were powerful and fanatical Nazis. Born had to leave Germany, and in 1936 became professor at the University of Edinburgh, returning to Germany after his retirement in 1954.

Dirac's quantum mechanics

Shortly after the appearance of Heisenberg's first paper on quantum mechanics, another important contribution was made by Paul Adrien Maurice Dirac (1902–1984). Dirac was trained as an electrical engineer at the University of Bristol, and being unable to obtain employment secured in 1923 a research scholarship to work for his Ph.D. degree at Cambridge under R. H. Fowler. In July, 1925, Heisenberg lectured at Cambridge and Dirac attended the lecture but later said that he had paid little attention to it. In September he saw the proofs of Heisenberg's paper and initially saw nothing useful in it. He soon changed his mind and quickly produced an alternative formulation, his first paper on quantum mechanics appearing late in 1925. The essential feature of his quantum mechanics, which is equivalent to Heisenberg's but more comprehensible, is that it is based on Hamilton's system of classical mechanics, transforming it by a number of postulates into a complete system of quantum mechanics.

Dirac's first paper on quantum mechanics was followed by many others of great importance. In 1926 he applied his methods to the hydrogen atom, and in papers that appeared in 1928 he combined his system of quantum mechanics with the theory of relativity. The resulting theory was an extremely powerful one which had far-reaching consequences. For example, Dirac was able to deduce that there are two quantum numbers for electron spin, a conclusion that had already been reached on the basis of experiment (Section 10.1). Dirac also predicted from his theory the existence of an elementary particle having the mass of the electron but a positive charge. This particle, the *positron*, was discovered in 1932 by the American physicist Carl David Anderson (1905–1991).

Dirac wrote an important book, *The Principles of Quantum Mechanics*, which first appeared in 1930 and exerted a wide influence. He shared the 1933 Nobel Prize in physics with Schrödinger.

Victor de Broglie: the wave properties of particles

Before Heisenberg and Dirac made their pioneering contributions, the idea had been put forward that, just as radiation has particle properties, particles might have wave properties. This idea was first clearly expressed by Louis Victor, Prince* de Broglie (1892–1987), in a series of papers that appeared in 1923 and in his doctoral thesis presented late in 1924 to the Université de Paris; this thesis was also published in 1925 in the *Annales de physique*. Louis de Broglie had first intended to become a civil servant and in 1910 obtained a licentiate in history. Later, partly under the influence of his elder brother Maurice, Duc de Broglie (1875–1960), who was a distinguished experimental physicist, Louis became interested in physics and philosophy, obtaining a licentiate in the physical sciences at the Sorbonne in 1913. He then worked towards his doctorate, specializing in fundamental problems of space, time and atomic structure. He was particularly impressed by the work of the American physicist Arthur Holly Compton (1892–1962) who, by observing the scattering of X-rays, obtained in 1923 further striking confirmation of Einstein's suggestion that radiation can exhibit particle properties.

In his papers and his doctoral thesis de Broglie suggested the converse hypothesis, that particles such as electrons can have wave properties. To obtain the wavelength of the wave associated with a particle he reasoned as follows. For a photon, according to Einstein's special theory of relativity, the energy E is the product of the momentum p and the speed of light c, which is its wavelength multiplied by its frequency ν:

$$E = pc = p\lambda\nu. \tag{10.21}$$

Since $E = h\nu$ it follows that

$$\lambda = \frac{h}{p} = \frac{h}{mc}. \tag{10.22}$$

For a beam of particles of mass m and velocity u de Broglie suggested that the corresponding expression would apply:

$$\lambda = \frac{h}{mu}. \tag{10.23}$$

---

* 'Prince' was a courtesy title conferred upon a de Broglie ancestor by Francis I of the Holy Roman Empire, and to which all descendants were entitled. After his elder brother Maurice died in 1960, Victor inherited the French title of Duc, which took precedence over the title of Prince.

He also approached the problem in a way that related to Bohr's atomic theory. Equation 10.16 rearranges to

$$2\pi r = n\frac{h}{mu},\qquad(10.24)$$

and $2\pi r$ is the circumference of the orbit. A wave of wavelength h/mu would thus fit into the orbit if $n = 1$; with $n = 2$, two waves would fit into the orbit, and so on. The proposal is thus mathematically equivalent to Bohr's quantum condition, but the approach to the problem is completely different.

If it had not been for the intervention of Einstein, de Broglie's doctoral thesis might have suffered a similar fate to that of Arrhenius 40 years earlier (Section 7.3). One of the examiners was Paul Langevin (1872–1946) who thought the work to be 'far-fetched'; however, at Langevin's suggestion de Broglie sent a copy to Einstein, who returned an enthusiastic endorsement which included the phrase 'He has lifted a corner of the great veil.' In spite of this all four examiners, another of whom was Perrin, remained sceptical of the work but awarded the degree. Van de Graaf (1901–1969), the inventor of the electrostatic generator that bears his name, was present as a student at the defence of the thesis, and later remarked 'Never had so much gone over the heads of so many.'

Other physicists, including Debye, Born and Heisenberg, were also initially sceptical of de Broglie's theory, and he did not receive his Nobel Prize until 1929, by which time the theory had been confirmed experimentally. The first experiments to support the theory involved the diffraction of electrons. According to the de Broglie equation 10.23, electrons accelerated by a potential of about 100 V have wavelengths similar to the interatomic spacing in crystals, and are therefore suitable for diffraction experiments. Clinton Joseph Davisson (1881–1958) had since 1919 been studying electron scattering in the laboratories of the American Telephone and Telegraph Company, and of the Weston Electric Company in New York. In January, 1927, with his colleague Lester Halbert Germer, he succeeded in demonstrating the diffraction of electrons by a single crystal of nickel. In May of the same year, at the University of Aberdeen, George Paget Thomson (1892–1975) and his research student Andrew Reid observed the diffraction of electrons by thin films. Ten years later Davisson and Thomson shared the Nobel Prize in physics for this work. It has been commented that J. J. Thomson was awarded the 1906 Nobel Prize for showing that the electron is a particle, while his son obtained a 1937 Nobel Prize for showing that it is a wave.

De Broglie's idea played no part in the first quantum-mechanical treatments of Born, Heisenberg and Dirac. It did, however, strongly influence the thinking of Schrödinger.

Schrödinger's wave mechanics

The quantum mechanics of Erwin Schrödinger (1887–1961) has been the most popular at least with chemists, since it provides a more easily visualized

Erwin Schrödinger

(1887–1961)

Schrödinger was born in Vienna, and received much of his early education from his father, a businessman of broad interests. From 1898 he attended the Akademisches Gymnasium in Vienna, and in 1906 went to the University of Vienna. He had hoped to be taught by Boltzmann, who committed suicide in that year, and instead he was greatly influenced by Boltzmann's successor Friedrich Hasenöhrl (1874–1915), a brilliant teacher whose career was cut short on the battlefield in World War I. Schrödinger received his doctorate of philosophy degree in 1910 for a thesis on the conduction of electricity on the surface of insulators.

After serving in World War I as an artillery officer Schrödinger became associate professor at the University of Stuttgart in 1921, and in the following year was appointed professor of theoretical physics at the University of Zürich. At first he worked mainly on problems related to the statistical theory of heat. On hearing in 1925 of Louis de Broglie's ideas on the wave nature of electrons he began his greatest work, his wave mechanics (Section 10.2), six papers on the subject appearing in 1926. In the following year he succeeded Max Planck as professor of theoretical physics at the University of Berlin.

Disapproving of the Nazi regime he left Germany in 1933 and spent some time at Oxford where F. A. Lindemann had been able to arrange temporary appointments for him as well as for several Jewish physicists. Schrödinger was elected a Fellow of Magdalen College, Oxford, and shortly after the election he heard that he was to share the 1933 Nobel Prize with Dirac. In 1936 he became professor at the University of Graz, but being out of favour with the Nazis because of his earlier departure from Berlin he was dismissed in 1938. In the following year he became Director of the Institute for Advanced Studies in Dublin, resigning the directorship in 1945 but retaining his professorship. He returned to the University of Vienna in 1956.

Although Schrödinger made scientific contributions of great importance, his judgement tended to be clouded by his highly emotional temperament. This often led to rather bizarre personal behaviour. It was a quarrel about the way in which his office was cleaned that led him to resign the Directorship of the Institute for Advanced Studies in Dublin. He tended to dress unconventionally, and as a result on several occasions had difficulty gaining admission to scientific meetings or to his own university lectures.

His amorous activities were equally unusual, and have been extensively documented in Walter Moore's excellent biography. His unpreposessing appearance, with his thick spectacles, hardly corresponds to the popular idea of a Lothario or Casanova, but such he was. In 1920 he married Annemarie Berthel, and although their relationship was punctuated by many stormy episodes they remained together until his death. Schrödinger had no children by his wife, but he had at least three illegitimate daughters and there would have been more children had it not been for abortions. He never had the son he longed for. Until he was in his sixties he was continuously involved in sometimes simultaneous liaisons, conducted openly and with his wife's knowledge; indeed she sometimes helped to arrange them. It does not appear that he ever succeeded in having two of his mistresses pregnant at the same time, but that was not for want of trying.

His wave mechanics was begun during one of these amorous adventures in late 1925, when he stayed at a resort with a mistress while his wife remained in Zürich, and it was perhaps not only the theory that was conceived at that time. The woman involved has not been identified, but it would seem that she deserved at least a Nobel Prize nomination as Best Supporting Actress.

When he was in Oxford, Schrödinger manoeuvered Lindemann into appointing a German colleague, and Lindemann was later infuriated to discover that the motive had been to form a liaison with the colleague's wife. This same woman, along with her daughter by Schrödinger, was later a component of a *ménage-à-trois* in Dublin, where Schrödinger made it clear that she and his wife were to be treated on the same footing. The child was brought up more by his wife than by her natural mother, and the two women shared the domestic duties.

In his early sixties Schrödinger began to suffer from the effects of excessive indulgence in tobacco, alcohol and amorous activity. His health and strength remained poor for several years, and when he died at the age of 73 the death certificate indicated no specific ailment but general aging of the heart and arteries; in layman's terms, he died of premature old age. Although his connection with the Roman Catholic Church had for many years been largely confined to attending the weddings and funerals of friends, to failure to practice birth control, and to membership of the Pontifical Academy of Sciences, he was buried in a Catholic cemetery at Alpbach in the Tirol.

References: A. Hermann, DSB, 12, 217–223 (1975); Heitler (1961); Moore (1989).

representation of atomic structure, in contrast to the rather formal approaches of Heisenberg and Dirac. Schrödinger, an Austrian by birth, succeeded Debye in 1921 as professor at the University of Zürich, and it was there that he did his original work in wave mechanics.

Schrödinger's wave mechanics evolved directly from de Broglie's ideas, which he had first thought to be 'rubbish' until he was persuaded otherwise. In the words of the physicist Hermann Weyl (1885–1955), Schrödinger obtained his inspiration for wave mechanics while engaged in a 'late erotic outburst in his life'. The work was done at a holiday resort towards the end of 1925, and his paper appeared in three parts early in 1926, with three additional papers later in that year. He did not derive his wave equation, but proceeded by analogy with the equations used in Maxwell's electromagnetic theory for wave motion in ordinary radiation. His wave equation is a differential equation involving a wave function or *eigenfuction* ψ, and an energy E known as an *eigenvalue*. Quantum restrictions are not introduced arbitrarily, but appear as a direct consequence of the wave equation, for which no solution is possible unless the energy has one of a number of permitted values.

At first Heisenberg and Dirac were quite critical of Schrödinger's wave mechanics. Born approved of it from the start, and in 1926 provided a physical interpretation of the wavefunction ψ. For a problem in atomic or molecular structure, ψ according to Born is related to the probability that an electron is present in a particular small region of space. If the function ψ is real (i.e., does not involve the square root of minus one), it is ψ^2 that represents the probability. If it is complex (does involve $\sqrt{-1}$) one forms the complex conjugate ψ^*, by changing the signs of terms involving $\sqrt{-1}$, and then it is $\psi\psi^*$ that represents the probability. When Born submitted his paper he thought that the probability was given by ψ rather than by ψ^2; he made the change when he corrected the proofs, so that this important relationship appears in a footnote.

The misgivings of Heisenberg and Dirac about Schrödinger's theory evaporated when later in 1926 Schrödinger proved that his and Heisenberg's formulations were mathematically equivalent. Starting with Heisenberg's equations, which involve matrices representing physical properties, Schrödinger showed that each physical property can be replaced by an appropriate operator, and that his wave equation was then obtained. Modern quantum-mechanical calculations are usually based on this procedure.

The Copenhagen interpretation

Heisenberg's uncertainty principle, and Born's interpretation of $\psi\psi^*$ as representing a probability, are important components of what came to be called the Copenhagen interpretation of quantum mechanics, since Niels Bohr had much to do with formulating this point of view. The Copenhagen interpretation implies a lack of complete determinism, in that future events do not follow inevitably from past conditions, pure chance playing some role. Most physicists have accepted this interpretation, but Einstein and Schrödinger took strong exception to it.

Einstein's objection is summarized in his often-quoted statement that 'God does not play dice.' In a number of forceful but always friendly arguments with Bohr, Einstein tried to devise ways of circumventing the uncertainty principle, but Bohr was always able to show that he was in error—sometimes by invoking the theory of relativity!

Schrödinger was an intensely emotional person, and he found the Copenhagen interpretation deeply distressing. With regard to Born's probability explanation, Schrödinger would say: 'I can't believe that an electron hops about like a flea.' In one exchange with Bohr he said that since people were giving these interpretations to his wave mechanics 'I regret having been involved in this thing.' To Schrödinger an electron has wave properties and is not to be regarded as a particle darting about. In his view electronic orbitals are to be considered in terms of wave properties, not as the movement of particles.

10.3 Chemical bonding

The way in which atoms join together to form molecules is one of the most important problems with which chemists are concerned. Modern theories of chemical bonding are based on quantum mechanics, but many important ideas that were developed earlier played a significant role in the application of quantum mechanics to chemical bonding. The subject is a vast one, and several excellent historical accounts have been published; only a very brief outline can be given here. Table 10.3 lists some of the more important contributions that have been made, and includes several monographs that were especially helpful in leading chemists to an understanding of this important matter. Modern textbooks are often historically unreliable; chemistry texts often give the credit for the early contributions entirely to G. N. Lewis, while physics books almost invariably give it to Niels Bohr, with no mention of Lewis. Both these men made valuable contributions, but so did a number of others.

Early electronic theories of valency

Since about the middle of the nineteenth century chemists have found it useful to consider chemical bonding in terms of the concept of *valency* or *valence*, words that derive from the Latin *valentia*, which means strength or capacity. The valency of an atom is the number of bonds that it can form with other atoms. The idea of valency appears in a rudimentary form in the work of Berzelius, but it was Edward (later Sir Edward) Frankland (1825–1899) who in 1852 first presented the concept of valency as we understand it today. His contribution was important in leading to the periodic table proposed by Dmitri Ivanovich Mendeleev (1834–1907), who related the valency of an element to its position in his table.

Until the electron had been identified by J. J. Thomson in 1897 (Section 6.7), no useful interpretation of valency could be given. The first electronic theory of valency was put forward in 1904 by Richard Wilhelm Heinrich Abegg

TABLE 10.3 Key publications dealing with electronic theories of valency

Year	Author	Contribution
1904	Abegg	'Rule of eight'; electron loss and gain; rule of normal valency and contra-valency
1904	Thomson	Plum-pudding model of the atom; outer-shell configurations; rare-gas configurations
1907	Thomson	*The Corpuscular Theory of Matter*
1914	Thomson	Two-electron bond
1916	Kossel	Theory of ionic bonds
1916	Lewis	Bonding by electron pairs; the coordinate link
1918	Vegard	Electronic configurations deduced from X-ray spectra
1919–1921	Langmuir	Elaboration of Lewis's ideas
1921	Bury	Electronic configurations deduced from chemical evidence
1923	Lewis	*Valence and the Structure of Atoms and Molecules*
1927	Sidgwick	*The Electronic Theory of Valency*
1927	Heitler and London	Quantum-mechanical theory of the covalent bond
1928–1939	Pauling	Extension of Heitler–London theory to other types of bonds
1928	Hund	Molecular-orbital theory
1928	Mulliken	Molecular-orbital theory
1933	Sidgwick	*The Covalent Link in Chemistry*
1935	Pauling and Wilson	*Introduction to Quantum Mechanics, with Applications to Chemistry*
1939	Pauling	*The Nature of the Chemical Bond*

(1869–1910), who at the time was professor of chemistry at the University of Breslau, having previously been an assistant to Nernst. He suggested that the electron configurations in the recently discovered noble gases must be particularly stable, and that every element has both a positive and a negative valency or contra-valency, the sum of the two always being eight. He noted that the positive valency of an element corresponds to its group in the periodic table. A significant comment on Abegg's ideas was made later in the same year by Paul Drude, whose important work with Nernst on electrostriction was noted in Section 7.4. Drude suggested that

Abegg's positive valency number v, whether it is a normal or a contra-valency, signifies the number of loosely attached negative electrons in the atom; his negative valency number v' means that the atom has the power of removing v' negative electrons from other atoms, or at least of attaching them more firmly to itself.

These ideas of Abegg and Drude show remarkable foresight, in that they anticipated theories that were not formulated for some years. Both men might well have made further important advances had it not been for their untimely deaths; Drude, who had become editor of the *Annalen der Physik*, died suddenly and unexpectedly in 1906 in his forty-third year, and Abegg was killed in 1910 in a balloon accident.

During the first decade of the twentieth century J. J. Thomson was developing his famous 'plum-pudding' theory of the atom, in which he envisioned the electrons as embedded in a positively charged sphere. In the course of his work he arrived at a number of concepts which although not now used in their original form nevertheless made an important contribution to modern ideas about valency. He recognized that the electrons are arranged in shells surrounding the centre of the positive sphere, and that only the outermost electrons are concerned in chemical bonding. He concluded from the chemical periodicity that there must be a periodic repetition of the structures of the outermost electronic shells, and he realized that in the noble gases the outer shell are complete. Thomson had a greater appreciation of chemistry than most physicists, and he discussed at some length the chemical implications of his proposals.

In 1916 two important papers were published on the electronic theory of valency. One of them, 133 pages in length, was by the physicist Walther Ludwig Kossel (1888–1956) of the University of Munich. In this paper Kossel quoted Abegg's paper many times, and developed a theory of bonding in which there are complete transfers of electrons from one atom to another, the two being held together by electrostatic attraction. It was therefore a theory of polar bonds, although Kossel attempted to apply it to a large number of bonds which we now know not to be polar, realizing himself that some modification to his theory was required. The most important feature of his theory was his emphasis on the special significance of the electronic structures that are found in the noble gases, resulting in the tendency for other atoms to attain those particular structures.

The second 1916 paper on the subject was much shorter, 22 pages in length, but of more general significance. It was by G. N. Lewis, whose important work in thermodynamics was discussed in Section 4.4. In 1916 Lewis was at the University of California at Berkeley, but his theory had been first developed in a more rudimentary form, and presented to students, in 1902 when he was an instructor at Harvard. The important idea that Lewis introduced is that bonding can take place as the result of the sharing of electrons between two atoms. From the work of Abegg and Thomson he appreciated the importance of

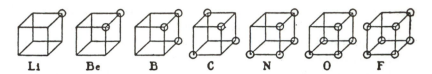

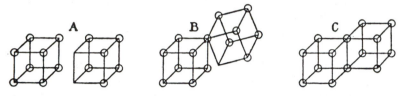

FIGURE 10.2 Diagrams given by G. N. Lewis (1923) to represent the electronic arrangements in some atoms and simple molecules.

groups of eight electrons ('octets'), which for convenience he represented as being stationed at the corners of cubes (Figure 10.2). Bonding could occur by an overlap of the edges of two cubes. He recognized that in some molecules both of the bonding electrons may originally have come from one of the atoms, an arrangement referred to later as a *coordinate link* or a *semi-polar bond*. With suitable modifications Lewis's model could also explain the tetrahedral arrangement of bonds emanating from a carbon atom.

In 1919 these ideas of Lewis were taken up with great enthusiasm by Irving Langmuir, who already had a wide reputation, having done much of his important work on solid and liquid surfaces (Chapter 9) that was later to bring him his Nobel Prize. Langmuir extended and to some extent modified Lewis's ideas, applying them to a wider range of compounds. He also suggested a number of the terms that are used today, such as *covalence* and *electrovalence*. He published a dozen papers on the subject between 1919 and 1921, and he also presented the theory in a number of lectures he gave in the United States and abroad. Unlike Lewis, who was somewhat shy and diffident and not a good lecturer, Langmuir was always eloquent and compelling—to the extent that people sometimes complained that they found themselves convinced by things that they knew to be wrong! As a result, although Langmuir always gave due credit to Lewis, the theory began to be known as the Lewis–Langmuir theory, which did not entirely please Lewis. In Britain, in fact, the theory was sometimes known as the Langmuir theory, which produced the following reaction from Lewis in a letter to W. A. Noyes in 1926:

. . . to persist, as they do in England, in speaking of the Langmuir theory of valence is inexcusable.

Lewis was also resentful of the indifference of most of his immediate colleagues, particularly when he was developing his theory at Harvard from 1902 onwards, and in 1929 he refused an honorary degree from Harvard.

In addition, Lewis was somewhat disappointed by the attitude of the physicists, who regarded his theory as unreasonably crude, and tended to favour Kossel's theory even though it was clearly inapplicable to covalent bonds. The physicists particularly objected to the fact that Lewis seemed to be regarding the electrons as fixed in certain positions, and that the electron-pair bond that he was proposing did not seem capable of binding the atoms together. In his turn Lewis, like many other chemists, was critical of the physicists' idea of completely mobile electrons, which seemed incapable of explaining the undoubted fact, established by strong chemical evidence, that molecules have definite shapes. Over the years both Lewis's and the physicists' concepts became modified and by 1923, when Lewis's book *Valence and the Structure of Atoms and Molecules* appeared, his ideas were entirely consistent with the ideas that the physicists themselves were then putting forward.

The great importance of Lewis's theory is that it provided chemists with a valuable way of visualizing the electronic structures of atoms and molecules, and for practical purposes his ideas are still used today. Physicists have tended to ignore his contribution, and modern historical accounts by physicists often fail to mention it although they usually mention Kossel's work. A few physicists have appreciated what Lewis did, as illustrated by the following passage from *Wave Mechanics* (1945) by Walter Heitler, whose important quantum-mechanical work on the covalent bond is considered later:

Long before wave mechanics was known Lewis put forward a semi-empirical theory according to which the covalent bond between atoms was effected by the formation of pairs of electrons shared by each pair of atoms. We see now that wave mechanics affords a full justification of this picture, and, moreover, gives a precise meaning to these electron pairs: they are pairs of electrons with antiparallel spin.

To many it has seemed surprising that Lewis did not receive a Nobel Prize in view of his great contributions to the theory of bonding as well as to thermodynamics. The fact that his bonding theory in its original form seemed naive to the physicists may have been the determining factor. In the 1920s and 1930s the nominations of Niels Bohr were given much consideration by the Nobel Prize committee, and many of his nominees received the award. He nominated a number of physicists whose work was of special interest to chemists, and Langmuir specifically for his work on surfaces, but he never nominated Lewis.

An important aspect of valency theory relates to the arrangement of the electrons in the various atoms, and this was actively investigated during the second decade of the twentieth century. In 1913 a contribution of fundamental importance was made by Henry Gwyn Jeffreys Moseley (1887–1915), who was working in Rutherford's laboratories at the University of Manchester. Moseley measured the characteristic frequencies of the X-ray lines of a number of elements, discovering a relationship from which he was able to deduce the charge on the atomic nucleus and hence the number of electrons in the neutral atom. This number, the *atomic number*, is more or less the order of the elements when they are listed in order of increasing atomic weight. Moseley's brilliant career was cut short when he was killed in action at the battle of Gallipoli.

Further investigations of X-ray spectra were carried out by L. Vegard at the Physical Institute of Christiania (Oslo) in Norway. In 1918 he published an important paper in which he deduced from his results the electronic configurations of the atoms up to uranium. In the following year Langmuir, in the course of his extensive investigations and expositions of valence theory, deduced the electronic configurations of the elements from the chemical evidence. His conclusions were not, however, satisfactory, and an alternative proposal was put forward in 1921 by Charles Rugeley Bury (1890–1968), who was a lecturer at the University College of Wales at Aberystwyth. The scheme that he described succinctly in a mere seven pages is essentially the scheme to be found in modern introductory textbooks of chemistry and physics. He deduced from the chemical evidence that the electrons are arranged in successive layers containing 2, 8, 18 and 32 electrons. He gave a clear discussion of the electronic arrangements in the actinides and lanthanides, and even made some predictions (inevitably not quite correct) for the transuranic elements.

Bury's scheme was reproduced in *The Electronic Theory of Valency* by Nevil Vincent Sidgwick (1873–1952); this was an important book that first appeared in 1927 and which interpreted the chemical behaviour of the elements in terms of their electronic configurations. Sidgwick acknowledged the important contribution of Bury, but almost all subsequent accounts have failed to do so and Bury's name is now almost entirely forgotten. Many accounts of the electronic configurations give the credit to Bohr. In 1921 Bohr did write two letters to *Nature* on the electronic configurations, but he only considered the noble gases. In his Nobel Prize address, a translation of which was published in *Nature* in 1923, Bohr did mention Bury and included a scheme that is essentially Bury's, without making it clear that this is the case.

A striking example of the failure to acknowledge Bury's work is to be found in connection with the discovery of the element hafnium, of atomic number 72. When Bury wrote his paper hafnium had not been discovered, but he referred to the missing element 72 and predicted that it would not be a rare earth but would resemble zirconium. Bohr, knowing of this prediction, suggested to his assistants György Hevesy (1885–1966) and Dirk Coster (1889–1950) that they should look for the missing element in zirconium ores. They discovered it in 1922, in time for Bohr to announce its discovery in his Nobel Prize address in December of that year. Hevesy and Coster confirmed that hafnium is not a rare earth but an analogue of zirconium, but in their report of the discovery Bury was not mentioned.

Solving the Schrödinger equation

The new quantum mechanics has introduced important changes into the way one thinks about atoms and molecules. In Bohr's theory the electrons were regarded as moving in definite orbits about the atomic nucleus. According to quantum mechanics, however, and particularly in the light of Heisenberg's uncertainty principle, one should not think of orbits. There is an uncertainty in our knowledge of the path traversed by an electron around a nucleus, so that one cannot precisely specify the orbit. Another way of looking at the situation is to note that if a precise orbit were specified, the wavefunction ψ, which relates to the probability that an electron is in a particular region of space, would be zero at any position away from the orbit. The solutions of the Schrödinger equation show instead that ψ varies smoothly, having maximum values at certain positions but diminishing and approaching zero at large distances from the nucleus.

In quantum mechanics one replaces the word 'orbit' with *orbital*, a word introduced by R. S. Mulliken who defined it in a simple way by saying that an orbital is as much like an orbit as quantum mechanics permits. More precisely, the word 'orbital' refers to a wavefunction that relates to a single electron associated with an atom or molecule.

Most modern work in quantum chemistry is based on Schrödinger's wave mechanics, since it is more easily visualized and handled than the treatments of Heisenberg and Dirac. It is a comparatively easy matter to set up the appropriate Schrödinger equation for any atomic or molecular system. One first writes down the classical-mechanical equation in a particular form (usually that of W. Rowan Hamilton), and then uses certain postulates suggested by Schrödinger in order to transform the classical equation into the wave-mechanical one.

The next step is to obtain the mathematical solution to the wave equation, and it is here that difficulties arise. In the 1930s, soon after quantum mechanics had first been formulated, some scientists such as Dirac believed that all physical problems were essentially solved, and it was even sometimes suggested that there would be no further need for experimental work since everything could now be calculated from first principles. It was soon realized that this is by no means the case. For any except the simplest systems it is impossible to obtain analytical solutions to the Schrödinger equation, and it is then necessary to employ approximation techniques. These are usually extremely laborious and time-consuming even with the aid of the best modern computers, and sometimes the results are unreliable.

The only atoms that can be treated exactly are hydrogen-like atoms, which have a single electron. All other atoms, and all molecules, must be treated by the approximation methods, of which there are several. One procedure is to devise a plausible wavefunction and then to calculate the energy that corresponds to it. This is a relatively simple procedure, but since the wavefunction is not the correct one the energy calculated is in error.

Help is provided by a theorem derived in 1909 by Walter Ritz, who during his short life—he died at 34 from tuberculosis—accomplished much, mainly in the field of spectroscopy (see Section 6.5). As applied to the quantum-mechanical calculation of an energy, Ritz's theorem leads to the conclusion that the energy calculated from any wavefunction must be equal to or greater than the true energy. This theorem led Ritz to his so-called *variation principle*, the application of which to quantum mechanics is as follows. If one chooses a trial wavefunction having variable parameters one can obtain an expression for the energy as a function of these parameters. One can then vary the energy until the minimum energy is obtained, and this must be the best value that can be obtained with this particular type of wavefunction.

For example, for a diatomic molecule AB one could choose the wavefunction

$$\psi = \psi_A + \lambda\psi_B, \tag{10.25}$$

where λ is a variable parameter and ψ_A and ψ_B are reliable wavefunctions for the individual atoms A and B. The energy would be calculated as a function of λ, and by varying λ to obtain the lowest energy one knows that one has the best value. Many modern calculations of energy levels in molecules are based on this principle. To obtain satisfactory energies it is usually necessary to start with wavefunctions having a considerable number of variable parameters, and the calculations can then be very time-consuming.

One useful approximate method for atoms is the *central-field*, or *self-consistent field* method devised and employed by the British physicist Douglas Rayner Hartree (1897–1958). In this method the outermost electron in the atom is regarded as being in the electric field created by the nucleus and the remaining electrons. The method was improved by the Russian physicist Vladimir Alexandrovitch Fock (b. 1898), and the modified version is usually known as the *Hartree–Fock method*.

The first application of quantum mechanics to an atomic problem was made in 1926 by Pauli, who solved Heisenberg's matrix equations for the hydrogen atom, and showed that the theory explained the features of the hydrogen spectrum. He made no numerical calculations to allow quantitative comparison with experiment, and did not include electron spin, to which he made a reference in the final section of the paper. Shortly afterwards, in the first of his three 1926 papers on quantum mechanics, Schrödinger also gave a solution for the hydrogen atom. An interesting feature of his solution was that the distance from the nucleus corresponding to the maximum value of ψ^2 is the same as the radius of the orbit that had been deduced by Bohr.

The simplest molecular system is the hydrogen molecule ion, H_2^+, which consists of two protons and a single electron. An approximate solution for this species was first given in 1927 by the Danish physicist Öyvind Burrau, who used elliptical coordinates and obtained reliable energies by numerical methods.

The Heitler–London treatment: valence-bond theory

The hydrogen molecule, H_2, is of special importance since the bond in it is the prototype of all covalent bonds. In 1916 G. N. Lewis made a great advance with his suggestion that a single covalent bond is due to a pair of electrons held jointly by two atomic nuclei. This idea forms the basis of all treatments of the covalent bond.

The first satisfactory quantum-mechanical treatment of the hydrogen molecule was given in 1927 in a paper written jointly by the German physicist Walter Heinrich Heitler (1904–1981) and the German physicist Fritz Wolfgang London (1900–1964). Heitler had been a student at the University of Munich where he met London who was doing his research under the direction of Sommerfeld. By 1927 both had transferred to the University of Zürich and were working with Schrödinger. The essential feature of their treatment of the hydrogen molecule was that a wavefunction for the pair of electrons was constructed from the wavefunctions of the separated atoms, which had been obtained exactly by Schrödinger in 1926. Suppose that $\psi_A^{(1)}$ is the wavefunction for an electron designated 1 on a hydrogen nucleus designated A, and that $\psi_B^{(2)}$ is the corresponding wavefunction for electron 2 associated with nucleus B. For the two electrons in the molecule Heitler and London first considered the wave-function $\psi_A^{(1)} \psi_B^{(2)}$, but realized that this was unsatisfactory since it implied that electron 1 was more related to nucleus A than to B, and 2 more related to B than to A. They avoided this difficulty by interchanging the electrons so as to obtain $\psi_A^{(2)} \psi_B^{(1)}$, and then took the sum or difference of the two functions:

$$\psi = \psi_A^{(1)} \psi_B^{(2)} \pm \psi_A^{(2)} \psi_B^{(1)}. \qquad (10.26)$$

From this function they then obtained the energy at various separations between the nuclei. They found that this treatment accounted for about two-thirds of the actual binding between the atoms. Considering the simplicity of the treatment this is quite good agreement, and improvement can be achieved by introducing additional terms and applying the variation principle. The Heitler–London treatment of the hydrogen molecule, and its extensions to other molecules, are often called *valence-bond theory*.

An important conclusion of the Heitler–London treatment is that the binding energy can be regarded as consisting of two types of energy. One is the *coulombic energy*, which roughly corresponds to the energy that would be obtained in a classical calculation. The other, the *exchange energy*, has no counterpart in classical mechanics, and arises in the quantum-mechanical treatment as a result of the fact that the electrons, being indistinguishable, can be interchanged. The exchange energy is much larger than the coulombic energy, so that a purely classical calculation does not lead to anything like the correct binding energy. The results of the Heitler–London treatment of the hydrogen atom are shown in Figure 10.3. The binding due to the coulombic energy alone is seen to be much less than the true binding energy. Even when the exchange energy is included the calculated energies are higher than the true energies, as is required by the variation principle.

In 1928 London extended the Heitler–London treatment to the H_3 species, an unstable intermediate in the reaction between a hydrogen atom and a hydrogen molecule. Although London's formula involved serious approximations, it

Walter Heinrich Heitler

(1904–1981)

Heitler was born in Karlsruhe, Germany, and became interested in science, particularly chemistry, at an early age. After school he went first to the Technische Hochschule in Karlsruhe, and later spent a year at the University of Berlin where he attended lectures by Einstein, Planck, von Laue, Nernst and Haber. He then moved to the University of Munich, where the professor of physics was Sommerfeld; however he carried out his research under K. F. Herzfeld on concentrated solutions. After obtaining his Ph.D. degree he worked for a period with Bjerrum in Copenhagen on ions in solution.

Being particularly interested in the new quantum mechanics Heitler then went to the University of Zürich, where Schrödinger was professor. Fritz London was also at Zürich, and the two developed their theory of the hydrogen molecule (Section 10.3), their paper on the subject appearing in 1927. Max Born then offered Heitler an assistantship at Göttingen, which he had no hesitation in accepting. For the next few years his main work was on group theory, and on applications of quantum mechanics to chemical problems.

In 1933 he felt it wise to leave Germany on account of his Jewish ancestry, and his first appointment was at the University of Bristol, where N. F. Mott was the professor of theoretical physics. There he began to work on quantum electrodynamics and radiation theory. His book *The Quantum Theory of Radiation* was first published in 1936; it had two later editions and was translated into several languages.

proved useful in early treatments of chemical reactions in terms of potential-energy surfaces (Section 8.3).

Both Heitler and London had distinguished scientific careers, and both had to leave their native countries under pressure from the Nazis. Heitler went first to the University of Bristol and then to the Dublin Institute for Advanced Studies of which Schrödinger was the Director; in 1949 he went to the University of Zürich as Director of the Institute of Theoretical Physics. London went to Duke University in Durham, North Carolina, where he remained to the end of his life.

Following the formulation of the Heitler–London theory in 1927, much effort went into extending it to other types of bonds. No contribution in this area of research was greater than that of Linus Carl Pauling (1901–1994). After obtaining his Ph.D. degree in 1925 at the California Institute of Technology, Pauling went in the spring of 1926 to work with Sommerfeld at the University

In 1940, after the fall of France, Heitler was interned for three months, as was the case with all 'enemy aliens' in Britain. In 1941 he went as an assistant professor to the Dublin Institute of Advanced Studies, of which Schrödinger had been made the first Director in 1939; he was promoted to full professor in 1943, and he succeeded Schrödinger as Director in 1946. He continued his work on radiation and also investigated nuclear forces. At Dublin he was noted for the clarity and high quality of his lectures. For the benefit of students of chemistry he gave an introductory course on wave mechanics and its application to the chemical bond. This course formed the basis of a small book, *Elementary Wave Mechanics*, which was published in 1945 and gave an unusually clear account of the subject.

In 1949 he became Director of the Institute for Theoretical Studies at the University of Zürich, where he worked particularly on philosophical problems. In the 1960s and 1970s he published a number of books and articles on the relationship of science to humanism and religion, and these have been of great interest not only to philosophers but to theologians and the general public. During his later years he was a member of the Swiss Reformed Church.

Reference: Mott (1982).

of Munich, where he became friendly with Heitler and London who were soon to develop their theory. Before going to Europe Pauling had heard a lecture by Max Born on the matrix mechanics that he and Heisenberg had developed, and on arriving in Europe Pauling attended Sommerfeld's first lectures on quantum mechanics. Pauling quickly became familar with the new quantum-mechanical theories, and he recognized that the Schrödinger formulation was most suitable for chemists. He had a particular interest in the X-ray analysis of the structure of crystals, and he prepared an important paper in which he estimated, from quantum mechanics, the sizes of ions and demonstrated that his estimates were consistent with the interionic distances obtained from the X-ray analysis of crystal structures; this paper appeared in the *Proceedings of the Royal Society* in 1927. He later published a number of other papers on ionic crystals and related matters.

Pauling was particularly impressed by the Heitler–London treatment and he

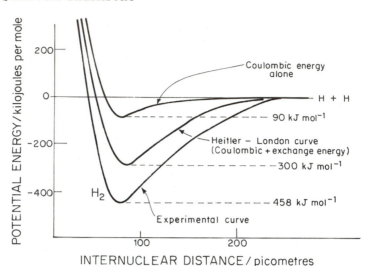

FIGURE 10.3   Results obtained by the Heitler–London treatment of the hydrogen molecule.

at once began to extend it to bonds involving atoms more complex than hydrogen atoms. In 1927 he took up an appointment at the California Institute of Technology where he has spent his entire career. During the 1930s he published a series of papers of great importance having the general title 'The nature of the chemical bond'. In these papers he dealt with many molecules that were too complex for a rigorous quantum-mechanical treatment, introducing approximation methods that allowed a number of general principles to be deduced. By use of these procedures he showed that the tetrehedral arrangement of bonds about a carbon atom can be interpreted in terms of a hybridization of orbitals. In an important paper with G. W. Wheland that appeared in 1933 he also gave a treatment of aromatic molecules on the basis of resonance between different structures, the wave function being a linear combination of those for the individual structures.

In 1935 Pauling and Edgar Bright Wilson (b. 1908) of Harvard prepared a monograph, *Introduction to Quantum Mechanics, with Applications to Chemistry*. This book gives a particularly clear, detailed and authoritative exposition of the fundamental principles of quantum mechanics, particularly in the Schrödinger formulation, and discusses many application to chemical problems. Besides dealing with hybridization and resonance it includes a section on the construction of potential-energy surfaces for chemical reactions (Section 8.3). For many years this was the book used by physical chemists to gain a knowledge of quantum mechanics and its chemical applications. In 1939 Pauling published another book of great importance, *The Nature of the Chemical Bond, and the Structure of Molecules and Crystals*, based on a series of lectures he had given at

Cornell University in 1937. This book was dedicated to G. N. Lewis, for whom Pauling always had a great admiration. It, too, exerted a wide influence, providing a convenient summary of the structural work that Pauling had done during the previous few years, as well as much unpublished material.

Molecular orbitals

An alternative treatment of molecules is the method of molecular orbitals. One way of looking at this method is to imagine that in a diatomic molecule AB the two nuclei are brought together, resulting in a 'united atom'. The treatment of this united atom can then be taken as a first approximation for the AB molecule. Whereas in the Heitler–London method the electrons forming a single bond are treated as a pair, in the method of molecular orbitals the electrons are imagined to be fed one by one into the molecule.

Suppose that the wavefunction for electron 1 on an isolated atom A is $\psi_A^{(1)}$, and that for the same electron on atom B the wavefunction is $\psi_B^{(1)}$. A molecular orbital for electron 1 on the molecule AB might then be written as

$$\psi_{AB}^{(1)} = \psi_A^{(1)} + \psi_B^{(1)}. \tag{10.27}$$

A similar expression applies to a second electron, so that for the pair of electrons one could use the wavefunction

$$\psi_{AB} = [\psi_A^{(1)} + \psi_B^{(1)}] \, [\psi_A^{(2)} + \psi_B^{(2)}]. \tag{10.28}$$

The procedure for constructing a molecular orbital as a linear combination of atomic orbitals (LCAO) was first suggested in 1929 by John Edward (later Sir John) Lennard-Jones (1894–1954), who was then at the University of Bristol.

Whereas the valence-bond method was created in the one paper by Heitler and London, the method of molecular orbitals evolved over a period of years. The first papers relevant to the method were descriptive and intuitive and were mainly concerned with the interpretation of molecular spectra. More analytical and quantitative treatments were developed, based on the mathematics of the new quantum mechanics.

The main credit for the molecular-orbital theory is usually given, entirely appropriately, to Hund and Mulliken, but mention should be made of some others who made significant contributions. Lennard-Jones's work has already been referred to. Gerhard Herzberg's important contributions to the understanding of molecular spectra were made using what is essentially the method of molecular orbitals. In 1929 he first explained chemical bonding in a simple and convincing manner in terms of bonding and antibonding electrons, an antibonding electron counterbalancing the effect of a bonding electron.

Although Hund and Mulliken never published jointly they were close friends from 1925 onwards, and regularly exchanged their ideas through visits and correspondence; there was always close cooperation, and never any rivalry, between them. Friedrich Hund (b. 1896) was for many years at Göttingen, for a time as Born's research assistant. From 1926 to 1927 he was with Niels Bohr in

Copenhagen and it was there that he prepared some of his important papers on molecular orbitals. From 1929 to 1946 he was professor of theoretical physics at Leipzig. After that he was at the University of Jena, also in East Germany, serving as its Rector for a time until he was dismissed, apparently because he did not do what was expected for a visiting Russian dignitary. In 1951 he and his family managed to leave East Germany with just their suitcases and went to the University of Frankfurt; six months later, with uncharacteristic cooperation, the East German authorities sent the rest of their possessions on to them. In 1956 Hund became professor of theoretical physics at Göttingen, and is now a professor emeritus there. He is particularly remembered for 'Hund's rules', concerned with electron spin. His pioneering paper on molecular orbitals appeared in 1928, and was followed by many other publications in the same and other fields. When Mulliken was awarded his Nobel Prize in 1966 he expressed regret that he had not shared it with Hund. This would indeed have been appropriate, since Hund's contributions were substantial and of lasting importance.

Robert Sanderson Mulliken (1896–1986) received his Ph.D. degree in chemistry at the University of Chicago in 1921. After appointments at Harvard and New York University, where he did much of his early work on molecular-orbital theory, he was at the University of Chicago until his death, with a dual appointment at Florida State University from 1964 to 1971. In 1924 he made a study of the spectrum of boron monoxide, BO, which directly confirmed the conclusion of Einstein and Stern that there is a zero-point energy. It was only later, with the development of quantum mechanics, that the zero-point energy could be properly understood. Mulliken's first paper on molecular orbitals was published in 1928 at almost exactly the same time as Hund's initial paper on the subject. In many subsequent papers Mulliken interpreted spectra in terms of the theory.

Another who made important contributions to molecular–orbital theory was Erich Hückel (1896–1980). Hückel took his doctorate in physics at Göttingen under Debye, whom he followed in 1922 to the Technische Hochschule of Zürich; there he and Debye developed their theory of strong electrolytes (Section 7.2). Later Hückel held a rather unsatisfactory position at the Technische Hochschule of Stuttgart, where he carried out most of his work on molecular structure. From 1937 until his death he was at the University of Marburg.

Hückel's work was influenced to some extent by his brother Walter Hückel (1895–1973), a distinguished organic chemist, and also by Hund. In a series of papers published from 1930 onwards Hückel treated benzene and other aromatic molecules in terms of molecular-orbital theory. It was he who suggested the idea of σ and π bonds, although it was Hund who suggested the notation. Hückel's work was of great practical importance, since it clarified the whole problem of the chemical behaviour of more complicated chemical systems.

At first there was considerable rivalry between the protagonists of the valence-bond (Heitler–London) method and the molecular-orbital method. Later it was realized that when improvements are made to the methods they approach each other. For more complex molecules, such as the aromatic systems treated by Hückel, molecular-orbital theory is usually the more convenient. For the most satisfactory understanding of molecules it is best to look at them from both points of view.

Suggestions for further reading

For general reviews of both the old quantum theory and quantum mechanics see

E. U. Condon, Sixty years of quantum physics, *Physics Today*, **15**, 37–49 (October, 1962).

W. H. Cropper, *The Quantum Physicists and an Introduction to their Physics*, Oxford University Press, 1970.

G. Gamow, *Thirty Years that Shook Physics*, Doubleday, New York, 1966; Dover Publications, New York, 1985.

B. Hoffmann, *The Strange Story of the Quantum*, Harper, New York, 1947; Dover Publications, New York, 1959.

M. Jammer, *The Conceptual Developments of Quantum Mechanics*, McGraw-Hill, New York, 1966.

For accounts of the quantum theories of Planck and Einstein see

M. Jammer, *op. cit.*, Chapter 1 (pp. 1–61).

M. J. Klein, Max Planck and the beginnings of the quantum theory, *Arch. Hist. Exact Sci.*, **1**, 459–479 (1961).

M. J. Klein, Planck, entropy, and quanta: 1901–1906, in *The Natural Philosopher*, Blaisdell, New York, 1963, Vol. 1, pp. 83–108.

M. J. Klein, Einstein's first paper on quanta, in *The Natural Philosopher*, Blaisdell, New York, 1963, Vol. 2, pp. 59–86.

A. Pais, *Subtle is the Lord: the Science and the Life of Albert Einstein*, Clarendon Press, Oxford, 1982.

A. Pais, *Inward Bound: Of Matter and Forces in the Physical World*, Clarendon Press, Oxford, 1986.

L. Rosenfeld, La première phase de l'évolution de la théorie des quanta, *Osiris*, **2**, 149–196 (1936).

For accounts of Bohr's theory of the atom see

Blanca L. Haendler, Presenting the Bohr atom, *J. Chem. Education*, **59**, 372–376 (1982).

M. Jammer, *op. cit.*, Chapter 2, pp. 62–88.

A. Pais, *Niels Bohr's Times, in Physics, Philosophy, and Polity*, Clarendon Press, Oxford, 1991.

For accounts of the development of quantum mechanics see

M. Jammer, *op. cit.*, Chapters 4–7.

M. Jammer, *The Philosophy of Quantum Mechanics: The Interpretations of Quantum Mechanics in Historical Perspective*, Wiley, New York, 1974.

M. J. Klein, Einstein and the wave-particle duality, in *The Natural Philosopher*, Blaisdell, New York, 1964, Vol. 3, pp. 3–49.

M. J. Klein, The essential nature of the quantum hypothesis, in *Paul Ehrenfest*, North-Holland Publishing Co., Amsterdam, 1970, Vol. 1, Chapter 10 (pp. 217–263).

W. C. Price, S. S. Chissick and T. Ravensdale (Eds.), *Wave Mechanics: The First Fifty Years*, Butterworths, London, 1973.

B. L. van der Waerden (Ed.), *Sources of Quantum Mechanics*, North-Holland Publishing Co., Amsterdam, 1967; Dover Publications, New York, 1968. This book reproduces many of the early papers on quantum mechanics, up to but not including Schrödinger's wave mechanics. The historical introduction by van der Waerden (pp. 1–59) is particularly useful.

For discussions of the electronic interpretation of the periodic table see

Davies, Mansel, Charles Rugeley Bury and his contributions to physical chemistry, *Arch. Hist. Exact Sci.*, **36**, 75–90 (1986).

Davies, Mansel, C. R. Bury, L. Vegard, and the electronic interpretation of the periodic table. A note, *Arch. Hist. Exact Sci.*, **41**, 185–187 (1990).

Pais, A., *Niels Bohr's Times*, op. cit. This book gives an excellent account of Bohr's work, but makes no mention of the important contributions of Lewis, Langmuir, Vegard and Bury to the electronic theory of valency.

For accounts of the development of theories of chemical bonding see

W. B. Jensen, Abegg, Lewis, Langmuir, and the octet rule, *J. Chem. Education*, **61**, 191–200.

D. J. Klein and N. Trinajstic, Valence-bond theory and chemical structure, *J. Chem. Education*, **57**, 633–637 (1990).

R. E. Kohler, The origin of G. N. Lewis's theory of the shared pair bond, *Hist. Stud. Phys. Sci.*, **3**, 343–376 (1971).

R. E. Kohler, Irving Langmuir and the octet theory of valence, *Hist. Stud. Phys. Sci.*, **4**, 39–87 (1972).

R. E. Kohler, The Lewis–Langmuir theory of valence and the chemical community, 1920–1928, *Hist. Stud. Phys. Sci.*, **6**, 431–468 (1975).

R. S. Mulliken, Molecular scientists and molecular science: some reminiscences, *J. Chem. Phys.*, **43**, S2–S11 (1965).

R. S. Mulliken, Spectroscopy, molecular orbitals, and chemical bonding, *Science*, **157**, 13–24 (1967). This is Mulliken's Nobel Prize address.

R. S. Mulliken, *Life of a Scientist*, Springer-Verlag, 1989. This book, edited by B. J. Ransil, includes an introduction by F. Hund, an outline of Mulliken's career, and a complete list of his publications.

C. A. Russell, *The History of Valency*, Leicester University Press, 1971.

M. D. Saltzman, The bonds of conformity: W. A. Noyes and the initial failure of the Lewis theory in America, *J. Chem. Education*, **61**, 119–123 (1984).

J. W. Servos, *Physical Chemistry from Ostwald to Pauling: The Making of a Science in America*, Princeton University Press, 1990. This book includes accounts of the work of Lewis and Pauling on the chemical bond.

J. C. Slater, Molecular-orbital and Heitler–London methods, *J. Chem. Phys.*, **43**. S11–S17 (1965).

A. N. Stranges, Reflections on the electron theory of the chemical bond: 1900–1925, *J. Chem. Education*, **61**, 185–190 (1984).

APPENDIX

Scientific periodicals

Below are listed the main periodicals cited in the references (pp. 361–418), with some details about them. The city given is the one in which publication was begun. The year shown is the one in which the publication was founded; the first issue may have appeared in a later year.

The abbreviations used in the references are indicated below by **bold-italic** type. Square brackets, e.g. [*Mém*], are used below to show that in the abbreviated form the item has been transferred to another position for convenience. To avoid undue complexity, minor changes in title have sometimes been ignored.

*Abhandl*ungen des Könglige (later Könglige *Preuss*ische) *Akad*emie der *Wiss*enschaft in *Berlin*. 1804; it was formerly *Mém. Acad. roy. sci. Berlin*.

Ambix. Cambridge, England. 1937.

*Amer*ican *J*ournal of *Phys*ics. Woodbury, NY 1940.

*Amer*ican *J*ournal of *Sci*ence. New Haven, Connecticut. ('Silliman's Journal'). 1818.

*Ann*alen *der Chem*ie *und Pharm*acie. Leipzig and Heidelberg. 1840; it succeeded Ann. der Pharm. Vols. 33 (1840)—168 (1873). In 1870 it was renamed Justus Liebig's Annalen der Chemie und Pharmacie. Leipzig. Vols. 169 (1870)—172 (1874). In 1875 it became:

Justus Liebig's *Ann*alen *der Chem*ie, Weinheim. Vols. 173 (1875)—766 (1972); volume numbers were discontinued from 1973. In 1979 the journal was renamed *Liebig's Annalen der Chemie*, with no volume numbers.

*Ann*alen *der Pharm*acie. Lengo, Heidelberg. 1832—Vol. 32 (1839). It was succeeded by *Ann. der Chem. und Pharm.*

(Gilbert's) Annalen der Physik. Halle. 1799—Vol. 60 (1819). It was succeeded by:

*(Gilbert's) Ann*alen *der Phys*ik und *physikal*ische *Chem*ie. Leipzig. Vols. 61 (1819)—76 (1824). It was succeeded by:

*(Poggendorff's) Ann*alen *der Phys*ik *und Chem*ie. Leipzig Ser. 2, Vols. 1 (1824)—160 (1878). It was succeeded by:

*(Wiedemann's) Ann*alen *der Phys*ik *und Chem*ie. Ser. 3, Vols. 1 (1877)— 69 (1899). It was succeeded by:

Annalen der Physik. Leipzig. Ser. 4, Vols. 1 (1900)—87 (1928); Ser. 5, Vols. 1 (1929)—43 (1943); Ser. 6. Vols. 1 (1947)—20 (1957).

Annales de chimie. Paris. 1789. It continued from 1816 as:

Annales de chimie et de physique. Ser. 2, Vol. 1, 1816. In 1914 this was split into the following two journals:

Annales de chimie. Ser. 9. Vol. 1, 1914.

Annales de physique, Ser. 9, Vol. 1, 1914.

Annals of Science. London. 1936

Archive for the History of Exact Sciences. Berlin. 1960.

Berichte der Bunsengesellschaft für physikalische Chemie. Weinheim. 1963. It was successor to *Z. Elektrochem.*

Berichte den Deutschen chemischer Gesellschaft. Berlin. 1868.

Bericht über die zur Bekenntmachung geigneten Verhandlung der Könglische Preussische Akademie der Wissenschaften zu Berlin. 1836. It was discontinued in 1855 and succeeded by *Monatsber. Akad. Wiss. Berlin.*

Biochemical Journal. London, 1906.

British Association for the Advancement of Science Reports. 1831.

Bulletin de la société chimique de Paris (later de France). Paris. 1858.

Chemical Reviews. Baltimore. 1924.

Chemisch Weekblad. The Hague. 1903.

(Crell's) Chemische Annalen für die Freunde der Naturlehre, Arzneygelahrtheit, Haushaltungskunst und Manufacturen. Helmstädt/Leipzig. 1784. It was successor to two other journals founded by Lorenz Krell, and was discontinued in 1803.

Comptes rendus hébdomadaires des sciénces. Académie des Sciénces. Paris. 1835.

Denkschriften der Bayerische Akademie der Wissenschaften. Munchen. 1808.

Discussions of the Faraday Society. London. 1947. In 1972 this became *J. Chem. Soc. Faraday Disc.*

Elektrochemische Zeitschrift. Berlin. 1890; it ceased publication in 1922.

Ergebnisse der exacten Naturwissenschaften. Berlin. 1922; it ceased publication in 1965.

Forhandlinger. Videnskabs-Selskabet i Christiania. 1858.

Historical Studies in the Physical Sciences. Berkeley, California. 1968.

Isis, Philadelphia. 1913.

Journal de chimie physique. Paris. 1903. It continued from 1939 (Vol. 36) as Journal de *chimie physique* et de physico-chimie biologique.

[Rozier's] Journal *de phys*ique, de *chim*ie, et *d'hist*oire *nat*urelle. Paris ('Rozier's Journal'). 1794. It was successor to a journal founded by Rozier in 1771, and ceased publication in 1823.

Journal *de physique* théoretique et appliquée. Paris. 1872. In 1919 it became the Journal de *phys*ique et de *radium*.

Journal *der Physik*. Halle & Leipzig. 1790; it became (*Gilbert's*) *Ann. der Physik* in 1799.

Journal für *prakt*ische *Chem*ie. Leipzig. 1834.

Journal of the *Amer*ican *Chem*ical *Soc*iety. Washington, DC. 1879.

Journal of *Chem*ical *Phys*ics. New York. 1933.

Journal of the *Chem*ical *Soc*iety. London. 1862. It was successor to the Memoirs of the Chemical Society, founded in 1858. There were no volume numbers after Vol. 127 (1925).

Journal of the *Chem*ical *Soc*iety. *Faraday Trans*actions. 1972. It was successor to *Trans. Faraday Soc.*

Journal of the *Franklin Inst*itute. Philadelphia. 1929.

Journal of *Phys*ical *Chem*istry. Ithaca, NY 1896

Journal of *Phys*ics: published by the Academy of Sciences of the *USSR* Moscow. 1939.

Journal of the *Roy*al *Inst*itute of *Chem*istry. London. 1943. It continued the Journal of the Institute of Chemistry, and in 1965 was absorbed into *Chemistry in Britain.*

[J] *Roy*al *Institution* of Great Britain. The Journal of Science and the Arts. London. 1816; it ceased publication in 1831.

Journal of the *Soc*iety of *Chem*ical *Ind*ustry. London. 1882; it continued from 1951 as the Journal of *Appl*ied *Chem*istry.

*Kong*līge *Danske Videnskabe*rs *Selsk*ab. Copenhagen. 1814.

[Mém] *Acad*émie *roy*ale des *sci*ences. Histoire . . . avec les mémoires. *Berlin.* 1745. It was successor to a journal published in Latin and founded in 1710; in 1804 it became the *Abhandl. Akad. Wiss. Berlin.*

[Mém] Histoire de l'*acad*émie *roy*ale des *sci*ences. Avec les mémoires de mathematiques et de physique pour la même année. Paris. 1699. It ceased publication in 1797 and was succeeded by:

[Mém] Institut national *de France*. Mémoires. Paris. 1796.

[Mém. phys. chem] La *soc*ieté *d'Arceuil*. Mémoires de physique et de chimie. Paris. 1807; it ceased publication in 1817.

*Mém*oires des *soc*iétées *sav*antes et littéraires de la république français. *Paris.* 1801. Only two volumes were issued, in 1801 and 1802.

*Mitt*eilungen der *naturforsch*enden *Ges*ellschaft in *Zürich*. 1847. It was discontinued after 1856.

*Monatsber*ichte der *Könglige Preuss*ische *Akad*emie der *Wiss*enschaften zu *Berlin*. 1856. It was successor to *Ber. Akad. Wiss. Berlin.*

*Nachr*ichten von der *Königle Ges*ellschaft der *Wiss*enschaften zu *Göttingen*. Mathematish-physikalische Klasse. 1894. It was discontinued after 1933.

Nature. London. 1869.

*Naturwiss*enschaftliche Rundschau. Braunschweig. 1886. It continued from 1913 as Die *Naturwiss*enschaften.

[Nicholson's] Journal of Natural Philosophy, Chemistry and the Arts. London. 1797; in 1814 it was absorbed into the Philosophical Magazine:

*Phil*osophical *Mag*azine, Comprehending the Various Branches of Science, Liberal and Fine Arts, Agriculture, Manufactures and Commerce. London. 1798. From 1827 new series have been issued, Volume 1 of each series being in the following years: [2], 1827, [3], 1833, [4], 1851, [5], 1875, [6], 1901, [7], 1926. [8], 1956

*Phil*osophical *Trans*actions. London 1665. In 1776 the title was changed officially to *Phil*osophical *Trans*actions of the Royal Society of London.

*Phys*ical *Rev*iew. Ithaca, NY 1893.

*Phys*ical *Rev*iew *Letters.* Woodbury, NY 1958.

*Physikal*ische *Z*eitschrift. Leipzig. 1819.

*Proc*eedings of the *Amer*ican *Acad*emy of *Arts* and *Sci*ences. Boston. 1846.

*Proc*eedings of the *Camb*ridge *Phil*osophical *Soci*ety. Cambridge, England. 1843.

*Proc*eedings of the *Chem*ical *Soc*iety. London. 1957. In 1965 this was absorbed by Chemistry in Britain; prior to 1957 the Society's Proceedings had been issued at irregular times.

*Proc*eedings of the *Nat*ional *Acad*emy of *Sci*ences. Washington, DC. 1863.

*Proc*eedings of the *Roy*al *Institution.* London. 1851.

*Proc*eedings of the *Roy*al *Soc*iety. London. 1832.

*Proc*eedings of the *Roy*al *Soc*iety of *Edinburgh.* 1832.

*Rec*euil des *trav*aux *chim*iques des *Pays-Bas.* Leiden. 1882.

*Rev*iews of *Mod*ern *Phys*ics. Minneapolis. 1930.

Science. New York. 1883.

*Sitsungsber*ichte der *Niederrhein*ischen *Ges*ellschaft für Natur- und Heilkunde. *Bonn.* 1818.

*Sitzungsber*ichte der *Preuss*ische *Akad*emie der *Wiss*enschaften. *Berlin.* 1882.

*Sitzungsber*ichte der *Akad*emie der *Wiss*enschaften. *Wien.* Mathematisch-Naturwissenschaftliche Klasse. 1848.

*Trans*actions of the *Conn*ecticut *Acad*emy of *Arts* and *Sci*ences. Hamden, Connecticut. 1866.

*Trans*actions of the *Electrochem*ical *Soc*iety. New York. 1931.

*Trans*actions of the *Faraday Soc*iety. London. 1905. In 1972 this became *J. Chem. Soc. Faraday Trans.*, in two series, I and II, which were merged in 1990.

*Trans*actions of the *Roy*al *Soc*iety of *Edinburgh*. 1783.

*Verhandel*ingen der *Konink*ijke *Akad*emie van *Wet*enschappen. *Amsterdam*. 1892.

*Verhandl*ungen der *Deut*schen *physikal*ischen *Ges*ellschaft. Berlin. 1882.

*Verhandl*ungen der *naturforsch*enden *Ges*ellschaft in *Basel*. 1880.

*Z*eitschrift für *anorgan*ische *Chem*ie. Hamburg. 1892.

*Z*eitschrift für *Elektrochem*ie. Leipzig. 1894.

*Z*eitschrift für *Naturforsch*ung. Wiesbaden. 1946.

*Z*eitschrift für *Physik* und *Math*ematik. *Wien*. 1826; after some name changes, it ceased publication in 1842.

*Z*eitschrift für *physikal*ische *Chem*ie, Stöchiometrie und Verwandschaftslehre. Leipzig. 1887.

*Zhur*nal. *Russ*koe *Fiz*iko-*Khim*icheskoe *Obs*hchestvo. St Petersburg. 1869.

References and notes

For the complete titles of journals, and further information about them, see the Appendix (p. 356).

Abegg, R. (1904) Die Valenz und das periodische System (Valency and the periodic system), *Z. anorgan. Chem.*, **39**, 330–380.

Allmand, A. J. (1926) Einstein's law of photochemical equivalence, *Trans. Faraday Soc.*, **21**, 438–452. This was an introductory address at a Faraday Society Discussion on this subject.

Andrews, Thomas (1863) Andrews' work on critical phenomena was first announced on p. 328 of the third edition (1863) of W. A. Miller's *Chemical Physics*, which is Part 1 of Miller (1855–56–57).

Andrews, Thomas (1869) Bakerian Lecture—On the continuity of the gaseous and liquid states of matter, *Phil. Trans.*, **159**, 575–590. An excerpt is in Magie (1935).

Andrews, Thomas (1876) Bakerian Lecture—On the gaseous states of matter, *Phil. Trans.*, **166**, 421–449.

Andrews, The late Thomas (1887) On the properties of matter in the gaseous and liquid states under various conditions of temperature and pressure, *Phil. Trans.*, **178**, 45–56. This paper of Andrews, who died in 1885, was found among his papers by P. G. Tait and communicated to the journal by G. G. Stokes; it is believed to have been written several years previously.

Ångström, A. (1868) *Recherches sur le spectre solaire* (Research on solar spectra), Uppsala.

Ångström, A. (1871) Sur les spectres des gaz simples (On the spectra of simple gases) *Comptes rendus*, **73**, 369–373. English translation in *Phil. Mag.*, **42**, 395–399 (1871).

Archer, M. D., M. I. C. Ferreira, G. Porter and C. J. Tredwell (1976) Picosecond study of Stern–Volmer quenching of thionine by ferrous ions, *Nouveau J. de chimie*, **1**, 9–12.

Armstrong, H. E. (1885) *Brit. Assn. Reports*, p. 953.

Arrhenius, S. (1887) Über die Dissoziation der in Wasser gelösten Stoffe (On the dissociation of substances dissolved in water), *Z. physikal. Chem.*, **1**, 631–648. A translation of this paper appears in *The Foundations of the Theory of Dilute Solutions*, Alembic Club Reprints, No. 19, Oliver and Boyd, Edinburgh, 1929.

Arrhenius, S. (1887) Über das Leitungsvermögen der phosphoreszierenden Luft (On the induction of conductivity in phosphorescent air), (*Poggendorff's*) *Ann. der Phys. und Chem.*, [3], **32**, 545–572.

Arrhenius, S. A. (1889) Arrhenius's expression for the elevation of the boiling point is quoted in Beckmann (1889).

Arrhenius, S. (1889) Über die Reaktionsgeschwindigkeit bei der Inversion von Rohrzucker durch Säuren (On the rate of the inversion of cane sugar by acids), *Z. physikal. Chem.*, **4**, 226–248. A translation of the four pages of this paper that deal with temperature dependence is in Back and Laidler (1967).

Arrhenius, S. A. (1899) Zur Theorie der chemischen Reaktionsgeschwindigkeit (On the theory of chemical reaction rates), *Z. physikal. Chem.*, **28**, 317–335.

Arrhenius, S. (1919) Electrolytic dissociation, *Trans. Faraday Soc.*, **15**, 10–17.

Arrowsmith, P., F. E. Bartoszek, S. H. P. Bly, T. Carrington, P. E. Charters and J. C. Polanyi (1980) Chemiluminescence during the course of a reactive encounter: $F + Na_2 \rightarrow FNaNa^{**} \rightarrow NaF + Na$, *J. Chem. Phys.*, **73**, 5895–5897, and later publications.

Aston, F, W. (1919) A positive ray spectrograph, *Phil. Mag.*, [6], **38**, 707–714.

Aston, F. W. (1923) *Isotopes*, Longmans, Green & Co., London. In 1933 the book appeared in a revised and expanded form as *Mass Spectra and Isotopes*, Arnold, London.

Avogadro, A. (1811) Essaie d'une manière de déterminer les masses rélatives des molécules élémentaires des corps, et les proportions selon lesquelles elles entrent dans ces combinations (Essay on a method of determining the masses of the elementary molecules of substances, and the proportions in which they enter into these compounds), *J. de physique*, **73**, 58–76. Excerpts of this paper are included in Alembic Club Reprints, No. 4, Oliver and Boyd, Edinburgh, 1899, and in Leicester and Klickstein (1968), pp. 232–238.

Back, M. H., and K. J. Laidler (1967) *Selected Readings in Chemical Kinetics*, Pergamon Press, Oxford.

Bacon, Francis (1620) *Novum Organum* (The new Organum). *Organum* is the Latinized form of the title of Aristotle's work.

Balmer, J. J. (1885) Notiz über die Spektralinien des Wasserstoffs (Note on the spectral lines of hydrogen), *Verhand. naturforsch. Ges. Basel*, **7**, 548–560; *(Wiedemann's) Ann. der Phys. und Chem.*, [3], **25**, 80–87. An excerpt is in Magie (1935). Balmer's only other article on spectra is the following:

Balmer, J. J. (1897) Eine neue Formel für Spektralwellen (A new formula for spectra), *Verhand. Naturforsch. Ges. Basel*, **11**, 448–463; *(Wiedemann's) Ann. der Phys. und Chem.*, [3], **60**, 380–391. English translation in *Astrophys. J.*, **5**, 199–209 (1897).

Bancroft, Wilder D. (1927) *J. Phys. Chem.*, **31**, 635–638. This is a review of Hinshelwood's *Thermodynamics for Students of Chemistry* (1926).

Bawn, C. E. H., and G. Ogden (1934) Wave mechanical effects and the reactivity of hydrogen isotopes, *Trans. Faraday Soc.*, **30**, 432–443.

Beckmann, E. O. (1888) Über die Methode der Moleculargewichtbestimmung durch Gefrierpunktserniedrigung (On the method of determining molecular weights from the lowering of the freezing point), *Z. physikal. Chem.*, **2**, 638–645, 715–743. The first paper describes the Beckmann thermometer, with a diagram.

Beckmann, E. O. (1889) Bestimmung von Moleculargewichten aus Siede-punktserhöhungen (The determination of molecular weights from the elevation of the boiling point), *Z. physikal. Chem.*, **3**, 603–604.

Beckmann, E. O. (1889) Studien zur Praxis der Bestimmung des Moleculargewichts aus Dampfdruckerniedrigungen (Studies on the technique of determining molecular weights from the lowering of the vapour pressure), *Z. physikal. Chem.*, **4**, 532–552. Arrhenius's treatment of the lowering of the vapour pressure and the elevation of the boiling point is on pp. 550–551.

Becquerel, A. E. (1843) By this year Becquerel had made an excellent monochrome daguerreotype of the solar spectrum. An engraving made from it later appeared in Becquerel (1867) and is reproduced as Figure 6.4 of the present book.

Becquerel, A. E. (1845) By this year Becquerel had performed the remarkable feat of recording coloured solar spectra on daguerreotype plates. Examples still survive in the Conservatoire des Arts et Métiers in Paris and in the Science Museum in London. Hand-tinted engravings made by Becquerel from his daguerreotype plates are reproduced in Becquerel (1867) and in Buerger (1989).

Becquerel, A. E. (1854) Nouvelles recherches sur les impressions colorées produites lors de l'action chimique de la lumière, *Comptes rendus*, **39**, 63–67. Translation as New researches on the coloured impressions produced by the chemical action of light, *Humphrey's Journal of Photography*, 267–279 (December 15, 1854). The full paper, with the same title, appeared in *Ann. de chemie*, **42**, 81–106 (1854).

Becquerel, A. E. (1867) *La lumière, ses causes et ses effets*, Paris. Reproductions of some of the spectra shown in Becquerel's book, including tinted engravings, are to be found in Buerger (1989).

Beer, A. (1852) Bestimmung der Absorption des rothen Licht im farbingen Flussigkeiten (Estimation of the absorption of red light in coloured liquids), *(Poggendorff's) Ann. der Phys. und Chem.*, **86**, 78–88.

Behn, U. (1898) Über die specifische Wärme einiger Metalle bei tiefen Temperaturen (On the specific heats of some metals at low temperatures), *(Wiedemann's) Ann. der Phys. und Chem.*, [3], **66**, 237–244.

Bell, R. P. (1941) *Acid–Base Catalysis*, Clarendon Press, Oxford. The treatment of ionic strength effects in the light of transition-state theory is on pp. 28–31.

Berkeley, Earl of, and E. G. J. Hartley (1904) A method of measuring directly high osmotic pressures, *Proc. Roy. Soc., A*, **73**, 436–443. This was the first of several papers on osmotic pressure; for a complete list of Berkeley's publications see H. Hartley (1942).

Bernoulli, Daniel (1738) *Hydrodynamica; sive, de viribus et motibus fluidorum commentarii* (Hydrodynamics; otherwise, a commentary on forces and motion in fluids), Strasbourg. English translation by T. Carmody and H. Kobus, *Hydrodynamics*, Dover, New York, 1968. A translation of the section on kinetic theory is in Brush (1965; pp. 57–65).

Berthelot, M. (1862) Essai d'une théorie sur la formation des éthers (A suggested theory of the formation of ethers), *(Poggendorff's) Ann. der Phys. und Chem.*, [3], **66**, 110–128.

Berthelot, M., and P. L. de Saint-Gilles (1862) Recherches sur les affinités: De la formation et de la décomposition des éthers (Research on affinities: on the formation and decomposition of ethers), *Ann. de chim. et de pharm.*, **65**, 385–422, and several later papers.

Berthelot, M. (1883) *Sur la force des matières explosives d'après la thermochemie* (On the force of explosive substances according to thermodynamics), 2 vols.

Berthollet, C. L. (1803) *Essai de statique chimique* (Essay on chemical statics), 2 vols., Paris.

Berthoud, A. (1912) Théorie de l'influence de la température sur la vitesse des réactions chimiques (Theory of the effect of temperature on the rates of chemical reactions), *J. de chim. phys.*, **10**, 573–597.

Berthoud, A. (1913) Formule de Maxwell généralisée (Generalized Maxwell equation), *J. de chim. phys.*, **11**, 577–583.

Berzelius, J. J. (1836) Quelques idées sur une nouvelle force agissant dans les combinaisons des corps organiques (Some ideas about a new force involved in the combination of organic substances), *Ann. de chim. et de phys.*, **61**, 146–151, English translation of part of this article in Leicester and Klickstein (1952), pp. 265–268.

Bjerrum, N. (1909) A new form for the electrolytic dissociation theory, *Proc. Seventh International Congress of Applied Chemistry*, Section X, 55–60. Part of this paper is reproduced in Bjerrum (1949), pp. 56–57.

Bjerrum, N. J. (1911, 1912) Über die spezifische Wärme der Gaze (On the specific heat of gases), *Z. Elektrochem.*, **17**, 731–734 (1911); 18, 101–104 (1912). English translation in Bjerrum (1940), pp. 27–33.

Bjerrum, N. J. (1912) Über die ultraroten Absorptionsspektren der Gase (On the infrared absorption spectra of gases), *Nernst Festschrift*, Verlag von Knapp. English translation in Bjerrum (1949), pp. 34–40.

Bjerrum, N. J. (1914) Über ultrarote Spektren. II. Eine directe Messung der Grösse von Energie-quanten (On infrared spectra. II. A direct measurement of the size of the energy quanta), *Verhandl. Deut. physikal. Ges.*, **16**, 640–642. Part III: Die Konfiguration des CO_2-Moleküls und die Gesetze der intramolekularen Kräfte (The configuration of the carbon dioxide molecule and the laws of the intramolecular forces), *ibid.*, **16**, 737–753 (1914). English translation in Bjerrum (1949), pp. 41–55.

Bjerrum, N. (1916) De staerke Elektrolyters Dissociation (Dissociation of strong electrolytes), *Fysisk Tids.*, **15**, 59–73; this was a paper presented at the Sixteenth Scandinavian Meeting in Oslo. A German version appeared as Die Dissoziation der starker Elektrolyte, *Z. Elektrochem.*, **24**, 321–328. English translation in Bjerrum (1949), pp. 58–70.

Bjerrum, N. (1924) Zur Theorie der chemischen Reaktionsgeschwindigkeit (On the theory of chemical reaction rates), *Z. physikal. Chem.*, **108**, 82–100.

Bjerrum, N. (1926) Untersuchungen über Ionenassoziation (Investigations on ionic association), *Kgl. Danske Videnskab. Selskab. Copenhagen*, 7, No. 9, 1–48.

Bjerrum, Niels (1949) *Selected Papers*, Einar Munksgaard, Copenhagen.

Black, Joseph (1803) *Lectures on the Elements of Chemistry*, 2 vols., Edinburgh. This publication was prepared by Black's former student John Robison from notes made by Black and his students; Black himself published none of his work on heat. An excerpt dealing with heat is in Magie (1935).

Blagden, Charles (1788) Experiments on the effect of various substances in lowering the point of congelation of water, *Phil. Trans.*, 78, 277–312.

Bleaney B., and R. P. Penrose (1946) Ammonia spectrum in the 1 cm wavelength region, *Nature*, 339–340, and many other publications.

Bloch, F., W. W. Hansen and N. Packard (1946) Nuclear induction, *Phys. Rev.*, 69, 127.

Bodenstein, M. (1894) Über die Zersetzung des Jodwasserstoffgases in der Hitze (On the thermal decomposition of gaseous hydrogen iodide), *Z. physikal. Chem.*, 13, 56–127.

Bodenstein, M. (1897) Zersetzung des Jodwasserstoff im Licht (The photochemical decomposition of hydrogen iodide), *Z. physikal. Chem.*, 22, 23–33.

Bodenstein, M. (1899) Gasreaktionen im der chemischen Kinetik. II. Einfluss der Temperatur auf Bildung und Zersetzung von Jodwasserstoff (Gas reactions in chemical kinetics. II. Influence of temperature on the formation and decomposition of hydrogen iodide), *Z. physikal. Chem.*, 29, 295–314.

Bodenstein, M., and S. C. Lind (1907) Geschwindigkeit der Bildung des Bromwasserstoffe aus sein Elementen (Rate of formation of hydrogen iodide from its elements), *Z. physikal. Chem.*, 57, 168–192.

Bodenstein, M., and W. Dux (1913) Photochemische Kinetik des Chlorknallgasen (Photochemical kinetics of chlorine–hydrogen mixtures), *Z. physikal. Chem.*, 85, 297–328.

Bodenstein, M. (1913) Eine Theorie der photochemischen Reaktionsgeschwindigkeit (A theory of photochemical reaction velocity), *Z. physikal. Chem.*, 85, 329–397, *Verh. deut. physik. Ges.*, 15, 690–704 (1913). Bodenstein's steady-state hypothesis appears in this paper, six months after the hypothesis had been suggested by Chapman and Underhill.

Bodenstein, M. (1916) Die Vereinigung von Chlor und Wasserstoff (The combination of chlorine and hydrogen), *Z. Elektrochem.*, 22, 53–61.

Bodenstein, M., and H. Lütkemeyer (1924) Die photochemische Bildung von Bromwasserstoff und die Bildungsgeschwindigkeit der Brommolekul aus den atomen (The photochemical formation of hydrogen bromide and the rate of formation of bromine molecules from the atoms), *Z. physikal. Chem.*, 114, 208–236.

Bodenstein, M. (1927) Analysen der Zeitgesetze zusammengesetzter chemische Reaktionen (Analysis of the time laws of complex chemical reactions), *Ann. der Physik*, 82, 836–840. This was a reply to Skrabal's criticism of the steady-state hypothesis.

Bodenstein, M. (1931) Photochemical kinetics in gaseous systems, *Trans. Faraday Soc.*, **27**, 413–424.

Bodenstein, M., and C. Wagner (1929) Ein Vorschlag für die Bezeichnung der Lichtmenge in der Photochemie (A suggestion for the designation of quantity of light in photochemistry), *Z. physikal. Chem.*, *B*, **3**, 456–458.

Bohr, N. (1913) On the constitution of atoms and molecules, *Phil. Mag.*, [6], **26**, 1–25, 476–502, 857–875.

Bohr, N. (1913) Reports on Bohr's 1913 papers on the structure of the atom appeared in 'Physics at the British Association', *Nature*, **92**, 305 (1913) and in 'Problems of radiation (from our special correspondent)', *The Times*, 13 September 1913, p. 10.

Bohr, N. (1914) On the effects of electric and magnetic fields on spectral lines, *Phil. Mag.*, [6], **27**, 506–524.

Bohr, N. (1915) On the quantum theory of radiation and the structure of the atom, *Phil. Mag.*, [6], **30**, 394–415.

Bohr, N. (1921) Atomic structure, *Nature*, **107**, 104–107, 108, 208–209. In these two Letters to the Editor, Bohr discusses the electronic configurations of the noble gases only.

Bohr, N. (1923) The structure of the atom, *Nature*, **112**, 29–44. This was Bohr's Nobel Prize address, given on 11 December, 1922. The news of the discovery of hafnium reached Bohr just in time for him to announce it in this address, in which he made brief reference to the work of Vegard and Bury. Bohr had to improvise during the first part of his address, since he had left his notes and slides at his hotel.

Bohr, N. (1961) Rutherford Memorial Lecture, *Proc. Phys. Soc. London*, **78**, 1083–1115.

Boltzmann, L. (1866) Über die mechanische Bedeutung des zweiten Hauptsatzes der Wärmetheorie (On the mechanical significance of the second law of thermodynamics), *Sitzungsber. Akad. Wiss. Wien*, **53**, 195–220.

Boltzmann, L. (1868) Studien über des Gleichgewicht der lebendigen Kraft zwischen bewegten materiallen Punkten (Studies on the average distribution of kinetic energy in a system of material points), *Sitzungsber. Akad. Wiss. Wien*, **58**, 517–560.

Boltzmann, L. (1871) Analytische Beweis des zweiten Hauptsatzes der mechanischen Wärmetheorie aus den Sätzen über das Gleichgewicht der lebendigen Kraft (Analytical proof of the second law of thermodynamics from the principles governing the distribution of kinetic energy), *Sitzungsber. Akad. Wiss. Wien*, **63**, 712–732.

Boltzmann, L. (1871) Zur Priorität des Auffindung der Beziehung zwischen den zweiten Hauptsatz der mechanische Wärmetheorie und dem Principe der kleinsten Wirkung (On the priority in discovering the connection between the second law of thermodynamics and the principle of least motion), *(Poggendorff's) Ann. der Phys. und Chem.*, **143**, 211–230. In this paper Boltzmann claimed that his paper of 1866 was essentially the same as that of

Clausius (1871). Clausius (1872) conceded Boltzmann's priority, but in any case the argument was invalid!

Boltzmann, L. (1872) Weitere Studien über das Wärmegewicht unter Gasmoleculen (Further studies on the thermal equilibrium of gas molecules), *Sitzungsber. Akad. Wiss. Wien*, **66**, 275–370. English translation in Brush, (1965), Vol. 2, Selection 2, pp. 88–175.

Boltzmann, L. (1877) Über die Natur der Gasmolecüle (On the nature of the gas molecule), *Sitzungsber. Akad. Wiss. Wien*, [2], **74**, 553–560. Abstract in *(Poggendorff's) Ann. der Phys. und Chem.*, [2], **160**, 175–176 (1877), and an English abstract in *Phil. Mag.*, [5], **3**, 320 (1877).

Boltzmann, L. (1877) Bemerkungen über einige Probleme der mechanische Wärmetheorie (Remarks on some problems in thermodynamics), *Sitzungsber. Akad. Wiss. Wien*, [2], **75**, 62–100.

Boltzmann, L. (1877) Über die Beziehung eines allgemeine mechanische Sätzes zum zweiten Hauptsatze der Wärmetheorie (On the relation between a general mechanical theorem and the second law of thermodynamics), *Sitzungsber. Akad. Wiss. Wien*, [2], **75**, 67–73. Translation in Brush (1965), Vol. 2, pp. 188–193.

Boltzmann, L. (1877) Über die Beziehung zwischen den zweiten Hauptsatz der mechanischen Wärmetheorie und den Wahrscheinlichkeitsrechnung respektive den Sätzen über des Wärmegleichgewicht (On the relationship between the second law of thermodynamics and statistics in relation to heat theorems), *Sitzungsber. Acad. Wiss. Wien*, [2], **76**, 373–435. Abstract in *(Wiedemann's) Ann. der Phys. und Chem.*, [3], **3**, 166–174 (1879).

Boltzmann, L. (1884) Ableitung des Stefan'schen Gesetzes betreffend die Abhängigkeit der Wärmestrahlung von der Temperatur aus der elektromagnetischen Lichttheorie (Derivation of Stefan's law concerning the dependence of heat radiation on temperature, from the electromagnetic theory of radiation), *(Wiedemann's) Ann. der Phys. und Chem.*, [3], **22**, 291–294.

Boltzmann, L. (1896) Entgegnung auf die wärmetheoretischen Betrachtungen des Hrn. E Zermelo (Reply to Zermelo's remarks on the theory of heat), *(Wiedemann's) Ann. der Phys. und Chem.*, [3], **7**, 773–784.

Born, M. (1924) Über Quantenmechanik (On quantum mechanics), *Z. Physik*, **26**, 379–395. English translation in van der Waerden (1967).

Born, M., and P. Jordan (1925) Zur Quantenmechanik (On quantum mechanics), *Z. Physik*, **34**, 858–888. English translation in van der Waerden (1967).

Born, M., W. Heisenberg and P. Jordan (1926) Zur Quantenmechanik. II, *Z. Physik*, **35**, 557–615.

Born, M. (1926) Zur Quantenmechanik der Stossvorgänge (On the quantum mechanics of collision processes), *Z. Physik.*, **37**, 863–876; **38**, 803–827 (1926). Born at first thought that ψ represents the probability; the change to ψ^2 was made to the galley proofs of the first paper.

Bosanquet, R. H. M. (1877) Notes on the theory of sound, *Phil. Mag.*, [5], **3**, 271–278, 343–349, 418–424; **4**, 25–39, 125–136, 216–222.

Boscovich, Roger (1758) *Theoria Philosophiae Naturalis* (Theory of natural philosophy), Vienna, revised 1763. English translation by J. M. Child, *A Theory of Natural Philosophy*, Open Court Publishing Co., Chicago 1922, reprinted by MIT Press, Cambridge, Mass., 1966.

Boyle, Robert (1661) *The Sceptical Chymist: or Physico-Chymical Doubts and Paradoxes . . .* , London; Everyman edition, London, 1911.

Boyle, Robert (1662) *A Defence of the Doctrine Touching the Spring and Weight of the Air*, London. The first edition of this book appeared in 1660 but did not contain a statement of the pressure–volume relationship, which first appeared in the second edition of 1662; a third edition appeared in 1682. An excerpt dealing with the gas law is in Magie (1935).

Boynton, H. (1948) *The Beginnings of Modern Science*, W. J. Black, Roslyn, NY.

Brandsma, W. F. (1922) Equilibria and reaction velocities, *Chem. Weekblad.*, **19**, 318–322.

Brandsma, W. F. (1928) Reaction velocities. II, *Chem. Weekblad.*, **47**, 94–104.

Brandsma, W. F. (1929) Reaction velocities. III, *Chem. Weekblad.*, **48**, 1205–1218.

Brewster, David (1834) Observations on the lines of the solar spectrum, and on those produced by the earth's atmosphere, and by the action of nitrous acid gas, *Trans. Roy. Soc. Edinburgh*, **12**, 519–530.

Briers, F., D. L. Chapman and E. Walters (1926) The influence of the intensity of illumination on the velocity of photochemical changes, *J. Chem. Soc.*, 562–569. This describes for the first time the rotating-sector method.

Briers, F., and D. L. Chapman (1928) The influence of the intensity of illumination on the velocity of the photochemical union of bromine and hydrogen, and the determination of the mean life of a postulated catalyst, *J. Chem. Soc.*, 1802–1811. This paper presented an application of the rotating-sector method.

Brönsted, J. N. (1922) Zur Theorie der chemischen Reaktionsgeschwindigkeit (On the theory of chemical reaction rates), *Z. physikal. Chem.*, **102**, 169–207.

Brönsted, J. N. (1923) Einige Bemerkungen über den Begriff der Säuren und Basen (Some remarks on the definition of acids and bases), *Rec. trav. chim. Pays-Bas*, **42**, 718–728.

Brönsted, J. N., and V. K. La Mer (1924) The activity coefficients of ions in very dilute solution, *J. Amer. Chem. Soc.*, **46**, 555–573.

Brooks, P. R. (1988) Spectroscopy of transition region species. *Chem. Rev.*, **88**, 407–428.

Brooks, P. R. See also Odiorne, Brooks and Kasher (1971).

Browne, Walter R. (1879) *The Mechanical Theory of Heat*, Macmillan, London; this was an edited English translation of Clausius (1865–1867).

Brush, S. G. (1965) *Kinetic Theory, Vol. 1. The Nature of Gases and of Heat*, Pergamon Press, Oxford.

Brush, S. G. (1967–1968) Foundations of statistical mechanics, *Arch. Hist. Exact Sci.*, **4**, 145–183.

Buerger, Janet E. (1989) *French Daguerreotypes*, University of Chicago Press.

Bull, T. H., and P. B. Moon (1954) A mechanical method for the activation of fast reactions, *Disc. Faraday Soc.*, **17**, 54–57.

Bunsen, R. W. (1853) Untersuchungen über die chemische Verwandtschaft (Investigations of chemical change), *Ann. der Chem. und Pharm.*, **85**, 137–155. This paper was concerned with gaseous explosions.

Bunsen, R., and H. E. Roscoe (1855) Photochemische Untersuchungen (Photochemical investigations), *(Poggendorff's) Ann. der Phys. und Chem.*, **96**, 373–394, and many later papers.

Bunsen, R. (1857) *Gasometrische Methoden* (Gasometric methods), Vieweg & Sohn, Brunswick. This book brought gas analysis to a new level of precision, and describes Bunsen's investigations on gaseous explosions, including his measurements of flame speeds.

Bunsen, R., and G. Kirchhoff (1860, 1861) Chemische Analyse durch Spectralbeobachten (Chemical analysis by spectrum observations), *(Poggendorff's) Ann. der Phys. und Chem.*, **110**, 161–189; 113, 337–381. This article in two parts, which gives the first description of the Bunsen–Kirchhoff spectroscope, contains coloured engravings of spectra, but no diagram of the spectroscope. An English version of the paper, with the spectra and including a diagram (reproduced in Figure 6.2 of the present book), appears in *Phil. Mag.*, **20**, 89–109 (1861); **22**, 329–349, 498–510 (1861). An early Bunsen–Kirchhoff spectroscope, believed to have been given by Bunsen to Professor Charles Daubeny, professor of chemistry at Oxford, is to be seen in the Museum of the History of Science, Old Ashmolean Building, Oxford.

Burrau, Ö. (1927) Berechnung des Energiewertes des Wasserstoffmolekel-Ions (Calculation of the energy value of the ionized hydrogen molecule in its normal state), *Kgl. Danske Vid. Selskab.*, **7**, No. 14, 1–18. More accurate treatments were later given by E. A. Hylleras, *Z. Physik*, **71**, 739–763 (1931) and by G. Jaffe, *Z. Physik*, **87**, 535–544 (1934).

Bury, C. R. (1921) Langmuir's theory of the arrangement of electrons in atoms and molecules, *J. Amer. Chem. Soc.*, **43**, 1602–1609.

Butler, J. A. V. (1932) The mechanism of overvoltage and its relation to the combination of hydrogen atoms at metal electrodes, *Trans. Faraday Soc.*, **28**, 379–382.

Butler, J. A. V. (1936) Hydrogen overvoltage and the reversible hydrogen electrode, *Proc. Roy. Soc., A*, **157**, 423–433.

Buys-Ballot, C. H. D. (1858) Über die Art von Bewegung, welche wir Wärme und Electricität Nennen (On the type of motion that we call heat and electricity), *(Poggendorff's) Ann. der Phys. und Chem.*, [2], **103**, 240–259.

Cannizzaro, S. (1858) Sunto di un corso di filosofia chimica (Sketch of a course in chemical philosophy), *Nuevo Cimento*, **7**, 321–366. Translations are in Ostwald's *Klassiker*, No. 30, and in Alembic Club Reprints, No. 18, Edinburgh, 1910; a translated abstract appears in Leicester and Klickstein (1968), pp. 407–417.

Cardwell, D. S. L. (1971) *From Watt to Clausius*, Cornell University Press, NY.

Carnot, S. (1824) *Réflexions sur la puissance motrice du feu et sur les machines propre à déveloper cette puissance* (Reflections on the motive power of fire and on machines designed to develop that power), Bachelier, Paris. Original copies of this book are now very rare, but facsimile editions are available. An English translation by R. H. Thurston, *Reflections on the Motive Power of Heat* (1897), has been reprinted many times, for example in E. Mendoza (Ed.), *Reflections on the Motive Power of Fire by Sadi Carnot and other papers on the Second Law of Thermodynamics* by E. Clapeyron and R. Clausius, New York, 1960. Excerpts are in Magie (1935).

Chance, B. (1940) The accelerated flow method for rapid reactions, *J. Franklin Institute*, **229**, 455–476, 613–640, 737–760. The stopped-flow method has been further developed, for example by Gibson and Milmes (1964).

Chapman, D. L. (1899) On the rate of explosion in gases, *Phil. Mag.*, [5], **47**, 90–104.

Chapman, D. L., and C. H. Burgess (1905) Note on the cause of the period of chemical induction in the union of hydrogen and chlorine, *Proc. Roy. Soc.*, **74**, 400 and many subsequent publications. For a review of Chapman's work on this reaction see Laidler (1988), pp. 272–276.

Chapman, D. L., and H. E. Jones (1910) The homogeneous decomposition of ozone in the presence of oxygen and other gases, *J. Chem. Soc.*, **97**, 2463–2477.

Chapman, D. L., and H. E. Jones (1912) Decomposition of dry ozone, *J. Chem. Soc.*, **99**, 1811–1819.

Chapman, D. L., and L. K. Underhill (1913) The interaction of chlorine and hydrogen. The influence of mass, *J. Chem. Soc.*, **103**, 496–508. This paper includes the first statement of the steady-state hypothesis.

Chapman, D. L. (1913) A contribution to the theory of capillarity, *Phil. Mag.*, [6]. **25**, 475–481.

Chapman, D. L., see also Briers.

Chapman, [Mrs] M. C. C. (1924) The first law of photochemistry, *J. Chem. Soc.*, **125**, 1521–1526.

Christiansen, J. A. (1919) On the reaction between hydrogen and bromine, *Kgl. Danske Videnskab. Selsk.*, **1**, No. 14, 1–17.

Christiansen, J. A. (1921) Reaktionskinetishe Studier (Studies in reaction kinetics), Ph.D. thesis, University of Copenhagen, Vilhelm Priors Kgl. Hofboghandel, Copenhagen. The thesis was published in October, 1921. Important features in it are the treatment of chain reactions and of unimolecular reactions, more detailed than that of Lindemann (1921).

Christiansen, J. A., and H. A. Kramers (1923) Über die Geschwindigheit chemischer Reaktionen (On the rate of chemical reactions), *Z. physikal. Chem.*, **104**, 451–469.

Clapeyron, B. P. E. (1834) Mémoire sur la puissance motrice de la chaleur (Memoir on the motive power of heat), *J. de l'École polytechnique*, **14**, 153–190.

An English translation, by R. Taylor, appeared in *Scientific Memoirs Selected from the Transactions of Foreign Academies of Science and Learned Societies and from Foreign Journals*, **1**, 347–376 (1837); this appears also in the book edited by Mendoza (see Carnot, 1824).

Clausius, R. (1850) Über die bewegende Kraft der Wärme und die Gesetze welche sich deraus für die Wärmelehre selbst ableiten lessen (On the moving force of heat and the laws of heat that may be deduced therefrom), *(Poggendorff's) Ann. der Phys. und Chem.*, [2], **79**, 368–397, 500–524. All of Clausius's papers on thermodynamics are included in Clausius (1865–1867), and in translation in Hirst (1887). Excerpts from this 1850 paper are in Magie (1935).

Clausius, R. (1854) Über eine veränderte Form des zweiten Hauptsätzes der mechanischen Wärmetheorie (On a modified form of the second fundamental theorem in the mechanical theory of heat), *(Poggendorff's) Ann. der Phys. und Chem.*, [2], **93**, 481–506; English translation in Browne (1879) and Hirst (1887).

Clausius, R. (1857) Über die Art der Bewegung, welche wir Wärme nennen (On the type of motion that we call heat), *(Poggendorff's) Ann. der Phys. und Chem.*, [2], **100**, 353–380. An English translation of this appeared in *Phil. Mag.*, **14**, 108–127 (1857) and is reproduced in Brush (1965), pp. 111–134.

Clausius, R. (1857) Über die Elektricitätsleitung in Elektrolyten (On the conduction of electricity through electrolytes), *(Poggendorff's) Ann. der Phys. und Chem.*, **101**, 338–360. The idea that ions exist free in solution is sometimes referred to as the Williamson–Clausius hypothesis, but this expression does not seem justified. In papers in 1851 and 1852 the British chemist Alexander William Williamson (1824–1904) suggested the dissociation of substances but not into charged species, and his proposal was in no way related to the conductances of solutions.

Clausius, R. (1858) Über die mittlere Länge der Wege, welche bei Moleculärbewegung gasförmige Körper von den einzelnen Molecülen zurückgelegt werden, nebst einigen anderen Bemerkungen über die mechanische Wärmetheorie (On the mean lengths of the paths traversed by the individual molecules in gases, together with additional comments on the mechanical theory of heat), *(Poggendorff's) Ann. der Phys. und Chem.*, [2], **105**, 239–258. An English translation by F. Guthrie appears in *Phil. Mag.*, **17**, 81–91 (1859) and is reproduced in Brush (1965), pp. 135–147.

Clausius, R. (1862) Über die Wärmeleitung gasförmige Körper (On the thermal conductivities of gases), *(Poggendorff's) Ann. der Phys. und Chem.*, [2], **115**, 1–56. This paper pointed out serious numerical errors in Maxwell (1860). For example, in one calculation Maxwell had treated hours as if they were seconds, and as a result of this and other errors estimated the thermal conductivity of copper to be 10^7 times that of air; the factor should have been 7×10^3.

Clausius, R. (1862) On the application of the equivalence of transformations to the internal work of a mass of matter, *Phil. Mag.*, [4], **24**, 81–97, 201–213. This was a translation, submitted by Clausius himself, of a paper in *Mitt. physikal. Ges. Zürich*, **7**, 48–95 (1862).

Clausius, R. (1865) Über verschiedener für die Anwendung bequeme Formen der Haupt gleichungen der mechanishen Wärmetheorie (On several convenient forms of the fundamental equations of the mechanical theory of heat), *(Poggendorff's) Ann. der Phys. und Chem.*, [2], **125**, 353–399; English translation in Hirst (1887).

Clausius, R. (1865–1867) *Abhandlungen über die mechanische Wärmetheorie* (Treatise on thermodynamics), 2 vols., Brunswick. An edited English translation of this, by Walter R. Browne, appeared as *The Mechanical Theory of Heat*, Macmillan, London, 1879. Another edited translation, by T. Archer Hirst, appeared as *The Mechanical Theory of Heat with its Applications to the Steam Engine and to the Physical Properties of Bodies*, London, 1887.

Clausius, R. (1870) Über einen auf die Wärme andwendbaren mechanische Satz (On a mechanical theorem applicable to heat), *Sitzungsber. Niederrhein. Ges. Bonn*, 114–119. An English translation appears in *Phil. Mag.*, [5], **40**, 122–127 (1890) and is reproduced in Brush (1965), Selection 11, pp. 172–178.

Clausius, R. (1871) Über die Zurückführung der zweiten Hauptsatzes der mechanischen Wärmetheorie auf allgemeine mechanische Prinzipien (On the derivation of the second law of thermodynamics from general mechanical principles), *(Poggendorff's) Ann. der Phys. und Chem.*, [2], **142**, 433–461. This paper was at once followed by Boltzmann (1871) claiming priority for the mechanical explanation for the second law, a claim that was accepted in Clausius (1872).

Clausius, R. (1872) Bemerkungen zu der Prioritätreclamation des Hrn. Boltzmann (Remarks on the priority claim of Herr Boltzmann), *(Poggendorff's) Ann. der Phys. und Chem.*, [2], **144**, 265–274. In this paper Clausius conceded Boltzmann's claim to priority for the mechanical explanation.

Cohen, I. Bernard (1964) Newton, Hooke and 'Boyle's law' (discovered by Power and Towneley), *Nature*, **204**, 618–621.

Collins, P. (1975) Humphry Davy and heterogeneous catalysis, *Ambix*, **22**, 205–217.

Compton, A. H. (1923) A quantum theory of the scattering of X rays by light electron, *Phys. Rev.*, **21**, 483–503. Compton's results were initially of considerable controversy, for example at the December, 1923, meeting of the American Physical Society, and at the August, 1924, meeting in Toronto of the British Association for the Advancement of Science. Maurice de Broglie was one of the first to confirm Compton's results, a fact that is significant in connection with his brother Victor's interest in the work.

Condon, E, U. (1962) Sixty years of quantum physics, *Physics Today*, **15**, 37–49 (October, 1962).

Conway, B. E., and M. Salomon (1964) Studies on the hydrogen evolution reaction down to −150 °C and the role of proton tunnelling, *J. Chem. Phys.*, **41**, 3169–3177. See also Salomon and Conway (1965).

Coriolis, G. G. de (1829) *Du calcul et de l'éffet des machines; ou, Considérations sur l'emploi des moteurs et sur leur évaluation* (On the design and the effect of machines; or, considerations regarding the use and evaluation of engines), Paris.

Crookes, W. (1861) On the existence of a new element, probably of the sulphur group, *Phil. Mag.*, [4], **21**, 301–305.

Crookes, W. (1879) The Bakerian Lecture—On the illumination of lines of molecular pressure, and the trajectory of molecules, *Phil. Trans.*, **170**, 135–164. This gives a summary of his work on the radiometer, and on electrical discharges. Reproduced in part in Magie (1935).

Cuvier, G. (1801) Rapport sur la galvanisme fait â l'Institut national (Report on galvanism made to the National Institute), *[Rozier's] J. de phys. de chim. et d'hist. nat.*, **52**, 318–321; Rapport sur les expériences galvaniques (Report on galvanic experiments), *Mém. soc. sav. Paris*, **1**, 132–137 (1801).

Daguerre, L. J. M. (1839) The first public announcement of Daguerre's invention was made by F. Arago, with support from J. B. Biot, at a meeting of the Académie des Sciences on 7 January 1839; the report, in *Comptes rendus*, **8**, 4–6 (1839), is a very brief account with no technical details. An English translation appeared in the Literary Gazette, No. 1150, p. 74 (2 February 1839). During the next few months further reports were presented by Arago, some with reference to Talbot's priority claims (see Talbot, 1839).

Dalton, John (1805) Theory of the absorption of gases by water, etc., *Memoirs of the Literary and Philosophical Society of Manchester*, [2], **1**, 271–287. This paper, Dalton's first on his atomic theory and containing his first table of atomic weights, was read to the Society in 1803, and is reprinted, together with other papers, in *Foundations of the Atomic Theory: Comprising Papers and Extracts by John Dalton, William Hyde Wollaston, M.D., and Thomas Thomson, M.D. (1802–1808)*, Alembic Club Reprints, No.2, Edinburgh, 1899. This paper was followed by others; much of Dalton's work on the atomic theory was included in the following publication:

Dalton, John (1808–1827) *A New System of Chemical Philosophy*, Manchester. Part 1 appeared in 1808, Part II in 1810 and Vol. II, Part I in 1827; no more was published. Facsimile reprint, London, 1853.

Daniell, J. F. (1836) On voltaic combinations, *Phil. Trans.*, **126**, 107–124, 125–130.

Daniell, J. F. (1839) *An Introduction to the Study of Chemical Philosophy*, London; second (and final) edition, 1843.

Daniell, J. F. (1842a) On the constant voltage battery, *Phil. Mag.*, **20**, 294–303. In this paper Daniell maintained that Grove's cell, described in Grove (1839), had been inspired by Daniell (1836); this was disputed by Grove.

Daniell, J. F. (1842b) Sixth letter on voltaic combinations: addressed to

Michael Faraday, Esq., DCL, FRS, *Phil. Trans.*, **132**, 137–156. Another polemic against Grove.

Daniell, J. F., and W. A. Miller (1844) Additional researches on the electrolysis of secondary compounds, *Phil. Trans.*, **134**, 1–19; *Phil. Mag.*, [3], 25, 175–188.

Daub, E. E. (1970) Entropy and dissipation, *Hist. Stud. Phys. Sci.*, **2**, 321–354.

Davidson, N. R., R. Marshall, A. E. Larsh and T. Carrington (1951) Direct observation of the rate of recombination of iodine atoms, *J. Chem. Phys.*, **19**, 1311.

Davisson, C. J., and L. H. Germer (1927) Diffraction of electrons by a crystal of nickel, *Phys. Rev.*, **30**, 705–740. Their technique was similar to von Laue's method of X-ray diffraction, in contrast to the Thomson and Reid technique.

Davy, Humphry (1802) See Wedgwood, Thomas, and H. Davy (1902).

Davy, Humphry (1808) The Bakerian Lecture: On some new phenomena of chemical change produced by electricity, particularly the decomposition of the fixed alkalies, and the exhibition of the new substances which constitute their bases; and on the general nature of alkaline bodies, *Phil. Trans.*, **98**, 1–44. The lecture was delivered on 20 November 1806, The paper is included in Davy (1839–1840), Vol. 5, pp. 1–56, and in Davy (1901).

Davy, Humphry (1808) Electrochemical researches on the decomposition of the earths; with observations on the metals obtained from alkaline earths, and on the amalgam procured from ammonia, *Phil. Trans.*, **98**, 333–370.

Davy, Humphry (1817) Some new experiments and observations on the combustion of gaseous mixtures, and an account of a method of preserving a continued light in mixtures of inflammable gases and air without flame, *Phil. Trans.*, **107**, 77–85. For a discussion see Collins (1975).

Davy, Humphry (1818) *On the Safety Lamp for Coal Mines with Some Researches on Flame*, R. Hunter, London.

Davy, Humphry (1839–1840) *The Collected Works of Sir Humphry Davy*, edited by his brother, John Davy, MD, FRS, 9 vols. London.

Davy, Humphry (1901) *The Decomposition of the Fixed Alkalies and Alkaline Earths*, Alembic Club Reprint No. 6, Edinburgh. This is a reprint of periodical articles that appeared in 1807 and 1808.

Dawson, H. M., and F. Powis (1913) The catalytic activity of acids. Evaluation of the activities of the hydrogen ion and the undissociated acid, *J. Chem. Soc.*, **103**, 2135–2244, and many later papers.

de Broglie, L. (1923) Ondes et quanta (Waves and quanta) *Comptes rendus*, **177**, 610–624, and other short communications: *Comptes rendus*, **177**, 548–550, 630–632.

de Broglie, L. (1924) *Thèses . . . pour obtenir la grade de docteur ès sciences physiques . . .*, Masson et Cie, Paris. Published also as Recherches sur la theorie des quanta (Researches on the theory of quanta), *Ann. de physique*, **3**, 22–128. An English translation of part of this is in G. Ludwig, *Wave Mechanics*, Pergamon Press, New York, 1968.

Debye, P. (1912) Zur Theorie der spezifischen Wärme (On the theory of specific heat), *Ann. der Physik*, [4], **39**, 789–839.

Debye, P. J. W., and E. Hückel (1923) Zur Theorie der Elektrolyte (On the theory of electrolytes), *Physikal. Z.*, **24**, 185–206, 305–325.

Debye, P. (1942) Reaction rates in ionic solutions, *Trans. Electrochem. Soc.*, **82**, 265–271.

Dempster, A. J. (1918) New method of positive ray analysis, *Phys. Rev.*, [2], **11**, 316–325.

Desaguliers, J. T. (1719) *A System of Experimental Philosophy*, London.

Desaguliers, J. T. (1734, 1744) *A Course in Experimental Philosophy*, 2 vols., London.

Descartes, René (1637) *Discours de la méthode pour bien conduire la raison et chercher la verité dans les sciences* (A discourse on the method of best reasoning and of seeking the truth in the sciences), Leiden. English translation by P. J. Olscamp in *Discourse on method, optics, geometry and meteorology*, Bobbs-Merril, Indianapolis, 1965.

Descartes, René (1644) *Principia Philosophiae* (Principles of philosophy), Amsterdam.

Dewar, J. (1905) Studies with the liquid hydrogen and ice calorimeters, *Proc. Roy. Soc., A*, **76**, 325–340.

Dirac, P. A. M. (1925) The fundamental equations of quantum mechanics, *Proc. Roy. Soc., A*, **109**, 642–653. Reprinted in van der Waerden (1967).

Dirac, P. A. M. (1926) Quantum mechanics and a preliminary investigation of the hydrogen atom, *Proc. Roy. Soc., A*, **110**, 561–569. Reprinted in van der Waerden (1967).

Dirac, P. A. M. (1928) The quantum theory of the electron, *Proc. Roy. Soc., A*, **117**, 610–624, 118, 351–361 (1928).

Dirac, P. A. M. (1930) *The Principles of Quantum Mechanics*, Clarendon Press, Oxford. There were several later editions, for example the fourth edition in 1958.

Dixon, H. B. (1884) Conditions of chemical change in gases. Hydrogen, carbonic oxide and oxygen, *Phil. Trans.*, **175**, 617–684.

Dixon, H. B. (1893) Bakerian Lecture—The rate of explosion in gases, *Phil. Trans.*, **184**, 97–188. The chronograph of the type designed by F. J. Smith [Jervis-Smith] is described in this paper.

Dixon, H. B. (1903) On the movements of the flame in the explosion of gases, *Phil. Trans.*, **200**, 315–352.

Döbereiner, J. W. (1813) His hydrogen–air lamp, ignited catalytically by spongy platinum, was described, from personal communications, by P. L. Dulong and L. J. Thénard, *Ann. de chim. et de phys.*, **23**, 440–443, 24, 380–387.

Draper, J. W. (1841) On some analogies between the phaenomena of the chemical rays, and those of radiant heat, *Phil. Mag.*, [3], **19**, 195–210.

Draper, J. W. (1843) On the rapid dithionizing power of certain gases and vapours, and on an instantaneous method of producing spectral appearances,

Phil, Mag., [3], **22**, 161–165. This paper describes the photography of spectra. A daguerreotype, in monochrome, of the solar spectrum, taken by Draper in 1842 and at one time in the possession of Sir John Herschel, is now in the National Museum of Photography, in Bradford, England.

Draper, J. W. (1843) On a new system of inactive tithonographic spaces in the solar spectrum analogous to the fixed lines of Fraunhofer, *Phil. Mag.*, [3], **22**, 360–364. This paper compares the lines obtained in the solar spectrum with those obtained by Fraunhofer. It includes engravings made from a daguerreotype plate.

Draper, J. H. (1843) Description of the tithonometer, an instrument for measuring the chemical force of the indigo-tithonic rays, *Phil. Mag.*, [3], **23**, 401–415. This paper describes a 'tithonometer' (photometer) based on the reaction between hydrogen and chlorine; it was the prototype of the improved photometer later devised by Bunsen and Roscoe (1855).

Draper, J. W. (1844) *Of the forces that produce the organization of plants*, New York. This pamphlet, published privately, describes the first successful photography of a spectrum obtained with a ruled-glass diffraction grating.

Draper, J. W. (1845) On the interference spectrum, and the performance of the tithonic rays, *Phil. Mag.*, [3], **26**, 465–478. It was in this paper that Draper demonstrated that it is the absorbed rays that bring about chemical action (the 'Grotthuss-Draper law').

Draper, J. W. (1847) On the production of light by heat, *Phil. Mag.*, [3], **30**, 345–360. In this paper Draper stated qualitatively the spectroscopic principles that were presented mathematically by Kirchhoff in 1860.

Draper, J. W. (1847) On the production of light by chemical action, *Phil. Mag.*, [3], **32**, 100–114. This paper describes the spectra of flames and discusses their origin.

Draper, J. W. (1857) On the diffraction spectrum—Remarks on M. Eisenlohr's recent experiments, *Phil. Mag.*, [4], **13**, 153–156.

Draper, J. W., On the measurement of the chemical action of light, *Phil. Mag.*, [4], **14**, 161–164. This paper describes a tithonometer (photometer) based on the photolysis of ferric oxalate.

Draper, J. W. (1874) Early contributions to spectrum photography and photochemistry, *Nature*, **10**, 243–244. In this Letter to the Editor Draper referred to the daguerreotype of a spectrum that he had produced in 1842 and published in 1844, and made the justifiable claim that his paper of 1847, entitled 'On the production of light by heat', had contained all the qualitative ideas later expressed mathematically by Kirchhoff in 1860.

Drude, P., and W. Nernst (1894) Über Elektrostriction durch freie Ionen, *Z. physikal. Chem.*, **15**, 79–85.

Drude, P. (1904) Optische Eigenschaften und Elektronentheorie (Optical properties and electron theory), *Ann. der Physik*, [4], **14**, 677–725.

Dulong, P. L., and A. J. Petit (1819) Sur quelques points importants de la théorie de la chaleur (On some important points in the theory of heat), *Ann. de chim. et de phys.*, **10**, 395–413. An excerpt is in Magie (1935).

Dutrochet, R. J. H. (1824) *L'agent immédiat du mouvement vital dévoilé dans sa nature et dans son mode d'action, chez les végétaux et chez les animaux* (The agency responsible for movement revealed in nature and its mode of action, in plants and animals), Paris.

Dutrochet, R. J. H. (1828) *Nouvelles recherches sur l'endosmose et l'exosmose...* (New researches on endosmosis and exosmosis), Paris. Dutrochet also published numerous papers on the same subject in the *Annales de chemie*.

Ehrenfest, P. (1911) Welche Züge der Lichtquantenhypothese spielen in der Theorie der Wärmestrahlung eine wesentliche Rolle? (What type of light-quantum hypothesis plays an important role in the theory of heat radiation?), *Ann. der Physik*, [4], **36**, 91–118.

Ehrenfest, Paul and Tatiana (1911) *Begriffliche Grundlagen der statistischen Auffassung in der Mechanic* (The conceptual foundations of the statistical approach in mechanics), *Encyklopädie der mathematische Wissenschaften*, Vol. 4, Part 32, Tenbruer, Leipzig. English translation by M. J. Moravsic, Cornell University Press, 1959, Dover Publications, Mineola, New York, 1991.

Eigen, M. (1954) Methods for investigation of ionic reactions in aqueous solutions with half lives as short as 10^{-9} s, *Disc. Faraday Soc.*, **17**, 194–205.

Einstein, A. (1902) Kinetische Theorie des Wärmesgleichgewicht und des zweiten Hauptsatzes der Thermodynamik (Kinetic theory of thermal equilibrium and the second law of thermodynamics), *Ann. der Physik*, [4], **9**, 417–433.

Einstein, A. (1905) Über einen die Erzeugung und Verwandlung des Lichtes betreffenden heuristischen Gesichtspunkt (On a heuristic viewpoint concerning the emission and transformation of light), *Ann. der Physik*, [4], **17**, 132–148. A translation of this paper appears in A. B. Arons and M. B. Peppard, *Amer. J. Phys.*, **33**, 367–374 (1965).

Einstein, A. (1905) Die von der molekularkinetischen Theorie der Wärme geforderte Bewegung von in ruhenden Flüssigkeiten suspendierten Teilchen (On the movement of small particles suspended in a stationary liquid, according to the kinetic molecular theory of heat), *Ann. der Physik*, [4], **17**, 549–560. English translations of this and of Einstein's four other papers on Brownian movement, by A. D. Cowper with notes by R. Fürth, are in *Investigations on the Theory of the Brownian Movement*, Methuen, London, 1926, reprinted by Dover Publications, New York, 1956.

Einstein, A. (1906) Zur Theorie der Brownschen Bewegung (On the theory of the Brownian movement), *Ann. der Physik*, [4], **19**, 371–381.

Einstein, A. (1906) Eine neue Bestimmung des Molecüldimensionen (A new determination of molecular dimensions), *Ann. der Physik.*, [4], **19**, 289–306. Corrections to this paper are in *Ann. der Physik*, [4], **34**, 591–592 (1911); in the Methuen and Dover translations these corrections are incorporated in the text of the original article.

Einstein, A. (1907) Theoretische Bemerkungen über die Brownsche Bewegung (Theoretical observations on the Brownian movement). *Z. Elektrochem.*, **13**, 41–42.

Einstein, A. (1907) Die Plancksche Theorie der Strahlung und die Theorie der specifische Wärme (Planck's radiation theory and the theory of specific heats, *Ann. der Physik*, [4], **22**, 180–190.

Einstein, A. (1908) Elementare Theorie der Brownschen Bewegung (Elementary theory of the Brownian movement), *Z. Elektrochem.*, **14**, 235–239. This paper was written especially for chemists, at the suggestion of R. Lorentz.

Einstein, A. (1912) Thermodynamische Begründung des photochemischen Äquivalentgesetzes (Thermodynamic basis of the law of photochemical equivalence), *Ann. der Physik*, [4], **37**, 832–838.

Einstein, A. (1912) Nachtrag zu meinen Arbeit: Thermodynamische Begründung . . . (Supplement to my article on the thermodynamic basis . . .), *Ann. der Physik*, [4], **38**, 881–884, *J. de physique*, **3**, 272–282 (1913).

Einstein, A., and O. Stern (1913) Einige Argumente für die Annahme eine molekular Agitation bein absoluten Nullpunkt (Some arguments in support of the assumption of molecular vibration at the absolute zero), *Ann. der Physik*, [4], **40**, 551–560.

Einstein, A. (1916) Zur Quantentheorie der Strahlung (On the quantum theory of radiation), *Mitteilungen der physikal. Ges. Zürich*, **18**, 47–62; reprinted in *Physikal. Z.*, **18**, 1121–128 (1917). English translation in van der Waerden (1967)

Eley, D. D., and E. K. Rideal (1941) The catalysis of the parahydrogen conversion by tungsten, *Proc. Roy. Soc., A*, **178**, 429–451.

Elster, J., and H. Geitel (1889) Notiz über die Zerstreuung der negativen Elektricität durch das Sonnen- und Tageslicht (Note on the scattering of negative electricity by sunlight and daylight), *(Wiedemann's) Ann. der Phys. und Chem.*, [3], **38**, 40–41 and later papers.

Erdey-Gruz, T., and M. Volmer (1830) Zur Theorie der Wasserstoffüberspannung (On the theory of hydrogen overvoltage), *Z. physikal. Chem.*, **150**, 203–213.

Erdey-Gruz, T., and H. Wick (1932) Zur Frage der Wasserstoffüberspannung (On the question of hydrogen overvoltage), *Z. physikal. Chem.*, **162**, 53–62.

Erdey-Gruz, T., and H. Wick (1932) Die Abscheidungsspannung des Quecksilbers an Fremdelectroden (The deposition potential of mercury on foreign electrodes), *Z. physikal. Chem.*, **162**, 63–7C.

Eucken, A. (1930) *Lehrbuch der chemischen Physik* (Textbook of chemical physics), Akademische Verlags Gesellschaft, Leipzig, pp. 58–59.

Euler, L. (1750) Recherche sur l'origine des forces (Research on the origin of forces), *Mém. Acad. Sci. Berlin*, **6**, 419–447. In this paper Euler criticized Newton's use of the expression *vis inertiae* (force of inertia).

Euler, L. (1750) Découverte d'un nouveau principe de méchanique (Discovery of a new principle of mechanics), *Mém. Acad. Sci. Berlin*, **6**, 185–217. This paper was the first to recognize as a fundamental mechanical law the statement that force is mass times acceleration.

Evans, M. G., and M. Polanyi (1935) Some applications of the transition state method to the calculation of reaction velocities, especially in solution, *Trans. Faraday Soc.*, **31**, 875–895.

Evans, M. G., and M. Polanyi (1937) On the introduction of thermodynamic variables into reaction kinetics, *Trans. Faraday Soc.*, **33**, 448–452.

Eyring, H., and F. Daniels (1930) The decomposition of nitrogen pentoxide in inert solvents, *J. Amer. Chem. Soc.*, **52**, 1472–1492.

Eyring, H., and M. Polanyi (1931) Über einfachs Gasreaktionen (On simple gas reactions), *Z. physikal. Chem.*, B, **12**, 279–311. A translation of the first part, dealing with the construction of the potential-energy surface, is in Back and Laidler (1969).

Eyring, H. (1935) The activated complex in chemical reactions, *J, Chem. Phys.*, **3**, 107–115. See also Wynne-Jones and Eyring (1935) which later in the same year gave a more general formulation, in terms of thermo-dynamics.

Eyring, H., S. Glasstone and K. J. Laidler (1939) Application of the theory of absolute reaction rates to overvoltage, *J. Chem. Phys.*, **7**, 1053–1065.

Eyring, H. See also Gershinowitz and Eyring (1935).

Faraday, M. (1833) Experimental researches in electricity. Third Series. 7. Identities of electricities derived from different sources, *Phil. Trans.*, **123**, 23–54. Reprinted in Faraday (1852). Faraday's many papers are conveniently classified in Scott (1964).

Faraday, M. (1834) Experimental researches in electricity, Sixth Series. 12. On the power of metals and other solids to induce the combination of gaseous bodies, *Phil. Trans.*, **124**, 55–76.

Faraday, M. (1844) A speculation concerning electric conduction and the nature of matter, *Phil. Mag.*, **24**, 136–144; reprinted in Volume II of Faraday (1830, 1844, 1852).

Faraday, M. (1852) *Experimental Researches in Electricity*, Francis and Taylor, London, earlier editions 1830 and 1844; reprinted by Dover Publications, New York, 1964. There is also an Everyman's Library edition which includes selections: J. M. Dent, London, and E.P.Dutton, New York, 1912 with many later reprints.

Fick, A. (1855) Über Diffusion, *(Poggendorff's) Ann. der Phys. und Chem.*, [2], **94**, 59–86.

Fock, V. (1930) Näherungsmethode zur Lösung des quantenmechanischen Mehrkörperproblems (Approximate methods for the solution of quantum-mechanical many-body problems), *Z. Physik*, **61**, 126–148.

Fontenelle, Bernard Le Bouvier de (1686) *Entretiens sur la pluralité des mondes* (A discourse on the plurality of worlds), Paris.

Fowler, R. H. (1932) Theories of adsorption of gases. Quantum mechanics of the reversible cell and electrolysis, *Trans. Faraday. Soc.*, **28**, 368–378.

Fowler, R. H., and T. E. Sterne (1932) Statistical mechanics with special reference to the vapour pressures and entropies of crystals, *Rev. Mod. Phys.*, **4**, 635–722.

Fowler, R. H. (1936) *Statistical Mechanics. The Theory of the Properties of Matter in Equilibrium*, Cambridge University Press.

Fowler, R. H., and E. A. Guggenheim (1939) *Statistical Thermodynamics. A Version of Statistical Mechanics for Students of Physics and Chemistry*, Cambridge University Press. Their modification of the formula for the activation energy given by Tolman (1920) is on pp. 491–506.

Fox Talbot, see Talbot, W. H. F.

Fraenkel, G. (1986) Nuclear magnetic resonance line-shape analysis of re-organizing systems, in F. Bernasconi (Ed.), *Investigation of Rates and Mechanisms of Reactions*, Wiley, New York, Part II, pp. 547–605.

Frank-Kamenetsky, D. A. (1940) Conditions for the applicability of Bodenstein's method in chemical kinetics, *J. Phys. Chem. U.S.S.R.*, **14**, 695–700. By 'Bodenstein's method' he meant the steady-state hypothesis, actually first suggested by Chapman and Underhill (1913).

Frankland, E. (1852) On a new series of organic bodies containing metals, *Phil. Trans.*, **142**, 417–444.

Franklin, Benjamin (1751) *Experiments and Observations on Electricity, made at Philadelphia in America*, E. Cave, London. This 90-page book was put together by Peter Collinson with whom Franklin had an extensive correspondence.

Franklin, Benjamin (1774) Of the stilling of waves by means of oil, *Phil. Trans.*, **64**, 445–460.

Fraunhofer, J. von (1817, 1821) Bestimmung des Brechungs- und Farbenzerstreuungs- Vermögens verschiedenen Glasarten, in Bezug auf die Vervollkommung achromatische Fernröhre (Determination of the refractive and colour dispersion powers of different kinds of glass, in relation to the perfection of achromatic telescopes), *Denkschr. Akad. Wiss. Munchen*, **5**, 193–226 (1817); **8**, 1–76 (1821).

Fraunhofer, J. von (1823) Kurzer Bericht von den Resultaten neuen Versuche über die Gesetze des Licht, und die Theorie derselben (A short report on the results of new experiments concerning the laws of light, and theories relating to them), *(Gilbert's) Ann. der Phys. und physikal. Chem.*, **74**, 337–378.

Freind, John (1709) *Praelectiones Chymicae*, London; English translation, *Chymical Lectures*, London, 1712.

Fulhame, Mrs [Elizabeth] (1794) *A Essay on Combustion with a View to a New Art of Dying and Painting, wherein the Phlogistic and Antiphlogistic Hypotheses are Proved Erroneous*, published by the author and printed by J. Cooper, Bow Street, Covent Garden, London, 182 pages; the Preface is dated 5 November 1794. Chapter VIII is entitled 'Reduction of Metals by Light'. There is a detailed abstract of the book in a commentary on it by J. F. Coindet, *Ann. de chimie*, **26**, 58–85 (1798).

Galilei, Galileo (1590, 1600) *On Motion and On Mechanics*, translated by I. E. Drabkin and Stillman Drake, Univ. of Wisconsin Press, Madison, 1960; the first work, *De motu*, was written in about 1590, the second, *Le mecchaniche*, in about 1600. Excerpts are included in Magie (1935).

Galilei, Galileo (1638) *Discorsi e dimonstrazioni matematiche intorno a due nueve scienze, attenenti alla Mecanica & i Movimenti Locali* (Discourses and mathematical demonstrations concerning two new sciences), Leiden; English translation by S. Drake, *Two New Sciences*, Univ. of Wisconsin Pres, Madison, 1974.

Garber, Elizabeth (1969) James Clerk Maxwell and thermodynamics, *Amer. J. Phys.*, **37**, 146–155.

Gay-Lussac, J. L. (1802) Sur la dilatation des gaz et des vapeurs (On the expansion of gases and vapours), *Ann. de chimie*, **43**, 137–175 (1802). English version in *Nicholson's Journal*, **33**, 207–216, 257–267 (1802) and German version in *(Gilbert's) Ann. der Physik*, **12**, 257–291 (1803). An excerpt is in Magie (1935), pp. 165–172.

Gay-Lussac, J. L. (1809) Mémoire sur la combinaison des substances gazeuses, les unes avec les autres (Memoir on the combination of gaseous substances with each other), *Mém. phys. chem. soc. Arcueil*, **2**, 207–234. English translations of parts of this are in *Foundations of the molecular theory; comprising papers and extracts by John Dalton, Joseph-Louis Gay-Lussac and Amadeo Avogadro (1808–1811)*, Alembic Club Reprints, No. 4, Edinburgh, 1819. An excerpt from the paper is in Leicester and Klickstein (1968).

Gershinowitz, H., and H, Eyring (1935) The theory of trimolecular reactions, *J. Amer. Chem. Soc.*, **57**, 985–991.

Gibbs, J. W. (1873) Graphical methods in the thermodynamics of fluids, *Trans. Conn. Acad. Arts Sci.*, **2**, 382–404. Gibbs's papers are all reproduced in *The Scientific Papers of J. Willard Gibbs* (1906), the thermodynamic papers in Vol. 1.

Gibbs, J. Willard (1873) A method of graphical representation of the thermodynamic properties of substances by means of surfaces, *Trans. Conn. Acad. Arts Sci.*, **2**, 382–404.

Gibbs, J. Willard (1876, 1878) On the equilibrium of heterogeneous substances, *Trans. Conn. Acad. Arts Sci.*, **3**, 108–248, 343–524.

Gibbs, J. Willard (1902) *Elementary Principles in Statistical Mechanics Developed with Special Reference to the Rational Foundations of Thermodynamics*, New Haven.

Gibbs, J. Willard (1906) *The Scientific Papers of J. Willard Gibbs* (Ed. H. A. Bumsted and R. G. Van Name), 2 vols. New York, reprinted 1961. The thermodynamic papers are in Vol. 1, the papers on statistical mechanics in Vol. 2.

Gibson, Q. H., and L. Milmes (1964) Apparatus for rapid and sensitive spectrophotometry, *Biochem. J.*, **91**, 161–171.

Goodwin, H. M. (Ed., 1899) *The Fundamental Laws of Electrolytic Conduction*, New York.

Gordon, J. P., H. G. Zeiger and C. H. Townes (1954, 1956) Molecular microwave oscillator and new hyperfine structure in the microwave spectrum of NH_3, *Phys. Rev.*, **95**, 282–284; The maser—new type of microwave amplifier, frequency standard and spectrometer, *Phys. Rev.*, **99**, 1264–1274 (1956). See also Schawlow and Townes (1958).

Gouy, G. (1888) Note sur le mouvement Brownien (A note on Brownian movement), *J. de physique*, [2], **7**, 561–564.

Gouy, G. (1910) Sur la constitution de la charge électrique à la surface d'un électrolyte (On the constitution of the electric charge on the surface of an electrolyte), *J. de physique*, [4], **9**, 457.

Graham, T. (1833) On the law of diffusion of gases, *Phil. Mag.*, **2**, 175–190, 269–276, 351–358.

Graham, T. (1846) On the motion of gases, *Phil. Trans.*, **136**, 573–631.

Graham, T. (1849) On the motion of gases. Part II, *Phil. Trans.*, **139**, 349–391.

Graham, T. (1850) Bakerian Lecture—On the diffusion of liquids, *Phil. Trans.*, **140**, 1–46.

Graham, T. (1866) On the absorption and dialytic separation of gases by colloidal septa, *Phil. Trans.*, **156**, 399–439.

Greene, E. F., R. W. Roberts and J. Ross (1960) Variation of a reaction cross section with energy, *J. Chem. Phys.*, **32**, 940–941.

Grimm, H. G., H. Ruf and H. Wolff (1931) Über den Zusammenhang zwischen Molekülbau und Reaktionsgeschwindigkeit bei der Vereinigung von Triäthylamin und Åthyljodid in verschieden Lösingsmittel (On the relationship between molecular structure and rate of reaction in the combination of triethylamine and ethyl iodide in different solvents, *Z. physikal. Chem.*, *B*, **13**, 301–315.

Gronwall, T. H., V. K. La Mer and K. Sandved (1928) The influence of the so-called higher terms in the Debye–Hückel theory of solutions of strong electrolytes, *Physikal. Z.*, **29**, 358–393; see also La Mer, Gronwall and Greiff (1931).

Grotthuss, C. J. D. von (1805, 1806) *Mémoire sur la décomposition de l'eau et des corps, qu'elle tient en dissolution, à l'aide de l'électricité galvanique* (Memoir on the decomposition of water and of substances it holds in solution, by means of galvanic electricity), Rome, 1805, Milan, 1806; *Ann. de chimie*, 58, 54–74 (1806); English version in Phil. Mag. 25, 330–339 (1806).

Grotthuss, C. J. D. von (1819) Über die chemische Wirksamheit des Lichts und der Elektricität . . . (On the chemical action of light and of electricity), *(Gilbert's) Ann. der Phys. und der physikal. Chem.*, **61**, 50–74.

Grove, W. R. (1839) On voltaic series and the combination of gases by platinum, *Phil. Mag.*, **14**, 127–130. This gives a brief account of the gas battery; further details are in papers that appeared in 1842, 1843 and 1845.

Grove, W. R. (1839) On a new voltaic combination, *Phil. Mag.*, **20**, 294–303. This cell had zinc in H_2SO_4 and platinum in HNO_3.

Grove, W. R. (1839) On a small voltaic battery of great energy; some observations on voltaic combinations and forms of arrangement; and on the inactivity of a copper positive electrode in nitro-sulphuric acid, *Phil. Trans.*, **15**, 287–293.

Grove, W. R. (1842) Remarks on a letter of Prof. Daniell in the Philosophical Magazine, *Phil. Mag.*, **21**, 333–335. Reply to Daniell (1842a).

Grove, W. R. (1842) On a gaseous voltaic battery, *Phil. Mag.*, **21**, 417–420.

Grove, W. R. (1843) On the subject of Prof. Daniell's last communication, *Phil. Mag.*, **22**, 32–35. Reply to Daniell (1842b).

Grove, W. R. (1843) On the gas voltaic battery, *Phil. Trans.*, **133**, 91–112.

Grove, W. R. (1845) On the gas voltaic battery, *Phil. Trans.*, **135**, 351–361.

Grove, W. R. (1846) *The Correlation of Physical Forces*, London; 6th edition, with reprints of many of Grove's papers, 1874.

Grove, W. R. (1852) On the electro-chemical polarity of gases, *Phil. Trans.*, **142**, 87–101.

Grove, W. R. (1858) On the electrical discharge and its stratified appearance in rarified media, *Proc. Roy. Institution*, **23**, 5–10.

Guggenheim, E. A. (1933) *Modern Thermodynamics by the Methods of Willard Gibbs*, Methuen, London.

Guggenheim, E. A. (1956) More about the laws of reaction rates and of equilibrium, *J. Chem. Education*, **33**, 544–545.

Guldberg, C. M., and P. Waage (1864) Studier over affiniteten (Studies on affinity), *Forhand. Videnskabs-Selskab. Christiania*, **35**, 35–40, 92–94, 111–120; reproduced as a pamphlet in French in 1867 (see below).

Guldberg, C. M., and P. Waage (1867) *Etudes sur les affinitées chimiques* (Studies on chemical affinities) Christiania, 1867. This and other publications by Guldberg and Waage are reprinted in Ostwald's *Klassiker* (1899), No. 104, and translated excerpts are in Leicester and Klickstein (1968). For discussions of the Guldberg and Waage work see Lund and Hassel (1964), Mysels (1956), Guggenheim (1956) and Lund (1965).

Guldberg, C. M. (1870) Sur la loi des points de congélation de solutions salines (On the law for the freezing points of salt solutions), *Comptes rendus*, **70**, 1349–1352.

Guldberg, C. M., and P. Waage (1879) Chemische Affinität (Chemical affinity), *J. prakt. Chem.*, **19**, 69–114; reproduced in Ostwald's *Klassiker* (1899), No. 104.

Gurney, R. W. (1931) The quantum mechanics of electrolysis, *Proc. Roy. Soc., A*, **134**, 137–154.

Gurney, R. W. (1932) The quantum mechanics of electrochemistry. II, *Proc. Roy. Soc., A*, **136**, 378–391.

Haas, A. E. (1910) Über die elektrodynamische Bedeutung des Planck'schen Strahlgesetzes und über eine neue Bestimmung des elektrischen Elementarquantums und der Dimensionen des Wasserstoffatoms (On the electrodynamic significance of Planck's radiation law, and on a new significance for the quantum of electricity and the dimensions of the hydrogen atom), *Sitzungsber. Akad. Wiss. Wien*, **119**, 119–144. Niels Bohr, in his classic 1913 papers on the hydrogen atom, referred to this work of Haas, but in his 1961 Rutherford Memorial Lecture, through a lapse of memory, he said that he only learnt of the work later.

Hales, Stephen (1727) *Vegetable Staticks*, London.

Hales, Stephen (1733) *Haemostaticks*, London.

Hamilton, W. Rowan (1834) On a general method in dynamics, in which the study of the motions of all free systems of attracting or repelling points is reduced to the search and differentiation of one central relation or characteristic function, *Phil. Trans.*, **124**, 247–308.

Hammett, L. P. (1940) *Physical Organic Chemistry*, McGraw-Hill, New York. The 'Hammett equation' is discussed on pp. 184–199.

Harcourt, A. Vernon, and W. Esson (1865) On the laws of connexion between the conditions of a chemical change and its amount, *Phil. Trans.*, **156**, 193–222 (1865); *Proc. Roy. Soc.*, **14**, 470–474 (1865). Excerpts are in Back and Laidler (1967).

Harcourt, A. Vernon, and W. Esson (1895) On the connexion between the conditions of a chemical change and its amount. Part III. Further researches on the reaction of hydrogen dioxide and hydrogen iodide, *Phil. Trans.*, *A*, **186**, 817–895. This was their Bakerian Lecture to the Royal Society.

Harcourt, A. Vernon (1912) On the variation with temperature of the rate of a chemical change, *Phil. Trans.*, *A*, **212**, 187–204. This paper has an extensive mathematical appendix, pp. 193–204, written by W. Esson.

Harned, H. S. (1918) Notes on neutral salt catalysis. I. The role of the solvent in neutral salt catalysis in aqueous solution, *J. Amer. Chem. Soc.*, **40**, 1461–1481, and later papers.

Hartree, D. R. (1928) The wave mechanics of an atom with a non-coulombic central field, *Proc. Camb. Phil. Soc.*, **24**, 89–110, 111–132, 426–437.

Hartridge, H., and F. J. W. Roughton (1923) A method of measuring the velocity of very rapid chemical reactions, *Proc. Roy. Soc.*, *A*, **104**, 376–394.

Heisenberg, W. (1925) Über quantentheoretische Umdeutung kinematischer Beziehungen (On a quantum-theoretical interpretation of kinematic and mechanical relationships), *Z. Physik*, **33**, 879–893. Translation in van der Waerden (1967).

Heisenberg, W. (1927) Über den anschaulichen Inhalt der quantentheoretischen Kinematik und Mechanik (On the intuitive content of quantum-theoretical kinematics and mechanics), *Z. Physik*, **43**, 172–198.

Heitler, W., and F. London (1927) Wechselwirkung neutrale Atome und homopolare Bindung nach der Quantenmechanik (Interaction of neutral atoms and homopolar bonds according to quantum mechanics), *Z. Physik*, **44**, 455–472. See also London (1928).

Helmholtz, H. (1847) *Über die Erhaltung der Kraft* (On the conservation of force [energy]) G. Reiner, Berlin. This 53-page book is based on an address to the Physical Society of Berlin given on 23 July, 1847. The book is subdivided as follows: Introduction; I. The conservation of *vis viva*; II. The principle of conservation of energy; III. The application of the theory to mechanical theorems; IV. The energy equivalent of heat; V. The energy equivalent of electrical processes; VI. The energy equivalent of magnetism and electromagnetism. In the above list of contents the word Kraft has been translated as energy. For a detailed discussion of the publication see Y. Elkana, Helmholtz's'Kraft': an illustration of concepts in flux, *Hist. Stud. Phys. Sci.*, **2**, 263–298 (1970). However, this discussion is somewhat laboured: the point can be made very simply—Helmholtz used *Kraft* in two senses; in the general sense of any property producing some effect (including force as now defined), and as energy as now defined. Excerpts from a translation by John Tyndall are in Magie (1935).

Helmholtz, H. von (1858) Über Integrale der hydrodynamischen Gleichungen welche den Wirbelbewegungen entsprechen (On the integrals of the hydrodynamical equations that give rise to vortex motion), *(Crelle's) J. für die reine und angew. Math.*, **55**, 25–55. A free translation of this, made by P. G. Tait and approved by Helmholtz, appeared in *Phil. Mag.*, [4], **33** (Supplement), 485–512 (1867). This translation includes a note by Kelvin, The translational velocity of a circular vortex ring, pp. 511–512.

Helmholtz, H, (1877) Über galvanische Ströme, verursacht durch Concentrations-Unterschiede; Folgerungen aus der mechanischen Wärmetheorie (On galvanic currents, caused by concentration differences; conclusions from the mechanical theory of heat), *Monatsber. Preuss. Akad. Wiss. Berlin*, 713–726. English version in *Phil. Mag.*, [5], **5**, 348–358 (1878). Reproduced in Ostwald's *Klassiker*, No. 124. 1902.

Helmholtz, H. (1879) Studien über electrische Grenzschichten (Studies on electrical double layers), *(Wiedemann's) Ann. der Phys. und Chem.*, **7**, 337–382.

Helmholtz, H. von (1881) On the modern development of Faraday's conception of electricity, *J. Chem. Soc.*, **39**, 277–304; reproduced in *Faraday Lectures, 1869–1928*, London, 1928.

Helmholtz, H. (1882) Helmholtz's suggestion of a dropping mercury electrode is on pp. 30–35 of A. König, *(Wiedemann's) Ann. der Phys. und Chem.*, **16**, 1–38.

Helmholtz, H. (1882) Zur Thermodynamik chemischer Vorgange (On the thermodynamics of chemical processes), *Sitzungsber. Preuss. Akad. Wiss. Berlin*, **1**, 23–39).

Henry, William (1803) Experiments on the quantity of gases absorbed by water at different temperatures and under different pressures, *Phil. Trans.*, **93**, 29–42, 274–276.

Herapath, John (1821) Tables of temperature, and a mathematical development of the causes and laws of phenomena which have been adduced in support of hypotheses of 'calorific capacity', 'latent heat', etc., *Annals of Philosophy*, [2], **2**, 50, 89, 201, 256, 363, 434; **3**, 16.

Herapath, John (1847) *Mathematical Physics; or the Mathematical Principles of Natural Philosophy; with a development of the causes of heat, Gaseous Elasticity, Gravitation and other great phenomena of Nature*, Whittaker & Co. and (Herapath's) Railway Journal Office, London.

Herschel, John F. W. (1823) On the absorption of light by coloured media, and on the colours of the prismatic spectrum exhibited by various flames, with an account of a ready mode of determining the absolute dispersive power of any medium, by direct experiment, *Trans. Roy. Soc. Edinburgh*, **9**, 445–460. This is one of the earliest articles on spectral analysis, the first being Melvill (1752).

Herschel, Sir John F. W. (1839) Note on the art of photography, or the application of the chemical rays of light to the purposes of pictorial representation, *Proc. Roy. Soc.*, **4**, 131–133. This was a brief report, based on the reading of the paper at the 14 March 1839, meeting of the Royal Society. A full paper of the same title was submitted by Herschel to the *Philosophical Transactions* but was withdrawn, probably because he had been able to improve his procedures as described in some of his later papers. The manuscript of this withdrawn paper, after being missing for over a century, has been discovered by Larry Schaaf and reproduced by him, with a commentary, in *History of Photography*, **3**, 47–60 (1979).

Herschel, Sir John F. W. (1840) On the chemical action of the rays of the solar spectrum on preparations of silver and other substances, both metallic and non-metallic, and on some photographic processes, *Proc. Roy. Soc.*, **4**, 205–210. This was a brief report based on presentations to the Royal Society in February and March, 1840. The paper includes photography in the infrared. The full paper appeared in *Phil. Trans.*, **130**, 1–59 (1840).

Herschel, Sir John F. W. (1842) On the action of the rays of the solar spectrum on vegetable colours, and on some new photographic processes, *Proc Roy. Soc.*, **4**, 397–398, read to the Royal Society on 17 November 1842. This paper describes the 'cyanotype', or blueprint. The work is described in greater detail in *Phil. Trans.*, **132**, 181–214, and this paper is reprinted in *Phil. Mag.*, [3]. **22**, 5–21, 107–116, 135, 170–180, 246–252, 505–511. Some of Herschel's original coloured photographs of spectra are held by the Royal Society.

Herschel, W. (1800) Experiments on the refrangibility of the visible rays of the sun, *Phil. Trans.*, **92**, 284–292.

Hertz, H. (1887) Über einem Einfluss das ultravioletten Lichtes auf die elektrische Entladung (On the effect of ultraviolet light on an electric discharge), *(Wiedemann's) Ann. der Phys. und Chem.*, [3], **32**, 545–572.

Hertz, H. (1888) Über Inductionserscheinung hervorgerufen durch die electrischen Vorgänge in Isolatoren (On induction phenomena brought about by electrical processes in insulators), *(Wiedemann's) Ann. der Phys. und Chem.*, [3], **34**, 273–285.

Herzberg, G. (1929) Zum Aufbau der zweiatomigen Moleküle (On the structure of diatomic molecules), *Z. Physik*, **57**, 601–630.

Herzberg, G., and D. A. Ramsay (1950) *Disc. Faraday Soc.*, **9**, 80–81. This was a contribution to the general discussion at the meeting.

Herzfeld, K. F. (1919) Zur Theorie der Reaktionsgeschwindigkeit in Gazen (On the theory of reaction rates in gases), *Ann. der Physik*, [4], **59**, 635–666, *Z. Elektrochem.*, **25**, 301–304.

Herzfeld, K. F., and F. O. Rice (1928) Dispersion and absorption of high frequency sound waves, *Phys. Rev.*, [2], **31**, 691–695.

Herzfeld, K. F., and T. A. Litovitz (1959) *Absorption and Dispersion of Ultrasonic Waves*, Academic Press, New York.

Heyrovsky, J. (1922) Electrolysis with a dropping mercury cathode, *Phil. Mag.*, [6], **45**, 303–315 (1923).

Heyrovsky, J., and M. Shikata (1925) Researches with the dropping mercury cathode, *Rec. trav. chim. Pays-Bas*, **44**, 488–513. In this paper the word 'polarography' was first suggested.

Higgins, William (1789) *A Comparative View of the Phlogistic and Antiphlogistic Theories, and Inductions, to which is annexed an Analysis of the Human Calculus, with Observations on its Origin . . .* , J. Murray, London; second edition, 1791; this second edition is reproduced in facsimile in Wheeler and Partington (1960).

Higgins, William (1814) *Experiments and Observations on the Atomic Theory*, Longman, Hurst, Rees, Orme and Brown, London; reproduced in facsimile in Wheeler and Partington (1960).

Hinshelwood, C. N., and H. C. Tingey (1922) The catalytic decomposition of formic acid on surfaces of platinum and silver, *J. Chem. Soc.*, 1668–1676, and many other papers on surface reactions appearing between 1922 and 1929. This work is reviewed in more detail in Laidler (1988), pp. 261–263.

Hinshelwood, C. N. (1927) On the theory of unimolecular reactions, *Proc. Roy. Soc., A*, **113**, 230–233 (1927). The first part of this article is reproduced in Back and Laidler (1967).

Hinshelwood, C. N., and H. W. Thompson (1928) The kinetics of the combination of hydrogen and oxygen, *Proc. Roy. Soc., A*, **118**, 170–175.

Hinshelwood, C. N., and H. W. Thompson (1929) The mechanism of the homogeneous combination of hydrogen and oxygen, *Proc. Roy. Soc., A*, **122**, 610–621.

Hirst, T. Archer (1887) *The Mechanical Theory of Heat with its Application to the Steam Engine and to the Physical Properties of Bodies*, London; this is an edited English translation of Clausius (1865–1867). This volume has useful mathematical appendices which the Browne translation (1879) lacks.

Hittorf, J. W. (1853–1859) Über die Wanderung der Ionen während der Elektrolyse (On the movement of ions during electrolysis), *(Poggendorff's) Ann. der Phys. und Chem.*, [2], **89**, 177–211; **98**, 1–33 (1856); **103**, 1–56 (1858); **106**, 337–411, 513–586 (1859). These papers are reproduced in Ostwald's *Klassiker* (1891), Nos. 21 and 23. An English translation of the first paper is in Goodwin (1899). Selections from the papers appear in Leicester and Klickstein (1952).

Hooke, Robert (1665) *Micrographia; or, Some Physiological Descriptions of Minute Bodies Made by Magnifying Glasses*, London. This book includes Hooke's theory of combustion, an excerpt of which is in Leicester and Klickstein (1952), pp. 48–53.

Hooke, Robert (1677) *Lampas: or, Descriptions of some Mechanical Improvements of Lamps and Waterpoises. Together with some other Physical and Mechanical Discoveries*, Printed for John Martyn, Printer to the Royal Society, at the Bell in St Paul's Church-Yard, London. This book describes some extensions of his theory of combustion that was in the *Micrographia*.

Horstmann, A. F. (1868) Zur Theorie der Dissociations-Erscheinungen (On the theory of dissociation phenomena), *Ber. Deut. chem. Ges.*, **1**, 210–215. Some of Horstmann's papers are included in Ostwald's *Klassiker* (1899) No. 104, and in van't Hoff's *Abhandlungen* (1903). Since Horstmann's work was done before the concept of free energy had been introduced his treatments were somewhat cumbrous, but essentially correct.

Horstmann, A. F. (1871) Zur Theorie der Dissociation (On the theory of dissociation), *Ber. Deut. Chem. Ges.*, **4**, 635–639.

Horstmann, A. F. (1873) Theorie der Dissociation (Theory of dissociation), *(Poggendorff's) Ann. der Phys. und Chem.*, **170**, 192–210.

Hückel, E. (1931–1933) Quantentheoretische Beiträge zum Benzolprobleme (Quantum-mechanical treatment of the benzene problem), *Z. Physik*, **70**, 204–286, **72**, 310–337 (1931); **76**, 628–648 (1932); **83**, 632–668 (1933).

Hückel, E. (1934) Theory of free radicals of organic chemistry, *Trans Faraday Soc.*, **30**, 40–52.

Huggins, W., and W. A. Miller (1862–1863) Note on the lines in the spectra of some of the fixed stars, *Proc. Roy. Soc.*, **12**, 444–445. No photographs of spectra were reported in this paper.

Huggins, W., and W. A. Miller (1864) On the spectra of some of the fixed stars, *Phil. Trans.*, **154**, 413–436. This paper describes attempts to photograph the spectra, but the clock used to drive the telescope was not precise enough to allow the spectral lines to be clearly distinguished.

Huggins, W., and W. A. Miller (1866) On the spectrum of a new star in Corona Borealis, *Proc. Roy. Soc.*, **15**, 146–149. This paper included hand-drawn engravings of spectra, but no photographs.

Huggins, W., and W. A. Miller (1866) Note on the spectrum of the variable star α Orionis, *Monthly Notices of the Royal Astronomical Society*, **26**, 215–218.

Huggins, W. (1880) On the photographic spectra of stars, *Phil. Trans.*, **171**, 669–690. By this time Huggins was using a clock that drove the telescope sufficiently accurately to allow sharp spectral photographs to be taken, and in the later work he used dry plates instead of wet collodion. The paper includes drawings made from the photographs of a number of stars.

Hund, F. (1927) *Linien Spektren und das periodische Systeme der Elemente* (Line spectra and the periodic system of the elements), Springer, Berlin.

Hund, F. (1928) Zur Deutung der Molecülspectren. IV (On the significance of molecular spectra. IV), *Z. Physik*, **51**, 759–795.

Hund, F. (1928) Quantenmechanik und chemische Bindung (Quantum mechanics and chemical bonding), *Z. Elektrochem.*, **34**, 437–442.

Hund, F. (1929) Chemical binding, *Trans. Faraday Soc.*, **25**, 646–648.

Hund, F. (1931–1932) Zur Frage der chemischen Bindung (On the question of chemical bonding), *Z. Physik*, **73**, 1–30, 565–577; **74**, 429–430.

Huygens, C., (1669) Regles du mouvement dans le rencontre des corps (Laws of motion in the collision of bodies), *Phil. Trans.*, **4**, 925–928.

Huygens, C., (1673) *Horologium Oscillatorium sive de Motu Pendulorum ad Horologia Aptato Demonstrationes Geometricae* (The pendulum clock, or geometrical demonstrations of the motion of pendulums), F. Muguet, Paris. This book, mainly about clocks, discusses many principles of mechanics. Excerpts are in Magie (1935)

Javan, A., W. R. Bennett and D. R. Herriott (1961) Population inversion and continuous optical maser oscillation in a gas discharge containing a He–Ne mixture, *Phys. Rev. Letters*, **6**, 106–110.

Jeans, J. H. (1904) *Dynamical Theory of Gases*, Cambridge University Press, and later editions.

Jeans, J. H. (1905) On the partition of energy between matter and radiation, *Phil. Mag.*, [6], **10**, 91–98. Jeans's contribution to the Rayleigh–Jeans formula was in only one sentence: 'It seems to me that Lord Rayleigh has introduced an unnecessary factor 8 by counting negative as well as positive values of his integrals.'

Jeans, J. H. (1905) A comparison between two theories of radiation, *Nature*, **72**, 293–294; this was a Letter to the Editor.

Jeans, J. H. (1914) *Report on Radiation and the Quantum Theory*, Physical Society, London. In this report Jeans gave his approval to the quantum theory.

Jeans, J. H. (1940) *Introduction to the Kinetic Theory of Gases*, Cambridge University Press, and later editions. The quantum theory is not treated in this book, even in later editions, although the specific heats of gases at low temperatures are discussed. To Jeans, 'kinetic theory' apparently meant 'classical kinetic theory'.

Jones, Catharine M., and W. C. McC. Lewis (1920) Studies in catalysis. Part XIV. The mechanism of the inversion of sucrose, *J. Chem. Soc.*, **117**, 1120–1133, and later papers.

Jones, W. J., A. Lapworth and H. M. Lingford (1913) The influence of water on the partial pressures of hydrogen chloride above its alcoholic solutions, *J. Chem. Soc.*, **103**, 252–263.

Joule, J. P. (1841) On the heat evolved by metallic conductors of electricity, and in the cells of a battery during electrolysis, *Phil. Mag.*, **19**, 260–277. In this paper Joule established that the heat is proportional to the square of the current and the first power of the potential (C^2V). An excerpt is in Magie (1935).

Joule, J. P. (1843) On the production of heat by voltaic electricity, *Proc. Roy. Soc.*, **4**, 280–281: this was a brief abstract, communicated by P. M. Roget (of *Thesaurus* fame), then Secretary of the Royal Society.

Joule, J. P. (1843) On the calorific effects of magnetoelectricity and the mechanical value of heat, *Phil. Mag.*, [3], **23**, 263–276, 347–355, 435–443.

Joule, J. P. (1850) On the mechanical equivalent of heat, *Phil. Trans.*, **140**, 61–82. This was Joule's major paper on the subject. It was communicated by Michael Faraday, Joule himself then not being a Fellow of the Royal Society. The paper contains references to Rumford, but not to Mayer. Excerpts are in Magie (1935).

Joule, J. P. (1852) On the heat disengaged in chemical combinations, *Phil. Mag.*, **3**, 481–504.

Joule, J. P., and W. Thomson (1854, 1862) On the thermal effects of fluids in motion, *Phil. Trans.*, **144**, 321–364 (1854); **152**, 579–589 (1862). These papers were concerned with the 'Joule–Thomson effect'.

Joule J. P. (1884–1887) *The Collected Papers of James Prescott Joule*, 2 vols. Taylor & Francis, London.

Kammerlingh Onnes, H. (1911) Further experiments with liquid helium, *Communications from the Physical Laboratory of the University of Leiden*, No. 122b; earlier work was described in No. 118b.

Kammerlingh Onnes, H. (1967) Investigations into the properties of substances at low temperatures, which have led, among other things, to the preparation of liquid helium, in *Nobel Lectures, 1901–1921*, Elsevier, New York, pp. 306–336.

Kasper, J. V. V., and G. C. Pimentel (1965) HCl chemical laser, *Phys. Rev. Letters*, **14**, 352–354.

Kassel, L. S. (1928) Studies in homogeneous gas reactions. I, *J. Phys. Chem.*, **32**, 225–242.

Kassel, L. S. (1932) *The Kinetics of Homogeneous Gas Reactions*, Chemical Catalog Company, New York.

Keill, John (1701) *Introductio ad Veram Physicam, Oxford; English translation, An Introduction to Natural Philosophy*, London, 1720. Note the translation of veram physicam (literally, true physics) as natural philosophy; at the time the English word 'physics' was ambiguous, as it might have referred to the science of physic, i.e., medicine.

Keill, J. (1708) On the laws of attraction and other principles of physics, *Phil. Trans.*, **27**, 97.

Kelvin, Lord, see Thomson, W.

Kenney-Wallace, G. A., E. L. Quitevis and E. F. G. Templeton (1985) Ultra-fast laser spectroscopy: a probe of the photodynamics of chemical intermediates, *Rev. Chem. Intermediates*, **6**, 197–235, and many other publications.

Kirchhoff, G. (1860) Über die Fraunhofer'schen Linien (On the Fraunhofer lines), *(Poggendorff's) Ann. der Phys. und Chem.*, **109**, 148–150. An English translation of part of this paper, made by G. G. Stokes, appeared as 'On the simultaneous emission and absorption of rays of the same definite refrangibility', *Phil. Mag.* [4], **19**, 193–197 (1860). An excerpt is in Magie (1935).

Kirchhoff, G. (1860) Über des Verhältniss zwischen dem Emissionsvermögen der Körper für Wärme und Licht (On the relation between the radiative and absorptive powers of different bodies for heat and light), *(Poggendorff's) Ann.*

der Phys. und Chem., **109**, 275–301. English translation in *Phil. Mag.*, [4], **20**, 1–21 (1860).

Kirchhoff, G., and R. Bunsen (1861) Chemische Analyze durch Spektral-beobachtungen (Chemical analysis by spectral observations), *(Poggendorff's) Ann. der Phys. und Chem.*, **113**, 337–381. English translation in *Phil. Mag.*, [4], **22**, 229–309, 498–510.

Kirchhoff, G. (1863) Contributions towards the history of spectrum analysis and of the analysis of the solar atmosphere, *Phil. Mag.*, [4], **25**, 250–263. This paper presents Kirchhoff's point of view regarding the priorities for the various contributions in this field. He gives particular credit to Herschel, Talbot and W. Allen Miller for their contributions to spectral analysis. He discusses Balfour Stewart's work on the principles of spectroscopy, but makes no mention of Draper.

Klein, M. J. (1970) Maxwell, his demon, and the second law of thermodynamics, *American Scientist*, **58**, 84–97.

Knott, C. G. (1911) *Life and Scientific Work of Peter Guthrie Tait*, Cambridge University Press.

Kohlrausch, F. (1868) Über Galvanische Widerstandbestimmung flüssiger Leiter (On the determination of the electrical resistances of solutions), *Nachr. Ges. Wiss. Göttingen*, 415–420 (1868).

Kohlrausch, F. (1869) Über die Gültigkeit der Ohm'schen Gesetz für Elektro-lyten (On the validity of Ohm's law for electrolytes), *Nachr. Ges. Wiss. Göttingen*, 14–16.

Kohlrausch, F., and W. A Nippoldt (1869) Über der Gültigkeit des Ohm'schen Gesetz für Elektrolyten und eine numerische Bestimmung des Leitung-widerstandes der verdünnsten Schwefelsäure durch alternirande Ströme (On the validity of Ohm's law for electrolytes, and a numerical estimation of the electrical resistance of dilute sulphuric acid with alternating currents), *(Poggendorff's) Ann. der Phys. und Chem.*, **138**, 28–298, 370–390; English summary in *Phil. Mag.*, **40**, 227–229 (1870).

Kohlrausch, F., and O. Grotrian (1874) Das elektrische Leitungsvermögen der Chlor-Alkalien und alkalinischen Erden, sowie der Salpetersäure in wässerigen Lösungen (The electrical conductivity of alkali chlorides, alkaline earth chlorides and nitric acid in aqueous solutions), *Nachr. Ges. Wiss. Göttingen*, 405–408; *(Poggendorff's) Ann. der Phys. und Chem.*, **154**, 1–14, 215–239 (1875); English version in *Phil. Mag.*, **49**, 417–425 (1875).

Kohlrausch, F. (1878) Über das elektrische Leitungsvermögen des Wassers und einiger anderen schlechter Leiter (On the electrical conductivity of water and some other poor conductors), *(Poggendorff's) Ann. der Phys. und Chem., Ergänz.*, **8**, 1–16.

Kohnstamm, P., and F. E. C. Scheffer (1911) Thermodynamic potential and velocities of reaction, *Verhandel. Konink. Akad. Wet. Amsterdam*, **13**, 789–804.

Kooij, D. M. (1893) Über die Zersetzung des gasförmigen Phosphorwasserstoffs (On the decomposition of gaseous phosphine), *Z. physikal. Chem.*, **12**, 155–161.

Kossel, W. (1916) ber Molekülbildung als Frage des Atombaus (Molecular structure in relation to atomic constitution), *Ann. der Physik*, [4], **49**, 229–362.

Krönig, A. K. (1856) *Grundzüge einer Theorie der Gäze* (Fundamentals of the theory of gases). This was first published as a small book by Hayn, Berlin, and reprinted in *(Poggendorff's) Ann. der Phys. und Chem.*, [2], **99**, 315–322 (1856).

Lagrange, J. L. (1788) *Méchanique analytique*, Paris.

Laidler, K. J. (1988) Chemical kinetics and the Oxford college laboratories, *Arch. Hist. Exact Sci.*, **38**, 197–283.

Lambert, J. H. (1760) *Photometria: sive, de mensura et gradibus luminis, colorum et umbrae* (Light measurements relating to the intensity and variation of light, for colours and shadows), Augsburg.

La Mer, V. K., T. H. Gronwall and L. J. Greiff (1931) The influence of the higher terms of the Debye–Hückel theory in the case of unsymmetrical valence type electrolytes, *J. Phys. Chem.*, **35**, 2245–2288.

Langmuir, I. (1912) Chemically active modification of hydrogen, *J. Amer. Chem. Soc.*, **34**, 1310–1325.

Langmuir, I. (1915) Dissociation of hydrogen into atoms. II. Calculation of the degree of dissociation and the heat of formation, *J. Amer. Chem. Soc.*, **37**, 417–458.

Langmuir, I. (1916) The constitution and fundamental properties of solids and liquids. I. Solids, *J. Amer. Chem. Soc.*, **38**, 2221–2295. This paper includes a derivation of the 'Langmuir adsorption isotherm' and a discussion of some surface reactions in terms of it.

Langmuir, I. (1917) The constitution and fundamental properties of solids and liquids. II. Liquids, *J. Amer. Chem. Soc.*, **39**, 1845–1906. The film balance is described, with a diagram.

Langmuir, I. (1918) Adsorption of gases on plane surfaces of glass, mica and platinum, *J. Amer. Chem. Soc.*, **40**, 1361–1403.

Langmuir, I. (1919) The arrangement of electrons in atoms and molecules, *J. Amer. Chem. Soc.*, **41**, 868–934.

Langmuir, I. (1920) Radiation as a factor in chemical action, *J. Amer. Chem. Soc.*, **42**, 2190–2205; also a comment at a Faraday Society Discussion: *Trans. Faraday Soc.*, **17**, 600–601 (1922).

Langmuir, I. (1921) The mechanism of the catalytic action of platinum on the reactions $2CO + O_2 = 2CO_2$ and $2H_2 + O_2 = 2H_2O$, *Trans. Faraday Soc.*, **17**, 621–654.

Laplace, P. S. (1796) *Exposition du système du monde*, 2 vols., Paris; English translation by H. H. Harte, *The System of the World*, 2 vols., Dublin, 1830. A version of this book was prepared by Mary Somerville (1831).

Laplace, P. S. (1799–1825) *Traité de méchanique céleste*, 5 vols. in 4, Paris; English translation by H. H. Harte, *A Treatise of Celestial Mechanics*, 2 vols. Dublin, 1822–27.

Lapworth, A. (1908) An examination of the conception of hydrogen ions in catalysis, salt formation, and electrolytic conduction, *J. Chem. Soc.*, **93**, 2187–2203. See also Jones, Lapworth and Lingford (1913).

Laszlo, P. (1979) Fast kinetics studied by NMR, *Progress in Nuclear Magnetic Resonance Spectroscopy*, **13**, 257–270.

Lavoisier, A., and P. S. Laplace (1780) *Mém. Acad. roy. sci.*, 355–408.

Le Chatelier, H. L. (1884), Sur un énoncé général des lois des équilibres chimiques (On a general formulation of the laws of chemical equilibrium), *Comptes rendus*, **99**, 786–789; an English translation of part of this is in Leicester and Klickstein (1968).

Le Chatelier, H. L. (1888) Sur les lois d'équilibre chimique (On the laws of chemical equilibrium), *Comptes rendus*, **106**, 355–357.

Leclanché, G. (1868) *Les Mondes*, **16**, 532.

Leicester, H. M., and H. S. Klickstein (1968) *A Source Book in Chemistry*, Harvard University Press, Cambridge, Mass.

Leibniz, G. W. (1686) Brevis demonstratio erroris mirabilis Cartesii (Short demonstration of a memorable error of Descartes), *Acta Eruditorum*, **5**, 561–563. An excerpt is in Magie (1935).

Leibniz, G. W. (1695) Specimen Dynamicum (A dynamical model), *Acta Eruditorum*, **14**, 145–157.

Lenard, P. (1900) Erzeugung von Kathodenstrahlen durch ultraviolettes Licht (Production of cathode rays by ultraviolet light), *Ann. der Physik*, [4], **2**, 359–375.

Lenard, P. (1902) Über die lichtelektrische Wirkung (On the photoelectric effect), *Ann. der Physik*, [4], **8**, 145–198.

Lennard-Jones, J. E. (1929) The electronic structures of some diatomic molecules, *Trans. Faraday Soc.*, **25**, 668–686.

Lewis, G. N. (1900) A new conception of thermal pressure and a theory of solutions, *Proc. Amer. Acad.*, **36**, 145–168. German version as 'Eine neue Auffassung vom thermischen Drucke und eine Theorie der Lösungen', *Z. physikal. Chem.*, **35**, 343–368 (1900).

Lewis, G. N. (1901) The law of physico-chemical change, *Proc. Amer. Acad.*, **37**, 49–69. German version as 'Das Gesetz physiko-chemische Vorgänge', *Z. physikal. Chem.*, **38**, 205–226 (1901).

Lewis, G. N. (1907) Outlines of a new system of thermodynamic chemistry, *Proc. Amer. Acad.*, **43**, 259–293. German version as 'Umriss einer neuen Systems der chemischen Thermodynamik', *Z. physikal. Chem.*, **61**, 129–165 (1907).

Lewis, G. N. (1910) The use and abuse of the ionic theory, *Z. physikal. Chem.*, **70**, 212–219.

Lewis, G. N. (1912) The activities of the ions and the degree of dissociation of strong electrolytes, *J. Amer. Chem. Soc.*, **34**, 1631–1648.

Lewis, G. N. (1916) The atom and the molecule, *J. Amer. Chem. Soc.*, **38**, 762–785.

Lewis, G. N., and M. Randall (1921) Activity coefficients of strong electrolytes, *J. Amer. Chem. Soc.*, **43**, 1112–1154.

Lewis, G. N., and M. Randall (1923) *Thermodynamics and the Free Energy of Chemical Substances*, McGraw-Hill, New York. There were a number of later editions.

Lewis, G. N. (1923) *Valence and the Structure of Atoms and Molecules*, Chemical Catalog Co., New York.

Lewis, G. N. (1926) The conservation of photons, *Nature*, **118**, 874–875. As this title implies, Lewis's concept of a photon, as a particle that is conserved, is different from the modern one.

Lewis, W. C. McC. (1916) Studies in catalysis. Part V. Quantitative expressions for the velocity, temperature coefficient, and effect of the catalyst, from the point of view of the radiation hypothesis, *J. Chem. Soc.*, **109**, 796–815. In this and many of his other papers Lewis interpreted kinetic results in terms of the radiation hypothesis. Although this proved to be wrong these contributions of Lewis are valuable, as they can easily be reinterpreted in terms of the more satisfactory theories.

Lewis, W. C. McC. (1917) Studies in catalysis. Part VII. Heat of reaction, equilibrium constant, and allied quantities, from the point of view of the radiation hypothesis, *J. Chem. Soc.*, **111**, 457–469.

Lewis, W. C. McC. (1918) Studies in catalysis, Part IX. The calculation in absolute measure of velocity constants and equilibrium constants in gaseous systems, *J. Chem. Soc.*, **113**, 471–492. This paper gives Lewis's classic application of the kinetic theory to the rate constant of a reaction. Excerpt in Back and Laidler (1969). pp. 36–40.

Lewis, W. C. McC. (1921) The radiation hypothesis of chemical reactivity, *Trans. Faraday Soc.*, **17**, 573–587.

Lewis, W. C. McC., see also C. M. Jones and Lewis (1920).

Linde, C. (1896) Erzielung niedrigster Temperaturen (The achievement of the lowest temperatures), *(Wiedemann's) Ann. der Phys. und Chem.*, **57**, 328–332.

Lindemann, F. A. (1922) *Trans. Faraday Soc.*, **17**, 598–599. Reproduced in Back and Laidler (1967). Lindemann's brief oral presentation to the Faraday Society was on 28 September 1921.

Liveing, G. D., and J. Dewar (1879) On the spectrum of sodium and potassium, *Proc. Roy. Soc.*, **29**, 398–402, and many later papers, collected in:

Liveing, G. D., and Sir J. Dewar (1915) *Collected Papers on Spectroscopy*, Cambridge University Press.

Locke, John (1690) *Essay concerning Human Understanding*.

Lockyer, J. N. (1874) Atoms and molecules spectroscopically considered, *Nature*, **10**, 69–71, 89–90.

Lodge, O. (1893) On the connection between aether and matter, *British Assn. Reports*, **61**, 688.

London, F. (1928) Zur Quantentheorie der homopolaren Valenzzahlen (On the quantum theory of homopolar valence numbers), *Z. Physik*, **46**, 455–477.

London, F. (1928) In *Probleme der modernen Physik*, Sommerfeld Festschrift, F. Hirtzel, Leipzig.

London, F. (1931) Über einige Eigenschaften und Anwendungen der Molekularkräfte (On some properties and applications of molecular forces), *Z. physikal. Chem.*, B, **11**, 222–251.

London, F. (1937) The general theory of molecular forces, *Trans. Faraday Soc.*, **33**, 8–26.

Lowry, T. Martin (1923) The uniqueness of hydrogen, *J. Soc. Chem. Ind.*, **42**, 43–47.

Lund, E. W., and O. Hassel (1964) Guldberg and Waage and the law of mass action, in *The Law of Mass Action: A Centenary Volume*, Universitats forlajet, Oslo, 1964, pp. 37–46.

Lund, E. W. (1965) Guldberg and Waage and the law of mass action, *J. Chem. Education*, **42**, 548–550.

Magie, W. F. (1935) *A Source Book in Physics*, McGraw-Hill, New York.

Maiman, T. H. (1960) Stimulated optical radiation in ruby, *Nature*, **187**, 493–494.

Marcelin, André (1931) *Solutions Superficielles: Fluides à deux dimensions et stratifications monomoléculaires*, Les Presses Universitaire de France, Paris.

Marcelin, René (1910) Sur la méchanique des phénomènes irréversibles (On the mechanism of reversible processes), *Comptes rendus*, **151**, 1052–1055.

Marcelin, R. (1914) Expression des vitesses de transformation des systèmes physico-chimiques en fonction de l'affinité (Equation for the rates of transformation in physico-chemical systems as a function of the affinity), *Comptes rendus*, **158**, 116–118.

Marcelin, R. (1914) Influence de la temperature sur les vitesses de transformation des systèmes physico-chimiques, (Influence of temperature on rates of transformation in physico-chemical systems), *Comptes rendus*, **158**, 407–408.

Marcelin, R. (1914) Méchanique des phénomènes irreversibles à partir de la règle de distribution de Boltzmann–Gibbs (Dynamics of irreversible processes from the standpoint of the Boltzmann–Gibbs distribution law), *J. de chim. phys.*, **12**, 451–460.

Marcelin, R. (1915) Contribution à l'étude de la cinetique physico-chimique (Contribution to the study of physico-chemical kinetics), *Ann. de physique*, [9], **3**, 120–184, 185–231. The first part of this paper is concerned with theoretical kinetics, the second with experimental studies on rates of evaporation and sublimation, with the use of novel techniques.

Marcelin, R. (1918) Étude experimentelle sur le developpement des cristaux (An experimental study of the development of crystals), *Ann. de physique*, **10**, 185–188. This paper and the following one are posthumous papers, prepared by his brother André Marcelin, from notes made by René before he was killed in action in September, 1914.

Marcelin, R. (1918) Structure des cristaux et lames très minces; nouvelle détermination expérimentale des grandeurs moléculaires (Structure of crystals and very thin films; a new experimental determination of molecular dimensions), *Ann. de physique*, [9], **10**, 189–194.

Marcus, R. A. (1952) Unimolecular dissociations and free radical combination reactions, *J. Chem. Phys.*, **20**, 359–368.

Mariotte, Edmé (1679) *Essaie de la Nature de l'Air*, Paris. An excerpt dealing with the pressure–volume relationship is in Magie (1935).

Maxwell, J. Clerk (1860) Illustrations of the dynamical theory of heat, *Phil. Mag.*, **19**, 19–32, **20**, 21–37. Reprinted in part in Brush (1965).

Maxwell, J. Clerk (1860) On the results of Bernoulli's theory of gases as applied to their internal friction, their diffusion, and their conductivity for heat, *British Assn. Reports*, 30th Meeting, Part 2, 15–16. This publication is not included in the collected papers (1890).

Maxwell, J. Clerk (1866) The Bakerian Lecture: On the viscosity or internal friction of air and other gases, *Phil. Trans.*, **156**, 249–268.

Maxwell, J. Clerk (1867) On the dynamical theory of gases, *Phil. Trans.*, **157**, 49–88. Reprinted in Brush (1966), Vol. 2, pp. 23–87.

Maxwell, J. Clerk (1871) *Theory of Heat*, Longmans, Green & Co., London; 4th edition, greatly revised, 1875; 7th edition, 1883; 11th edition with notes by Lord Rayleigh, 1894. Simple accounts of the entropy concept are found in the 4th and later editions, for example in the 7th edition on pp. 162–165. See also Elizabeth Garber (1969).

Maxwell, J. Clerk (1873) *A Treatise on Electricity and Magnetism* (Ed. W. D. Niven), 2 vols., Oxford University Press. This book was the culmination of a series of papers on electromagnetic theory, the first appearing in 1855.

Maxwell, J. Clerk (1873) Molecules, *Nature*, **8**, 437–441 (1873).

Maxwell, J. Clerk (1873) In a letter to P. G. Tait, dated 1 December, 1873, reprinted in Knott (1911), Maxwell derided the ideas of Clausius and Boltzmann, who were regarding the second law as a consequence of the laws of mechanics. The letter is signed dp/dt; for a discussion of the equation $dp/dt = JCM$ on which this designation is based see Klein (1970).

Maxwell, J. Clerk (1875) On the dynamical evidence of the molecular constitution of bodies, *Nature*, **11**, 357–359, 374–377.

Maxwell, J. Clerk (1875) Atom, *Encyclopaedia Britannica*, 9th Edition. This article contains a lengthy section on the vortex atom.

Maxwell, J. Clerk (1877) The kinetic theory of gases, *Nature*, **16**, 342–346.

Maxwell, J. Clerk (1878) Tait's Thermodynamics, *Nature*, **17**, 257–259.

Maxwell, J. Clerk (1879) On Boltzmann's theorem on the average distribution of energy in a system of material points, *Trans. Camb. Phil. Soc.*, **12**, 547–570.

Maxwell, J. Clerk (1890) *The Scientific Papers of James Clerk Maxwell* (Ed. W. D. Niven), Cambridge University Press. Reprinted by Dover, New York, 1965. This compilation is unsatisfactory in two respects: it omits some important publications, and it fails to specify the references completely, not giving the dates of publication.

Mayer, J. R. (1842) Bemerkungen über die Kräfte der unbelebten Natur (Notes on the forces of inanimate nature), *Ann. der Chem. und Pharm.*, **42**, 233. Mayer had previously submitted a paper entitled 'Über die quantitative und qualitative Bestimmung der Kräfte' (On the quantitative and qualitative estimation of forces) to the *Ann. der Physik und Chemie*, but its editor, Poggendorff, rejected it; the paper that appeared in the *Ann. der Chem. und Pharm.* was an improved version, and was accepted by Liebig.

Mayer, J, R. (1845) *Die organische Bewegung in ihrem Zusammenhang mit dem Stoffwechsel* (Living motion in its relationship with chemical change), privately published, Heilbronn.

Mayer, J. R. (1867) *Die Mechanik der Wärme* (The mechanics of heat), Stuttgart.

McLeod, H. (1884) Apparatus for measurements of low pressures of gas, *Phil. Mag.*, **48**, 110–113.

Mellor, J. W. (1904) *Chemical Statics and Dynamics*, Longmans, Green & Co., London.

Melvill, T. (1752) Observations on light and colours, *Edinburgh Physical and Literary Essays*, **2**, 35.

Menschutkin, N. A. (1890) Über die Affinitätskoeffizienten der Alkylhaloide und die Amine (On the affinity coefficients [rate constants] for [the reaction of] alkyl halide and amine). *Z. physikal. Chem.*, **6**, 41–57.

Meyer, J. Lothar (1864) *Die modernen Theorien der Chemie* (The modern theories of chemistry), Breslau.

Michelson, A. A. (1892) On the application of interference methods to spectroscopic measurements, *Phil. Mag.*, [5], **32**, 338–346.

Miller, W. A. (1845) Experiments and observations on some cases of lines in the prismatic spectrum produced by the passage of light through coloured vapours and gases, and from certain coloured flames, *Phil. Mag.*, [2], **27**, 81–91. This work had been presented at the Cambridge meeting of the British Association for the Advancement of Science in 1845. Miller was perhaps the first to publish detailed drawings of flame and absorption spectra.

Miller, W. A. (1855–56–57) *Elements of Chemistry: Theoretical and Practical*, John W. Parker & Son, London. The book is in three volumes, of which the subtitles are *Part 1. Chemical Physics, Part 2. Inorganic Chemistry*, and *Part 3, Organic Chemistry*. It is interesting to see this threefold division of chemistry as early as 1855. There were five later editions; the 6th edition (Longmans, Green, Reader and Dyer, 1877–80) was revised and greatly expanded after Miller's death in 1870 by Herbert McLeod (1841–1923), still remembered particularly for the 'McLeod gauge'. The book was reprinted in the United States. Miller's book was a successor to Daniell's *An Introduction to the Study of Chemical Philosophy* (1845). It was used as a standard textbook for examination purposes at King's College, London, at Oxford, and presumably at other British universities. It was in the 3rd edition of *Part 1. Chemical Physics* (1863) that Thomas Andrews' work on critical phenomena was first announced. Andrews also helped McLeod with parts of the 6th edition.

Miller, W. A. (1862) On spectrum analysis, *Chemical News*, **5**, 201–203, 214–218.

Miller, W. A. (1862) On the photographic transparency of various bodies, and on the photographic effects of metallic and other spectra obtained by means of the electric spark, *Phil. Trans.*, **152**, 861–887; the paper is reproduced in J. Chem. Soc., 19, 59–88 (1862). This paper includes a diagram of the apparatus used to photograph spectra, and excellent photographs of the spectra of a number of salts. The work was presented at the Manchester meeting of the British Association for the Advancement of Science in March, 1891, and a brief report appeared in *British Assn. Reports*, 1861, Part 2, 87–88.

Miller, W. A. (1867) Spectrum analysis, with its applications to astronomy, *Chemical News*, **15**, 259–263, 276–278, 285–287; **16**, 8–11, 20–22, 47–49, 71–73. This was an account of a series of four illustrated lectures given by Miller at the Royal Institution in 1867.

Millikan, R. A. (1910) A new modification of the cloud method of determining the elementary charge and the most probable value of that charge, *Phil. Mag.*, [6], **19**, 209–228.

Millikan, R. A. (1913) On the elementary electrical charge and the Avogadro constant, *Phys. Rev.*, [2], **2**, 109–143.

Millikin, G. A. (1936) Photometric methods of measuring the velocity of rapid reactions. III. A portable micro-apparatus applicable to an extended range of reactions, *Proc. Roy. Soc., A*, **155**, 277–292.

Milner, S. R. (1912) Virial of a mixture of ions, *Phil. Mag.*, [6], **23**, 551–578; **25**, 742–751 (1913); **35**, 214–220 (1918).

Milner, S. R. (1918) The effect of interionic forces on electrolysis, *Phil. Mag.*, [6], **35**, 352–364 (1918).

Milner, S. R. (1919) Note on the effect of interionic forces in electrolytes, *Trans. Faraday Soc.*, **15**, 148–151.

Mitscherlich, E. (1842) Chemischer Zersetzung und Verbindung vermittelst Contactsubstanzen (Chemical decomposition and combination brought about by contact substances) *Ann. der Chem. und Pharm.*, **44**, 186–204.

Mitscherlich, E. (1842) Über die chemische Zersetzung und Verbindung mittelst Contactsubstanzen (On chemical decomposition and combination brought about by contact substances), *(Poggendorff's) Ann. der Phys. und Chem.*, **55**, 209–229.

Moelwyn-Hughes, E. A., and C. N. Hinshelwood (1932) The kinetics of two bimolecular reactions in solution and in the vapour phase, *J. Chem. Soc.*, 230–240. For a review of the work of Hinshelwood and Moelwyn-Hughes see Laidler (1988) and Shorter (1985).

Moelwyn-Hughes, E. A. (1932) The kinetics of certain bimolecular reactions in solution, *Chem. Rev.*, **10**, 241–264.

Moseley, H. G. J. (1913) The high-frequency spectra of the elements, *Phil. Mag.*, [6], **26**, 1024–1034.

Moseley, H. G. J. (1914) The high-frequency spectra of the elements, Part II, *Phil. Mag.*, [6], **27**, 703–713. In 1913 Moseley returned to Oxford from Manchester and carried out his later experiments in the Electrical Laboratory, which is now part of the Clarendon Laboratory. Some of his apparatus, including X-ray tubes and a spectrometer, and samples of metals and alloys, are to be seen in the Museum for the History of Science, Old Ashmolean Building, Oxford.

Mulliken, (1924) Isotope effects in the band spectrum of carbon monoxide and silicon hydride, *Nature*, **113**, 423–424. A related paper is in *Nature*, **114**, 349–350 (1924).

Mulliken, R. S. (1925) The isotope effect in band spectra. II. The spectrum of boron monoxide, *Phys. Rev.*, **25**, 259–294.

Mulliken, R. S. (1928) The assignment of quantum numbers for electrons in molecules, *Phys. Rev.*, **32**, 186–222, 761–772.

Mulliken, R. S. (1929) Band spectra and atomic nuclei, *Trans. Faraday Soc.*, **25**, 534–645.

Mulliken, R. S. (1935) Electronic structures of polyatomic molecules and valence. VI. On the method of molecular orbitals, *J. Chem. Phys.*, **3**, 375–378. This paper introduced the expression 'linear combination of atomic orbitals' (LCAO), the idea being due to Lennard-Jones (1929).

Mysels, K. J. (1956) The laws of reaction rates and of equilibrium, *J. Chem. Education*, **33**, 178–179.

Nernst, W. (1888) Zur Kinetik der in Lösung befindlichen Körper. I. Theorie der Diffusion (On the kinetics of substances in solution. I. Theory of diffusion), *Z. physikal. Chem.*, **2**, 613–637.

Nernst, W. (1889) Zur Theorie unkehrbarer galvanische Elemente (On the theory of reversible electrodes), *Sitzungsber. Preuss. Akad. Wiss. Berlin*, 83–85. This paper proposed the hydrogen electrode as the standard and expressed the emf of a concentration cell.

Nernst, W. (1889) Die elektromotorische Wirksamkeit der Ionen (The electromotive action of ions), *Z. physical. Chem.*, **4**, 129–181. This paper gives the equation for the emf of a cell other than a concentration cell.

Nernst, W. (1893) *Theoretische Chemie vom Standpunkte der Avogadroschen Regel und der Thermodynamik* (Theoretical chemistry from the standpoint of Avogadro's principle and of thermodynamics), Leipzig. English edition translated by C. S. Palmer, *Theoretical Chemistry*, London, 1st edition, 1895.

Nernst, W. (1894) Dielektrizitätskonstante und chemische Gleichgewicht (Dielectric constant and chemical equilibrium), *Z. physikal. Chem.*, **13**, 531–536.

Nernst, W. (1894) Zur Dissociation des Wassers (On the dissociation of water), *Z. physikal. Chem.*, **14**, 155–156.

Nernst, W. (1894) Methode zur Bestimmung von Dielektrizitätskonstanten (Method of determining dielectric constants), *Z. physikal. Chem.*, **14**, 622–663.

Nernst, W. (1894) See Drude and Nernst (1894).

Nernst, W. (1897) Die elektrolytische Zersetzung wässerige Lösungen (The electrolysis of aqueous solutions), *Ber. Deut. chem. Ges.*, **30**, 1547–1563.

Nernst, W. (1901) Zur Theorie der Lösungen (On the theory of solutions), *Z. physikal. Chem.*, **38**, 487–500. This paper includes a treatment of solubility products.

Nernst, W. (1904) Theorie der Reaktiongeschwindigkeit in heterogenen Systemen (Theory of reaction rates in heterogeneous systems), *Z. physikal. Chem.*, **47**, 52–55. This paper includes a treatment of the Nernst diffusion layer.

Nernst, W. (1906) Über die Berechnung chemische Gleichgewichte aus thermische Messungen (On the calculation of chemical equilibria from thermal measurements), *Nachr. Ges. Wiss. Göttingen*, 1–40. Nernst's heat theorem is discussed in detail in W. Nernst, *Die theoretischen und experimentellen Grundlagen des neuen Wärmesatzes*, Krapp, Halle, 1918; this appeared in translation as *The New Heat Theorem*, Methuen, London, 1926.

Nernst, W. (1911) Zur Theorie der spezifischen Wärme und über die Anwendung der Lehre von den Energiequanten auf physikalisch-chemische Fragen überhaupt (On the theory of specific heats and the application of the laws of energy quanta to physico-chemical problems in general), *Z. Elektrochem.*, **17**, 265–275.

Nernst, W., and F. A. Lindemann (1911) Spezifische Wärme und Quantentheorie (Specific heats and quantum theory), *Z. Elektrochem.*, **17**, 817–827.

Nernst, W. (1914) Über die Anwendung des neuen Wärmesatzes auf Gase (On the application of the new heat theorem to gases), *Z. Elektrochem.*, **20**, 357–360.

Nernst, W. (1918) Zur Anwendung des Einsteinschen photochemichschen Aequivalentgesetzes (On the application of Einstein's law of photochemical equivalence), *Z. Elektrochem.*, **24**, 335–336. In this paper Nernst proposed his chain mechanism for the hydrogen–chlorine reaction.

Nernst, W. (1918) *Die theoretische und experimentellen Grundlagen des neuen Wärmesatzes*, Knapp, Halle. Translation as *The New Heat Theorem*, Methuen, London (1926) and Dover Publications, New York (1969).

Neumann, Carl G. (1875) *Vorlesungen über die mechanische Theorie der Wärme* (Lectures on the mechanical theory of heat), Leipzig.

Newhall, Beaumont (1980) *Photography: Essays and Images*, Museum of Modern Art, New York. This book contains reproductions of some of the pioneering articles on photography.

Newton, Isaac (1672) A new theory about light and colours, *Phil. Trans.*, **6**, 3075–3087.

Newton, Isaac (1687) *Philosophae Naturalis Principia Mathematica*, The Royal Society, London. For translations and commentaries see *Sir Isaac Newton's Mathematical Principles of Natural Philosophy and His System of the World*, translated by A. Motte, revised and annotated by F. Cajori, Univ. of California

Press, Berkeley, 1934; A. Koyne and I. B. Cohen (Eds.), *Newton's Philosophae Naturalis Principia Mathematica*, 2 vols., Cambridge Univ. Press, 1972; and D. T. Whiteside (Ed.), *The Mathematical Papers of Isaac Newton*, Cambridge Univ. Press, 1974. Excerpts from Newton's *Principia* which deal with his theory of gases are reproduced in Brush (1965), Vol. 1, Selection 2, pp. 292–294. Excerpts dealing with mechanics are in Magie (1935).

Newton, Isaac (1710) *De Rerum Acidorum* (Concerning the Nature of Acids).

Newton, Isaac (1704) *Opticks*, Innys, London. Fourth edition (1730) reproduced in facsimile by G. Bell and Sons, London, 1931, and by Dover Publications, New York, 1952, 1979; the Dover edition has a preface by I. Bernard Cohen. Newton's theory of matter was first presented in the second edition, which appeared in 1717. Newton, always highly sensitive to criticism, postponed the publication of *Opticks* until a year after Robert Hooke's death (1703), since Hooke had made a strong attack on Newton's paper of 1672.

Nicholson, J. (1800) Account of the new electrical or galvanic apparatus of Sig. Alex. Volta, and experiments performed with the same, *Nicholson's Journal*, **4**, 179–187. This paper appeared in July, 1800, before the publication of the paper by Volta (1800). Excerpts from Nicholson's paper are included in Magie (1935) and in Boynton (1948).

Nollet, J. A. (1748) Recherches sur les causes du bouillonnement des liquides (Research on the causes of the boiling of liquids), *Mém. acad. roy. sci.*, 57–104.

Norrish, R. G. W., and G. Porter (1949) Chemical reactions produced by very high light intensities, *Nature*, **164**, 658.

Norrish, R. G. W., and G. Porter (1952) Spectroscopic studies of the hydrogen–oxygen explosion initiated by the flash photolysis of nitrogen dioxide, *Proc. Roy. Soc., A*, **210**, 439–460.

Odiorne, T. J., P. R. Brooks and J. V. Kasher (1971) Molecular beam reaction of K with HCl: effect of vibrational excitation of HCl, *J. Chem. Phys.*, **55**, 1980–1982.

Onnes, H. Kamerlingh, see Kamerlingh Onnes.

Onsager, L. (1926) Zur Theorie der Elektrolyte (On the theory of electrolytes), *Physikal. Z.*, **27**, 388–392.

Onsager, L. (1927) Report on a revision of the conductivity theory, *Trans. Faraday Soc.*, **23**, 341–349.

Ostwald, W. (1877–1880) Volum-chemische und optische-chemische Studien (Volumetric and optical studies on chemical systems), *J. prakt. Chem.*, [2], **16**, 385–423 (1877); [2], **18**, 328–371 (1878); [2], **22**, 305–322 (1880).

Ostwald, W. (1883–1888) Studien zur chemischen Dynamik (Studies in chemical dynamics), *J. prakt. Chem.*, [2], **27**, 1–39; [2], **28**, 449–495 (1883); [2], **29**, 385–408 (1884); [2], **31**, 307–317 (1885); [2], **35**, 112–121 (1887); *Z. physikal. Chem.*, **2**, 127–147 (1888).

Ostwald, W. (1884) Studien zur chemischen Dynamik (Studies in chemical dynamics), *J. prakt. Chem.*, **29**, 385–400. In this paper Ostwald called attention to Wilhelmy's work.

Ostwald, W. (1884) Notiz über das elektrische Leitungsvermögung der Säuren (Note on the electrical conductivities of salts), *J. prakt. Chem.*, [2], **30**, 93–95. This short paper was written after Ostwald received a copy of Arrhenius's thesis.

Ostwald, W. (1885–1887) *Lehrbuch der allgemeinen Chemie* (Textbook of general chemistry), Akademische Verlagsgesellschaft, Leipzig, 2 vols. A second and greatly enlarged edition of this, in 2 volumes and three parts, but never completed, was published between 1891 and 1902. This work was commonly known as 'der grosse Ostwald', to distinguish it from 'der kleine Ostwald' (1889). At the time the words 'allgemeine Chemie' (general chemistry) and 'physikalische Chemie' (physical chemistry) were used interchangeably in both German and English.

Ostwald, W. (1887) Studien zür Kontäktelektrizität (Studies on contact electricity), *Z. physikal. Chem.*, **1**, 583–630. This paper deals with the dropping mercury electrode.

Ostwald, W. (1888) Zur Theorie der Lösungen (On the theory of solutions), *Z. physikal. Chem.*, **2**, 36–37 (1888). A translated excerpt is in Leicester and Klickstein (1952).

Ostwald, W. (1888) Studien zur chemischen Dynamik. Über oxidations und reduktions Vorgänge (Studies in chemical dynamics. On oxidation and reduction processes), *Z. physikal. Chem.*, **2**, 127–147. The idea of half-life was introduced in this paper.

Ostwald, W. (1888) Über die Dissociationstheorie der Elektrolyte (On the dissociation theory of electrolytes), *Z. physikal. Chem.*, **2**, 270–283.

Ostwald, W. (1880) Zur Dissociationstheorie der Elektrolyte (On the dissociation theory of electrolytes), *Z. physikal. Chem.*, **3**, 588–602.

Ostwald, W. (1889) *Grundriss der allgemeinen Chemie*, Leipzig; this was commonly known as 'der kleiner Ostwald'. This book was translated into English by James Walker as *Outlines of General Chemistry*, Macmillan, London, 1890; a later edition was translated by W. W. Taylor, and published under the same title by Macmillan in 1912. These books are remarkable in containing not a single mathematical equation.

Ostwald, W. (1891) *Klassiker der exacten Wissenschaft* (Classics of the exact sciences), Wilhelm Engelmann, Leipzig.

Ostwald, W. (1900) Über Oxidationen mittels freien Sauerstoffs (On oxidations brought about by free acids), *Z. physikal. Chem.*, **34**, 248–252. Coupled reactions were considered in this paper.

Ostwald, W. (1901) Uber Katalyse (On catalysis), *Z. Elektrochem.*, **7**, 995–1066; Catalysis, *Nature*, **65**, 522–526 (1902), the latter being a slightly abridged translation of the German article.

Ostwald, W. (1902) Über Katalyse (On catalysis), *Physikal. Z.*, **3**, 313–322.

Ostwald, W. (1904) Faraday Lecture: Elements and compounds, *J. Chem. Soc.*, **85**, 506–522. His definition of 'chemical dynamics' is on p. 508.

Ostwald, Wolfgang (1917) *Introduction to Theoretical and Applied Colloid Chemistry*, translated by M. H. Fischer, John Wiley, New York.

Paneth, F., and W. Hofeditz (1929) Über die Darstellung von freien Methyl (On the preparation of the free methyl radical), *Ber. Deut. chem. Ges., B,* **62,** 1335–1347.

Pauli, W. (1923) Über die Gesetzmässigkeiten des anomalen Zeemaneffekts (On the regularities of the anomalous Zeeman effect), *Z. Physik,* **16,** 155–164.

Pauli, W. (1926) Über das Wasserstoffspektrum vom Standpunkt der neuen Quantenmechanik (On the hydrogen spectrum from the standpoint of the new quantum mechanics), *Z. Physik,* **36,** 336–363.

Pauling, L. (1927) The theoretical prediction of the physical properties of many-electron atoms and ions. Mole fraction, diamagnetic susceptibility and extension in space, *Proc. Roy. Soc., A,* **114,** 181–211. This and the following paper were written when Pauling was working with Sommerfeld at the University of Munich.

Pauling, L. (1927) The sizes of ions and the structure of ionic crystals, *J. Amer. Chem. Soc.,* **49,** 765–790.

Pauling, L. (1928) The shared-electron chemical bond, *Proc. Nat. Acad. Sci.,* **14,** 359–362.

Pauling, L. (1928) The application of the quantum mechanics to the structure of the hydrogen molecule and hydrogen molecule ion and to related problems, *Chem. Revs.,* **5,** 173–213.

Pauling, L. (1931) The nature of the chemical bond. Application of results obtained from the quantum mechanics and from a study of the paramagnetic susceptibility to the structure of molecules, *J. Amer. Chem. Soc.,* **53,** 1367–1400.

Pauling, L. and G. W. Wheland (1933) The nature of the chemical bond. V. The quantum-mechanical calculation of the resonance energy of benzene and naphthalene and the hydrocarbon free radicals, *J. Chem. Phys.,* **1,** 363–374.

Pauling, L., and E. Bright Wilson (1935) *Introduction to Quantum Mechanics, with Applications to Chemistry,* McGraw-Hill, New York.

Pauling, L. (1939) *The Nature of the Chemical Bond, and the Structure of Molecules and Crystals: An Introduction to Structural Chemistry,* Cornell University Press, Ithaca, N.Y. This book was dedicated to G. N. Lewis.

Pelzer, H., and E. Wigner (1932) Über die Geschwindigkeitkonstante von Austauschreaktionen (On the rate constants of substitution reactions), *Z. physikal. Chem., B,* **15,** 445–471.

Perman, E. P., and R. H. Greaves (1908) The decomposition of ozone by heat, *Proc. Roy. Soc., A,* **80,** 353–369.

Perrin, J. B. (1895) Nouvelles propriétés des rayons cathodiques (New properties of cathode rays), *Comptes rendus,* **121,** 1130–1134. An excerpt is in Magie (1935).

Perrin, J. B. (1908) L'agitation moléculaire et le mouvement brownien (Molecular agitation and Brownian movement), *Comptes rendus,* **146,** 967–970.

Perrin, J. B. (1908) L'origine du mouvement brownien (The origin of the Brownian movement), *Comptes rendus*, **147**, 530–532.

Perrin, J. B. (1909) Mouvement brownien et constantes moléculaires (Brownian movement and molecular constants), *Comptes rendus*, **149**, 477–479.

Perrin, J. B. (1909) La mouvement brownien de rotation (Brownian movement of rotation), *Comptes rendus*, **149**, 549–551.

Perrin, J. B. (1909) Mouvement Brownien et réalité moléculaire (Brownian movement and molecular reality), *Ann. de chim. et de phys.*, **18**, 5–114. English translation by F. Soddy, *Brownian Movement and Molecular Reality*, Taylor and Francis, London, 1910. This book was for the most part an account of work previously published in the *Comptes rendus*.

Perrin, J. B. (1911) Les grandeurs moléculaires (nouvelles mesures (Molecular dimensions—new measurements), *Comptes rendus*, **152**, 1380–1382.

Perrin, J. B. (1913) *Les atomes*, Paris. English translation by D. L. Hammick, *Atoms*, Constable, London, 1916.

Perrin, J. B. (1919) Matière et lumière. Essai de synthése de la méchanique chimique (Matter and light. Essay on the foundations of chemical kinetics), *Ann. de physique*, **11**, 5–108. The argument regarding unimolecular reactions is on p. 9.

Perrin, J. B. (1922) Radiation and chemistry, *Trans. Faraday Soc.*, **17**, 546–572.

Pfaundler, P. (1867) Beiträge zur chemischen Statik (Contribution to chemical statics), *(Poggendorff's) Ann. der Phys. und Chem.*, [2], **131**, 55–85.

Pfaundler, P. (1872) Elementare Ableitung den Grundgleichung der dynamischen Gastheorie (Elementary derivation of the fundamental equation of gas dynamics), *(Poggendorff's) Ann. der Phys. und Chem.*, [2], **144**, 428–438.

Pfeffer, W. F. P. (1877) *Osmotische Untersuchungen, Studien zur Zellenmechanik* (Osmotic investigations, studies in cell physiology), Leipzig. Translation in *Harper's Scientific Memoirs* (Eds. J. S. Ames and H. C. Jones), New York, 1899.

Pierce, G. W. (1925) Piezoelectric crystal oscillators applied to the precision measurement of the velocity of sound in air and CO_2 at high frequencies, *Proc. Amer. Acad. Arts Sci.*, **60**, 271–302.

Planck, M. (1887) Über die molekulare Konstitution verdunnter Lösungen (On the molecular constitution of dilute solutions), *Z. physikal. Chem.*, **1**, 577–582.

Planck, M. (1899) Über irreversible Strählungsvörgange (On irreversible radiation processes), *Sitzungsber. Preuss. Akad. Wiss. Berlin*, 440–480.

Planck, M. (1900) Über eine Verbesserung der Wienschen Spektralgleichung (On an improvement of Wien's radiation law), *Verhandl. Deut. physikal. Ges.*, **2**, 202–204. This is a report of Planck's comment at the meeting of the German Physical Society on 10 October, following Kurlbaum's presentation of his work with Rubens.

Planck, M. (1900) Zur Theorie des Gesetzes der Energieverteilung im Normal Spektrum (On the theory of the energy distribution law of the normal spectrum), *Verhandl. Deut. physikal. Ges.*, **2**, 237–245. This paper was presented at the meeting of the German Physical Society on 14 December 1900.

Pockels, Agnes (1891) Surface tension, *Nature*, **43**, 437–439. This consists of a translation of a letter from Fräulein Pockels to Lord Rayleigh, with a note by Rayleigh.

Poincaré, H. (1890) Sur le problème des trois corps et les équations de dynamique (On the three-body problem and the equations of dynamics), *Acta Mathematica*, **13**, 1–270. Extracts from this paper are included in Brush (1965), Vol. 2, 194–202.

Poisson, S. D. (1811) Mémoire sur la distribution de l'éléctricitè à la surface des corps conducteurs, *Mém. Inst. de France*, 1–92, 163–274. Translations of brief accounts of this work appeared in *Annals of Philosophy*, **1**, 152–156 (1813), and **3**, 391–392 (1814).

Polanyi, J. C. (1959) Energy distribution between reagents and products of atomic reactions, *J. Chem. Phys.*, **31**, 1338–1351.

Polanyi, J. C. (1961) Proposal for an infrared maser dependent on vibrational excitation, *J. Chem. Phys.*, **34**, 347–348.

Polanyi, J. C. See also Arrowsmith *et al.* (1980).

Polanyi, M. (1920) Reaktionsisochore und Reaktionsgeschwindigkeit vom Standpunkte der Statistik (Reaction isochore and reaction velocity from the standpoint of statistics), *Z. Elektrochem.*, **26**, 49–54.

Polanyi, M., H. Beutler and S. von Bogdandy (1926) Über Luminescenz hochverdünnter Flammen, *Naturwiss.*, **14**, 164–165, and many later papers; the work is summarized in:

Polanyi, M. (1932) *Atomic Reactions*, Ernest Benn, London.

Polanyi, M., see also Evans, M. G., and M. Polanyi.

Porter, G. (1950) Flash photolysis and spectroscopy. A new method for the study of free radical reactions, *Proc. Roy. Soc., A*, **200**, 284–290.

Porter, G. (1950) The absorption spectroscopy of substances of short life, *Disc. Faraday Soc.*, **9**, 60–69.

Porter, G., and J. I. Steinfeld (1966) Giant-pulse-laser photolysis of phthalocynanine vapor, *J. Chem. Phys.*, **45**, 3456–3457.

Porter, G., and M. R. Topp (1968) Nanosecond flash photolysis and the absorption spectra of excited singlet states, *Nature*, **220**, 1228–1229.

Porter, G., and M. R. Topp (1970) Nanosecond flash photolysis, *Proc. Roy. Soc., A*, **315**, 163–184.

Porter, G. (1976) see Archer *et al.*

Priestley, Joseph (1777) *Disquisitions Relating to Matter and Spirit*, London.

Purcell, E. H., R. V, Pound and N. S. Torrey (1946) Resonant absorption by nuclear magnetic moments in a solid, *Phys. Rev.*, **69**, 37–38.

Raman, C. V., and K. S. Krishnan (1928) A new type of secondary radiation, *Nature*, **121**, 501–502.

Raman, C. V. (1928) A change in wave-length in light scattering, *Nature*, **121**, 619.

Raman, C. V., and K. S. Krishnan (1928) The optical analogue of the Compton effect, *Nature*, **121**, 711.

Rankine, W. J. M. (1850) Abstract of a paper on the hypothesis of molecular vortices, and its application to the mechanical theory of heat, *Proc. Roy. Soc. Edinburgh*, **2**, 275–286.

Rankine, W. J. M. (1851) On the centrifugal theory of elasticity as applied to gases and vapours, *Phil. Mag.*, **2**, 509–542.

Rankine, W. J. M. (1853) On the mechanical action of heat especially in gases and vapours, *Trans. Roy. Soc. Edinburgh*, **20**, 147–190. This and the preceding paper were read before the Royal Society of Edinburgh on 4 February, 1850.

Rankine, W. J. M. (1859) *Manual of the Steam Engine*, C. Griffith, London.

Raoult, F. M. (1882) Loi générale de congélation des dissolvants (General law for the freezing of solutions), *Comptes rendus*, **95**, 1030–1033.

Raoult, F. M. (1887) Loi générale des tensions de vapeur des dissolvants (General law for the vapour pressure of solutions), *Comptes rendus*, **104**, 1430–1433.

Rayleigh, Lord (1899) Investigations in capillarity, *Phil. Mag.*, [5], **48**, 331–337.

Rayleigh, Lord (1900) Remarks upon the law of complete radiation, *Phil. Mag.*, [5], **49**, 539–540.

Rayleigh, Lord (1905) The dynamical theory of gases and of radiation, *Nature*, **72**, 54–55. This is a letter to the Editor.

Regnault, V. (1840) Recherches sur la chaleur spécifique des corps simples et composés (Research on the specific heats of simple and complex bodies), *Ann. de chim. et de phys.*, **73**, 5–72 (1840).

Reicher, L. T. (1885) Über die Geschwindigkeit der Verseifung (On the velocity of saponification), *Ann. der Chemie*, **228**, 257–287; **232**, 103–114 (1886); **238**, 276–286 (1887).

Rice, F. O., H. C. Urey and R. N. Washburn (1928) The mechanism of homogeneous gas reactions. I. The effect of black body radiation on a molecular beam of nitrogen pentoxide, *J. Amer. Chem. Soc.*, **50**, 2402–2412.

Rice, F. O. (1931) The thermal decomposition of organic compounds from the standpoint of free radicals, I. Saturated hydrocarbons, *J. Amer. Chem. Soc.*, **53**, 1959–1972.

Rice, F. O., and B. L. Evering (1932) The thermal decomposition of organic compounds from the standpoint of free radicals. II. Experimental evidence of the decomposition of organic compounds into free radicals, *J. Amer. Chem. Soc.*, **54**, 3359–3543.

Rice, F. O., and A. L. Glasebrook (1933) The free methyl radical, *J. Amer. Chem. Soc.*, **55**, 4329–4330.

Rice, F. O., and K. F. Herzfeld (1934) The thermal decomposition of organic compounds from the standpoint of free radicals. VI. The mechanism of some chain reactions, *J. Amer. Chem. Soc.*, **56**, 284–287.

Rice, F. O., and A. L. Glasebrook (1934) The thermal decomposition of organic compounds from the standpoint of free radicals. XI. The methylene radical, *J. Amer. Chem. Soc.*, **56**, 2381–2383.

Rice, J. (1915) *British Assn. Report*, 397–398, This brief report states some interesting conclusions about critical complexes, but is unfortunately so brief that it is impossible to know what procedure was followed.

Rice, O. K., and H. C. Ramsperger (1927) Theories of unimolecular reactions at low pressures, *J. Amer. Chem. Soc.*, **49**, 1617–1629. Most of this article is reproduced in Back and Laidler (1967).

Rideal, E. K. (1939) A note on a simple molecular mechanism for homogeneous catalytic reactions, *Proc. Camb. Phil. Soc.*, **35**, 130–132. See also Eley and Rideal (1941).

Ritter, J. W. (1801) *(Gilbert's) Ann. der Physik*, **7**, 527. Reprinted in *Physische-chemische Abhandlung in chronologische Ordnung*, Vol. 2, Leipzig, 1806.

Ritter, J. W. (1801) Galvanische Versuche über die chemische Natur des Wassers (Galvanic investigations on the chemical nature of water), *(Crell's) Chemische Ann.*, **1**, 41–63; *Mém. Soc. Sav. Paris*, **2**, 288–296 (1801).

Ritz, W. (1908) Über ein neues Gesatz der Serienspektren (On a new law of series spectra), *Phyzikal. Z.*, **9**, 521–529. English version in *Astrophys. J.*, **28**, 237–243 (1908).

Ritz, W. (1909) Über eine neue Methode zur Lösung gewisser Variationsprobleme der mathematischen Physik (On a new method of solution of certain variational problems in mathematical physics), *Journal für die reine und angewandte Mathematik (Crelle's Journal)*, **135**, 1–61.

Roozeboom, H. W. B. (1901) *Die heterogenen Gleichgewichte vom Standpunkt der Phasenlehre* (Heterogeneous equilibrium from the standpoint of the phase rule), Vieweg, Braunschweig.

Rubens, H., and F. Kurlbaum (1900) Über die Emission langwelliger Wärmestrahlen durch den schwarzen Körper bei verschiedenen Temperaturen (On the long-wave emission of heat radiation by black bodies at different temperatures), *Sitzungsber. Akad. Wiss. Berlin*, 929–941.

Rumford, Benjamin Count of [Benjamin Thompson] (1798) An enquiry concerning the chemical properties that have been attributed to light, *Phil. Trans.*, **88**, 449–468. This paper refers to Mrs Fulhame's work.

Rutherford, E. (1897) The velocity and rate of recombination of the ions of gases exposed to Röntgen radiation, *Phil. Mag.*, **44**, 422–440.

Rydberg, J. R. (1890) Recherches sur la constitution des spectres d'émission des élements chimiques (Research on the structure of the emission spectra of the chemical elements), *Kung. Svenska vetenskaps. hand.*, **23**, No. 11, 155 pp. English summary, 'On the structure of the line spectra of the chemical elements', *Phil. Mag.*, [5], **29**, 331–337 (1890). In the paper in the Swedish journal Rydberg claimed that he had been using Balmer's type of equation long before the appearance of Balmer's paper.

Salet, G. (1875) Sur les spectres doubles (On double spectra), *J. de physique*, **4**, 225–227.

Salet, G. (1875) Sur les spectres multiples (On multiple spectra), *Comptes rendus*, 485–489, and later publications.

Salomon, M., and B. E. Conway (1965) Chemical and quantum-mechanical effects in electrochemical proton discharge, and the kinetics at low temperatures, *Disc. Faraday Soc.*, **39**, 223–238.

Schawlow, A. L., and C. H. Townes (1958) Infrared and optical masers, *Phys. Rev.*, **112**, 1940–1949.

Scheffer, F. E. C., and W. F. Brandsma (1926) Reaction velocities, *Rec. trav. chim. Pays-Bas*, **45**, 522–534.

Schrödinger, E. (1926) Quantizierung als Eigenwertproblem (Quantization as an eigenvalue problem), *Ann. der Physik*, **79**, 361–376, 489–527; **80**, 437–490. Translations of these and other papers, by J. Shearer and W. Deans, are in *Collected Papers on Wave Mechanics by E. Schrödinger*, Blackie, Glasgow, 1928.

Schrödinger, E. (1926) Über das Verhältnis der Heisenberg-Born-Jordanschen Quantenmechanik zu der meinen (On the relationship between the Heisenberg–Born–Jordan quantum mechanics and mine), *Ann. der Physik*, **79**, 734–756.

Schuster, A. (1879) On harmonic ratios in the spectra of gases, *Nature*, **20**, 533.

Schuster, A. (1881) On harmonic ratios in the spectra of gases, *Proc. Roy. Soc.*, **31**, 337–347.

Schuster, A. (1887) Experiments on the discharge of electricity through gases, *Proc. Roy. Soc., A*, **42**, 371–379.

Scott, William T. (1964) A bibliographic reference table for Faraday's papers on electricity, in *The Natural Philosopher*, Blaisdell, New York, Vol. 3, pp. 75–95.

Semenov, N. N. (1927) Die Oxydation des Phosphordampfes bei neidrigen Drucken (The oxidation of phosphorus vapour at low pressures), *Z. Physik*, **46**, 109–131. A portion of this paper appears in translation in Back and Laidler (1967), pp. 127–153. Prior to Semenov's work on the phosphorus oxidation, J. Chariton and Z. Walta (*Z. Physik*, **39**, 547–556 (1926) had reported a lower limit for the explosion, but their work was not entirely convincing; Semenov's investigation, based on that of Chariton and Walta, was more thorough.

Seneca, Lucius Annaeus (*c.*63 AD) *Quaestiones Naturales* (Enquiries of nature), Book 1, Chapters 2–8. A translation of this book appears in J. Clarke, *Physical Science in the Time of Nero, being a Translation of Quaestiones Naturales*, Macmillan, London, 1910.

Shank, C. V. (1986) Investigation of ultrafast phenomena in the femtosecond time domain, *Science*, **233**, 1276–1280.

Shorter, J. A. (1985) The British school of physical organic chemistry, *Chemtech*, April, 1985, 252–256.

Siedentopf, H. and R. Zsigmondy (1903) Über Sichtbarmachung und Grössenbestimmung ultramikroskopischer Teilchen, mit besonderer

Anwendung auf Goldrubingläser (On the visualization and size-estimation of ultramicroscopic particles with particular application to ruby glass containing colloidal gold particles), *Ann. der Physik*, [4], **10**, 1–39.

Simon, F. E. (1927) Zum Prinzip von der Unerreichbarkeit des absoluten Nullpunktes (The principle of the unattainability of absolute zero), *Z. Physik*, **14**, 806–809.

Simon, F. E. (1930) Fünfundzwanzig Jahre Nernstscher Wärmesatz (Twenty-five years of Nernst's heat theorem) *Ergebn. exakt. Naturwiss.*, 222 (1930).

Skrabal, A, (1927) Zur Deutung des Zeitgesetzes der Bildung des Bromwasserstoffes aus seinen Elementen (On the significance of the equation for the rate of formation of hydrogen bromide from its elements), *Ann. der Physik*, [4], **82**, 138–142. This paper contains a criticism of the steady-state hypothesis, answered by Bodenstein (1927).

Smith, F. J. [Jervis-Smith] (1889) An experimental investigation of the circumstances under which a change of the velocity of the propagation of the ignition of an explosive gaseous mixture takes place in closed and open vessels. Part I. Chronographic measurements, *Proc. Roy. Soc.*, **45**, 451–452. Several other papers over the next few years are concerned with his chronograph.

Smoluchowski, M. (1905) Zur kinetischen Theorie der Brownschen Molecularbewegung und der Suspensionen (On the kinetic theory of Brownian movement and of suspensions), *Ann. der Physik*, [4], **21**, 756–780.

Smoluchowski, M. (1914) Studien über Molekularstatistik von Emulsionen und deren Zusammenhang mit der Brownschen Bewegung (Studies on the molecular statistics of emulsions and on their connection with the Brownian movement), *Sitzungsber. Akad. Wiss. Wien, IIa*, **123**, 2381–2405.

Smoluchowski, M. (1916) Drei Vorträge über Diffusion, Brownsche Molecularbewegung und Koagulation von Kolloidteilschen (Three reports on diffusion, Brownian movement and the coagulation of colloid particles), *Physikal. Z.*, **17**, 557–571, 585–599.

Smoluchchowski, M. von (1917) Versuch einer mathematischen Theorie der Koagulationskinetic kolloider Lösungen (An attempt at a mathematical theory of the kinetics of coagulation of colloidal solutions), *Z. physikal. Chem.*, **92**, 129–168.

Somerville, Mary (1831) *The Mechanism of the Heavens*, John Murray, London.

Somerville, Mary (1834) *On the Connexion of the Physical Sciences*, John Murray, London.

Somerville, Mary (1836) Expériences sur la transmission des rayons chimiques du spectre solaire, à travers différents milieux (Experiments on the transmission of chemical rays of the solar spectrum across different media), *Comptes rendus*, **3**, 473–476. This was an abstract of a letter from Mrs Somerville to Arago, who presented the work to the Académie des Sciences.

Somerville, Mary (1846) On the action of the rays of the spectrum upon vegetable juices, *Phil. Trans.*, **136**, 111–121. This was an extract from a letter from Mrs Somerville to Sir John Herschel; she was then residing in Italy on account of her husband's health, and the letter was sent from Rome, where perhaps the work was done.

Sommerfeld, A. (1916) Zur Theorie der Balmerschen Serie (On the theory of the Balmer series), *Münchener Berichte*, 425–458.

Sommerfeld, A. (1916) Zur Quantentheorie der Spektrallinien, *Ann. der Physik*, [4], **51**, 1–94, 125–167.

Sommerfeld, A. (1919) *Atombau und Spektrallinien*, Vieweg, Braunschweig; *Atomic Structure and Spectral Lines* (a translation by H. L. Brose from the third, 1922, German edition), Methuen, London, 1923.

Stark, J. (1908) Weiter Bemerkungen über die thermische und chemische Absorption im Bandenspektrum (Further remarks on thermal and chemical absorption in band spectra), *Physikal. Z.*, **9**, 889–894.

Stark, J. (1908) Über die zerstäubende Wirkung des Lichtes und die optische Sensibilisation (On the disintegrating action of light, and optical sensitisation), *Physikal. Z.*, **9**, 894–900. For a comment on Stark's priority over Einstein for the law of photochemical equivalence, see Allmand (1926).

Stark, J. (1913, 1914) Beobachtungen über den Effekt des elektrischen Feldes auf Spektrallien (Observations on the effect of an electric field on spectral lines), *Sitzungsber. Akad. Wiss. Berlin*, 932–946; *Ann. der Physik*, **43**, 965–1047 (1914).

Stefan, J. (1879) Über die Beziehung zwischen der Wärmestrahlung und der Temperatur (On the relationship between heat radiation and temperature), *Sitzungsber. Akad. Wiss. Wien*, **79**, 391–428.

Stern, O., and W. Gerlach (1922) Der experimentalle Nachweis des magnetischen Moments der Silberatome (The experimental detection of the magnetic moment of the silver atom), *Z. Physik*, **8**, 110–111.

Stern, O. (1924) Zur Theorie des elektrolytischen Doppelschicht (The theory of the electrical double layer), *Z. Elektrochem.*, **30**, 508–516.

Stewart, Balfour (1858) An account of some experiments on radiant heat, involving an extension of Prévost's law of exchange, *British Assn. Reports*, **27**, 23–24; *Trans. Roy. Soc. Edinburgh*, **22**, 1–20 (1857–61). Reproduced in D. B. Brace (Ed.), *The Laws of Radiation and Absorption: Memoirs by Prévost, Stewart, Kirchhoff, and Kirchhoff and Bunsen*, New York, 1901.

Stokes, G. G. (1852) On the change of refrangibility of light, *Phil. Trans.*, **142**, 463–562. Excerpt in Magie (1935).

Stokes, G. G. (1864) On the reduction and oxidation of the colouring matter of the blood, *Proc. Roy. Soc.*, **13**, 355–364.

Stoney, G. Johnstone (1868) The internal motions of gases compared with the motions of waves of light, *Phil. Mag.*, **36**, 132–141.

Stoney, G. Johnstone (1868) On the cause of the interrupted spectra of gases, *Phil. Mag.*, **41**, 291–296.

Stoney, G. Johnstone (1881) On the physical units of nature, *Phil. Mag.*, **11**, 381–389; this paper was presented to the British Association in 1874.

Stoney, G. Johnstone (1891) On the cause of double lines and of equidistant satellites in the spectra of gases, *Trans. Roy. Dublin Soc.*, **4**, 563–608. The word 'electron' is suggested on p. 583.

Stoney, G. Johnstone (1892) The line spectra of the elements, *Nature*, **46**, 29, 126, 222, 268–269.

Stoney, G. Johnstone (1894) Of the 'electron' or atom of electricity, *Phil. Mag.*, [5], **38**, 418–420.

Sutherland, W. (1902) Ionization, ionic velocities, and ionic sizes, *Phil. Mag.*, [6], **3**, 161–177.

Sutherland, W. (1906) The molecular constitution of aqueous solutions, *Phil. Mag.*, [6], **12**, 1–20.

Sutherland, W. (1907) Ionization in solution and two new types of viscosity, *Phil. Mag.*, [6], **14**, 1–35.

Tafel, J. (1905) Über die Polarisation bei kathodische Wasserstoffentwicklung (On polarization during cathodic evolution of hydrogen), *Z. physikal. Chem.*, **50**, 641–712.

Talbot, W. H. F. (1826) Some experiments on coloured flames, *(Brewster's) Edinburgh J. Science*, **5**, 77–82. This paper is one of the earliest contributions to spectral analysis, the earliest being Melvill (1752).

Talbot, W. H. F. (1834) Facts relating to optical science. No. 2, *Phil. Mag.*, [3], **4**, 112–114, 289–290.

Talbot, W. H. F. (1836) Facts relating to optical science. No. 3, *Phil. Mag.*, [3], **9**, 1–4.

Talbot, H. H. F. (1839) Some account of the art of photogenic drawing, or the process by which natural objects may be made to delineate themselves without the aid of the artist's pencil, *Proc. Roy. Soc.*, **4**, 120–121. This paper was read to the Royal Society on 31 January 1839; an expanded version with the same title appeared in *Phil. Mag.*, **14**, 196–211 (1839).

Talbot, W. H. F. (1839) At a meeting of the Académie des Sciences on 4 February 1939, Arago announced that he and Biot had received letters from Talbot claiming priority over Daguerre. Arago supported Daguerre and a letter of reply to Talbot was tabled. This is reported in *Comptes rendus*, **8**, 170–174 (1839), and later in the year there were several other reports of the same kind, with further correspondence between Talbot, Arago and Biot.

Talbot, H. H. F. (1839) An account of the process employed in photogenic drawing, *Proc. Roy. Soc.*, **4**, 124–126. This paper was read to the Royal Society on 21 February 1839, and gave sufficient details to allow the procedures to be repeated by others.

Talbot, W. H. F. (1839) *Some Account of the Art of Photogenic Drawing*, R. & J. E. Taylor, London. Reproduced in Newhall (1980).

Talbot, W. H. F. (1841) An account of some improvements in photography, *Proc. Roy. Soc.*, **4**, 312–315.

Talbot, H. Fox (1844-1846) *The Pencil of Nature*, published in six instalments by Longman, Brown, Green and Longmans, London. Facsimile edition limited to 250 copies, in 7 volumes, published by Hans P. Kraus, New York, 1989. Volume 1 is an introduction by Larry J. Schaaf. The designation 'H. Fox Talbot' is unusual; Talbot usually signed himself 'H. F. Talbot' or 'Henry F. Talbot'.

Talbot, W. H. F. (1869–1872) Note on the early history of spectrum analysis, *Proc. Roy. Soc. Edinburgh*, **7**, 461–466.

Taylor, E. H., and S. Datz (1955) Study of chemical reaction mechanisms with molecular beams: the reaction of K with HBr, *J. Chem. Phys.*, **23**, 1711–1718.

Taylor, H. S. (1925) A theory of the solid catalytic surface, *Proc. Roy. Soc.*, *A*, **108**, 105–111.

Taylor, H. S. (1926) Photosensitization and the mechanism of chemical reactions, *Trans. Faraday Soc.*, **21**, 560–568.

Taylor, H. S. (1931) The activation energy of adsorption processes, *J. Amer. Chem. Soc.*, **53**, 578–597.

Thénard, L. J. (1813) Résultats d'expériences sur le gaz ammoniac (The results of experiments on gaseous ammonia), *Ann. de chimie*, **85**, 61–64.

Thénard, L. J. (1818) Observations sur l'influence de l'eau dans le formation des acides oxigénés (Observations on the effect of water on the formation of oxyacids), *Ann. de chim. et de phys.*, [2], **9**, 314–317.

Thompson, Benjamin, see Rumford.

Thomson, G. P., and A. Reid (1927) Diffraction of cathode rays by a thin film, *Nature*, **119**, 890.

Thomson, G. P. (1927) The diffraction of cathode rays by thin films of platinum, *Nature*, **120**, 802.

Thomson, G. P. (1928) Experiments on the diffraction of cathode rays, *Proc. Roy. Soc.*, *A*, **117**, 600–609.

Thomson, G. P. (1961) Early work on electron diffraction, *Amer. J. Phys.*, **29**, 821–825.

Thomson, J. J. (1890) On the electricity of drops, *Nature*, **42**, 295, 614.

Thomson, J. J. (1897) Cathode rays, *Phil. Mag.*, [5], **44**, 293–316. Reproduced in part in Magie (1935).

Thomson, J. J. (1897) On the electricity of drops, *Phil. Mag.*, [5], **45**, 341–358.

Thomson, J. J. (1898) On the charge of electricity carried by the ions produced by Röntgen rays, *Phil. Mag.*, [5], **47**, 528–545.

Thomson, J. J. (1898) *The Discharge of Electricity through Gases*, Constable, London.

Thomson, J. J. (1899) On the masses of the ions in gases at low pressure, *Phil. Mag.*, [5], **48**, 547–567.

Thomson, J. J. (1907) On rays of positive electricity, *Phil. Mag.*, [6], **13**, 561–575.

Thomson, J. J. (1910) Rays of positive electricity, *Phil. Mag.*, [6], **20**, 752–767.

Thomson, J. J. (1913) *Rays of Positive Electricity and their Application to Chemical Analysis*, Longmans, Green & Co., London.

Thomson, J. J. (1914) The forces between atoms and chemical affinity, *Phil, Mag.*, [6], **27**, 757–789.

Thomson, W. (1848) An account of Carnot's theory of the motive power of heat, with numerical results deduced from Regnault's experiments on steam, *Trans. Roy. Soc. Edinburgh*, **16**, 541–574. Reprinted in Thomson (1882).

Thomson, W. (1848) On an absolute thermometric scale founded on Carnot's theory of the motive power of heat, calculated from Regnault's observations, *Proc. Camb. Phil. Soc.*, **1**, 66–70. Reprinted in *Phil. Mag.*, [3], **33**, 313–317 (1848) and in Thomson (1882).

Thomson, W. (1851) On the mechanical theory of electrolysis, *Phil. Mag.*, [4], **2**, 429–444. Reprinted in Thomson (1882).

Thomson, W. (1852) On a universal tendency in nature to the dissipation of mechanical energy, *Proc. Roy. Soc. Edinburgh*, **3**, 139–142. Reprinted in *Phil. Mag.*, [4], **4**, 304–306 and in Thomson (1882). Excerpts are in Magie (1935).

Thomson, W. (1852) On the dynamical theory of heat, with numerical results deduced from Mr Joule's equivalent of a thermal unit, and M. Regnault's observations on steam, *Trans. Roy. Soc. Edinburgh*, **20**, 261–288. Reprinted in *Phil. Mag.*, [4], **4**, 8–21, 105–117 (1852) and in an expanded form in Thomson (1882).

Thomson, W. (1867) On vortex atoms, *Proc. Roy. Soc. Edinburgh*, **6**, 94–105 (1867). Reprinted in *Phil. Mag.*, [4], **34**, 15–24 (1867).

Thomson, W., and P. G. Tait (1867) *Treatise on Natural Philosophy*, Vol. 1, Clarendon Press, Oxford; Cambridge University Press, 1878, 1883. This book was referred to as 'T and T''; its promised Vol. 2 never appeared.

Thomson, W., (1874) The kinetic theory of the dissipation of energy, *Proc. Roy. Soc. Edinburgh*, **8**, 325–334.

Thomson, Sir W. (1882) *Mathematical and Physical Papers*. Cambridge University Press. This is a collection of all of Kelvin's published papers.

Thomson, W. (1901) [now Lord Kelvin] Nineteenth century clouds over the dynamical theory of heat and light, *Phil. Mag.*, [6], **2**, 1–40.

Thomson, W. (1902) [now Lord Kelvin] Aepinus atomized, *Phil. Mag.*, [6], **3**, 257–283.

Thomson, W. (1904) [now Lord Kelvin] Plan of an atom to be capable of storing an electrion [sic] with enormous energy for radioactivity, *Phil. Mag.*, [6], **10**, 695–698. In this and the previous paper Kelvin proposed an entirely new model of the atom to accommodate the electrons.

Thomson, W. (1904) *The Baltimore Lectures on Molecular Dynamics and the Wave Theory of Light*, Cambridge University Press. Reprinted as *Kelvin's Baltimore Lectures and Modern Theoretical Physics: Historical and Philosophical Perspectives* (Eds. R. Kargon and P. Achinstein), MIT Press, Cambridge, Mass., 1987; this volume also contains a number of essays on Kelvin's work in relation to modern physics. The lectures were given at Johns Hopkins University in 1884.

Tilden, W. A. (1903), The specific heats of metals and the relation of specific heats to atomic weight, *Phil. Trans.*, **201**, 37–43.

Tolman, R. C. (1920) Statistical mechanics applied to chemical kinetics, *J. Amer. Chem. Soc.*, **43**, 2506–2528.

Tolman, R. C. (1938) *The Principles of Statistical Mechanics*, Clarendon Press, Oxford.

Townsend, J. S. (1897) On electricity in gases and the formation of clouds in charged gases, *Proc. Camb. Phil. Soc.*, **9**, 244–258.

Townsend, J. S. (1898) Electrical properties of newly prepared gases, *Phil. Mag.*, [5], **45**, 125–151.

Trautz, M. (1916) Des Gesetz der Reaktionsgeschwindigkeit und der Gleichgewichte in Gasen (The laws of reaction velocity and of equilibrium in gases), *Z. anorgan. Chem.*, **96**, 1–28.

Trautz, M. (1918) Das Gesetz der thermochemischen Vorgänge (Zusammenfassung) und das der photochemischen Vörgange (The law of thermochemical processes (Summary) and of photochemical processes), *Z. anorgan. Chem.*, **102**, 81–129. The radiation hypothesis is discussed in this paper, but was much more clearly presented in W. C. McC. Lewis (1916, 1917 and 1918) and in Perrin (1919, 1922). Trautz had previously referred to the radiation hypothesis, but not very explicitly, in *Z. für wissenschaftliche Photographie*, **2**, 217 (1904) and **4**, 160–183 (1906).

Truesdell, C. (1971) *The Tragicomedy of Classical Thermodynamics*, Springer-Verlag, Vienna and New York.

Uhlenbeck, G. E., and S. Goudsmit (1925) Ersetzung der Hypothese vom unmechanischen Zwang durch eine Forderung bezüglich des inneren Verhaltens jedes einzelnen Elektrons (Replacement of the hypothesis of the nonmechanical stress by a postulate concerning the intrinsic behaviour of every single electron), *Naturwiss.*, **13**, 953–954. Their expression 'nonmechanical stress' refers to a proposal that Bohr had made (*Ann. der Physik*, [4], **71**, 228–288 (1923)) to explain the anomalous Zeeman effect.

van der Waals, J. D. (1873) *Over decontiuiteit van den gasen vloeisofoestand*, Sijthoff, Leiden. Translation as 'The continuity of the liquid and gaseous states of matter', by R. Threlfall and J. F. Adair, in *Physical Memoirs*, Vol. 1, Part 2, Taylor & Francis, for the Physical Society, London, 1890.

van't Hoff, J. H. (1874) *Voorstel tot uitbreiding der tegenvoordig in de scheikunde gebruikte structuur-formules im de ruimte; benevens een daarmee samenhangende opmerking omtrent het verband tusschen optisch actief vermogen en chemische constituie van organische verbindungen* (Proposal for an extension into space of the formulae now in use in chemistry; together with a related remark on the relation between the optical rotatory power and the chemical composition of organic compounds), Utrecht. A French translation is in Archives néerlandaises des sciences exactes et naturelles, 9, 445–454 (1874). There later appeared an English translation in *The Foundations of Stereo Chemistry; Memoirs by Pasteur, van't Hoff, Le Bel, and Wislicenus*, New York, 1901.

van't Hoff, J. H. (1884) *Études de dynamique chimique*, F. Muller, Amsterdam. At the time the word 'dynamics' was taken to include chemical equilibrium as well as chemical kinetics, both of which are covered in the book. A second edition, now in German, appeared as J. H. van't Hoff and E. J. Cohen, *Studien zur chemischen Dynamik*, Frederik Muller & Co., Amsterdam, 1896. At the same time there was an English translation of the second edition, made by van't Hoff's former student Thomas Ewen (1876–1955), then at the Yorkshire College of Science in Leeds: *Studies in Chemical Dynamics*, published in 1896 jointly by F. Muller and by Williams & Norgate, London.

van't Hoff, J. H. (1886) L'équilibre chimique dans les systèmes gazeux, ou dissous à l'état dilué (Chemical equilibrium in gaseous systems, or in dilute solutions), *Archives néerlandaises des sciences exactes et naturelles*, **20**, 239–302. The same material that was in this paper also appeared, almost word for word, in the Swedish journal *Svenska vetenskaps-academiens Handlingar*, **21**, 3–41, 42–49, 50–58 (1886). The section on pp. 42–49, entitled 'Une propriété général de la matière diluée' (A general property of dilute systems), was not in the *Archives* paper.

van't Hoff, J. H. (1887) Die Rolle des osmotischen Drücke in der Analogie zwischen Lösungen und Gasen (The role of osmotic presure in the analogy between solutions and gases), *Z. physikal. Chem.*, **1**, 481–508, English translation in *The Foundations of the Theory of Dilute Solutions*, Alembic Club Reprints, No. 19, Oliver & Boyd, Edinburgh, 1929, pp. 5–42.

van't Hoff, J. H. (1898–1901) *Vorlesungen über theoretische und physikalische Chemie*, 4 vols., Brunswick. An English translation by R. A. Lehrfeld appeared in 1899–1900 as *Lectures on Theoretical and Physical Chemistry*, 3 vols. Edward Arnold, London. Vol. 3 is concerned with chemical dynamics, the definition of which is on p. 9.

van't Hoff, J. H. (Ed.) (1903) *Abhandlungen zur thermodynamische chemische Vorgange* (Accounts of progress in chemical thermodynamics), Wilhelm Englemann, Leipzig.

van't Hoff, J. H. (1905) *Zur Bildung der ozeanische Salzablagerungen*. Erstes Heft (On the formation of oceanic salt deposits. First part), Vieweg, Braunschweig.

van't Hoff, J. H. (1912) *Untersuchungen über die Bildungsverhältnisse der ozeanischen Salzablagerungen insobesonere des Stassfurter Salzlagers* (Investigations on the composition of oceanic salt deposits especially in the Stassfurt deposits), edited by H. Precht and E. Cohen, Akademische Verlags, Leipzig. This is a collection of articles that appeared in *Sitzungsber. Preuss. Akad. Wiss. Berlin*.

Varignon, P. (1700) Manière générale de déterminer les forces, les vitesses, les espaces et les temps (General method of determining forces, speeds, distances and times), *Mém. acad. roy. sci.*, 22–27. This paper makes for the first time the statement that force is mass times acceleration.

Vegard, L. (1918) The X-ray spectra and the constitution of the atom, *Phil. Mag.*, [6], **35**, 293–326.

Volta, Alexander (1800) On the electricity excited by the mere contact of conducting substances of different kinds. In a letter from Mr Alexander Volta, in the University of Pavia, to the Rt. Hon. Joseph Banks, Bart., K. B., P. R. S. Read June 26, 1800, *Phil. Trans.*, **92**, 403–431. This letter, which was written in French and dated 29 March 1800, is reproduced in facsimile following an article by G. Sarton, The discovery of the electric cell (1800), *Isis*, **15**, 124–138 (1931). An English translation of the letter appears in E. C. Watson, *Amer. J. Phys.*, **13**, 399–406 (1945), and parts of the letter in translation appear in Magie (1935) and in Boynton (1948).

Walker, Revd Robert (1851) *Textbook of Mechanical Philosophy*, John Henry Parker, Oxford. This was an official textbook of physics, written for Oxford's Honour School of Natural Science, which had been created in 1850.

Warburg, E. (1918) Über den Energieumsatz bei photochemische Vörgange in Gasen. VII. Photolyse des Jodwasserstoffs (On the energy change in photochemical processes in gases. VII. Photolysis of hydrogen iodide). *Sitzungsber. Preuss. Akad. Wiss.*, 300–317.

Waterston, J. J. (1846) On the physics of media that are composed of free and perfectly elastic molecules in a state of motion, *Proc. Roy. Soc., A*, **5**, 604 (Abstract only).

Waterston, J. J. (1893) *Phil. Trans., A*, **183**, 1–79. The title is the same as for the abstract (see previous item). This paper, the Abstract, and Rayleigh's introduction, are reproduced in *The Collected Scientific Papers of John James Waterston* (Ed. J. S. Haldane), Oliver & Boyd, Edinburgh, 1928, and in *The World of Physics* (Ed. J. H. Weaver), Simon & Schuster, New York, 1987, Vol. 1.

Watson, Revd Richard (1770) Experiments and observations on various phoenomena attending the solution of salts, *Phil. Trans.*, **60**, 325–354.

Watson, Revd Richard (1781–1787) *Chemical Essays*, 5 vols., Cambridge University Press. This book, which had a number of later editions, is a collection of Watson's books and papers.

Weber, H. F. (1872) Die specifische Wärme des Kohlenstoffs (The specific heat of carbon), *(Poggendorff's) Ann. der Phys. und Chem.*, [2], **147**, 311–319.

Weber, H. F. (1875) Die specifische Wärme der Elemente Kohlenstoff, Bor und Silicium (The specific heats of the elements carbon, boron and silicon), *(Poggendorff's) Ann. der Phys. und Chem.*, [2], **154**, 367–423, 553–582.

Wedgwood, Thomas, and H. Davy (1802) An account of the method of copying paintings upon glass, by the agency of light upon nitrate of silver. Invented by T. Wedgwood, Esq. With observations by H. Davy, *J. Roy. Institution*, **1**, 170–174. This paper is reproduced in Newhall (1980).

Wenzel, C. F. (1777) *Lehre von der Verwandtschaft der Körper* (Laws of the transformation of substances), Dresden.

Wheeler, T. S., and J. R. Partington (1960) *The Life and Work of William Higgins, Chemist (1783-1825)*, Pergamon Press, Oxford.

Whewell, W. (1832) *First Principles of Mechanics*, Cambridge University Press.

Whewell, W. (1840) *The Philosophy of the Inductive Sciences, Founded upon their History*, 2 vols., London.

Whewell, W. (1841) *The Mechanics of Engineering*, Cambridge University Press.

Wicke, E., and M. Eigen (1952) Über den Einfluss des Raumbedarfs von Ionen, in wässerigen Lösung auf ihre Verteilung in elektrischen Feld und ihre Aktivitätskoefficienten (On the influence of the effective sizes of ions in aqueous solution on their distribution in an electric field and their activity coefficients), *Z. Elektrochem.*, **56**, 551–561.

Wicke, E., and M. Eigen (1953) Thermodynamische Eigenschaften konzentrierten wässrige Elektrolytlösungen (Thermodynamic properties of concentrated aqueous solutions in aqueous solution), *Z. Elektrochem.*, **57**, 319–330.

Wicke, E., and M. Eigen (1953) Raumbedarf und Aktivitätkoefficienten stärker Elektrolyten in wässrigen Lösung (Space requirements and activity coefficients of strong electrolytes in aqueous solution), *Z. Naturforsch.*, *A*, **82**, 161–167.

Wien, W. (1894) Temperatur und Entropie der Strahlung (Temperature and entropy of radiation), *(Wiedemann's) Ann. der Phys. und Chem.*, [3], **52**, 132–165 (1894).

Wien, W. (1896) Über die Energieverteilung im Emissionspectrum eines schwarzes Körpers (On the energy distribution in the emission spectrum of a black body), *(Wiedemann's) Ann. der Phys. und Chem.*, [3], **58**, 662–669. English translation in *Phil. Mag.*, [5], **43**, 214–220 (1987). Rayleigh (1900) referred to Wien's hypothesis as 'little more than a conjecture'. That Wien's formula (10.1) is consistent with Stefan's law may be seen as follows. Put $x = \beta\nu/T$, so that the integral $\alpha\int\nu^3 e^{-\beta\nu/T}\,d\nu$ becomes $\alpha(T^4/\beta^4)\int e^{-x}\,dx$. This integral is a number, so that the total radiation is proportional to T^4. The same result is obtained if $e^{-\beta\nu/T}$ is replaced by any function of ν/T, but it is not given by the Rayleigh–Jeans equation (10.2).

Wien, W. (1898) Untersuchungen über die electrische Entladung in verdünnter Gasen (Investigations on the electrical charges in attenuated gases), *(Wiedemann's) Ann. der Phys. und Chem.*, [3], **65**, 440–452.

Wigner, E. (1932) Über das Überschreiten von Potentialschwellen bei chemischen Reaktionen (On the penetration of potential-energy barriers in chemical reactions), *Z. physikal. Chem.*, *B*, **19**, 203–216. See also Pelzer and Wigner (1932).

Wilhelmy, L. (1850) Über das Gesetz, nach welchem die Einwirkung der Säuren auf den Rohrzucker stattfinden (On the law according to which acids act on cane sugar), *(Poggendorff's) Ann. der Phys. und Chem.*, [2], **81**, 413–433, 499–526. Reprinted in Ostwald's *Klassiker* (1891), No. 29. Translated excerpts in Leicester and Klickstein (1968), pp. 396–400.

Williamson, A. W. (1851) Über die Theorie der Aetherbildung (On the theory of esterification), *Ann. der Chem. und Pharm.*, **77**, 37–49. English version in *J. Chem. Soc.*, **4**, 229–338 (1852). In this paper Williamson discusses dissociation but not ionization; see the note on Clausius (1857).

Williamson, A. W. (1851–1854) Suggestions for the dynamics of chemistry derived from the theory of esterification, *Proc. Roy. Inst.*, **1**, 90–94.

Wilson, H. A. (1903) A determination of the charge on the ions produced in air by Röntgen rays, *Phil. Mag.*, [6], **5**, 429–441.

Windler, S. C. H. (1840) Über das Substitutionstheorie und die Theorie der Typen (On the substitution law and the theory of types), *(Liebig's) Ann. der Chem. und Pharm.*, **33**, 308–310. This was a satirical pastiche written by Wöhler; it is listed in the Index of the journal and in the Royal Society Catalogue. For a translation see R. E. Oesper, *The Human Side of Scientists*, University of Cincinnati Press, 1975.

Wollaston, W. H. (1802) A method of examining refractive and dispersive powers by prismatic reflection, *Phil. Trans.*, **92**, 365–380.

Wynne-Jones, W. F. K., and H. Eyring (1935) The absolute rate of reactions in condensed phases, *J. Chem. Phys.*, **3**, 492–502. Reprinted in Back and Laidler (1967).

Zavoisky, E. K. (1945) Spin-magnetic resonance in paramagnetic substances, *J. Phys. U.S.S.R.*, **9**, 245. This is a brief abstract in English.

Zeeman, P. (1897) On the influence of magnetism on the nature of light emitted by a substance, *Phil. Mag.*, [5], **43**, 226–239.

Zeeman, P. (1897) Doublets and triplets in the spectrum produced by external magnetic forces, *Phil. Mag.*, [5], **44**, 55–60, 255–259.

Zermelo, E. (1896) Über einen Satz der Dynamik und die mechanische Wärmetheorie (On a theorem of dynamics and the mechanical theory of heat), *(Wiedemann's) Ann. der Phys. und Chem.*, **57**, 485–494. English translation in Brush (1965), Vol. 2, Selection 7, pp. 207–217.

Zewail, A. H., and R. B. Bernstein (1988) Real-time laser femtochemistry, *Chem. Eng. News*, Nov. 7, 1988, 24–43.

Biographical Notes

Abegg, Richard (1869–1910)

B. Dantzig (Gdansk), Poland. He was professor at the Technische Hochschule in Breslau, and worked on a variety of problems in physical chemistry, including colligative and transport properties and solubility products. He made an important contribution to valency theory by distinguishing between 'normal valency' and 'contravalency' (Section 10.3). He was killed in a balloon accident.

Amontons, Guillaume (1663–1705)

B. Paris, France. Largely self-educated. He was employed on public works projects which gave him experience in practical mechanics. He became particularly interested in the design and improvement of instruments, and was the first to notice that the temperature of boiling water remains constant as heating is continued. He showed that the pressure of a gas varies linearly with the temperature at constant volume, and in 1703 he hinted at the idea of an absolute temperature scale and of the absolute zero.

Reference: J. Payen, DSB, 1, 138–139 (1970).

Andrews, Thomas (1813–1885)

B. Belfast, Ireland. *Educ.* University of Glasgow (under William Thomson), in Paris (under Dumas), and he then studied medicine in Dublin and Edinburgh (MD 1835). He was professor of chemistry at Queen's College, Belfast, from 1849 to 1879, and at the same time practised medicine. He is noted for his discovery of the critical behaviour of gases (Section 5.1), first announced in 1863 in Miller's *Chemical Physics* and discussed in a number of later publications.

Reference: E. L. Scott, DSB, 1, 160–161 (1970).

Ångstrom, Anders Jonas (1814–1874)

B. Lögdö, Sweden. *Educ.* University of Uppsala (doctorate in 1833). He was professor of physics at the University of Uppsala until his death. He made extensive studies of the spectra of the sun and the aurora borealis, and suggested the wavelength unit (10^{-10} m) that is known by his name.

Reference: C. O. Maier, DSB, 1, 166–167 (1907).

Arago, Dominique François Jean (1786–1853)

B. Paris, France. *Educ.* École Polytechnique, Paris. In 1809 he became professor of descriptive geometry at the École Polytechnique and also a member of the Institut de France. He was Director of the Paris Observatory and in 1825 became perpetual Secretary of the Académie des Sciences. He played a role of great importance in the development of French science, and in 1839 gave support to Daguerre's technique of photography (Section 8.4). He also had an active political career, serving for a period in the Chambre des Députés.

Reference: R. Hahn, DSB, 1, 200–203 (1970).

Armstrong, Henry Edward (1848–1937)

B. Lewisham, England. *Educ.* Royal College of Chemistry (under Frankland) and Leipzig (under Kolbe). He held positions at several of the London colleges that eventually became part of the University of London; his main appointment was at the Central Technical College. He carried out research in a variety of fields and exerted a wide influence, particularly by his advocacy of the heuristic methods of teaching science, which is based on the principle that people learn best what they teach themselves. He was ahead of his time in his concern about environmental problems and energy conservation. He had strong opinions, and was particularly scornful of Arrhenius's theory of electrolytic dissociation, preferring a hydrate theory (Section 7.2).

References: C. E. Ronneberg, DSB, 1, 288–289 (1970); Hartley (1971); Keeble (1941).

Avogadro, Amadeo, Count of Quaregno (1776–1856)

B. Turin, Italy; he succeeded to the title of Count of Quaregno, 1787. He received a doctorate in law in 1796 and practised law for three years. His main appointment was as professor of physics at the University of Turin. In 1811 he concluded that at a given temperature, pressure and volume, all gases contain the same numbers of molecules. He was one of the first to realize that the particles in a gas need not be atoms but may be combinations of atoms. His hypothesis was neglected until Cannizzaro publicized it in 1858.

Reference: M. P.Crosland, DSB, 1, 343–350 (1970).

Baker, Herbert Brereton (1862–1935)

B. Blackburn, Lancashire. *Educ.* Balliol College, Oxford. After some years as a schoolmaster he succeeded Vernon Harcourt as Dr Lee's Reader in chemistry at Christ Church. He was professor of chemistry at Imperial College, London, from 1912 to 1932. He made extensive studies of the effect of moisture on chemical change, and argued that 'Bakerian drying' prevents reaction from occurring.

References: Bowen (1970); Laidler (1988).

Balmer, Johann Jacob (1825–1898)

B. Lausen, Switzerland, *Educ.* in mathematics at Karlsruhe and Berlin; doctorate, Univ. of Basel, 1849. He taught at a girls' secondary school in Basel, 1859–

1898 and was part-time lecturer at the University of Basel from 1865 to 1890. He discovered the first spectral series (Section 6.5).

Reference: C. L. Maier, DSB, 1, 425–426 (1970).

Becquerel, Alexandre Edmond (1820–1891)

B. Paris. *Educ.* École Polytechnique and École Normale Supérieure. He was professor of physics at the Conservatoire des Arts et Métiers from 1852, and taught chemistry at the Société Chimique de Paris from 1860 to 1863. He was Director of the Muséum d'Histoire Naturelle from 1878. He investigated electricity, magnetism and optics, and did particularly important work on diamagnetism and luminescence. In 1840 he demonstrated that electric currents arise in certain light-induced reactions. In 1843 he took photographs of the solar spectrum and demonstrated the existence of Fraunhofer lines in the ultraviolet. His son, Antoine Henri Becquerel (1852–1908), shared the 1903 Nobel Prize in physics with the Curies, for his discovery of radioactivity.

References: J. B. Gough, DSB, 1, 555–556 (1970); Buerger (1989).

Berzelius, Jöns Jakob (1779–1848)

B. Vävesunda Sörgård, Sweden. *Educ.* University of Uppsala, where he studied medicine and acquired a profound knowledge of chemistry. He took a comprehensive view of science, and was a great systemist. He recognized the importance of Dalton's atomic theory early, and was the first to identify the phenomenon of catalysis, a word he coined himself as well as many other chemical terms.

Reference: H. M. Leicester, DSB, 2, 90–97 (1970); Hartley (1971); Bernhard (1989–90).

Bjerrum, Niels Jannisksen (1879–1958)

B. Copenhagen, Denmark. *Educ.* Univ. of Copenhagen (Ph.D., 1908). He was professor of chemistry at the Royal Veterinary and Agricultural College, Copenhagen, from 1914 to 1949. He worked on a wide range of problems in physics and physical chemistry, including kinetic theory, quantum theory, specific heats, spectra, Brownian movement and salt effects in kinetics. In 1909 he was the first to suggest that strong electrolytes are completely dissociated (Section 7.2), and in 1911 he was the first to give a satisfactory treatment of spectra and the specific heats of gases on the basis of quantum theory (Section 10.1). He also treated salt effects in solution (Section 8.5).

References: E. N. Hiebert, DSB, 2, 169–171 (1970); Jensen (1958–1959); Laidler (1986).

Blagden, Sir Charles (1748–1820)

B. Wooten-under-Edge, Gloucestershire, England. *Educ.* University of Edinburgh (MD, 1768). He was a medical officer in the British Army and then assistant to Henry Cavendish from 1782 to 1789. He was elected FRS in 1772 and was Secretary of the Royal Society from 1784. He discovered the relationship between the concentration and the lowering of the freezing point; although earlier found by Richard Watson this is usually known as Blagden's law. He was knighted in 1792. He spent much time in France and died in Arcueil.

Reference: E. L. Scott, DSB, 2, 186 (1970).

Bodenstein, Max (1871–1942)

B. Magdeburg, Germany. *Educ.* University of Heidelberg (Ph.D. 1893). He held various research appointments: at Göttingen with Nernst, and at Leipzig with Ostwald. He was professor at the University of Hannover from 1908 to 1923, when he succeeded Nernst at Berlin, remaining there until 1936. He worked almost entirely in kinetics, and was a remarkably skilful experimentalist. He made detailed studies of the thermal and photochemical reactions of hydrogen with the halogens (Section 8.4). In 1913 he was the first to suggest the idea of a chain reaction, and in the same year he independently suggested the steady-state hypothesis, shortly after it had been suggested by D. L. Chapman (Section 8.4).

Reference: E. Cremer, DSB, 15, 36–38 (1978); Cremer (1967).

Born, Max (1882–1970)

B. Breslau, Silecia (now Wroclaw, Poland). *Educ.* University of Göttingen (Ph.D. 1908). He was professor at the University of Berlin from 1915 to 1921 when he moved to Göttingen. He published the first paper on quantum mechanics in 1924 and then collaborated with Heisenberg and Jordan on various developments of the subject (Section 10.2). In 1936 he became professor of natural philosophy at the University of Edinburgh and won a Nobel Prize in physics in 1954 for his work on the quantum mechanics of collision processes.

Reference: A. Hermann, DSB, 17, 39–44 (1978).

Bosanquet, Robert Holford Macdowall (1841–1912)

Educ. Balliol College, Oxford. For a period he practised as a barrister, but from 1870 until his death he was a Fellow of St John's College, Oxford, where he established a small research laboratory. He carried out research on sound, particularly from organ pipes, and independently of Boltzmann gave the first satisfactory explanation for the specific heat of a diatomic gas (Section 5.3).

Boscovich, The Revd Roger Joseph (1711–1787)

B. Dubrovnik, Yugoslavia. He became a Jesuit priest and was professor of mathematics at the Collegium Romanum from 1740 and at the University of Pavia from 1763. He travelled widely in Europe and was elected FRS. His many books include *Philosophiae Naturalis Theoria* (1758); this gave his atomic theory, which exerted a wide influence in the nineteenth century, on Faraday and others.

References: Z. Markovic, DSB, 2, 326–332 (1970); Whyte (1961); Williams (1965).

Brandsma, Wiebold Frans (1892–1964)

B. Brummen, The Netherlands. *Educ.* in engineering at the Technische Universiteit Delft (Ph.D. 1925). He worked with Scheffer on theoretical kinetics, publishing on the subject from 1922 to 1929.

Brönsted, Johannes Nicolaus (1879–1947)

B. Varda, Denmark. *Educ.* University of Copenhagen (doctorate in 1908). He was professor of physical chemistry at the University of Copenhagen from 1908 until his death, his main work being on the thermodynamics and kinetics of electrolyte solutions. In 1923 he proposed definitions of acids and bases in terms of proton transfer (Section 8.5). In 1924, with La Mer, he used the Debye–Hückel theory to relate the activity coefficient of an ion to the ionic strength (Section 8.5). In 1922 he proposed the kinetic activity factor to explain ionic strength effects in kinetics, and in 1924 proposed the 'Brönsted relationships' between catalytic activity and acid or base strength (Section 8.5).

References: A. S. Veibel, DSB, 2, 498–499 (1970); Christiansen (1948–1949); Laidler (1986).

Bury, Charles Rugeley (1890–1968)

B. Henley-on-Thames, England. *Educ.* Trinity College, Oxford, and Göttingen. From 1913 to 1943 he was a lecturer at the University College of Wales at Aberystwyth, and from 1943 to 1952 he was on the staff at Imperial Chemicals Limited at Billingham. He was the first to deduce, from chemical evidence, the electronic configurations of the elements up to uranium; he made predictions for the transuranic elements (Section 10.4). He also worked on colloidal micelles, and on colour and chemical constitution.

References: Davies (two articles in 1986).

Butler, John Alfred Valentine (1899–1977)

B. Gloucestershire, England. *Educ.* University of Birmingham (D.Sc. 1927). From 1922 to 1939 he was at the University College of Swansea and at the University of Edinburgh, working on various aspects of electrochemistry, including overvoltage (Section 7.5). Subsequently he worked at the Courtauld Institute of Biochemistry and the Chester Beatty Research Institute, working on physico-chemical aspects of biological systems.

Reference: Mayneord (1979).

Cannizzaro, Stanislao (1826–1910)

B. Palermo, Italy. *Educ.* University of Palermo (in medicine). He was professor of chemistry at Genoa, Palermo and Rome. He was created a Senator in 1872 and later served as Vice-President of the Italian Senate. He discovered the 'Cannizzaro reaction' in 1853, and in 1858 he played an important role by giving a clear exposition of Avogadro's hypothesis.

Reference: H. M. Leicester, DSB, 3, 45–47 (1970).

Carlisle, Sir Anthony (1768–1840)

B. Durham, England. He studied surgery as an apprentice in York, Durham and London. He became highly skilled in medicine and surgery and had a fashionable practice in London. He was appointed surgeon to the Prince Regent (later George IV), was knighted in 1821, and became President of the College of Surgeons in 1829. He lectured on anatomy and wrote treatises on physiology, his activities being mainly in medicine and surgery, but he is now chiefly remembered for his electrolysis of water, with William Nicholson, in 1800.

Reference: A. Thackray, DSB, 3, 67–68 (1971).

Chance, Britton (b. 1913)

B. Wilkes-Barre, Pennsylvania. *Educ.* University of Pennsylvania (Ph.D. 1940). He has been professor of biophysics at the University of Pennsylvania from 1949. He devised the stopped-flow technique for the study of fast reactions (Section 8.5) and has made important investigations on a variety of biological reactions.

Christiansen, Jens Anton (1888–1969)

B. Velje, Denmark. *Educ.* Polytekniske Laereanstalt, Copenhagen, Carlsberg Laboratory, and University of Copenhagen (Ph.D. 1921). He taught at the University of Copenhagen from 1919 to 1959, being professor of inorganic chemistry from 1931 to 1948 when he succeeded Brönsted as professor of physical chemistry. In 1919 he proposed a mechanism for the hydrogen-bromine reaction (Section 8.4), and his 1921 Ph.D. thesis gave a treatment of unimolecular reactions (Section 8.5). He made a number of other important contributions to chemical kinetics, for example to acid-base catalysis (Section 8.6).

References: Ballhausen (1969); Bell (1970); Laidler (1986); Veibel (1970).

Clapeyron, Benoit Pierre Émile (1799–1864)

B. Paris, France. *Educ.* École Polytechnique. He taught for some years in St Petersburg, Russia, returning to France in 1830 after the July Revolution. He was an engineer of some distinction, specializing in bridges and locomotives. His only paper in pure science is notable for his restatement of Carnot's ideas in the language of the calculus (Section 4.1), and for his expression of vapour pressure as a function of temperature; this relationship was later improved by Clausius and is known as the Clapeyron–Clausius equation.

References: M. Kerker, DSB, 3, 286–287 (1971); Cardwell (1971).

Crookes, Sir William (1832–1919)

B. London, England. *Educ.* Royal College of Chemistry under A. W. Hofmann whose assistant he was from 1850 to 1854. Under the influence of Faraday, Wheatstone and Stokes he became more interested in physical aspects of chemistry. In 1854 he became Superintendent of the meteorological department of the Radcliffe Observatory at Oxford, and from 1855 to 1856 he taught chemistry at the College of Science at Chester. Subsequently he held no appointment but was a freelance scientist, earning his living by consulting and various commercial activities. He was a highly skilled experimentalist. He was a pioneer in the application of photography to scientific problems, particularly spectroscopy, and through spectroscopy discovered thallium in 1861 (Section 6.3). In 1875 he invented the radiometer, and his investigations of it led to his discovery that the cathode rays in a discharge tube are deflected in a magnetic field (Section 6.7). He developed an important theory of the origin of the elements. In 1859 he founded the weekly journal *Chemical News*, which he edited until 1906. He was much interested in spiritualism and made many investigations of mediums. He was knighted in 1897 and served as President of the Royal Society from 1913 to 1915.

References: W. H. Brock, DSB, 3, 474–482 (1971); d'Albe (1923).

Dalton, John (1766–1844)

B. Eaglesfield, near Cockermouth, Cumberland, the son of a Quaker weaver; he was largely self-educated. In 1781 he became a teacher in a boarding school in Kendall, and in 1793 was appointed a teacher of mathematics and science at New College, Manchester; he later supported himself by private tutoring. He made extensive studies in meteorology and biology, and investigated colour blindness ('Daltonism'). He carried out extensive studies of gases and discovered the law of partial pressures that is named after him. His great contribution was his atomic theory, first announced in 1803 and published in 1805 (Section 5.2). His book *A New System of Chemical Philosophy* appeared in three parts from 1808 to 1827 and was never completed.

References: A. Thackray, DSB, 3, 537–547 (1971); Greenaway (1966); Patterson (1970).

Dampier, Sir William Cecil (1867–1952)
(William Cecil Dampier-Whetham)

B. London, England as W. C. Dampier-Whetham, but changed his name to Dampier, that of his mother's family, after succeeding to a baronetcy. *Educ.*

Trinity College, Cambridge, of which he was a Fellow from 1891 to the end of his life. His earlier research was on the electrochemistry of solutions, and he is particularly remembered for his work on ionic mobilities (Section 7.2). He made a lasting contribution to agriculture through the Agricultural Research Council, of which he was the first Secretary. His *History of Science* first appeared in 1929 and went through several editions and translations into many languages.

Reference: Taylor (1954).

Daniell, John Frederic (1790–1845)

B. London, England. *Educ.* privately, mainly in the classics. He went into business and came in touch with scientific problems. In 1831 he was appointed the first professor of chemistry at King's College, London, where he was very active and influential. He devised many electrochemical cells (Section 7.1) and did important work in spectroscopy. He served as Secretary of the Royal Society from 1839 until his death, which occurred at a meeting of the Council of the Society.

Reference: A. Thackray, DSB, 3, 556–558 (1971).

Davy, Sir Humphry (1778–1829)

B. Penzance, Cornwall. He had little education, and became apprenticed to an apothecary. At the age of 20 he became Superintendent of the Pneumatic Institution at Clifton, near Bristol, begun by Thamas Beddoes. He later joined the newly founded Royal Institution in 1802 and became its professor of chemistry and Director. He did pioneering work in a variety of fields. He used electrolysis to isolate, for the first time, a number of elements. He became President of the Royal Society in 1820, but after 1823 he spent most of his time abroad, and died in Geneva.

References: D. M. Knight, DSB, 3, 598–604 (1971); Hartley (1966); Levere (1980); Russell (1959).

Dawson, Harry Medforth (1876–1939)

B. Leeds, Yorkshire. *Educ.* Yorkshire College of Science, Leeds, whose degrees were then granted by the University of London. He carried out research in Germany, mainly with van't Hoff, but also at Giessen, Leipzig and Breslau; his Ph.D. was from Giessen. He was demonstrator in chemistry at the Yorkshire College (which became incorporated in the University of Leeds in 1904), and became professor there in 1920. He worked extensively on the equilibrium and kinetic properties of solutions of electrolytes, and was the first to establish general acid–base catalysis (Section 8.7).

Reference: Whytlaw-Gray & Smith (1940); Laidler (1986).

Debye, Peter Joseph Wilhelm Debye (1884–1966)

B. MaasFricht, The Netherlands. *Educ.* University of Munich (doctorate in 1910); he first studied electrical engineering, then physics. He taught at the

Universities of Zürich, Leipzig and Berlin, and in 1940 moved to Cornell University where he later became head of the chemistry department. He made pioneering contributions in a variety of fields, including X-ray analysis, dipole moments and, with Hückel, the theory of strong electrolytes (Section 7.2).

References: C. P. Smyth, DSB, 3, 617–621 (1971); Davies (1967. 1970).

Desaguliers, The Revd John Theophilus (1683–1744)

B. La Rochelle, France. *Educ*. Christ Church, Oxford. He lectured for a time at Oxford and then in London. He was appointed to a number of church livings, and carried out experiments on heat, optics and electricity, some at Newton's suggestion. He was elected FRS in 1714 and became the Society's Curator of Experiments. He also did much work in applied science.

Reference: A. R. Hall, DSB, 4, 43–46 (1971); Stokes (1927).

Dewar, Sir James (1842–1923)

B. Kinkardine-on-Forth, Scotland. *Educ*. University of Edinburgh. After some appointments in Edinburgh he became Jacksonian professor of natural experimental philosophy at Cambridge in 1875. In 1877, finding the Cambridge laboratory facilities inadequate, he accepted the dual appointment of Fullerian professor at the Royal Instituton, where he did most of his research; he did, however, collaborate at Cambridge with Liveing in spectroscopy (Section 6.6). His most important work was on low temperatures, for which he constructed double-walled vessels in common use today. In 1891 he liquified oxygen, in 1898 hydrogen, and in 1899 he produced solid hydrogen having achieved a temperature of 14 K. He was knighted in 1904.

References: A. B Costa, DSB, 4. 78–81 (1971); Armstrong (1926, 1928).

Dirac, Paul Adrien Maurice (1902–1984)

B. Bristol, England. *Educ*. University of Bristol (degree in electrical engineering, 1921), St John's College, Cambridge. He proposed a mathematical formulation of quantum mechanics in 1925, and developed it in later years (Section 10.2), He was professor of mathematics at Cambridge from 1932 to 1969, and was professor of physics at Florida State University from 1971. He shared the 1933 Nobel Prize in physics with Schrödinger.

Reference: Dalitz and Peierls (1986).

Dixon, Harold Baily (1852–1930)

B. London. *Educ*. Christ Church, Oxford; he first studied classics but was unsuccessful and then studied chemistry under Vernon Harcourt. He was a Fellow of Trinity College, Oxford, 1875–1887, and in 1887 succeeded Sir Henry Roscoe as professor of chemistry at Owens College, Manchester (later the University of Manchester). He carried out comprehensive studies of explosions in gases, and investigated the effect of water on reactions in the gas phase (Section 8.5).

References: A. B. Costa, DSB, 4, 130; Baker & Bone (1931, 1932); Bowen (1970); Laidler (1988).

Döbereiner, Johann Wolfgang (1780–1849)

B. Hof an der Saale, Germany. He was apprenticed to an apothecary, and later taught at a technical college which was part of the University of Jena. His demonstration of spontaneous ignition of a hydrogen–air mixture in the presence of spongy platinum (Section 8.6) led to the manufacture of many highly popular lighters, but brought him no financial return since he had not taken out a patent. He was one of the first to notice certain regularities in the table of the elements ('Döbereiner's triads').

References: E. Farber, DSN, 4, 133–135 (1971)

Dux, Walter (1889–1987)

B. Hildesheim, Germany. *Educ.* Technischen Hochschule of Hannover (Dr Ing. 1913). In Hannover he worked with Bodenstein on the reaction between hydrogen and chlorine, and was much concerned with the development of the concept of a chain reaction (Section 8.4). He remained in Hannover until 1936 when he and his family had trouble with the Nazis, who expropriated the family business, and he lived for the rest of his long life in St Margarets-on-Thames, a suburb of London.

Eigen, Manfred (b. 1927)

B. Bochum, Germany. *Educ.* University of Göttingen, Germany Ph.D. 1951). He became a member of the staff of the Max-Planck Insität in Göttingen, becoming its Director. He developed relaxation techniques, such as temperature-jump, and applied them to many processes. He shared the 1967 Nobel Prize in chemistry with Porter and Norrish.

Esson, William (1839–1916)

B. Carnoustie, Forfarshire, Scotland. *Educ.* St John's College, Oxford. He became a fellow and tutor in mathematics of Merton College, Oxford, and in 1897 was appointed Savilian professor of geometry. He collaborated for over half a century with Vernon Harcourt on the analysis of kinetic results (Sections 8.1 and 8.2).

Reference: F. Szabadvary, DSB, 4, 411–412 (1971); Elliott (1917).

Evans, Meredith Gwynne (1904–1952)

B. Atherton, Lancashire. *Educ.* University of Manchester, where he became a lecturer. He was professor at the University of Leeds from 1939 to 1948 when he succeeded Michael Polanyi at the University of Manchester. He worked on a wide range of kinetic problems, especially polymerizations and oxidations. He collaborated with Polanyi on reactions of sodium atoms with halides, and on transition-state theory in 1935.

References: H. W. Melville (1953).

Ewan, Thomas (1868–1955)

B. Manchester. *Educ.* Owens College, Manchester, and the University of Munich (Ph.D. 1890). From 1893 to 1894 he worked with van't Hoff at the University of Amsterdam on oxidation reactions. From 1894 to 1896 he was demonstrator in chemistry at the Yorkshire College, Leeds, where he translated the second edition of van't Hoff's book, which appeared as *Studies in Chemical Dynamics*. From 1896 he held various industrial positions, first with the Aluminium Company, and then with the Cassel Cyanide Company.

Reference: Faulkner (1957).

Fizeau, Armand Hippolyte Louis (1819–1896)

B. Paris. *Educ.* He began medical studies at the Collège Stanislas but abandoned them and then studied at the Collège de France under Victor Regnault; he also attended lectures at the École Polytechnique and studied at the Paris Observatory under François Arago. He became interested in daguerreotypy from the beginning (1839) and improved the technique by gold toning and in other ways. He devised methods of converting a daguerreotype into an engraving without harming the original. In 1845, with Foucault, he took the first photograph of the sun, and in 1849 he made the first experimental determination of the velocity of light.

References: J. B. Gough, DSB, 5, 18–21 (1972); Buerger (1989).

Foucault, Jean Bernard Léon (1819–1868)

B. Paris. *Educ.* He began to study medicine at the École de Médicine, but abandoned medical studies and became assistant to Alfred Donné at the École. He became a science reporter for the *Journal des débats* and set up a laboratory in his home. He was interested in daguerreotypy from the beginning (1839) and took microphotographs for Donné. In 1845, with Fizeau, he took the first photograph of the sun. He invented a pendulum-clock device to keep a telescope pointing to the sun or a star, and incidentally, in 1851, demonstrated the earth's rotation ('Foucault's pendulum'). He was awarded the degree of docteur ès sciences physiques in 1853 for comparing the velocity of light in air and water.

Reference: H. J. Burstyn, DSB, 5, 84–87 (1972); Buerger (1989).

Fraunhofer, Joseph (1787–1826)

B. Staubing, Germany. He went into business in Munich as a maker of optical instruments. He made many spectroscopic investigations and mapped many of the lines that have come to be called the Fraunhofer lines. In 1823, while continuing his business activities, he became Director of the physics museum of the Bavarian Academy of Sciences in Munich.

Reference: R. V. Jenkins, DSB, 8, 142–144 (1972).

Freind, John (1675–1728)

B. Northamptonshire. *Educ.* Christ Church, Oxford (MA, MB, MD). He lectured at Oxford and carried out research in support of Newton's theory of matter. He was elected FRS in 1712. He served as physician to the English forces and then practised medicine in London. He was later physician to the Royal children and then to Queen Caroline.

Reference: Marie Boas Hall, DSB, 5, 156–157 (1972).

Galvani, Luigi (1727–1789)

B. Bologna, Italy. *Educ.* University of Bologna, where he later became professor of anatomy and obstetrics. He carried put research in physiology and anatomy, and is best remembered for work on the twitching of the muscles of a frog (Chapter 7).

Reference: T. M. Brown, DSB, 5, 267–269 (1972).

Gay-Lussac, Joseph-Louis (1778–1850)

B. St Léonard, Haute Vienne, France. *Educ.* École Polytechnique under Berthollet and others. He was professor of physics at the Sorbonne from 1808. In 1831 he was elected to the Chambre des Députés and elevated to the Chamber of Peers in 1834. He discovered the law of expansion of gases in 1802 and the law of combining volumes in 1809. In 1808 he and Thénard discovered boron.

Reference: M. P. Crosland, DSB, 5, 317–327 (1972).

Gilbert, Sir William (1540–1603)

B. Colchester, England. *Educ.* Cambridge and in Italy. Fellow of St John's College, Cambridge, from 1561, he later practised medicine in London and became President of the College of Physicians. He was granted a special pension by Queen Elizabeth to carry out research. He did pioneering work in electricity and magnetism (Chapter 7), and in his *Novum Organum* (1620) Francis Bacon used Gilbert's work to exemplify his inductive method.

Reference: S. Kelly, DSB, 5, 396–401 (1972).

Gouy, Louis Georges (1854–1926)

B. Vals-des-Bains, Ardèche, France. Little is known of his education; most of his career was as a professor at the Université de Lyon. He worked on a wide variety of topics, in particular on problems relating to the propagation of light. In 1888 he investigated Brownian motion and concluded that it is due to molecular bombardment (Section 9.1). In 1810 he proposed a theory of the diffuse double layer at a surface (Section 7.5).

Reference: J. B. Gough, DSB, 5, 483–484 (1972).

Graham, Thomas (1805–1869)

B. Glasgow, Scotland. *Educ.* University of Glasgow (MA 1826). He was professor of chemistry at the Andersonian College (now the Royal Technical College), Glasgow, from 1830 to 1837, and from 1837 to 1854 at University College, London. He was Master of the Mint from 1854. His earlier work, in Glasgow, was on the phosphates and arsenates. He formulated his law of diffusion in gases in 1833 and after 1837 studied diffusion in liquids (Chapter 9); this led him to his classic work on colloid chemistry, of which he is considered the founder. He participated in the formation of the Chemical Society of London and was its first President.

Reference: G. B. Kauffman, DSB, 5, 492–495 (1972).

Gray, Stephen (*c.* 1666–1736)

B. Canterbury, England. He had little education, first following his father's occupation as a dyer in Canterbury; later he moved to London and began scientific investigations, on the design of microscopes and other matters. In 1732 he published the results of experiments on electricity (Chapter 7), and in that year was elected FRS.

Reference: J. L. Heilbron, DSB, 5, 515–517 (1972).

Grimm, Hans Georg (1887–1958)

B. Hamburg, Germany. *Educ.* University of Munich (Ph.D., 1912). In 1924 he became professor of physical chemistry at the University of Würzburg. His main interest was in electronic theories of valency and their application to structural problems in inorganic chemistry. In 1931, with Ruf and Wolff, he published an important paper in chemical kinetics, giving activation energies for a Menschutkin reaction in a variety of solvents (Section 8.6).

Reference: Hofmann (1958)

Grotthuss, Theodor Christian Johann Dietrich von (1785–1822)

B. Leipzig. *Educ.* Leipzig, Naples, and Rome, and at the École Polytechnique in Paris. While in Italy, in 1805, he presented his theory of electrolysis (Section 7.1). Later he worked at the family farm on various scientific problems, principally on photochemistry. In 1809 he discovered that for photochemical action the light must be absorbed (Chapter 6); this was rediscovered by Draper in 1841 and known as the Grotthuss–Draper law. He also investigated combustion, gypsum deposits, meteorites, and the kinetic theory of gases. He suffered from an hereditary disease which caused him much suffering, and he committed suicide at the age of 37.

References: J. F. Stradins, DSB, 5. 558–559 (1972); Stradins (1964).

Grove, Sir William Robert (1811–1896)

B. Swansea, Wales. *Educ.* Brasenose College, Oxford. He became a barrister but carried out scientific work and was professor of experimental philosophy at the London Institution from 1841 to 1846. He served as a Secretary of the Royal Society from 1847 to 1849. He later returned to the practice of law, became a QC in 1853, and in 1856 was one of the counsel who defended the notorious William Palmer, a surgeon, who was convicted of murder by poisoning. He became a judge in 1871 and was knighted in 1872. He devised several electrochemical cells, including some fuel cells (Section 7.3). His book *On the Correlation of Physical Forces* (1846) was an important contribution to the law of conservation of energy (Section 4.2).

References: E. L. Scott, DSB, 5, 559–561 (1972); Webb (1961).

Hales, The Revd Stephen (1677–1761)

B. Kent. *Educ.* Benet's College (now Corpus Christi), Cambridge. He was ordained deacon in 1709 and was perpetual curate (vicar) of Teddington from then until his death, carrying out his scientific work there. He applied Newton's matter theory to a variety of problems in combustion, respiration and plant physiology of which he is the acknowleged founder. Elected FRS 1717; Copley Medal, 1739; Hon. DD, Oxford, 1733; foreign associate, Académie des Sciences, 1753.

Reference: H. Guerlac, DSB, 6, 35–48 (1972).

Hammett, Louis Plack (1894–1987)

B. Wilmington, Delaware. *Educ.* Harvard and Columbia (Ph.D. 1923). He taught at Columbia from 1920 to 1961, when he became professor emeritus, continuing his interest in science for many years. He was one of the founders of the field of physical organic chemistry, a term first used in the title of his book *Physical Organic Chemistry*, which appeared in 1940. He is particularly remembered for his work on linear free-energy relationships, referred to in Section 8.6.

Reference: Shorter (1990).

Heisenberg, Werner Karl (1901–1976)

B. Würzburg, Germany. *Educ.* University of Munich, under Sommerfeld (Ph.D. 1923). He became an assistant to Born in Göttingen in 1922 and also spent some time with Bohr in Copenhagen. In 1925 he developed his quantum mechanics based on matrices (Section 10.2), and in 1927 he proposed his uncertainty principle (Section 10.2). From 1927 he was professor at Leipzig, and in 1933 received the Nobel Prize for physics. During World War II he directed the German atomic energy programme, but opposed the oppressive activities of the Nazis.

References: D. C. Cassidy, DSB, 17, 394–403 (1990); Mott & Peierls (1977).

Herapath, John (1790–1867)

B. Bristol. Largely self-educated. He first worked in his father's business and later became a teacher of mathematics. His paper on kinetic theory appeared in 1821 in the *Annals of Philosophy* after rejection by the Royal Society (Section 3.3). He published later papers and a book, *Mathematical Physics*, and edited the *Railway Journal*.

Reference: S. G. Brush, DSB, 6, 291–293 (1972).

Herschel, Sir John Frederick William, Bart. (1792–1871)

B. Hawkshurst, Kent, the only son of Sir William Herschel. *Educ.* St John's College Cambridge (Senior Wrangler and Smith's Prizeman in 1813). He was elected Fellow of St John's and FRS in 1813, but held no paid position from 1816 until 1850, when he became Master of the Mint. He did pioneering work in photography (Sections 6.1 and 8.4) and in astronomy. He was knighted in 1831 and created a baronet in 1838.

Reference: D. S. Evans, DSB, 6, 323–328 (1972); Schaaf (1992).

Herschel, Friedrich Wilhelm (Sir William) (1738–1822)

B. Hannover, Germany; after the French occupation he escaped to England in 1757. He first supported himself by teaching, conducting and composing music, becoming organist of the Octagon Chapel in Bath in 1766. Later he became interested in astronomy and began to grind lenses and mirrors and construct telescopes. He soon came to be recognized as a leading astronomer, and in 1782 George III granted him a pension in return for living at Slough, near Windsor, and showing the heavens to members of the royal family as requested. He discovered the planet Uranus, and was the first to detect infrared radiation from its heating effect (Chapter 6). He was knighted in 1816.

References: M. A. Hoskin, DSB, 6, 328–336 (1972); Lovell (1968).

Herzfeld, Karl Ferdinand (1892–1978)

B. Vienna, Austria. *Educ.* Universities of Vienna, Zürich and Göttingen. He was *Privatdozent* at the University of Munich from 1920 to 1926. He was professor of physics at Johns Hopkins University, Baltimore, 1926–1936, and at the Catholic University of America, Washington, DC, 1936–1961. He worked on a variety of problems in chemical physics, including electrolyte solutions (with Heitler), energy transfer and ultrasonics. He suggested a mechanism for the hydrogen–bromine reaction (Section 8.4), and with F. O. Rice, chain mechanisms for organic decompositions (Section 8.5).

Heyrovsky, Jaroslav (1890–1967)

B. Prague, Czechoslovakia. *Educ.* University of Prague for one year and then, in 1910, at University College, London, where he studied under William Ramsay, W. C. McC. Lewis and F. G. Donnan; he was then at Charles University (Ph.D. in 1918). He worked first at London University and from 1921 at Charles University. In 1950 he became the first Director of the Central Polarographic Institute in Prague. He was awarded the 1956 Nobel Prize in chemistry for his development of polarography.

References: M. Teich, DSB, 6, 370–376 (1972); Butler & Zuman (1964); Müller (1941); Sherman (1990).

Hittorf, Johann Wilhelm (1824–1914)

B. Bonn, Germany. *Educ.* University of Berlin (doctorate in 1846). He held appointments at the University of Bonn and the Royal Academy of Münster. He worked on the allotropic forms of selenium and phosphorus, electric discharges through gases, and the passivity of metals. He measured concentration changes during electrolysis and introduced the idea of transport numbers (Section 7.2).

Reference: O. J. Drennan, DSB, 6, 438–440 (1972).

Hooke, Robert (1635–1703)

B. Isle of Wight. *Educ.* Christ Church, Oxford, and acted as assistant to Robert Boyle from about 1655 to 1662. He was Curator of Experiments to the Royal Society from 1661 to 1703, and was elected FRS in 1663; he was Secretary to the Royal Society from 1677. An experimentalist of great skill, he designed the air pump first used by Boyle. He was a distinguished microscopist, his book *Micrographia* (1665) containing remarkable drawings of microscopic observations. He did important work in various aspects of physics and astronomy, and on combustion and respiration. He engaged in bitter controversies with Newton.

References: R. S. Westfall, DSB, 6, 481–488 (1972); Andrade (1950).

Horstmann, August Friedrich (1842–1929)

B. Mannheim. *Educ.* Heidelberg (Ph.D. 1865). He was professor of theoretical chemistry from 1865, and did some of the earliest work in chemical thermodynamics (Section 4.4).

Reference: O. J. Drennan, DSB, 6, 519–520 (1972).

Hückel, Erich (1896–1980)

B. Charlottenburg, Germany. *Educ.* Göttingen (doctorate in 1921). In 1922 he became an assistant to Debye in Zürich and developed with him the theory of strong electrolytes (Section 7.2). Later he held appointments at Stuttgart and

Marburg. He carried out important work on the quantum mechanics of unsaturated and aromatic compounds (Section 10.4).

References: Hartmann & Longuet-Higgins (1982); Hückel (1965); Sunko & Trinajstic (1981).

Jordan, Ernst Pascual (1902–1980)

B. Hamburg, Germany. *Educ.* Technische Hochschule, Hannover, and the University of Göttingen. After 1925 he collaborated with Max Born and Heisenberg on matrix mechanics (Section 10.2). In 1929 he obtained an appointment at the University of Rostock, and in 1944 succeeded Max von Laue as Director of the Institute of Theoretical Physics at the University of Berlin. He was active in politics, and before the war held views somewhat sympathetic to the Nazi regime; from 1953 to 1961 he was a member of the German Bundestag. He was professor at the University of Hamburg from 1947 until his retirement in 1971.

Reference: K. von Meyenn, DSB, 17, 448–454.

Keill, John (1671–1721)

B. Edinburgh. *Educ.* Edinburgh and Balliol College, Oxford. From 1712 to 1721 he was Savilian professor of astronomy at Oxford. He taught the Newtonian principles and developed the idea that matter consists of attracting particles (Section 3.2).

Reference: D. Kubrin, DSB, 7, 275–277 (1973).

Kirchhoff, Gustav Robert (1824–1887)

B. Königsberg, Prussia (now Kaliningrad, Russia). *Educ.* University of Königsberg. After an appointment at Breslau he was at Heidelberg from 1854, where he worked on spectroscopy with Bunsen (Section 4.1). In the 1860s he established important principles of spectroscopy and introduced the idea of a black body (Section 6.2).

Reference: L. Rosenfeld, DSB, 7, 379–383 (1973).

Kohlrausch, Friedrich Wilhelm Georg (1840–1910)

B. Rinteln, Germany. *Educ.* Universities of Marburg, Erlangen and Göttingen (doctorate from Göttingen in 1863). He held appointments at the Universities of Göttingen (1866–1870), Zürich (1870–1871), Würzburg (1875-1888), Strassbourg (1888-1894), and the Physikalische Technische Reichsanstalt at Charlottenburg (from 1894). He was noted for experimental work of high precision on a variety of problems, including elasticity, magnetism and the conductivities of electrolytes (Section 7.2).

Reference: O. J. Drennan, DSB, 7, 449–450 (1973).

Kohnstamm, Philip Abraham (1875–1951)

B. Bonn, Germany. *Educ.* University of Amsterdam (Ph.D. 1901). He succeeded van der Waals as professor of physics at the University of Amsterdam in 1908, and beginning in about 1910 he worked on the theory of reaction rates, partly in collaboration with Scheffer and Brandsma; they suggested the idea of a Gibbs energy of activation (Section 8.3). In 1817 Kohnstamm's interest shifted to pedagogy; in 1930 he was appointed professor of pedagogy at the University of Utrecht, and from 1938 he was professor of pedagogy at Amsterdam.

Reference: Anon. (1970).

Lambert, Johann Heinrich (1728–1777)

B. Mulhouse, Alsace. He had little formal education and his early career was as a secretary and teacher; in 1765 he became a member of the Prussian Academy of Sciences in Berlin. His main work was in mathematics, astronomy and philosophy. His realization that the intensity of light varies exponentially as it passes through an absorbing medium (Section 6.4) is basic to work in colorimetry and spectrophotometry.

Reference: C. J. Scribe, DSB, 7, 595–600 (1972).

Lapworth, Arthur (1872–1941)

B. Galashiels, Scotland. *Educ.* Mason College and the City and Guilds of London Institute. In 1909 he went to the University of Manchester, becoming professor in 1913. He made important contributions to the knowledge of mechanisms of reactions in solution (Section 8.6).

References: J. Challenor, DSB, 8, 31–34 (1973) ; Robinson (1947); Saltzman (1972).

Lavoisier, Antoine Laurent (1743–1794)

B. Paris. *Educ.* Collège des Quatre Nations (usually known as Mazarin College). In 1768 he was elected a member of the Académie des Sciénces and, in order to finance his researches, became one of the *fermiers-génénaux*, concerned with the collection of taxes. In 1776 he was appointed Director of the government powder mills. His researches covered a wide range, including chemistry, meteorology and geology, and he was much interested in practical applications. He did important work with Laplace on calorimetry, but remained convinced that heat is one of the chemical elements. His most important work in chemistry relates to the nature of the atmosphere and to combustion. He reinterpreted results obtained by Joseph Priestley and others, and carried out carefully designed experiments which overthrew the old phlogiston theory, which he replaced with the modern theory of combustion. He did much to transform chemistry into a more quantitative field, placing emphasis on the use of the balance. His famous *Traité élémentaire de chimie*, published in 1789, exerted a wide influence on the development of the subject. For his activities as a fermier-général he was guillotined in 1794.

References: H. Guerlac, DSB, 8, 66–92; this article gives numerous references to other biographical material.

Lewis, Gilbert Newton (1875–1946)

B. West Newton, Massachesetts. *Educ.* University of Nebraska, and Harvard (Ph.D. 1897). He later worked with Ostwald and Nernst, and after a period at the Massachusetts Institute of Technology he went in 1912 to the University of California where he remained until his accidental death in his laboratory. He clarified and expanded Gibbs's thermodynamics, and in 1900 to 1907 he introduced and developed the concepts of fugacity and activity (Section 4.4). He made important contributions to atomic and molecular theory, and introduced the idea of electron pairs in bond formation (Section 10.4). He also developed a theory of acids and bases that is more general than that of Brönsted (Section 8.6). Lewis never received a Nobel Prize, but many physical chemists feel that he well deserved one.

References: R. E. Kohler, DSB, 8, 289–294 (1973); Servos (1990). The January, February and March, 1984, issues of *J. Chem. Education* (Vol. 61) contain many articles about G. N. Lewis.

Lewis, William Cudmore McCullach (1885–1956)

B. Belfast, Northern Ireland. *Educ.* Royal University of Ireland (now Queen's University, Belfast), University of Liverpool (under F. G. Donnan); he spent a year at Heidelberg with Georg Bredig. In 1913 Lewis succeeded Donnan as professor of physical chemistry at Liverpool, retiring in 1948. Much of his research was on the application of physical chemistry to colloidal and biological systems He was one of the first to apply quantum theory to chemical problems. In 1917 he applied simple kinetic theory to calculate a rate constant for a second-order reaction (Section 8.3). He wrote an influential book, *A System of Physical Chemistry*, first published in 1916.

References: Bawn (1958); King & Laidler (1984).

Lindemann, Frederick Alexander, 1st Viscount Cherwell (1886–1957)

B. Baden-Baden, Germany, of parents who normally resided in England. *Educ.* Darmstadt Hochschüle and the Institüt für physikalische Chemie, Berlin, where he worked with Nernst (Ph.D. 1910). He was professor of experimental philosophy at Oxford from 1919 to 1955, and was Paymaster-General in Winston Churchill's cabinet, from 1942 to 1945 and from 1951 to 1953. He worked on a variety of problems in physics, and played an important role in the overthrow of the radiation hypothesis, suggesting an alternative theory of unimolecular gas reactions (Section 8.5).

References: R. V. Jones, DSB, 8, 368–369 (1973); King & Laidler (1984); Thomson (1958).

Liveing, George Downing (1827–1924)

Educ. St John's College, Cambridge, of which he became a Fellow in 1853. He was professor of chemistry at Cambridge from 1861 to 1908, and after James Dewar arrived at Cambridge in 1875 the two collaborated on extensive investigations in spectroscopy. Liveing was a Fellow of the Chemical Society for 71 years, and he carried on research until at the age of 97 he died as the result of an accident.

Reference: Haydock (1925).

Lockyer, Sir Joseph Norman (1836–1920)

B. Rugby, England. He did not attend a university but became a civil servant. He was an amateur astronomer, becoming a lecturer in astronomy at the Normal School of Science (renamed in 1890 the Royal College of Science). From 1887 he was professor of astronomical physics at the Normal School and Director of the Solar Physics Laboratory at South Kensington. He made important contributions to astronomy and spectroscopy, being the first to recognize helium in the solar spectrum (Section 6.3). He founded the journal *Nature* in 1869 and edited it for 50 years (Section 2.1). He was knighted in 1897.

References: H. Dingle, DSB, 8, 440–442 (1973); Lockyer & Lockyer (1928); Meadows (1972).

Lodge, Sir Oliver Joseph (1851–1940)

B. Staffordshire, England. *Educ.* University of London (doctorate in 1877). He was professor of physics at the University of Liverpool from 1881 to 1900 and Principal of the University of Birmingham from 1900 to 1919. He worked in particular on electromagnetic radiation, including radio waves. Lodge played an active role in the British Association, often being called upon to resolve controversial issues. He devised a method of measuring transport numbers (Section 7.1). He was knighted in 1902.

Reference: C. Süsskind, DSB, 8, 443–444 (1973); Gregory & Ferguson (1941).

Lomonosov, Mikhail Vasilevich (1711–1765)

B. Denisovka (now called Lomonosov), near Archangel, Russia. *Educ.* Univ. of Marburg. He was professor of chemistry at the University of St Petersburg from 1745. He anticipated Lavoisier in opposing the phlogiston theory, and recognised the nature of heat and light. He did important work in astronomy, and is regarded as the founder of modern Russian science. He also wrote poems and plays.

Reference: B. M. Kedrov, DSB, 8, 467–472 (1973).

London, Fritz (1900–1954)

B. Breslau, Germany (now Wroclaw, Poland). *Educ.* Universities of Bonn, Frankfurt and Munich (doctorate in 1921 at Munich on the theory of knowledge). His first interest was philosophy, but in 1925 he returned to Munich to work under Sommerfeld. In 1927 he and Heitler produced their treatment of the hydrogen molecule (Section 10.4), and in the following year he extended the treatment to the H_3 system. At about the same time he developed a quantum-mechanical treatment of dispersion forces. In 1939 he became professor at Duke University in Durham, North Carolina, where he worked on superconductivity and related problems.

Reference: C. W. F. Everett & W. M. Fairbank, DSB, 8, 473–479 (1973).

Lubbock, Sir John William, 3rd Baronet (1803–1865)

B. London, England. *Educ.* Trinity College, Cambridge. He was by profession a banker, and spent his leisure time in scientific pursuits. He did important work on the theory of annuities, on the tides, and on the calculation of planetary orbits. He twice served as Vice-President and Treasurer of the Royal Society, and played an important role in the regrettable rejection of Waterston's paper on kinetic theory (Section 5.3). His son Sir John Lubbock (1834–1913), the 4th Baronet and later Lord Avebury, was distinguished as an entomologist, anthropologist, botanist and geologist.

Reference: S. H. Dieke, DSB, 8, 850 (1973).

Marcelin, René (1885–1914)

B. Cagny, Seine-et-Oise, France. *Educ.* Faculté des Sciences de Paris (Ph.D. 1914). He worked on the theory of reaction rates from the thermodynamic and statistical points of view. In 1914 he described a reaction for the first time in terms of the motion of a point in phase space, and so introduced the idea of a potential-energy surface (Section 8.3). Marcelin was killed in the first few days of World War I.

References: Laidler (1986); Mysels (1986).

Mariotte, Edmé (*c.* 1620–1684)

Little is known of Mariotte's personal life; he may have been a Roman Catholic priest. He played an important role in the Académie des Sciénces, and in the development of experimental science in France. He published a number of influential books.

Reference: M. S. Mahoney, DSB, 9, 114–122 (1974).

Mayer, Julius Robert (1814–1878)

B. Heilbronn, Württemberg (now Baden-Württemberg), Germany. *Educ.* Univ. of Tübingen, MD 1838. A paper of 1842 establishes his priority for the principle of conservation of energy (Section 4.1); Joule's later conclusion was more convincing. A bitter priority dispute affected Meyer's mental stability, but he received recognition in his later life, being, for example, awarded the Copley Medal of the Royal Society in 1871.

Reference: R. S. Turner, DSB, 9, 235–240 (1974).

Melvill, Thomas (1726-1753)

B. possibly in Glasgow, Scotland. *Educ.* University of Glasgow (in divinity). He carried out pioneering experiments in which he examined the spectra of flames into which various chemicals had been introduced, and was the founder of flame spectroscopy. He died in Geneva at the age of 27.

Reference: J. D. North, DSB, 9, 266–267 (1974).

Miller, William Hallowes (1801–1880)

B. Llandovery, Carmarthenshire, Wales. *Educ.* St John's College, Cambridge, of which he became a Fellow in 1829. His main work was in mathematics and crystallography, and he is still remembered for the 'Miller indices'. In 1832 he succeeded William Whewell as professor of mineralogy at Cambridge.

Reference: D. McKie, DSB, 9, 392–393 (1974).

Milner, Samuel Roslington (1875–1958)

B. Yorkshire, England. *Educ.* University College, Bristol (doctorate from the University of London in 1899). From 1900 he taught in the department of physics at the University of Sheffield, finally as professor of physics. He worked on a variety of problems, including spectroscopy, electromagnetic theory, tensor calculus and electrolyte theory. He formulated a detailed treatment of interionic forces in solution (Section 7.2).

Reference: Allibone & Clarke (1959).

Moseley, Henry Gwyn Jeffreys (1887–1915)

B. Weymouth, England. *Educ.* Trinity College, Oxford. He worked in Rutherford's department at the University of Manchester from 1910 to 1914, and from the X-ray spectra of elements determined their atomic numbers (nuclear charges). He was killed in action at Gallipoli.

References: J. L. Heilbron, DSB, 9, 542–545 (1974); Moseley (1974).

Nicholson, William (1753–1815)

B. London. He went to sea with the East India Company and later became a commercial agent of Josiah Wedgwood, the potter; he was also a waterworks engineer and a science writer. He wrote a successful *Introduction to Natural Philosophy* (1781), founded *Nicholson's Journal* in 1797, and in 1808 compiled a *Dictionary of Practical and Theoretical Chemistry*. With Carlisle, he electrolysed water in 1800 (Section 7.1).

Reference: A. Thackray, DSB, 10, 107–109 (1974).

Norrish, Ronald George Wreyford (1897–1978)

B. Cambridge, England. *Educ.* Emmanuel College, Cambridge (Ph.D. 1924). He was fellow of Emmanuel from 1924 to the end of his life, and professor of physical chemistry at Cambridge from 1937 to 1965. He worked mainly on the kinetics of reactions in the gas phase, but also did some work on solution reactions, especially oxidations and polymerizations. He and George Porter developed the technique of flash photolysis (Section 8.4), sharing with Manfred Eigen the 1967 Nobel Prize for chemistry.

Reference: Dainton & Thrush (1981); Laidler (1993).

Noyes, Arthur Amos (1866–1936)

B. Newburyport, Massachusetts. *Educ.* Massachusetts Institute of Technology and Leipzig (under Ostwald; Ph.D. 1890). He taught at MIT until 1919, being its acting President from 1907 to 1909. In 1919 he moved to the Throop College of Technology in Pasadena, California, which later changed its name to the California Institute of Technology. He was a competent research chemist and a great teacher of chemistry, playing an important role in the introduction of research in physical chemistry to the United States.

References: L. Pauling, DSB, 10, 156–158 (1974); Pauling (1958).

Paneth, Friedrich Adolf (1887–1958)

B. Vienna, Austria. *Educ.* Universities of Munich, Glasgow and Vienna (Ph.D. 1910). He held professorships at the Universities of Berlin (until 1929), Königsberg (until 1933) and Durham (1939–1953); from 1953 he was Director of the Max Planck Institute at Mainz. His research was in radiochemistry and inorganic chemistry. In 1929 he discovered the free methyl radical by means of the lead mirror technique (Section 8.5).

Reference: E. G. Spittler, DSB, 10, 288–289 (1974).

Pauling, Linus Carl (1901–1994)

B. Portland, Oregon. *Educ.* Oregon State Agricultural College, California Institute of Technology (Ph.D. 1925); postdoctoral work under Sommerfeld at the University of Munich. He has been professor of chemistry at the California Institute of Technology since 1927. His first work was on X-ray analysis. From about 1930 he was a pioneer in the development of quantum mechanics and in its application to chemical structure and behaviour (Section 10.4), developing the concepts of orbital hybridization and of resonance between alternative structures. In the 1950s he made important contributions to the understanding of proteins, suggesting that they have helical structures. He has also worked on the structures and properties of many other substances of biological import-ance. Much publicity resulted from his contention since 1970 that large doses of vitamin C are effective in the prevention of the common cold. He has received many honours, including the 1954 Nobel Prize in chemistry and the 1962 Nobel Prize for peace.

Reference: Servos (1990)

Perrin, Jean Baptiste (1870–1942)

B. Lisle, France. *Educ.* École Normale Supérieure, Paris. He was professor of physical chemistry at the Université de Paris from 1910 to 1940, when he emigrated to the United States, dying in New York. Beginning in the early years of the twentieth century he made important investigations of sedimentation and Brownian motion (Section 9.1) for which he was awarded the 1936 Nobel Prize for physics.

Reference: R.H. Stuewer, DSB, 10, 524–526 (1974).

Pfaundler, Leopold (1839–1920)

B. Innsbrück, Austria. *Educ.* Universities of Innsbrück, Munich and Paris (Ph.D. in physics at Innsbrück). He was professor of physics at Innsbrück from 1867 to 1891, when he became professor at the University of Graz. He carried out research over a wide range of physics and chemistry, and in 1867 and 1872 he applied kinetic theory to kinetic problems, taking into account the distribution of speeds (Section 8.3).

Reference: Root-Bernstein (1980).

Pickering, Percival Spencer Umfreville (1858–1920).

B. Pontefract, England. *Educ.* Balliol College, Oxford. He was lecturer in chemistry at Bedford College, London, from 1880 to 1887; he then carried out research on electrolyte solutions in a private laboratory he established in London. He was an outspoken critic of the theory of electrolytic dissociation. After 1896 he devoted himself to scientific fruit farming, in connection with which he carried out many valuable investigations.

Reference: 'A. H.' (1926); 'E. J. R.' (1921).

Polanyi, Michael (1891–1976)

B. Budapest, Hungary. *Educ.* University of Budapest (MD 1913, Ph.D. after 1918). He worked at the Institüt für physikalische Chemie in Berlin from 1920 to 1933 and was professor of physical chemistry at Manchester University from 1933 to 1948, when he became professor of social studies at Manchester. In 1931 he and Henry Eyring constructed a potential-energy surface for $H + H_2$ (Section 8.3). In the 1930s he carried out fundamental studies in gas kinetics which led to the later work in chemical dynamics (Section 8.7). In 1935 he and M. G. Evans developed transition-state theory at almost the same time as Eyring.

Reference: Wigner & Hodgkin (1977).

Powell, The Revd Baden (1796–1860)

B. Stamford Hill, England. *Educ.* Oriel College, Oxford. After ordination in 1820 he performed parochial duties until 1827 when he was appointed Savilian professor of geometry at Oxford. While vicar of Plumsted, Kent, he carried out experiments on the heating effect produced by infrared radiation. At Oxford he carried out experiments on the dispersion of light which supported the wave theory. A member of the Broad Church, he was a staunch advocate of the teaching of more science at universities. His son Robert Stephenson Smyth Baden-Powell (1857–1941), later Baron Baden-Powell, founded the scouting movement.

Reference: R. Fox, DSB, 11, 115–116 (1975).

Power, Henry (1623–1668)

B. Halifax, Yorkshire. *Educ.* Christ's College, Cambridge (MD 1665). He practised medicine in Halifax and investigated the circulation of the blood, the lymphatic system, microscopy and magnetism. He studied the pressure–volume relationships in gases from 1653 in Cambridge and Halifax, and with Richard Towneley at Towneley Hall. In 1663 he published *Experimental Philosophy*, the first book to describe metals under the microscope.

References: G. W. O'Brien, DSB, 11, 121–122 (1975); Cohen (1964); Cowles (1934).

Priestley, Joseph (1733–1804)

B. near Leeds, Yorkshire. *Educ.* at a Dissenting academy in Daventry. In 1755 he became a Unitarian clergyman and held various appointments. In the 1760s he began to study electricity and chemistry. He experienced hostility for his radical opinions, particularly for his support of the American colonists, and in 1794 he settled in Northumberland, Pennsylvania. In England and America he carried out chemical research, particularly on respiration and the properties of gases, but never abandoned the phlogiston theory. In 1795 he was offered, but declined, the professorship of chemistry at the University of Pennsylvania.

Reference: R. E. Schofield, DSB, 11, 139–147 (1975).

Rankine, William John Macquorn (1820–1872)

B. Edinburgh. *Educ.* University of Edinburgh (leaving without a degree). He practised as a civil engineer in Ireland and Scotland, and worked on various scientific problems being elected FRS in 1853. He was regius professor of civil engineering and mechanics at the University of Glasgow from 1855. He developed the vortex theory of the atom (Section 5.2) and gave a treatment of the second law of thermodynamics similar to that of Clausius (Section 4.3). He opposed the metric system and devised an absolute Fahrenheit scale.

References: E. M. Parkinson, DSB, 11, 291–295 (1975); Anon. (1873); Hutchison (1981); Raman (1973).

Raoult, François Marie (1830–1901)

B. Fournes, France. *Educ.* College of St Dié, where he also taught, and the University of Paris (docteur ès sciences physiques, 1863). He was professor of chemistry at the Faculté des Sciences de Grenoble from 1870. He worked in thermochemistry and electrochemistry and is particularly noted for his work on colligative properties, particularly the depression of the freezing point and vapour pressure ('Raoult's law').

Reference: L. I. Kuslan, DSB, 11, 297–300 (1975); van't Hoff (1902).

Rayleigh, 3rd Baron (John William Strutt) (1842–1919)

B. Langford Grove, near Maldon, England. *Educ.* Trinity College, Cambridge (Senior Wrangler and Smith's Prizeman, 1866); he was elected a Fellow of Trinity in 1867. He succeeded his father as 3rd baron in 1873, was Cavendish professor at Cambridge (succeeding Maxwell) from 1879 to 1884, and professor of natural philosophy at the Royal Institution from 1888 to 1905. Much of his scientific work was done at his family seat at Terling in Essex. He was interested in a wide range of problems in physics, especially wave theory. With William Ramsay he discovered argon in 1894, and he made important contributions to the understanding of surface films (Section 9.2). He had an unusually generous disposition, and was helpful to several other scientists. He won the 1904 Nobel Prize for physics, particularly for the discovery of argon.

References: R. B. Lindsay, DSB, 13, 100–107 (1976); Schuster (1921); Strutt (1934).

Regnault, Henri Victor (1810–1878)

B. Aix-la-Chapelle, France (now Aachen, Germany). *Educ.* École Polytechnique, Paris, and École des Mines, Paris. He was an assistant at the École Polytechnique to Gay-Lussac, whom he succeeded as professor of chemistry in 1840; Kelvin worked with him there in 1845–46. From 1854 he was Director of the famous porcelain factory at Sèvres. His main research was on the thermal properties of gases. He did much work on both scientific and artistic daguerreotypy.

References: R. Fox, DSB, 11, 353–354 (1975); Buerger (1989); Fox (1971).

Rice, Francis Owen (1890–1989)

B. Liverpool, England. *Educ.* Liverpool University (D.Sc. 1916). After appointments at New York University and Johns Hopkins University he was Chairman of the chemistry department at the Catholic University of America, Washington, DC, from 1938 to 1959; he was then Chairman of chemistry at Georgetown University from 1959 to 1962. With Urey and Washburn he first studied the kinetics of a reaction in a molecular beam (Section 8.7). He extended Paneth's lead-mirror technique for detecting free radicals, and with Herzfeld he devised mechanisms for organic decompositions (Section 8.5). He continued to do research until shortly before his death at the age of 99.

Rice, James (1874–1936)

B. Northern Ireland. *Educ.* Queen's College (now Queen's University), Belfast. After being a schoolmaster from 1901 to 1914 he taught at the University of Liverpool until his death, being associate professor from 1924 and Reader in theoretical physics from 1935. He was one of the first to recognize the importance of quantum theory in chemistry, and he developed Marcelin's idea of representing a chemical reaction in terms of the motion of a point in phase space.

References: King & Laidler (1984); Laidler & King (1983).

Rideal, Sir Eric Keighley (1890–1974)

B. Sydenham, England. *Educ.* Trinity Hall, Cambridge, and University of Bonn (Ph.D. 1912). He was visiting professor at the University of Illinois from 1919 to 1921. He was elected Fellow of Trinity Hall in 1921, and became professor of colloid physics at Cambridge in 1930, the title later becoming professor of colloid science. From 1946 to 1950 he was Director of the Davy–Faraday Laboratory at the Royal Institution. He was professor of chemistry at King's College, London, from 1950 to 1955. He worked on a wide range of topics including electrochemistry, colloid and surface chemistry, and chemical kinetics. He studied surface films, electrophoresis, and the kinetics of reactions on surfaces (Section 8.7). He was knighted in 1951.

References: W. H. Brock, DSB, 18, 738–743 (1990); Eley (1976).

Ritter, Johann William (1776–1810)

B. near Haynow, Silesia (now Chojnow, Poland). He was apprenticed to an apothecary, then studied and did research at the University of Jena. He was highly talented but presented eccentric interpretations and spent much time investigating the occult. He was the first to demonstrate, in 1801, the existence of the ultraviolet, by the blackening of silver chloride. In 1801 he repeated the electrolysis of water (Section 7.1). In 1802 he constructed a dry cell, and in 1803 a storage battery, both of these contributions being years ahead of their time (Section 7.3).

Reference: R. J. McRae, DSB, 11, 473–475 (1975).

Roget, Peter Mark (1779–1869)

B. London. *Educ.* University of Edinburgh. He became professor of physiology at the Royal Institution and published a number of treatises on physiology, electricity and magnetism. He discovered the persistence of vision, and was concerned in the establishment of the University of London. He was Secretary of the Royal Society from 1827 to 1848 and was involved in the rejection of Waterston's paper on the kinetic theory of gases (Section 5.3). He wrote his famous *Thesaurus of English Words and Phrases* (1852) in his retirement.

Roozeboom, Hendrik Willem Bakhuis (1854–1907)

B. Alkmaar, The Netherlands. *Educ.* University of Leiden (Ph.D. 1884). After working for some time in a butter factory he taught at the University of Leiden from 1878 until 1896 when he succeeded van't Hoff as professor of chemistry at Amsterdam. He was a pioneer in the application of Gibbs's phase rule.

Reference: H. A. M. Snelders, DSB, 11, 534–535 (1975).

Roscoe, Sir Henry Enfield (1833–1915)

B. Leatherhead, Surrey. *Educ.* University College, London (under Graham and Williamson). He then worked with Bunsen at Heidelberg on the photochemical hydrogen–chlorine reaction (Section 8.4). In 1857 he succeeded Edward Frankland as professor at Owens College, Manchester, where he was a highly successful teacher. He continued to collaborate with Bunsen after going to Manchester. He was knighted in 1884, and became a Privy Councillor in 1909.

Reference: R. H. Kargon, DSB, 11. 536–539 (1975).

Roughton, Francis John Worsley (1899–1972)

B. Kettering, England. *Educ.* Trinity College, Cambridge. In 1947 he succeeded Rideal as professor of colloid science at Cambridge, the title later becoming professor of biophysics. He developed flow methods for the study of fast reactions (Section 8.5) and made important contributions to the understanding of reactions of haemoglobin.

Reference: Gibson (1972).

Rydberg, Johannes Robert (1854–1919)

B. Lund, Sweden. *Educ.* University of Lund (doctorate in mathematics in 1875). He taught until his death at the University of Lund. In 1890 he generalized Balmer's formula for spectral series (Section 6.4).

References: C. L. Maier, DSB, 13, 42–44 (1976); Nepomucene (1960).

Scheffer, Frans Eppo Cornelius (1883–1954)

B. Veendam, The Netherlands. *Educ.* University of Amsterdam (Ph.D. 1909). After lecturing at the University of Amsterdam he became, in 1917, professor of analytical chemistry at the Technische Universiteit of Delft. He did some research in kinetics, some of it in collaboration with Kohnstamm and Brandsma (Section 8.3).

Reference: Snelders (1985).

Schuster, Sir Arthur (1851–1934)

B. Frankfurt, Germany; moved to Manchester, England in 1870. *Educ.* Owens College, Manchester, and Heidelberg (Ph.D. under Kirchhoff). He was professor of physics at Owens College from 1881. He worked on a range of problems, including spectroscopy (Section 6.6), electric discharges and X-rays.

References: R. H. Kargon, DSB, 12, 237–239 (1975); Simpson (1932–35, 1934).

Semenov, Nikolai Nikolaevich (1896–1986)

B. Saratov, Russia. *Educ.* University of St Petersburg (which when he graduated in 1917 had changed its name to the University of Petrograd). For a period he was at the Physico-Chemical Institute in Leningrad (formerly Petrograd and since 1991 St Petersburg again). In 1944 he became Director of the Institute for Chemical Physics at the Academy of Sciences in Moscow. He worked extensively in chemical kinetics and in 1927, independently of Hinshelwood, interpreted gaseous explosions in terms of branching chains (Section 8.5), sharing with Hinshelwood the 1956 Nobel Prize. He also developed the idea of degenerate branching, which accounts for cool flames.

Sidgwick, Nevil Vincent (1873–1852)

B. Oxford, England. *Educ.* Christ Church, Oxford, under Vernon Harcourt; after receiving first-class honours in Natural Science he proceeded to win first-class honours in Literae Humaniores ('Greats'). He spent a short time in Ostwald's laboratories in Leipzig and then went to Tübingen where he obtained a D.Sc. degree in 1901. He was then elected a Fellow of Lincoln College, Oxford, and resided in the College for the rest of his life. He did some research in solution kinetics, but his main contributions were his books, which were the result of a massive programme of inductive scholarship in which he collated a wide range of experimental results and theories. His book *The Electronic Theory of Valency*, published in 1927, was a classic which exerted a wide influence, as did his *The Covalent Link in Chemistry*, based on his lectures at Cornell University. In 1950 appeared his *The Chemical Elements and their Compounds*, which transformed inorganic chemistry from a mass of disconnected facts into a coherent whole.

References: Laidler (1988); Sutton (1958); Tizard (1954).

Skrabal, Anton (1877–1957)

B. Vienna. *Educ.* University of Vienna (DrTech, 1902). He taught at the University of Vienna for some years, and from 1917 to 1942 was professor of chemistry at the University of Graz. He did much work in kinetics, his book *Homogenkinetic* appearing in 1941.

References: Hüttig (1952); Schmid (1957).

Sommerfeld, Arnold Johannes Wilhelm (1868–1951)

B. Königsberg, East Prussia (now Kalingrad, Lithuania). *Educ.* University of Königsberg. In 1906 he succeeded Boltzmann as professor at the University of

Munich. In 1916 he extended Bohr's theory of the atom to allow for elliptical orbits (Section 10.1). He was not Jewish but bitterly opposed Nazi oppression, and was forced into retirement in 1940. He died after being struck by a car.

Reference: P. Forman and A. Hermann, DSB, 12, 525–532 (1975).

Stewart, Balfour (1828–1887)

B. Edinburgh, Scotland. *Educ.* Universities of St Andrews and Edinburgh. He was Director of the Kew Observatory from 1859 to 1870, and was then professor at Owens College, Manchester. In 1858 he formulated principles of emission and absorption of radiation.

Reference: D. M. Siegel, DSB, 13, 51–53 (1976).

Stokes, Sir George Gabriel, 1st Baronet (1819–1903)

B. Skreen, County Sligo, Ireland. *Educ.* Pembroke College, Cambridge (Senior Wrangler and Smith's Prizeman). He was elected Fellow of Pembroke in 1841 and was Lucasian professor at Cambridge from 1849 until his death; he became Master of Pembroke in 1902. He carried out theoretical and experimental investigations on a wide range of topics, including hydrodynamics, optics, gravity, sound, heat, solar physics, meteorology, chemistry and pure mathematics. He discovered fluorescence in 1852 (Section 6.1). He was created a baronet in 1889.

References: E. M. Parkinson, DSB, 13, 74–79 (1976); Rayleigh (1905).

Stoney, George Johnstone (1826–1911)

B. County Dublin, Ireland. *Educ.* Trinity College, Dublin. He was professor of natural philosophy at Queen's College, Galway, and from 1857 was Secretary to Queen's University, Dublin. In 1893 he moved to London to become active with the Royal Society. In 1874 he calculated the electric charge associated with an atom (Section 7.1), and in 1891 introduced the word 'electron'. He developed theories of spectra and suggested the idea of orbiting electrons (Section 6.5).

References: B. B. Kelham, DSB, 13, 82 (1976); 'J. J.' (1912).

Strutt, John William, 3rd Baron Rayleigh (1842–1919)

See Rayleigh.

Tafel, Julius (1862–1918)

B. Choindez, Canton Berne, Switzerland. *Educ.* Universities of Zürich, Munich and Erlangen. He was professor at the University of Würzberg, 1903–1910. He carried out elecrochemical syntheses, but began basic work in electrochemistry in about 1900 and arrived at the 'Tafel equation' in 1905 (Section 7.5). He suffered from tuberculosis and had to retire at the age of 48, finally taking his own life when he realized that he could no longer do useful work.

References: Emmert (1918); Müller (1969).

Talbot, William Henry Fox (1800–1877)

B. Dorsetshire, England. *Educ.* Trinity College, Cambridge. He had broad interests, including art and linguistics as well as science. He carried out some work in spectroscopy and in the 1830s invented a photographic technique which he first announced in 1839 (Section 8.4). He later brought about many improvements to his technique.

Reference: R. V. Jenkins, DSB, 13, 237–239 (1976); Schaaf (1992).

Taylor, Sir Hugh Stott (1890–1974)

B. St Helens, Lancashire. *Educ.* University of Liverpool (D.Sc. 1914). He worked in Arrhenius's laboratory on acid–base catalysis from 1912 to 1913, and with Bodenstein on the hydrogen–chlorine reaction from 1913 to 1914. He was professor at Princeton from 1914, becoming chairman of the department of chemistry and dean of graduate studies. He worked extensively on surface reactions, and proposed in 1925 the idea of active centres, demonstrating that adsorption can have an activation energy (Section 8.6). He remained a British subject and was knighted in 1953.

References: M. Stanley, DSB, 18, 898–899 (1990); Kemball (1975).

Thénard, Louis Jacques (1777–1857)

B. La Louptière, Aube, France (renamed La Louptière-Thénard after his death). He worked mainly on the isolation of new compounds, discovering boron (with Gay-Lussac) and hydrogen peroxide. He measured the rate of decomposition of hydrogen peroxide (Chapter 8). In 1832 he was elected to the Chambre des Députés, and was created a Baron in 1832.

Reference: M. P. Crosland, DSB, 13, 309–314 (1976).

Thomson, Sir George Paget (1892–1975)

B. Cambridge, England, the son of J. J. Thomson. *Educ.* Trinity College, Cambridge. He became a Fellow of Corpus Christi College, Cambridge, in 1914 and was later professor of physics at Aberdeeen and Imperial College, London, becoming Master of Corpus Christi College, Cambridge, in 1952. He and Clinton Davisson independently discovered electron diffraction, for which they shared the Nobel Prize in physics in 1937.

Reference: J. Hendry, DSB, 18, 908–912 (1990).

Thomson, Sir Joseph John (1856–1940)

B. Near Manchester, England. *Educ.* Owen's College, Manchester, and Trinity College, Cambridge. He became Cavendish professor of experimental physics at Cambridge in 1884, retaining that position until 1919 when he became research professor and Master of Trinity. He worked in a variety of fields, and is particularly noted for his work on cathode rays, in which he determined the mass and charge of the electron. He also worked on beams of positive particles (Section 10.3); he can be said to be the founder of mass spectrometry. He was awarded the Nobel Prize for physics in 1906 and is buried near Newton in Westminster Abbey.

Reference: J. L. Heilbron, DSB, 13, 362–372 (1976).

Tolman, Richard Chase (1881–1948)

B. West Newton, Massachusetts. *Educ.* Massachusetts Institute of Technology (Ph.D. 1910). He was professor of physical chemistry at the University of Illinois from 1916 to 1919, and at the California Institute of Technology from 1922 to 1948. His research covered a wide range of experimental and theoretical physics and chemistry, including statistical mechanics, relativity, cosmology and chemical kinetics. He interpreted the activation energy in 1921 (Section 8.3) and carried out experimental work on the kinetics of gas reactions.

Reference: J. R. Goodstein, 13, 429–430 (1976); King & Laidler (1984).

Towneley, Richard (1629–1668)

B. Towneley Hall, near Burnley, Lancashire, a member of a prominent Roman Catholic family. With Power he investigated the pressure–volume relationship in gases and in 1661 arrived at what has come to be called Boyle's law. He also worked in meteorology, especially on rainfall, and invented an improved micrometer.

References: C. Webster, DSB, 13, 444–445 (1976); Cohen (1964); Webster (1963, 1965).

Townsend, John Sealy Edward (1868–1957)

B. Galway, Ireland. *Educ.* Trinity College, Dublin. After five years of teaching mathematics, in 1895 he became a research student with J. J. Thomson in the Cavendish laboratory at Cambridge. In 1897 he made the first determination of the charge on the electron (Section 10.3). In 1900 he was appointed Wykeham professor of physics at Oxford.

References: T. J. Trenn, DSB, 13, 445–447 (1976); von Engel (1957).

Trautz, Max (1880–1961)

B. Karlsruhe, Germany. *Educ.* University of Karlsruhe. He was professor at the Universities of Heidelberg, Rostock and Münster, and worked in wide areas of physical chemistry, including thermodynamics, electrochemistry and chemical kinetics. He was one of the earliest proponents of the radiation theory of chemical reactions (Section 8.5), and he developed an early collision theory of chemical reactions (Section 8.3). His papers tended to be written rather obscurely and as a result have not received as much attention as they perhaps deserve.

References: Schwab (1955); Stranski (1961).

Van der Waals, Johannes Diderik (1837–1923)

B. Leiden, The Netherlands. *Educ.* University of Leiden. He was professor of physics at the University of Amsterdam from 1877 to 1908. He is famous for his equation for imperfect gases (Section 5.1) and for his 'law of corresponding states'. He was awarded the Nobel Prize for physics in 1910.

Reference: J. A. Prins, DSB, 14, 109–111 (1976).

Volmer, Max (1885–1965)

B. Marburg, Germany. *Educ.* University of Marburg. He was professor at the University of Marburg and later at the Technische Hochschule in Berlin-Charlottenburg. He worked on a range of problems particularly in chemical kinetics, including photochemistry and reactions occurring at solid surfaces.

References: Blumtritt (1985); Stranski (1961).

Volta, Count Alessandro Giuseppi Antonio Anastasis (1745–1827)

B. Como, Lombardi, Italy. *Educ.* at a Jesuit seminary. He first taught in a high school and carried out research, inventing the electrophorus, a hydrogen lamp, and an electrical condenser. From 1774 to 1804 he was professor of natural philosophy at the University of Pavia. He is best known for his work on the 'Voltaic pile' (Chapter 7).

Reference: J. L. Heilbron, DSB, 14, 69–82 (1976).

Warburg, Emil Gabriel (1845–1931)

B. near Hamburg, Germany. *Educ.* Heidelberg and University of Berlin (Ph.D. 1867). After an appointment at Strasbourg he was professor of physics at the University of Freiburg from 1876 to 1895, and after that at the University of Berlin. In 1905 he became President of the Physikalische-technische Reichanstalt. At Strasbourg with August Kundt (1839–1894) he confirmed in 1875 that gas viscosity and heat conduction are independent of pressure, and in 1876 that C_p/C_v for a monatomic gas is 5/3 (Section 5.3). Warburg was one of the founders of quantitative photochemistry, and worked mainly in this field in his later years. He confirmed the law of photochemical equivalence as applied to a primary process. His son, Otto Heinrich Warburg (1883–1970) received the 1931 Nobel Prize for physiology or medicine.

Reference: H. Ramser, DSB, 14, 170–172 (1976); Einstein (1922).

Waterston, John James (1811–1883)

B. Edinburgh. He attended lectures at the University of Edinburgh. He first practised as a civil engineer, and was Naval instructor to the East India Company in Bombay from 1839 to 1857; he then returned to Scotland. He published a number of papers on physical chemistry in the *Philosophical Magazine*. He submitted to the Royal Society an important paper on kinetic theory but it was rejected and was not published until 1892, after his death (Section 5.3).

References: S. G. Brush, DSB, 14, 184–186 (1976); Brush (1957, 1976).

Watson, The Revd Richard (1737–1816)

B. Heversham, Westmoreland. *Educ.* Trinity College, Cambridge. He was professor of chemistry at Cambridge from 1764 to 1771, and Regius professor of divinity at Cambridge from 1771; he was simultaneously Bishop of Llandaff from 1772 but resided at Windermere, carrying out research in chemistry but performing none of his episcopal duties. He discovered the relationship between freezing point and concentration (usually known as 'Blagden's law') in 1770. He wrote several books and articles on chemistry, particularly *Chemical Essays* (1781–87).

References: E. L. Scott, DSB, 14, 191–192 (1976); Coleby (1953); Millar & Millar (1985); Partington (1937).

Watt, James (1736–1819)

B. Greenock, Scotland. He was an instrument maker to the University of Glasgow from 1757 to 1766. In 1765 he designed the separate condenser for the Newcomen engine. From 1776 to 1774 he was a civil engineer in Glasgow, and then in Birmingham. In partnership with Matthew Boulton (1728–1809) he designed an improved steam engine. He also made important contributions to chemistry and physics; he established that water is a compound. His understanding of the working of the steam engine was of value to Carnot and others. He invented the first office duplicating machine in 1779.

References: H. Dorn, DSB, 14. 196–199 (1976); Carnegie (1913); Cardwell (1971); Robinson (1969–70).

Whetham, William Cecil Dampier—

See Dampier-Whetham

Wigner, Eugene Paul (b. 1902)

B. Budapest, Hungary. *Educ.* University of Berlin (doctorate in 1925). After teaching at the Universities of Berlin and Göttingen he became professor of mathematical physics at Princeton. He worked on many problems in theoretical physics, including the theory of the conservation of parity, and received the Nobel Prize for physics in 1963. In the 1930s he made important contributions to chemical kinetics, in terms of potential-energy surfaces and the theory of tunnelling (Section 8.3).

Wilhelmy, Ludwig Ferdinand (1812–1864)

B. Stargard, Pomerania (now Poland). He first studied and practised pharmacy, but sold the business in 1843 to study chemistry and physics at the Universities of Berlin, Giessen (now the Justus Liebig University) and Heidelberg (doctorate in 1846); he then studied with Regnault in Paris. He was *Privatdozent* at Heidelberg from 1849 to 1854 when he retired to private life. In 1850 he published pioneering work in chemical kinetics, on the rate of inversion of cane sugar, and proposed an empirical equation for its temperatutre dependence (Chapter 8).

Reference: S. J. Kopperl, DSB, 14, 359–360 (1976).

Wollaston, William Hyde (1766–1828)

B. Norfolk, England. *Educ.* Caius College, Cambridge, where he was a medical student (MD 1793) but also studied botany, chemistry and astronomy. He practiced medicine for a period and then devoted himself to research, covering a wide range of topics including the chemistry of platinum and other metals, crystallography, spectroscopy (Chapter 6), optics, muscular action, and the physiology of the ear. He discovered palladium and first isolated the amino acid cystine and other biological substances. He exerted a considerable influence, especially leading to the acceptance of Dalton's atomic theory.

Reference: D. G. Goodman, DSB, 14, 486–493 (1976).

References for biographies

References to the *Dictionary of Scientific Biography* (Ed. C. C. Gillispie), Charles Scribner, New York, 1970–present, are given below the individual biographies. Other useful sources of information about scientists are

J. R. Partington, *A History of Chemistry*, Macmillan, London, 1961; this includes biographical and other information about most of the people mentioned in this book.

Obituary Notices of Fellows of the Royal Society, Volumes 1–9 (1932–1954), which was succeeded by:

Biographical Memoirs of Fellows of the Royal Society, Vols. 1 (1955) to the present. Prior to 1932, obituaries of Fellows were published in the Proceedings of the Royal Society. A General Index to obituaries in Vols. 10–64 (1860–1899) of the Proceedings is to be found in Vol. 64, and for Vols. 65–75 (1900–1905) in Vol. 75.

Biographical Memoirs of the (U. S.) National Academy of Sciences.

J. C. Poggendorff biographisch-literarisches Handwörterbuch der exacten Naturwissenschaften, Academie Verlag, Berlin. This includes scientists of many nationalities.

Biografisch Woordenbock van Nederland, Elsevier, Amsterdam, for biographies of Dutch scientists.

Biographical Dictionary of Nobel Laureates in Chemistry, American Chemical Society and the National Foundation for the History of Chemistry, Washington, DC, 1993.

'A. H.' (1926) Percival Spencer Umfreville Pickering. 1858–1920, *Proc. Roy. Soc.*, **A**, 111, viii–xii.

Allibone, T. E., and J. R. Clarke (1959) Samuel Roslington Milner, *Biog. Memoirs F. R. S.*, **5**, 129–147.

Andrade, E. N. da C. (1950a) *Isaac Newton*, London.

Andrade, E. N. da C. (1950b) Wilkins Lecture: Robert Hooke, *Proc. Roy. Soc.*, **A**, 201, 439–473.

Anon. (1873) *Proc. Roy. Soc.*, **21**, i–iv. Obituary notice for Rankine.

Anon. (1882) *Proc. Roy. Soc.*, **33**, i–xvi. Obituary notice for Maxwell.

Anon. (1970) *Grote Winkler Prins Encyclopedie*, Elsevier, Amsterdam. Kohnstamm's biography, with portrait, is on pp. 215–216.

Archer, Mary D. (1989) Genesis of the Nernst equation, in *Electrochemistry*,

Past and Present (J. T. Stock and M. V. Orna, Eds.), American Chemical Society, Washington, DC.

Baker, H. B., and W. A. Bone (1931) Harold Baily Dixon, *J. Chem. Soc.*, 3349–3368.

Baker, H. B., and W. A. Bone (1932) Harold Baily Dixon, *Proc. Roy. Soc.*, A, **134**, i–xvii.

Ball, W. W. Rouse (1893, 1960) *A Short Account of the History of Mathematics*, Macmillan, London; reprinted by Dover Publications, 1960.

Ballhausen, C. J. (1969) Jens Anton Christiansen, Festskrift, University of Copenhagen, 237–239.

Bancroft, Wilder D. (1933) Wilhelm Ostwald, the great protagonist, *J. Chem. Education*, **10**, 539–542. 609–613.

Barnett, M. (1958) Sadi Carnot and the second law of thermodynamics, *Osiris*, **13**, 327–357.

Bawn, C. E. H. (1958) William Cudmore McCullach Lewis, 1885–1956, *Biog. Memoirs F. R. S.*, **4**, 193–203.

Bell, R. P. (1970) Jens Anton Christiansen, *Chemistry in Britain*, **6**, 491.

Bernhard, C. G. (1989–1990) *Berzelius—creator of the chemical language*, Saab-Scandia Griffin.

Blumtritt, Oskar (1985) *Max Volmer, 1885-1965. Eine Biographie* (65 pp.). Obtainable from Universitätsbibliothek, TU Berlin (Abt. Pubkicationen), Strasse des 17, Juni 135, D 1000 Berlin 12.

Boas, Marie (1958) *Robert Boyle and Seventeenth-Century Chemistry*, Cambridge University Press.

Bowen, E. J. (1958) David Leonard Chapman, *Biog. Memoirs F. R. S.*, **4**, 35–44.

Bowen, E. J. (1958) The Balliol–Trinity laboratories, Oxford, 1853–1940, *Notes & Records Roy. Soc.*, **25**, 227–239.

Bowen, E. J. (1967) Sir Cyril Hinshelwood, *Chemistry in Britain*, **3**, 534–536.

Boyer, C. B. (1943) History of the measurement of heat, *Scientific Monthly*, **57**, 442–452, 546–554.

Brush, S. G. (1957) The development of the kinetic theory of gases. II. Waterston, *Annals of Science*, **13**, 275–282.

Brush, S. G. (1958) The development of the kinetic theory of gases. III. Clausius, *Annals of Science*, **14**, 185–196.

Brush, S. C. (1965) *Kinetic Theory*, 2 Vols., Permagon Press, Oxford.

Brush, S. G. (1976) *The Kind of Motion that we call Heat*, North-Holland Publishing Co., Amsterdam.

Buerger, Janet E. (1989) *French Daguerreotypes*, University of Chicago Press.

Butler, J. A., and P. Zuman (1964) Jaroslav Heyrovsky, *Biog. Memoirs F. R. S.*, **13**, 167–191.

Cardwell, D. S. L. (1971) *From Watt to Clausius*, Manchester University Press and Cornell University Press.

Cardwell, D. S. L. (1983) The origin and consequences of certain of J. P. Joule's scientific ideas, in *Springs of Scientific Creativity* (Eds, R. Aris, H. T. Davis and R. H. Strewer), University of Minneapolis Press.

Carnegie, Andrew (1913) *James Watt*, Doubleday, Page & Co., New York.

Christiansen, J. A. (1948–1949) Johannes Nicolaus Brönsted, *Overs. Selsk. Virksomhed*, 57–59.

Cockcroft, J. D. (1963) Niels Henrik David Bohr, *Biog. Memoirs F. R. S.*, **9**, 37–53.

Cohen, J. Bernard (1964) Newton, Hooke and 'Boyle's law' (discovered by Power and Towneley), *Nature*, **204**, 618–621.

Coleby, L. J. M. (1953) Richard Watson, professor of chemistry in the University of Cambridge, 1764–71, *Annals of Science*, **9**, 101–123.

Conant, J. B., Robert Boyle'e experiments in pneumatics, in *Harvard Case Studies in Experimental Science* (Ed., J. B. Conant), Harvard University Press, 1957.

Cowles, T. (1934) Dr Henry Power, disciple of Sir Thomas Browne, *Isis*, **30**, 344–366.

Cremer, E. (1967) Max Bodenstein, *Ber. Deut. chem. Ges.*, **100**, xcv–cxxvi.

Crowther, J. C. (1935) *British Scientists of the Nineteenth Century*, Kegan Paul, Trench & Trubner. London.

'C. T.' [perhaps Charles Tomlinson] (1871) *Proc. Roy. Soc.*, **19**, xix–xxvi. Obituary notice for W. Allen Miller.

Dainton, F. S., and B. A. Thrush (1981) Ronald George Wreyford Norrish, *Biog. Memoirs F. R. S.*, **27**, 379–424.

d'Albe, E. E. Fournier (1923) *The Life of Sir William Crookes, O.M., F.R.S.*, London.

Dalitz, R. H., and Sir Rudolf Peierls (1986) Paul Adrien Maurice Dirac, *Biog. Memoirs F. R. S.*, **32**, 139–185.

Daub, E. E. (1970) Entropy and dissipation, *Hist. Stud. Phys. Sci.*, **2**, 321–354.

Davenport, D. A., and Kathleen M. Ireland (1989) The ingenious, lively, and celebrated Mrs Fulhame, *Bulletin of the History of Chemistry*, **5**, 37–42.

Davies, Mansel (1968) Peter J. W. Debye (1884–1966), *J. Chem. Education*, **45**, 467–473.

Davies, Mansel (1970) Peter Joseph Wilhelm Debye, *Biog. Memoirs F. R. S.*, **16**, 175–232.

Davies, Mansel (1986) C. R. Bury: his contributions to physical chemistry, *J. Chem. Education*, **63**, 741–743.

Davies, Mansel (1986) Charles Rugeley Bury and his contributions to physical chemistry, *Arch. Hist. Exact Sci.*, **36**, 75–90.

Derrick, M. Elizabeth (1982) Agnes Pockels, 1862–1935, *J. Chem. Education*, **59**, 1030–1031.

Dixon, H. B. (1920) A. G. Vernon Harcourt, 1834–1919, *Proc. Roy. Soc.*, **97**, vii–xi.

Donnan, F. G. (1912) *Proc. Roy. Soc., A*, **86**, xxxix–xliii. Obituary notice for van't Hoff.

Donnan, F. G. (1933) Ostwald Memorial Lecture, *J. Chem. Soc.*, 316–332.

Einstein, A. (1922) Emil Warburg als Forscher, *Naturwiss.*, **10**, 823–828. This article includes a list of publications.

'E. J. R.' (1921) Percival Spencer Umfreville Pickering, *J. Chem. Soc.*, **119**, 654–660.

Eley, D. D. (1976) Eric Keighley Rideal, *Biog. Memoirs F. R. S.*, **22**, 381–413.

Elliott, E. B. (1917) William Esson, *Proc. Roy. Soc., A*, **93**, lix–lvii.

Emmert, B. (1918) *Chemiker-Zeitung*, **42**, 481 (1918). Obituary notice for Tafel.

Eugster, H. D. (1971) The beginnings of experimental petrology, *Science*, **173**, 481–489. This article gives an account of van't Hoff's pioneering work in this field.

Everitt, C. W. F. (1983) Maxwell's scientific creativity, in *Springs of Scientific Creativity* (Eds R. Aris, H. T. Davis and R. H. Stuewer), University of Minnesota Press.

Faulkner, I. J. (1957) Thomas Ewen, 1868–1955, *Proc. Chem. Soc.*, 236.

Fitzgerald, G. F. (1891) *Proc. Roy. Soc.*, 48, i–viii. Obituary notice for Clausius.

Fox, R. (1969–70) Watt's expansive principle in the work of Sadi Carnot and Nicolas Clément, *Notes & Records Roy. Soc.*, **24**, 233–243.

Fox, R. (1971) *The Caloric Theory of Gases from Lavoisier to Regnault*, Clarendon Press, Oxford.

Fox, R. (1974) The rise and fall of Laplacian physics, *Hist. Stud. Phys. Sci.*, **4**, 89–136.

Garber, Elizabeth W. (1969) James Clerk Maxwell and thermodynamics, *Amer. J. Phys.*, **37**, 146–155.

Gibbs, J. Willard (1889) Rudolf Julius Emmanuel Clausius, *Proc. Amer. Acad. Arts Sci.*, **16**, 458–465.

Gibson, Q. H. (1972) Francis John Worsley Roughton, *Biog. Memoirs F. R. S.*, **19**. 563–582.

Giles, C. H., and S. D. Forrester (1971) The origins of the surface film balance (Part 2 of 'Studies in the early history of surface chemistry'), *Chemistry and Industry*, 2 January 1971, 43–53. This article contains much biographical information about Agnes Pockels and Lord Rayleigh, including portraits.

Goldman, M. (1982) *The Demon in the Aether: The Story of James Clerk Maxwell*, Paul Harris, Edinburgh.

Greenaway, F. (1966) *John Dalton and the Atom*, Cornell University Press.

Gregory, Sir Richard, and A. Ferguson (1941) Sir Oliver Lodge, *Obit. Notices F. R. S.*, **3**, 551–574.

Hall, Marie Boas, *Robert Boyle on Natural Philosophy*, Indiana University Press, 1965.

Hammick, D. L. (1959) David Leonard Chapman, *Proc. Chem. Soc.*, 101–103.

Hartley, Sir Harold (1966) *Humphry Davy*, Nelson, London.

Hartley, Sir Harold (1971) *Studies in the History of Chemistry*, Clarendon Press, Oxford.

Hartmann, H., and H. C. Longuet-Higgins (1982) Erich Hückel, *Biog. Memoirs F. R. S.*, **28**, 153–162.

Haycock, C. T. (1925) George Downing Liveing, *J. Chem. Soc.*, 2982–2984.

Heath, S. H. (1980) Henry Eyring, Mormon scientist, MA thesis, University of Utah.

Heath, S. H. (1985) The making of a physical chemist: the education and early researches of Henry Eyring, *J. Chem. Education*, **62**, 93–98.

Heilbron, J. L. (1974) *H. G. J. Moseley: The Life and Work of an English Physicist*, University of California Press.

Heitler, W. (1961) Erwin Schrödinger, *Biog. Memoirs F. R. S.*, **7**, 221–228.

Hey, D. H. (1955) Schools of chemistry in Great Britain and Ireland. XVIII. King's College, London, *J. Royal. Inst. Chem.*, **18**, 305–315. This article contains interesting information about J. F. Daniell, W. A. Miller and others.

Hiebert, E. N. (1971) The energetics controversy and the new thermodynamics, in D. H. D. Roller (Ed.), *Perspectives in the History of Science and Technology*, University of Oklahoma Press.

Hiebert, E. N. (1978) Nernst and electrochemistry, in G. Dubpernell and J. H. Westbrook (Eds.), *Selected Topics in the History of Electrochemistry*, The Electrochemical Society, Princeton, N. J.

Hiebert, E. N. (1982) Developments in physical chemistry at the turn of the century, in C. G. Bernhard, E. Crawford & D. Sörbom (Eds.), *Science, Technology and Society in the Time of Alfred Nobel*, Pergamon Press, Oxford.

Hirschfelder, J. O. (1982) Henry Eyring, 1901–1981, in *American Philosophical Society Year Book 1982*, Philadelphia.

Hofmann, U. (1958) Hans Georg Grimm zum 70 Geburtstag, *Ber. Bunsenges. physikal. Chem.*, **62**, 109–110.

Holleman, A. F. (1952) My reminiscences of van't Hoff, *J. Chem. Education*, **29**, 379–382.

Hückel, Anne (1965) Erich Hückel, *Nachr. Chem. Techn.*, **13**, 382–383. This is a biography written by his wife.

Hutchison, K. (1981) W. J. M. Rankine and the rise of thermodynamics, *British J. Hist. Sci.*, **14**, 1–26.

Hutchison, K. (1891), Rankine, atomic vortices and the entropy function, *Arch. Internationales d'Histoire des Sciences*, **31**, 72–134.

Hüttig, G. F. (1952) Anton Skrabal, *Oesterr. Chem. Z.*, **53**, 93.

Jensen, A. T. (1958–1959) *Overs. Selsk. Virksomhed*, 99. Obituary notice for Bjerrum.

'J. J.' (1912) *Proc. Roy. Soc.*, **86**, xx–xxxv. Obituary notice for Johnstone Stoney.

Keeble, Sir F. (1941) Henry Edward Armstrong, *Obit. Notices F. R. S.*, **3**, 229–245.

Kemball, C. (1975) Hugh Stott Taylor, *Biog. Memoirs F. R. S.*, **21**, 517–547.

Kerker, M. (1957) Sadi Carnot, *Scientific Monthly*, **85**, 145–149.

King, M. Christine (1981, 1982) Experiments with time, *Ambix*, **28**, 70–82, 29, 49–61.

King, M. Christine (1983) The chemist in allegory: Augustus Vernon Harcourt and the White Knight, *J. Chem. Education*, **60**, 177–180.

King, M. Christine (1984) The course of chemical change: the life and times of Augustus G. Vernon Harcourt (1834–1919), *Ambix*, **31**, 16–31.

King, M. Christine, and K. J. Laidler (1984) Chemical kinetics and the radiation hypothesis, *Arch. Hist. Exact. Sci.*, **30**, 45–86.

King-Hele, D. G. (Ed., 1988) Newton's *Principia* and its legacy, *Notes & Records Roy. Soc.*, **42**, 1–122.

Klein, M. J. (1969) Gibbs on Clausius, *Hist. Stud. Phys. Sci.*, **1**, 127–149.

Klein, M. J. (1970) Maxwell, his demon, and the second law of thermodynamics, *American Scientist*, **58**, 84–97.

Klein, M. J. (1983) The scientific style of Josiah Willard Gibbs, in *Springs of Scientific Creativity* (Eds., R. Aris, H. T. Davis and R. H. Stuewer). University of Minnesota Press.

Koyré, A., and I. B. Cohen (Eds., 1972) *Newton's Philosophiae Naturalis Principia Mathematica*, 2 vols., Cambridge University Press.

Laidler, K. J. (1984) The development of the Arrhenius equation, *J. Chem. Education*, **61**, 494–498.

Laidler, K. J. (1985) Chemical kinetics and the origins of physical chemistry, *Arch. Hist. Exact Sci.*, **32**, 43–75.

Laidler, K. J. (1985) René Marcelin (1885–1914): a short-lived genius of chemical kinetics, *J. Chem. Education*, **60**, 1012–1014.

Laidler, K. J. (1986) The development of theories of catalysis, *Arch. Hist. Exact Sci.*, **35**, 345–374.

Laidler, K. J. (1988) Chemical kinetics and the Oxford college laboratories, *Arch. Hist. Exact Sci.*, **38**, 197–283.

Laidler, K. J. (1993) in *Biographical Dictionary of Nobel Laureates in Chemistry*, American Chemical Society, Washington, DC.

Larmor, Sir Joseph (1908) William Thomson, Baron Kelvin of Largs, 1824–1907, *Proc. Roy. Soc., A*, **81**, iii–lxxvi.

Levere, T. H. (1980) Humphry Davy and the idea of glory, *Trans. Roy. Soc. Canada*, [4], **18**, 247–261 (1980).

Lockyer, T. Mary, and Winifred L. Lockyer (1928) *The Life and Work of Sir Norman Lockyer*, London.

Lovell, D. J. (1968) Herschel's dilemma in the interpretation of thermal radiation, *Isis*, **59**, 46–60.

MacCallum, T. W., and S. Taylor (Eds., 1938) *The Nobel Prize Winners and the Nobel Foundation*, 1901–1937, Central European Times Publishing Co., Zürich.

Macdonald, D. K. C. (1954) *Faraday, Maxwell, and Kelvin*, Doubleday & Co., New York.

McBride, W. A. E. (1987) J. H. van't Hoff, *J. Chem. Education*, **64**, 573–574.

Maddison, R. E. W. (1969) *The Life of the Honourable Robert Boyle, F.R.S.*, Taylor & Francis, London.

Maxwell, J. Clerk (1877) Scientific worthies: Hermann Ludwig Ferdinand Helmholtz, *Nature*, **15**, 389–391.

Mayneord, W. V. (1979) John Alfred Valentine Butler, *Biog. Memoirs F. R. S.*, **25**, 145–178.

Meadows, A. J. (1972) *Science and Controversy: A Biography of Sir Norman Lockyer*, Macmillan, London.

Melville, H. W. (1953) Meredith Gwynne Evans, *Obit. Notices F. R. S.*, **8**, 295–409.

Mendelsohn, K. (1973) *The World of Walther Nernst*, Macmillan, London.

Millar, Margaret, and I. T. Millar (1985) The chemist as biographer, *J. Chem. Education*, **60**, 365–370.

Moore, Ruth (1967) *Niels Bohr: the Man and the Scientist*, Hodder & Stoughton, London.

Moore, Walter (1989) *Schrödinger: Life and Thought*, Cambridge University Press.

More, Louis T. (1934) *Isaac Newton*, New York.

More, Louis T. (1944) *The Life and Work of the Honourable Robert Boyle*, Oxford University Press.

Mott, Sir Neville, and Sir Rudolf Peierls (1977) Werner Heisenberg, *Biog. Memoirs F. R. S.*, **23**, 213–251.

Mott, Sir Neville (1982) Walter Heinrich Heitler, *Biog. Memoirs F. R. S.*, **28**, 141–151.

Müller, K. (1969) Who was Tafel?, *J. Res. Inst. Catalysis, Hokkaido University*, **17**, 54–75.

Mysels, K. J. (1986) René Marcelin: experimenter and surface scientist, *J. Chem. Education*, **63**, 740.

Nepomucene, Sister St John (1960) Rydberg: the man and the constant, *Chymia*, **6**, 127–145.

Oesper, R. E. (1927) Robert Wilhelm Bunsen, *J. Chem. Education*, **4**, 431–439.

Oesper, R. E. (1975) *The Human Side of Scientists*, University of Cincinnati Press.

Partington, J. R. (1937) Richard Watson (1737–1816), *Chemistry and Industry*, **56**, 819–821.

Patterson, Elizabeth C. (1970) *John Dalton and the Atomic Theory*, New York.

Patterson, Elizabeth C. (1983) *Mary Somerville and The Cultivation of Science*, 1815–1840, Martinus Nijhoff.

Pauling, L. (1958) Arthur Amos Noyes: A biographical memoir, *Biog. Memoirs Nat. Acad. Sci.*, **31**, 322–346.

Porter, Sir George (1981) Michael Faraday—chemist (Faraday Lecture of the Chemical Society), *Proc. Roy. Inst.*, **53**, 90–99.

Raman, V. V. (1973) William John Macquorn Rankine, 1820–1872, *J. Chem. Education*, **50**, 274–276.

Rayleigh, Lord (1905) Sir George Gabriel Stokes, Bt., 1819–1903, *Proc. Roy. Soc.*, **75**, 199–216.

Reinold, A. W. (1896) *Proc. Roy. Soc.*, **59**, xvii–xxx. Obituary notice for Helmholtz.

Reynolds, Osborne (1892) Memoir of James Prescott Joule, *Memoirs & Proc. Manchester Lit. & Phil. Soc.*, [4], 6.

Robinson, E. (1969–1970) James Watt, engineer and man of science, *Notes & Records Roy. Soc.*, **24**, 221–232.

Robinson, R. (1947) Arthur Lapworth, *Obit. Notices F. R. S.*, **5**, 555–572.

Roller, D. (1957) The early developments of the concepts of temperature and heat, in J. B. Conant (Ed.), *Harvard Case Histories in Experimental Science*, Harvard University Press, Vol. 1, pp. 125–155.

Root-Bernstein, R. S. (1980) The Ionists: founding physical chemistry, Ph.D. Thesis, Princeton University.

Roscoe, H. E. (1900) Bunsen Memorial Lecture, *J. Chem. Soc.*, **77**, 513–554.

Rukeyser, Muriel (1942) *Willard Gibbs*, Doubleday, Doran, Garden City. NY; reprinted 1964.

Russell, C. A., The electrochemical theory of Sir Humphry Davy, *Annals of Science*, **15**, 1–25 (1959).

Saltzman, M. (1972) Arthur Lapworth: the genesis of reaction mechanisms, *J. Chem. Education*, **49**, 750–752.

Schaaf, L. J. (1990) The first fifty years of British photography: 1794–1844, in *Technology and Art: The Birth and Early Years of Photography* (Ed. Michael Pritchard), Royal Photographic Society Historical Group, Bath. This article includes an account of Elizabeth Fulhame's photoimaging experiments published in 1794.

Schaaf, L. J. (1992) *Out of the Shadows: Herschel, Talbot, and the Invention of Photography*, Yale University Press.

Schaffer, V. J. (1958) In Memoriam: Irving Langmuir—scientist, *J. Colloid Sci.*, **13**, 3–5.

Schmid, H. (1957) Anton Skrabal, *Oesterr. Chem. Z.*, **58**, 85, 285–289.

Schuster, Sir Arthur (1921) John William Strutt, Third Baron Rayleigh, *Proc. Roy. Soc.*, *A*, **98**, i–l.

Schwab, G. M. (1955) Zu Max Trautz' 75 Geburtstage, *Ber. Bunsenges. physikal. Chem.*, **59**, 139–140.

Servos, J. W. (1990) *Physical Chemistry from Ostwald to Pauling: The Making of a Science in America*, Princeton University Press. This book, particularly concerned with the development of physical chemistry in the United States, contains much information about Ostwald, van't Hoff, Arrhenius and Nernst as well as a number of American chemists.

Sherman, L. R. (1990) Jaroslav Heyrovsky (1890–1967), *Chemistry in Britain*, **26**, 1165–1168.

Shorter, J. (1980) A. G. Vernon Harcourt: A founder of chemical kinetics and a friend of 'Lewis Carroll', *J. Chem. Education*, **57**, 411–416.

Shorter, J. (1990) Hammett Memorial Lecture, *Prog. Phys. Org. Chem.*, **17**, 1–29.

Simpson, G. C. (1932–1933) *Obit. Notices F. R. S.*, **1**, 409–423. Obituary notice for Schuster.

Simpson, G. C. (1934) Sir Arthur Schuster, FRS, *Nature*, **134**, 595–597.

Snelders, H. A. M. (1985), in *Biografisch Woordenbock van Nederland*, Elsevier, Amsterdam. Biography of F. E. C. Scheffer.

Stokes, J. (1927) *The Life of John Theophilus Desaguliers*, Margate.

Stradins, J. P. (1964) The work of Theodor Grotthuss and the invention of the Davy safety lamp, *Chymia*, **9**, 125–145.

Stranski, I. N. (1961) Max Trautz, *Ber. Bunsenges. physikal. Chem.*, **56**, 401–402.

Stranski, I. N. (1965) Max Volmer, *Ber. Bunsenges. physikal. Chem.*, **69**, 755–756.

Strutt, R. J., Fourth Baron Rayleigh (1924) *Life of John William Strutt, Third Baron Rayleigh, O.M., F.R.S.*, London; 2nd edition, with additions, Madison, Wisconsin, 1968.

Sunko, D., and N. Trinajstic (1981) In Memoriam: Erich Hückel, *Croatica Chemica Acta*, **53**, xv–xvi. In English.

Sutton, L. E. (1958) Nevil Vincent Sidgwick, 1873–1952, *Proc. Chem. Soc.*, 310–319.

Taylor, G. I. (1954) *Obit. Notices F. R. S.*, **16**, 55–63. Obituary notice for Sir William Dampier (Dampier-Whetham).

Taylor, Sir Hugh (1958) Irving Langmuir, *Biog. Memoirs F. R. S.*, **4**, 167–184.

Thackray, A. (1970) *Atoms and Powers: An Essay on Newtonian Matter-Theory and the Development of Chemistry*, Harvard University Press.

Thompson, H. W. (1973) Cyril Norman Hinshelwood, *Biog. Memoirs F. R. S.*, **19**, 375–431.

Thompson, S. P. (1898) *Michael Faraday, His Life and Work*, London.

Thomson, G. P. (1958) Frederick Alexander Lindemann, Viscount Cherwell, *Biog. Memoirs F. R. S.*, **4**, 45–71.

Tizard, H. F. (1954) Nevil Vincent Sidgwick, 1873–1952, *Obit. Notices F. R. S.*, **9**, 237–258.

Tyndall, John (1869) *Faraday as a Discoverer*, London.

Urry, D. W. (1982) Henry Eyring (1901–1981): a 20th-century physical chemist and his models, *Mathematical Modelling*, **3**, 503–522.

Urry, D. W. (1982) Henry Eyring (1901–1981): a 20th-century architect of cathedrals of science, *Int. J. Quantum Chemistry, Quantum Biology Symposium*, No. 9. 1–3.

Van Doren, Carl (1938) *Benjamin Franklin*, New York. I. Bernard Cohen has commented in DSB that 'this is possibly the best biography of a scientist in English'.

Van Klooster (1952) Van't Hoff (1852–1911) in retrospect, *J. Chem. Education*, **29**, 376–379.

Van't Hoff, J. H. (1902) Raoult Memorial Lecture, *J. Chem. Soc.*, **81**, 969–981.

Veibel, A. S. (1970) Jens Anton Christiansen, *Overs. K. Dan. Vidensk. Selsk. Forh.*, 107.

von Engel, A. (1957) John Sealy Edward Townsend, *Biog. Memoirs F. R. S.*, **3**, 257–273.

Walker, J. (1913) Van't Hoff Memorial Lecture, *J. Chem. Soc.*, **103**, 1127–1143.

Walker, J. (1928) Arrhenius Memorial Lecture, *J. Chem. Soc.*, 1380–1401.

Webb, K. R. (1961) Sir Robert Grove (1811–1896) and the origins of the fuel cell, *J. Roy. Inst. Chem.*, **85**, 291–293.

Webster, C. (1963) Richard Towneley and Boyle's law, *Nature*, **197**, 226–228.

Webster, C. (1966) The discovery of Boyle's law and the concept of the elasticity of the air in the seventeenth century', *Arch. Hist. Exact Sci.*, **2**, 441–502.

Westfall, R. S. (1971) *Force in Newton's Physics*, London, 1971.

Westfall, R. S. (1980) *Never at Rest: A Biography of Isaac Newton*, Cambridge University Press.

Wheeler, L. P. (1951) *Josiah Willard Gibbs: The History of a Great Mind*, Yale University Press. This the official biography.

Wheeler, T. S., and J. R. Partington (1960) *The Life and Work of William Higgins, Chemist* (1763–1825), Pergamon Press, Oxford. The little that is known of Elizabeth Fulhame is on pp. 121–122.

Whitrow, G. J. (1989) Newton's role in the history of mathematics, *Notes & Records Roy. Soc.*, **43**, 71–92.

Whyte, L. L. (Ed. 1961) *Roger Joseph Boscovich, S.J., F.R.S., 1711–1787*, George Allen & Unwin, London.

Whytlaw Gray, R., and G. F. Smith (1940) Harry Medforth Dawson, *Obit. Notices F. R. S.*, **3**, 139–154.

Wigner, E. P., and R. A. Hodgkin (1977) Michael Polanyi, *Biog. Memoirs F. R. S.*, **23**, 413–448.

Williams, L. Pearce (1965) *Michael Faraday. A Biography*, Basic Books, New York; paperback reprint, Da Capo, New York, 1987.

Wilson, E. B. (1901) *Vector Analysis Founded upon the Lectures of J. Willard Gibbs*, Yale University Press.

Wilson, S. S. (1981) Sadi Carnot, *Scientific American*, **245**, 134–145 (August, 1981).

Index

Proper names, and page numbers relating largely to biographical information, are shown in **bold** type. Books and journals are shown in *italics*.